T0141324

PROCEEDINGS OF INTERNATIONAL SYMPOSIUM ON THE QINGHAI–XIZANG PLATEAU AND MOUNTAIN METEOROLOGY

MARCH 20–24, 1984   BEIJING, CHINA

PROCEEDINGS OF INTERNATIONAL SYMPOSIUM ON THE QINGHAI–XIZANG PLATEAU AND MOUNTAIN METEOROLOGY

MARCH 20–24, 1984   BEIJING, CHINA

CHINESE METEOROLOGICAL SOCIETY
STATE METEOROLOGICAL ADMINISTRATION
CHINESE ACADEMY OF SCIENCES

AMERICAN METEOROLOGICAL SOCIETY
NATIONAL OCEANIC AND ATMOSPHERIC
ADMINISTRATION
NATIONAL SCIENCE FOUNDATION

WORLD METEOROLOGICAL ORGANIZATION

SCIENCE PRESS, BEIJING
AMERICAN METEOROLOGICAL SOCIETY
BOSTON, MASSACHUSETTS
1986

Responsible Editor Xu Yigang

Preface

The International Symposium on the Qinghai-Xizang (Tibet) Plateau and Mountain Meteorology, held in Beijing was sponsored by the Chinese Meteorological Society, State Meteorological Administration, Chinese Academy of Sciences, American Meteorological Society, National Oceanic and Atmospheric Administration, National Science Foundation and World Meteorological Organization.

Its aim was to provide an opportunity for scientists both in China and abroad to summarize and exchange the results of their research in this field. Nearly a hundred scientists attended the symposium. Half were Chinese while the other half were well-known scholars from other 14 countries.

The exceptional level of support given to the symposium by various organizations reflects a recognition of both the importance of and the substantial advances in the study of meteorology of the Qinghai-Xizang (Tibet) Plateau and of mountain meteorology. On the one hand, China's climate is greatly affected by the Qinghai-Xizang Plateau, on the other hand, mountains of various scales on the earth can have not only a substantial effect on regional weather and climate, but also a significant and far-reaching effect on large-scale general circulation.

From May to August, 1979, a meteorological field observation experiment was carried out on the Qinghai-Xizang Plateau. In analyzing the observational data obtained in this experiment, Chinese meteorologists have been able to arrive at a variety of new conclusions.

In this description of the proceedings there are papers presented by the participants at the meeting, including those on the following important aspects of mountain meteorology: the field observation, the dynamic and thermal effects of the mountains on the general circulation, the results obtained through the numerical models with the large-scale topography and the circulation systems on the Plateau.

We highly appreciate the support given to the symposium by the various organizations mentioned above, without which this book would not have been possible. We also wish to express our thanks to the authors for their special efforts in writing these papers.

Beijing

December 1985   The editors

i

Contents

Session 1: Observation and Analysis (1)

THE QINGHAI-XIZANG PLATEAU METEOROLOGICAL EXPERIMENT (QXPMEX) MAY-AUGUST 1979

Tao Shiyan
Institute of
Atmospheric Physics,
Chinese Academy
of Sciences

Luo Siwei
Lanzhou Institute of
Plateau Atmospheric Physics,
Chinese Academy
of Sciences

Zhang Hongcai
Sichuan Meteorological Service

I. THE PRIMARY SCIENTIFIC OBJECTIVES

Many earlies studies have pointed out that the Qinghai-Xizang Plateau strongly influences the weather and climate of China, even the general circulation of the world. If one does not include the influence of the Qinghai-Xizang Plateau in the numerical simulation, the mean circulation and the characteristics of climate seem to conflict with the reality of the present situation, but after due consideration one can see that they are not so very incongruous after all. We have yet to understand well the exact reasons for the Plateau's influence, neither have we been able to conclude much from some of the important observational facts. For these reasons we have mounted the Qinghai-Xizang Plateau Meteorological Plateau Meteorological Experiment (QXPMEX) in May-August 1979, during the same period as the MONEX. The main objectives of the experiment were to investigate the thermal and dynamic effects of the Plateau on the general circulations in Eastern Asia, the radiation and heat budget on the surface on the Plateau and the weather on the Plateau under various synoptic conditions. The QXPMEX was divided into three phases: the experiment design phase (1978), the actual field observing phase (May-August, 1979) and the research phase (1980 and 1981). More than 100 scientists from various research institutes, provincial weather bureaus and universities participated in the experiment.

The cost of the experiment was about 3 million Yuan.

There are four specific tasks encompassed in the general objective of the experiment:

1) Field observation and analysis of the radiation and heat budget on the surface of the Qinghai-Xizang Plateau.

2) The Plateau's effect on the seasonal change of circulations in its surrounding areas.

3) Study of the weather systems on the Plateau and its surroundings in summer.

4) Numerical simulation experiments and theoretical studies on the effects of the Plateau on circulations and motion systems.

The detailed results of these investigations will be given in other papers at this symposium, though only some of the observational facts on the weather on the Plateau under various synoptic conditions will be presented here.

II. OBSERVATIONAL SYSTEMS IN QXPMEX

During the QXPMEX the experimental area and the distribution of the observational stations were planned as follows:

1. Experimental Area

The experimental area includes the whole Plateau, i.e. the region to the south of 40°N, to the north of 28°N, to the east of 80°E and to the west of 105°E. For studying the sub-synoptic scale systems (vortices and shearlines) around the Nagqu in the middle section of the Qinghai-Xizang Plateau and Qiangtang Plateau in its west part, two supplemental key regions of observation are established during the period that vortices and shearlines appeared. Region A is covered by 80°—90°E and 28°—37°N, and region B by 90°—95°E and 29°—34°N.

2. Observational Network

The density of the observational network is enhanced and the time of observation prolonged during the period of QXPMEX in the experimental area. There are 223 surface stations, 83 upper-air stations and 37 pilot balloon stations. In addition, there is a station for receiving the satellite photograph at Lhasa and two radar stations, one of which is located at Nagqu in the middle section of the Plateau and the other at Lhasa. In order to fill the blank area of observational data, six surface radiation measurement stations are set up and distributed on the Plateau in some typical regions with different vegetations.

III. THE WEATHER SYSTEMS ON THE PLATEAU
DURING THE QXPMEX

There are 4 main weather systems on the Plateau in summer, i.e. the vortex, shearline, moving anticyclone and cold air outbreak.

1. Vortex

The vortex on 500 hPa is one of the sub-synoptic scales and rain bearing weather systems on the Plateau in summer. Its composite structure and cause of formation shall be discussed in detail at this symposium and here we only mention it briefly.

The vortex occurred mainly in the middle and west Plateau and in total numbered was 54 during the QXPMEX. Figure 1 shows the distribution of vortices in QXPMEX. The vortex moves eastward in general along the shearline, and disappears in the east Plateau and along its edge. Only a few vortices can move eastward out from the Plateau.

Table 1 shows the eastward movement of vortices in the Qinghai-Xizang Plateau for May-August 1975—1982. It is seen from Table 1 that there are 54 vortices during QXPMEX, though only 7 moved out from the Plateau. 5 vortices disappeared at 102°—110°E and 2 vortices moved to the east of 110°E. This is similar to the mean situation for 1975—1982. Then there were 37 vortices per year, but only 9 vortices could move out from the Plateau. Seven disappeared at 102°—110°E and 2 moved to the east of 110°E.

The weather around the vortex is bad with frequent rainfall and low cloud as illustrated in the surface chart at 0600 GMT 14 July, 1979 (Fig. omitted). In the temperature and pressure field the vortex is very weak but its cyclonic circulation is strong. The vortex is shown clearly in the satellite picture. For example. at 0900 GMT 22 July, 1979 there are two vortices on the Plateau as shown in the

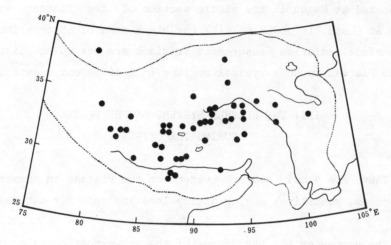

Fig. 1 The distribution of vortices on the 500 hPa chart
on the Plateau in May–August 1979.

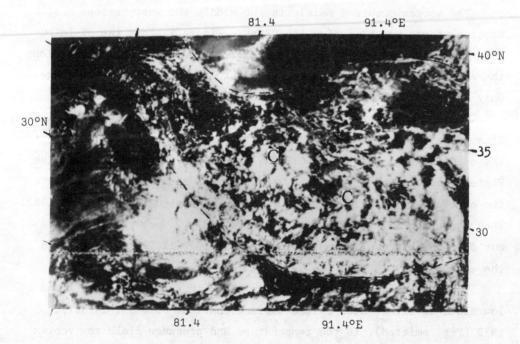

Fig. 2 The Tiros-N visible picture at 0900 GMT 22 July, 1979.

Table 1 The eastward movement of the vortices in
the Qinghai-Xizang Plateau.

Year	N	disappearing number of vortices at 102°—110°E				disappearing number of vortices to the east of 110°E			
		May	June	July	Aug	May	June	July	Aug
1975	42	0	3	3	0	1	0	0	0
1976	32	1	2	2	0	2	0	0	0
1977	36	1	1	1	0	2	1	0	0
1978	25	2	0	3	0	2	2	0	0
1979	54	2	2	1	0	1	1	0	0
1980	49	1	4	6	0	0	3	0	0
1981	28	4	6	3	0	0	0	0	0
1982	27	1	0	2	0	0	1	0	0
mean	37	1.5	2.3	2.6	0.4	1.0	1.0	0	0

N = total number of vortices

satellite picture marked C (Fig. 2).

The following six synoptic factors contribute to the creation of
a vortex: the large static instability near the surface, large barotropic instability, small vertical wind shear, large relative vorticity
large air-surface temperature difference and large relative humidity.

2. Shearline

The shearline is also one of the sub-synoptic scale and rain
bearing systems on the Plateau in summer. It is a quasi-stationary
front on the Plateau with no distinct discontinuity in temperature and
pressure gradient field. There is, however, a clear wind discontinuity
along the front. The shearline is shallow at about 2 km in depth; it is
located in warmer air with cyclonic vorticity along the line.

We performed a simple numerical simulation on the shearline on the
Plateau, using a two-layer p-σ primitive equation model, ideal topography (ellipsoid-shape, with top height about 5 km), ideal zonal
primitive flow field and ideal stationary heating field. But the size
and location of topography and the intensity of the heat source used

in the simulation is much similar to the reality of the Plateau.

It can be seen from Fig. 3 that owing to the dynamical effect of the Plateau a weak shearline appears over the east Plateau on the third day. It is then strengthened continuously, becoming a stationary shearline with nearly west-east direction after five and a half days. Fig.4

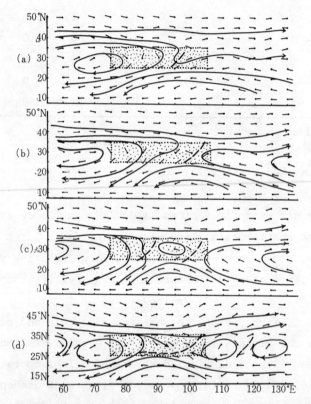

Fig. 3 The simulated streamfield on 500 hPa with pure dynamic effect of the Plateau.(a) 72h, (b) 96h, (c) 120h, (d) 132h.

shows the simulating streamfield on 500 hPa with both dynamical and thermodynamical effects of the Plateau. We can see that on the second day a shearline appears over the Plateau. Later it is enhanced and

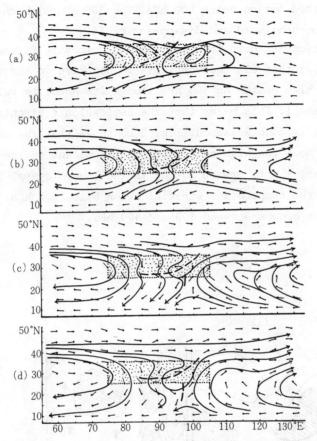

Fig. 4 The simulated streamfield on 500 hPa with both the
dynamical and thermodynamical effects of the Plateau.
(a) 48h, (b) 72h, (c) 96h, (d) 120h.

becomes stationary. So we arrive at the preliminary idea that the
shearline is mainly caused by the dynamical effect of the Plateau,
although the Plateau's thermodynamical effect also plays a strenthening role.

3. The Moving Anticyclone

Sometimes the 500 hPa subtropical high moves eastward from the
Iranian Highland and enters the Qinghai-Xizang Plateau, but, owing to
the intensive surface heating effect on the plateau, the anticyclone's
thermal structure below 2 km changes rapidly and the temperature lapse
rate becomes unstable. Thus the downward motion within the anticyclone

changes into upward motion below 400 hPa. The 500 hPa high on the
Plateau moves southeastward and disappears rapidly.

This situation is shown in the following composite figure. We can
see from Fig. 5(a) that all the vorticity is negative in the whole air
column over the high center on 500 hPa though its absolute value in the
lower layer is small. This means that the anticyclonic circulation
near the 500 hPa is much weakened. Fig. 5(b) shows that the original
downward motion near the high center is still dominant in the upper
layer but it changes into upward motion in the lower layers. This is
consistent with the temperature stratification below 400 hPa. A subsidence inversion occurs between 300 and 400 hPa.

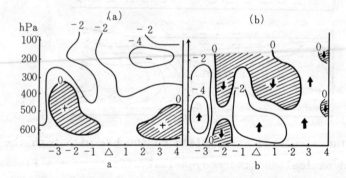

Fig. 5 East-west vertical profile of vorticity field (a)
$(10^{-5}s^{-1})$ and vertical velocity (b) $(10^{-4}hPa \cdot s^{-1})$
along the composite high pressure ridge-line on
the Plateau. The sign "$\triangle$" denotes the location of
the high center at the 500 hPa.

4. Cold Air Outbreak

Although the Plateau is under the control of subtropical weather
systems in summer, sometimes a deep westerly trough can extend southward
and enter into the southern part of the Plateau. It causes a cold wave
and a sharp temperature drop on the Plateau. On 30 July, 1979 a cold
air outbreak occurred on the Plateau with a deep westerly trough

extending southward to the southern part of the Plateau. The 24 hour temperature drop at 500 hPa is more than $16°C \cdot d^{-1}$ and on the surface is $15°C \cdot d^{-1}$ (Fig.6).

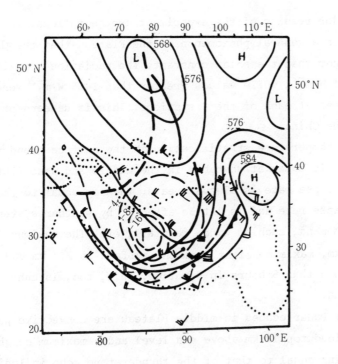

Fig. 6 500 hPa chart at 0000 GMT 30 July, 1979. The heavy
solid line indicates the trough line and the heavy
dashed line indicates that on the previous, the thin
solid line is the contour line, the thin dashed line
is 24h temperature change, " ➤ " is the surface
front and the dotted line denotes the Plateau.

IV. PUBLICATIONS AND MEETINGS

In the last four years two symposiums on the Qinghai-Xizang Plateau were held, specifically in 1980 and 1982. Two proceedings on these two symposiums have been issued and the third is to be published in the near future. In addition, two atlases of surface radiation budget and heat budget on the surface of the Plateau have also been published and the data of QXPMEX is now being prepared for publication.

V. THE MAJOR RESULTS OF THE EXPERIMENT

The major results of the experiment are as follows:

1) It is a general phenomenon on the Plateau that the global radiation is larger than the solar constant. The radiation's maximum value reaches 1521 $W \cdot m^{-2}$, while values greater than 1465 $W \cdot m^{-2}$ were found about 15 times at each of the 6 stations. This is unheard-of little known outside China.

2) The temperature stratification in the surface boundary layer below 1—2 km is somewhat unstable in the middle and west Plateau. The temperature lapse rate is always greater than or equal to the dry adiabatic lapse rate and is little affected by weather systems and weather phenomena, such as wind, rainfall, etc. The maximum intensity is -13.9°C/km, maximum height from the surface is 2.3 km and they are generally more than 6 hours in duration. This, too, is unheard-of outside China.

3) Most radar echoes in middle Plateau are convective with a mean height of cloud top 12 km above sea level and a maximum height reaching 18.2 km, being equal to that of the thunderstorm echo in India, but their horizontal dimension is smaller than in Beijing. This means that convective motion prevails in the Plateau, though the moisture is less than in Beijing.

4) The monthly and the ten day mean charts for the radiation budget and heat budget on the surface on the Plateau are compiled into an atlas.

5) The characteristics of the weather systems on the Plateau in summer are studied in more detail than before. By using composite analysis the structure of the vortices, shearlines and moving anticyclones is finally understood.

6) The distribution of the heat source of the atmosphere in Asia is evaluated using the data from the QXPMEX and MONEX. It is found that although on the Plateau there is a region of heat source, the center of heat source is the northeastern part of India and the northeastern

part of the Bay of Bengal with heating rate 7°C/day, on the Plateau the average heating rate is about 1°C/day.

7) The numerical simulation and theoretical studies on the influence of the Plateau on planetary circulations, on the summer monsoons in South Asia and East Asia and on the weather systems on the Plateau and those areas surrounding it have been completed and some preliminary results obtained.

In the area of Plateau meteorology, we suggest doing further research into the following problems:

1) It is important to study how the vortices on the Plateau move eastward out from the Plateau and how this affects the weather to the east of the Plateau.

2) We must study the interrelationships between the convections and the sub-synoptic motion systems on the Plateau and the large scale circulations in Asia.

3) Further diagnostical studies on the heat sources in Asia, including the numerical simulation on the feedback mechanism of heat source to the monsoon circulation in South and East Asia are needed.

4) The dynamic effect of the Plateau on the motion systems in Asia deserves to be studied in detail.

5) We must develop a good numerical model which will enable us to consider fully the thermal and dynamical effects of the Plateau.

6) Finally, we hope that it will be possible for a second TIPMEX to be carried out in the next five years.

Meteorological Measurement Systems in Mountainous Terrain

D. W. Beran

NOAA/ERL/WPL   Boulder, CO

1. Introduction

The complex atmospheric flows induced by mountainous terrain present
special measurement problems. The simplifying assumptions of horizontal
homogeneity and temporal continuity cannot be as readily applied as over
plains and water. Valleys that trap cold air, changing vegetation that alters
surface friction and radiation, and ridges that block low level flows are only
a few of the many added dimensions of meteorology in mountainous areas. Nor
does this terrain-induced complexity end at the top of the boundary layer or
over smoother surfaces in the lee of mountains. Atmospheric waves caused by
mountains are observed at great altitudes, and the effects of lee eddies can
be seen hundreds of kilometers downstream.

A first step toward understanding mountain meteorology is to observe the
flow and the thermodynamic properties of the atmosphere on scales that are
relevant to the problem. The key word here is scale. On very small scales or
where enough instruments can be deployed, conventional surface measurements
and upper-air soundings have proven adequate. As the scale of interest
increases, these methods become either very expensive or unfeasible, especially if greater upper-air temporal and spatial resolution is required.

Use of remote sensing devices over smoother surfaces has, in many cases,
proven to be an effective way of overcoming the limitations of conventional
sensors. This paper reviews a range of remote sensors and assesses their
applicability to the general area of mountain meteorology.

2. Remote Sensing of the Atmosphere

As the name suggests, remote sensing involves the measurement of a parameter at a point or points other than the instrument location. In general,
such techniques rely upon the interaction of waves (electromagnetic or
acoustic) with the atmosphere or the detection of naturally occurring radiance. One class of sensors (active) transmits a wave that interacts with the
atmosphere, then receives return signals that result from that interaction.
Another class (passive) monitors only those signals that are generated by some
physical mechanism within the atmosphere. Those interested in more information on these processes should see Derr (1972). The discussion here will be
confined to the meteorologically important measurements that can be derived
from both active and passive remote sensors.

In many cases, the ability of a remote sensor can be greatly enhanced by its manner of deployment. Fixed ground-based systems are, of course, effective only over the area within their inherent range. Aircraft mounting greatly extends the possible coverage, with penalties in the form of added operating costs, more complex data processing, and less temporal continuity. Satellite-based systems provide very broad coverage, again with a penalty of less spatial resolution and accuracy, and the ability to sense only a limited number of parameters. Not all sensors lend themselves to multiple forms of deployment. Physical size and the type of wave employed limit some systems to ground-based application only, reducing the range of parameters that can be measured.

For convenience the remote sensors discussed here are subdivided into broad ranges of the frequency spectrum in which they operate. Acoustic systems and electromagnetic systems (both microwave and optical) will be covered in that order.

3.   Acoustic Remote Sensing

The use of acoustics to remotely sense atmospheric structure started in the late 1960's with the experimental work of McAllister et al. (1969) in Australia and the theoretical analysis by Little (1969) in the U.S. Early experimental systems projected a sound pulse into the atmosphere, then detected the signal that was scattered from naturally occurring turbulence. Such monostatic devices have been variously called acoustic echo sounders, sodars, and echosondes. By carefully monitoring the time between sending the original pulse and receiving the returned signals, we can generate time-height plots of echo intensity. These plots reveal various structures in the atmosphere. These structures are associated with stronger scattering layers in regions where enhanced turbulence is present. Such layers are normally related with temperature inversions and wind shears.

The relatively slow moving acoustic waves suffer much greater attenuation than electromagnetic waves. The effective range of acoustic echo sounders is thus limited to only a kilometer. Further, the effective operating range is a function of the strength of turbulence. Because of this, acoustic systems tend to have an effective range that changes with the depth of the earth's boundary layer where turbulence is sufficient to produce a reliable echo.

The vertically pointing monostatic acoustic echo sounder has proven to be a useful system for monitoring the depth and oscillations of the earth's boundary layer. Figure 1 is an acoustic record showing the time history of boundary layer features in striking detail. This type of system produces a display showing the location of various features; however, it does not provide quantitative information on basic parameters such as wind, temperature, or moisture.

During the early 1970's, Beran et al. (1971a) demonstrated that monostatic acoustic systems could be used to measure the vertical (radial) wind speeds. Later, the antennas were tilted, and bistatic (transmitter and receiver antenna separated) systems that used the same Doppler principles were introduced (Beran and Willmarth, 1971b). While the equipment layout was somewhat more complex, this system provided profiles of the horizontal wind speed and direction in the first 600 m.

Both monostatic and bistatic Doppler systems have been used effectively in mountain meteorology experiments (Neff and King, 1984). They are ideal for the study of slow moving valley drainage winds where pollution transport is a prime concern. They are less useful when the measurement of high wind speeds is important. Strong winds interact with the antenna structure and produce eddies that saturate the sensitive transducers. This problem seriously limits the use of acoustic systems as general meteorological measuring devices. Because rather strong antenna sidelobes are unavoidable, care must also be exercised in siting systems in mountainous terrain. The sidelobe echo from nearby ridges or large structures will usually mask the weaker atmospheric returns.

Still, acoustic systems have an important role in mountain meteorology. They are one of the less expensive remote sensing systems commonly available today, they produce wind measurements to accuracies of $\sim \pm 1.0$ ms^{-1}, and they provide excellent range resolution, down to a few tens of meters. They are not, however, suited for all applications.

4. Electromagnetic Remote Sensing

a. Radar

The principle of operation of radars is similar to the acoustic systems described above except it is an electromagnetic wave that is transmitted and interacts with natural tracers in the atmosphere. The backscatter signal is, again, collected by a sensitive antenna. Non-Doppler weather radars employing wavelengths of less then 10 cm have been part of operational observing systems for many years. These systems rely upon scatter from natural hydrometeors and in some cases insects. They are very effective indicators of precipitation and have become the primary instrument for monitoring the position, movement, and strength of severe storms.

Doppler radars using these same wave lengths have been used for research and a 10 cm Doppler radar network (NEXRAD) is being implemented for operational use in the U.S. A single Doppler radar can measure the radial wind speeds out to ranges of a few hundred kilometers if hydrometeors of the proper size are present. Velocity azimuth display (VAD) techniques have been developed to interpret these radial velocities in terms of the general flow field.

More sophisticated wind measurements are possible when the radial velocity vectors from two or more Doppler radars separated by a known baseline distance are combined. Remarkable three-dimensional wind field measurements within convective storms and the earth's boundary layer have been produced with this dual Doppler technique, but the amount of data processing required is very large and complex. An example of wind fields measured by the dual Doppler technique over urban St. Louis, Missouri, is shown in Fig. 2 (Kropfli & Kohn, 1978).

The use of radar at wavelengths of less then 10 cm is critical to the investigation of precipitation and convective storm processes because of its sensitivity to hydrometeors. Powerful 10 cm Doppler systems can also measure clear-air winds, but generally only within the earth's boundary layer where the refractive index fluctuations provide strong echoes.

Chadwick and Gossard's (1983) review of clear-air radar remote sensing provides an excellent overview of such techniques, which started with the availability of special very sensitive radars in the late 1960's. These high

17

powered radars revealed such features as internal waves, breaking billows, and vortices. PPI scans showed such features as the doughnut-like appearance of convective cells in cross section.

One clear-air system transmits a frequency modulated continuous wave (FM-CW) (Richter, 1969) rather than the usual pulses. The FM-CW radar has several advantages over pulsed systems for atmospheric investigation. It can obtain very good range resolution (1.5 m) and has a very short minimum range (15 m), parameters that are determined by the pulse length in conventional pulsed radars. Because the FM-CW radar uses the difference between the frequency of the transmitted and returning signal to measure range to the target, it was long believed that FM-CW Doppler wind measurements would be impossible. This belief was dispelled in 1976 when a successful Doppler FM-CW system was demonstrated (Strauch et al., 1976, and Chadwick et al., 1976).

The effective range of present FM-CW radars is on the order of a few kilometers. Its excellent resolution and ability to measure winds make it an ideal instrument for boundary layer studies. Because it is more expensive to build than acoustic systems, its development has been slower. However, because it is not limited by strong surface winds it should eventually become the remote sensor of choice for general boundary layer observations.

A parallel development in clear-air radar sensing has been the exploitation of very large and powerful low frequency systems, originally developed for ionospheric probing, for measurements in lower regions of the atmosphere. Because they can provide wind measurements in the mesosphere, stratosphere, and troposphere, they are sometimes referred to as MST radars, but are more generally known as Profilers. These sytems operate in the VHF or UHF ranges. Several of them use the same 50 MHz frequency (VHF) found in the Jicamarca, Peru, ionospheric radar. These long wave lengths (~6 m) experience low attenuation, hence have excellent range capability, and do not interact with hydrometeors, allowing truly all-weather operation. The long wavelength and the need for a large power-aperture product combine to justify very large, relatively permanent antennas. The horizontal winds are measured by tilting the antenna beam a few degrees off vertical along two orthogonal directions. This is accomplished by phased arrays of coaxial cable suspended on posts covering areas as large as 100 meters on a side. Such antennas are inexpensive when compared with the large parobolic dishes required for shorter wavelength radars, but they are not easily transported.

In addition to the VHF units mentioned above, a UHF system (915 MHz) is now operating in Denver, Colorado. An example of the consistency of wind profiles from this unit is shown in Fig. 3. However, this higher frequency does interact with hydrometeors, and the resulting wind measurements can be contaminated by rain drop velocities. The UHF system provides better range resolution and lower minimum heights, although it gains these advantages at the expense of maximum range, non-all weather operation, and a costly parabolic antenna.

A 405 MHz Profiler is now under development. It combines the advantages of the 915 MHz and the 50 MHz systems while minimizing their weaknesses. This new system will have the capacity to essentially replace the rawinsonde and will greatly increase the temporal resolution of upper-air wind data because of its ability to operate continuously.

The 915 MHz system in Denver and four 50 MHz systems at other sites in Colorado form a small research network. They are sited on both sides of the main Rocky Mountain ridgeline through Colorado. This network has been used to analyze several storm systems and to support studies of pressure-height variability over mountains. A cross section analysis of the jet stream associated with a major storm system over the mountains is shown in Fig. 4. This cross section (Shapiro et al., 1984) used both Profiler and radiosonde data. Similar analysis of Profiler data taken between standard radiosonde launch times was used to track this major upper-air feature and its associated vorticity pattern hour by hour.

Wind Profilers provide a powerful new observing tool. Because they operate automatically, and high temporal resolution is achieved for long durations without the logistic problems and cost of radiosondes. With some special precautions they can readily be sited in mountain regions.

b. Microwave Radiometers

The previous section discussed active radar techniques that can be used to identify certain hydrometeor related features and to measure winds in clear

19

air. Passive microwave systems provide a method for direct measurement of thermodynamic parameters such as temperature and moisture.

Remote sensing by passive radiometers requires sensitive measurement of non-coherent electromagnetic energy that is emitted or scattered by the medium under observation. From the brightness of carefully chosen radiometric frequencies, we can infer temperatures at different altitudes in the atmosphere. Profiles of temperature can be derived from this information, mathematical inversion technique, surface temperature, and climatological soundings.

Radiometric measurements in the 50-60 GHz band, near the oxygen absorption line, have proven very effective for temperature profile measurements up to altitudes of about 500 hPa (Westwater and Decker, 1977). Radiometers operating in the 20-30 GHz band are used for measuring total water vapor and precipitable water.

Thermodynamic profiles measured by radiometers alone are generally smoother than those measured by a radiosonde (sharp inversions are not revealed). This characteristic is a result of the rather deep layer sensed by each frequency and the overall integrating effect of the technique. The resulting temperature profiles allow very precise determination of pressure heights, a parameter that is not affected by the inherent smoothing of the temperature profile.

It is possible to improve the radiometric profiles by adding information from other remote sensors. For example, because the wind Profiler uses an active technique that provides accurate range information and receives stronger return signals from inversions, it can determine the height of inversions. In addition, the gradients in the wind Profiler are related to the thermal structure of the atmosphere. When a radiometer temperature sounding is combined with this wind Profiler information, then very accurate, high resolution temperature profiles can be derived.

Radar remote sensing offers the greatest range of potential techniques for mountain meteorological measurements. The major limitation of such systems, especially the clear-air active devices, is their need for rather large antennas. This virtually dictates long term permanent siting. (Systems that depend upon hydrometeors as tracers are much smaller and have been used effectively from aircraft.) Antenna sidelobe interference is also a problem with active electromagnetic sensors. Careful site selection and improved data processing can eliminate most of these problems.

c.   Optical Remote Sensors

These devices generally operate in the visible portion of the frequency spectrum. Both passive and active techniques are employed. In addition, spectroscopic techniques could be used to determine the chemical make-up of the target, an advantage not found in the other remote sensing classes.

Turbulence in the clear-air atmosphere causes fluctuations in both phase and amplitude of an optical wave. The motion of turbulent eddies permits measurement of the transverse component of wind velocity between a light source and a receiver (Lawrence et al., 1972). The source can range from a star in space to an artificial light at the earth's surface, offering the possibility for transverse wind measurements along either a vertical or horizontal path. These line-of-site transverse wind sensors have proven extremely valuable for mountain meteorology applications such as measurement of drainage winds along a valley. Where used in a network to enclose a given area, they can provide direct measurement of the inflow/outflow which is related to low level convergence/divergence.

Lasers and laser radars (lidar) have made active optical sensing an important technique. Laser probes depend on the interaction of the laser (transmitted) energy with the atmosphere. This interaction attenuates the transmitted laser energy and introduces random fluctuations of amplitude and phase. Some of the attenuated laser energy appears as scattered light in the direction of a receiver, providing information on the atmosphere. As with other wave forms, the Doppler shift of the scattered light frequency is directly related to the component of wind speed along the beam in the volume where the scattering occurs. Natural aerosols provide the primary target for laser remote sensors.

Both continuous wave (CW) and pulsed coherent wave lasers are used to measure winds. The CW systems obtain range information by optically focussing the beam at some predetermined range. The range resolution for such systems varies from good near the instrument, to very poor as the range increases. Range resolution for pulsed systems is a function of the pulse length, and can be as small as a few hundred meters. Recently developed high power pulsed coherent CO_2 lasers, called TEA lasers, are being used as ground-based wind measuring systems. Ranges of 20 kilometers are common. These devices are being considered for use on board orbiting satellites where the motion of the satellite and an alternating forward and backward lidar scan will make it possible to directly measure the entire global wind field from space. This advanced capability will be invaluable for monitoring the effect of large mountain ranges on the global wind fields.

Passive infrared radiometers that operate like the microwave system described earlier are also available. The weighting functions for the ground-based microwave system and the satellite-based infrared devices peak at different altitudes. Thus, it is possible to combine the output from the two separate systems and to produce a significant improvement in the measured temperature profile.

Experiments using DIAL lasers and Raman scatter techniques to measure thermodynamic parameters have shown some degree of success. These methods are still very much in the development stage and are not readily available for practical application (Hardesty, 1984).

The very narrow beams projected by lasers offer distinct advantages for remote sensing. They can achieve very high cross-beam resolution and the annoying sidelobe problems associated with acoustic and radar devices are non-existent. These systems are also small enough to be readily portable and can be used on a variety of platforms. The expense of state-of-the-art laser systems and their inability to penetrate clouds containing large water droplets are two disadvantages that limit their use.

5.   Example of Application

The large range of remote sensing methods available for application to mountain meteorological problems makes it possible to sense the three basic meteorological parameters, wind, temperature, and moisture. In addition, in some special applications flux measurements and derived quantities such as divergence and vorticity can be obtained. Each application has its unique measurement requirements. Only when these requirements have been determined can the optimum sensor or combination of sensors be specified for a given experiment. Required spatial and temporal resolutions, accuracy, mobility,

22

and of course, cost and availability are all factors that must be balanced
before remote sensors are selected.

An excellent example of how remote sensors can be applied to solve a
problem in atmospheric measurement in mountainous terrain is the Atmospheric
Studies in Complex Terrain (ASCOT) project. The general thrust of this program is to investigate atmospheric boundary layer flow interactions with
complex underlying terrain. The primary motivation is to develop an understanding of these flows to help in the assessment of air quality.

Nocturnal drainage winds were chosen as the first phenomenon for investigation (Dickerson 1980). These winds are normally confined to relatively
small areas, on the order of 10-15 kilometers, and within about 1 kilometer in
the vertical direction. Doppler acoustic sounders and transverse optical wind
sensors were deployed along the valley under study. These instruments made it
possible to continuously monitor the vertical profile of the boundary layer
wind, the depth of the boundary layer, and the mean flow at various points
across the valley. The temporal continuity and density of measurements provided by the remote sensors in this experiment would have been very difficult
or impossible to achieve with conventional in-situ sensors.

References

Beran, D. W., C. G. Little, and B. C. Willmarth (1971a): Acoustic Doppler
    measurements of vertical velocities in the atmosphere. Nature, 230,
    160-162.

Beran, D. W., and B. C. Willmarth (1971b): Doppler winds from a bistatic
    acoustic sounder. Proc. 7th Int. Symp. on Remote Sensing of Environment,
    May 17-21, Univ. of Michigan.

Chadwick, R. B., and E. E. Gossard (1983): Radar remote sensing of the clear
    atmosphere - review and application. Proc. IEEE, 71, (6).

Chadwick, R. B., K. P. Moran, R. G. Strauch, G. E. Morrison, and
    W. C. Campbell (1976): Microwave radar wind measurements in the
    clear-air. Radio Sci., 11, 795-802.

Derr, V. E. (1972): Remote sensing of the troposphere. Superintendent of
    Documents, U.S. Govt. Printing office, Washington, D.C. 20402.

Dickerson, M. H., editor (1980): A collection of papers based on drainage
    wind studies in the geyser area of Northern California: Part I.
    Lawrence Livermore National Laboratory, Livermore, California. UCID-
    18884, ASCOT-80-7.

Hardesty, R. M. (1984): Coherent dial measurement of range-resolved water
    vapor concentration. To be published, Applied Optics, Aug. 1984.

Kropfli, R. A., and N. M. Kohn (1978): Persistent horizontal rolls in the
    urban mixed layer as revealed by dual-Doppler radar. J. Appl. Meteor.,
    17, 669-675.

Lawrence, R. S., G. R. Ochs, and S. F. Clifford (1972): The use of scintil-
    lations to measure average wind across a light beam. Appl. Opt., 11,
    239-243.

Little, C. G. (1969): Acoustic methods for the remote probing of the lower atmosphere. Proc. IEEE, 57, 571-578.

McAllister, L. G., J. R. Pollard, A. R. Mahoney, and P. J. R. Shaw (1969): Acoustic sounding - a new approach to the study of atmospheric structure, Proc. IEEE, 57, 579-587.

Neff, W. D., and C. W. King (1984): Studies of complex terrain flows using acoustic remote sensors. To be published as a tech. memo, NOAA/ERL, Boulder, Colorado 80303.

Richter, J. H. (1969): High-resolution tropospheric radar sounding. Radio Sci., 4, 1261-1268.

Shapiro, M. A., T. Hample, and D. van de Kamp (1984): Radar wind Profiler observations of fronts and jetstreams. To be published in Bull. AMS.

Strauch, R. G., W. C. Campbell, R. B. Chadwick, and K. P. Moran (1976): Microwave FM-CW Doppler radar for boundary layer probing. Geo. Rev. Meteor., 3, 193-196.

Westwater, E. R. and M. T. Decker (1977): Application of inversion to ground-based microwave remote sensing of temperature and water vapor profiles. In Atmospheric Remote Sounding, Academic Press, New York, 395-427.

25

DIAMOND D ACOUSTIC SOUNDER

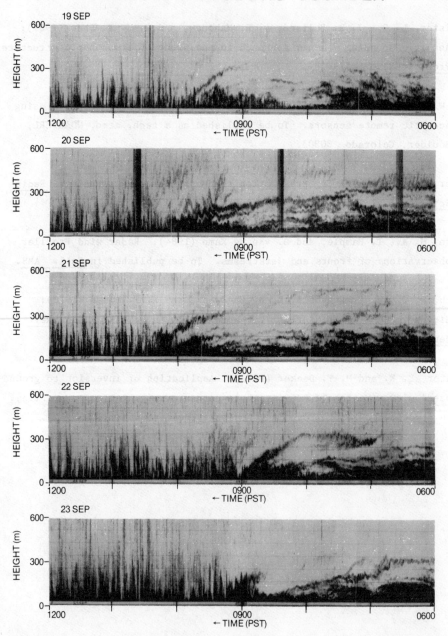

Figure 1.   Acoustic sounder record showing temperature structure
in the first 600 m on four consecutive days from
0600–1200 PST (right to left).

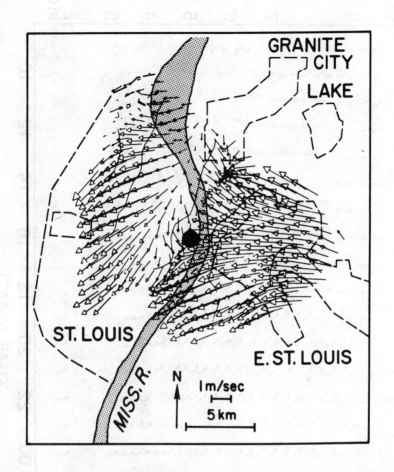

Figure 2.   Eddy wind field at 0.3 km AGL. Winds are obtained
from an average of 17 volume scans taken during a 70
min period ending at 1557 CDT. The mean wind over
the entire chaff cloud has been subtracted from these
data.

27

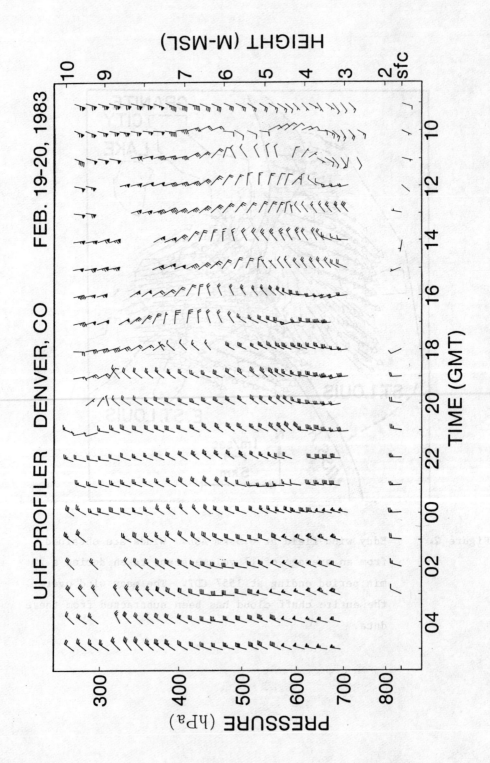

Figure 3.   Display of wind profiles from UHF Profiler.

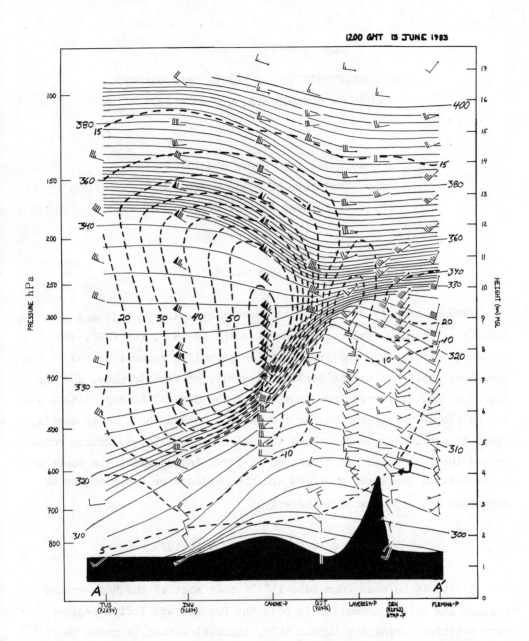

Figure 4.   Cross-section analysis of windspeed (ms^{-1}, dashed
            lines) and potential temperature (K, solid lines) at
            1200 GMT 13 June 1983.   Analysis is a composite of
            conventional rawinsonde soundings and radar wind
            profiles.   Profiler soundings are designated by the
            letter P at the horizontal axis.

SOME CHARACTERISTICS OF PLATEAU MONSOON AND
INDIAN SW MONSOON DURING SUMMER, 1979

Gao Youxi, Yuen Fumao and Li Ci
Lanzhou Institute of Plateau Atmospheric
Physics, Chinese Academy of Sciences

I. INTRODUCTION

Various anomaly charts are analysed here using data from a
variety of sources. These include the MONEX in 1979, the FGGE, and
pressure, temperature and precipitation anomalies as well as the 30 hPa
anomalies published by the Berlin University of West Germany, and the
sounding data collected from stations at the Qinghai-Xizang Plateau and
its neighbouring regions during the China's 1979 QXPMEX. These charts
focus in particular on the characteristics of the Plateau summer monsoon and the Indian southwest monsoon in 1979. In addition, the correlations between these two monsoons and the reasons for their causes of
formation are also discussed briefly.

II. THE WEAK SUMMER MONSOON IN 1979

1) It can be seen from Tab. 1[1,2] that most of the near-ground
pressur and height anomalies in both The Plateau and Indian region
were positive during May-August 1979, indicating that in these two
regions, the summer monsoon low systems in 1979 were weaker than those
in usual years.

2) According to the monthly mean chart of 600 and 500 hPa (Fig.
omitted) for the Plateau and Indian regions during May-August 1979,
the level the monsoon heat low is able to reach, taking the Plateau
region as the center, is lower than in normal years. From the mean

Tab. 1 The symbol contrast of pressure and height
anomalies from ground to 600 hPa level between the Plateau and the Indian regions.

		surface pressure	850 hPa	700 hPa	600 hPa
May	Plateau	+			−
	India	+	+	+	+
June	Plateau	+			+
	India	+	+	+	−
July	Plateau	+			+
	India	+	+	+	−
August	Plateau	+			−
	India	−	−	−	−

chart of years, the summer monsoon heat low may reach the vicinity of
400 hPa. Judging from the monthly mean chart of 500 hPa during June-
August of 1979, this heat low actually becomes a warm high. We can
also see from the monthly mean pressure charts that for the Indian
region, there are already Pressure highs at the 700 hPa level during
May-July, which indicates that the height of the Indian monsoon low in
1979 is still lower than in normal years.

3) Tab. 2[2] lists the meridional pressure gradients between
various stations. This table shows that at the 600 hPa level the west
monsoon component over the south Plateau during May-August in 1979 is
weaker than in other years, and that at the 500 hPa level, the west
monsoon completely disappears during the months of June to August. This
also indicates that the summer monsoon over the southern part of Plateau,
including the Indian region, is much weaker than in ordinary years.

III. THE LATE ONSET OF THE SUMMER MONSOON IN 1979

1) Tab. 3[2] shows the dates of the onset of the rainy season in
the Plateau and its neighbouring regions. In the summer of 1979, with
the exception of the westen part of Sichuan, the rainy season began

Table. 2 The intensity anomaly of the Plateau monsoon and
Indian monsoon (1979)

[pressure gredient unit: geo-potential meters]

		May		June		July		August	
		normal	1979	normal	1979	normal	1979	normal	1979
850 hPa	Bhubaneswar-Gauhati	19	12	21	-11	-12	-1	-16	-5
700 hPa	Bhubaneswar-Gauhati	25	-3	26	-2	-3	1	-12	-7
600 hPa	Bhubaneswar-Gauhati	30	-3	25	5	-10	7	-5	-4
	Gauhati-Nagqu	30	22	0	0	20	9	5	1
500 hPa	Bhubaneswar-Gauhati	40	25	14	0	-20	13	-17	3
	Gauhati-Nagqu	30	27	15	-26	5	-18	0	-35

Table. 3 The dates of the onset of the rainy season in the
Plateau and its neighbouring regions.

stations	1979	normal	difference (in days)	stations	1979	normal	difference (in days)
Zunyi	10/J	19/A	-52	Nagqu	6/J	19/M	-18
Guiyang	1/M	21/A	-19	Shiquanhe	1/J1	23/J	- 8
Bijie	10/J	4/M	-37	Golmud	16/J	23/A	-54
Kunming	1/J	13/M	-19	Chongqing	18/A	7/M	+19
Jinghong	8/M	3/M	-15	Yibin	30/A	18/M	+18
Tengchong	7/J	18/A	-40	Chengdu	30/A	28/M	+28
Lijiang	7/J	26/M	-12	Leshan	30/A	9/M	+ 9
Dêqên	7/J	15/A	-53	Kangding	1/M	11/M	+10
Nyingchi	16/M	26/A	-50	Barkam	18/A	11/M	+23
Lhasa	17/J	25/M	-20				

"+" denotes the days earlier than the normal;
"-" denotes the days later than the normal.
A: April, M: May, J: June, J1: July.

earlier than the mean date. The rainy seasons in the Plateau and its neighbouring Yunnan-Guizhou Plateau began 10—50 days later than in normal years.

2) Fig. 1[3] shows that in the northeastern part of India to the east of 90°E, the arrival of the SW monsoon is about one week later than in a normal year, which is consistent with the situation in eastern regions of the Plateau. The monsoon in the central part of India usually arrives on June 10, while in 1979, the date was closer to June 20, about 10 days later. This is also consistent with the onset of rainy season in the middle part of the Plateau.

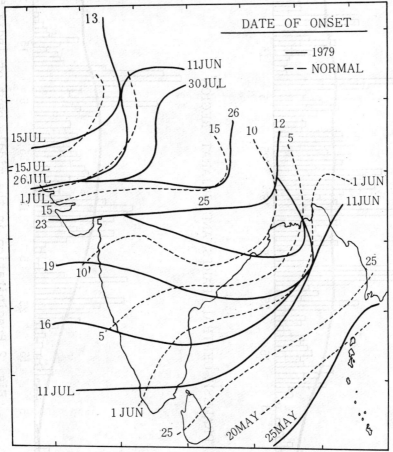

Fig. 1 The date of the onset of the Indian southwest
monsoon in 1979 (taken from reference [3]).

3) Fig. 2[3] shows the time-series curves for precipitation in the middle part of India from May 16 to September 30, 1979. From this

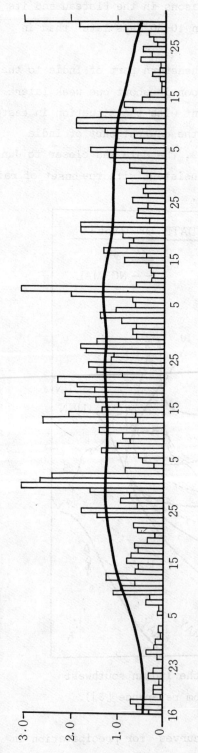

NORTHEAST INDIA

UTTAR PRADESH, CENTRAL PARTS AND GUJARAT REGION

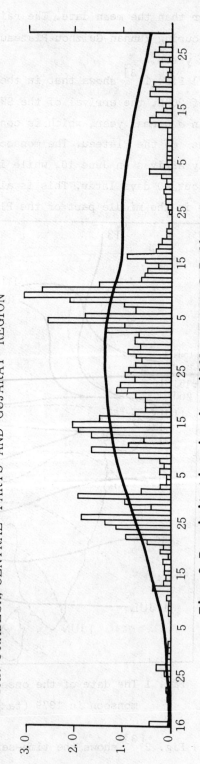

Fig. 2 Precipitation in the middle part of India from May 16 to September 30, 1979.

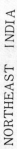

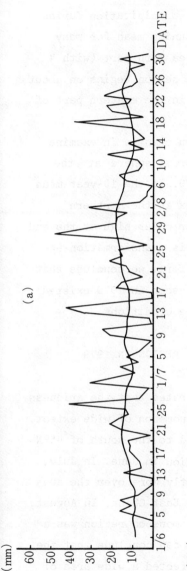

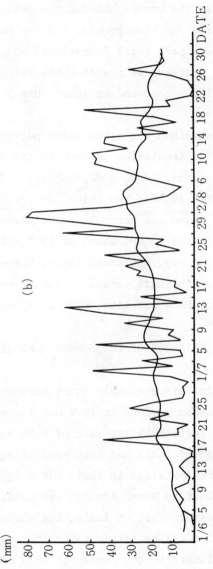

Fig. 3 The mean daily precipitation (in thin solid lines) from June to August, 1979 and the five-day running mean precipitation over a number of years.

(a) For the 4 stations in the eastern part of Plateau.

(b) For the 7 stations in the middle part of Plateau.

figure, we can ascertain that the burst of the southwest monsoon in
the middle part of India is normally about June 10, while in 1979
(see Fig. 2) the date on which precipitation reached the normal amount
was June 20, some 10 days later.

Fig. 3* gives the curves showing the daily precipitation during
June-August in 1979 and also present a 5 day running mean for many
years in the middle part (with 7 stations) and eastern part (with 4
stations) of the Plateau. We see that the rainy season begins on about
June 18 in the middle part and on about June 5 in the eastern part of
Plateau.

4) We also examined the circulation pattern changes in examine
the changes in the circulation pattern of the stratosphere at the
30hPa level(Fig.omitted) during May-August, 1979. In the 10-year mean
chart of 10 hPa and 30 hPa, the polar low vortex in the northern
hemisphere comes under the control of the summer polar high at the end
of April, while in the 30 hPa chart of 1979, this transformation appears at the end of May, one month later. Therefore, we conclude that
the phenomenon for the late coming of summer monsoon in 1979 existed
not only in troposphere but also in stratosphere conditions.

IV. THE ARIDNESS OF THE SUMMER MONSOON REGION IN 1979

1) The precipitation anomaly chart demonstrates that the aridness
of the summer monsoon regions in 1979 was a phenomenon of wide extent.
In May-June, all regions at the east of 60°E and to the south of 45°N
with the exception of India had less precipitation in June. In July,
there was less precipitation in India but slightly more over the area
from the Middle East to North Africa, excluding East China. In August,
except for the western part of India, the whole monsoon region was a
wide area characterized by little rain. Thus we can conclude that the
phenomenon of aridness in the summer of 1979 affected a wide area of
monsoon regions.

---

* Adopted from. Mr. Qu Zhang's Paper which is not published yet.

2) Table. 4[2] indicates that on the Plateau in the months of May, June and August, precipitation was of negative value, and that in India, there were few rains in May and July. In addition, from Tab. 4 demonstrates that the precipitation anomalies both on the Plateau and in India tend to be negative. Comparing with Tab. 4 the anomaly chart of temperature and humidity (Fig. omitted), we can see that in 1979 the blocking high in Europe developed well, the long-wave trough in the Balkhash lake was intensive, the cold-and-dry air was active, the influential latitude was low, and the maintaining period was long. Furthermore, from June to August, the precipitation anomaly symbols the Plateau were the opposite of those in India.

Tab. 4 The precipitation anomaly symbols for the Plateau and the middle and northern parts of the Indian regions during May-August, 1979.

	Plateau	India
May	-	-
June	-	+
July	+	-
August	-	+(-)*

* Indicating that there are few negative (or positive) anomalies in the particular area.

3) Tab. 5* shows the contrast between the symbols of humidity anomalies (dew point temperature) at different layers in the Plateau and Indian regions in May-August, 1979 and those symbols at the layers of 600—300 hPa. The dewpoint temperature departure for these two regions consisted of negative symbols and their humidities were lower than in other years. In comparing the height anomaly and the temperature anomaly, we suggest that the dry air over these two regions may originate from both polar and Iranian dry-and-cold air.

* Adopted from, Mr Gao Youxi, Yuen Fumao and Abudou Rousuli's paper which is not published yet.

Tab. 5 The dew-point temperature anomaly symbols for the
Plateau and Indian regions during May-August, 1979.

		600 hPa	500 hPa	300 hPa
May	Plateau	-	-	-
	India	-	-	-
June	Plateau	-	-	-
	India	-	-	-
July	Plateau	-	-	-(+)*
	India	-	-	-
August	Plateau	-	-	-
	India	-	-	

(* see Tab. 4).

V. THE RELATIONS BETWEEN THE PLATEAU SUMMER
MONSOON AND THE INDIAN SW MONSOON

The Plateau summer monsoon and the Indian south-west monsoon are
systems that both occur in the famous Asian monsoon regions. To a
certain extent, they are two independent systems, but there are certain
with definit relations between them. From Tab. 1, we see that their
pressure anomalies are mainly of same symbols on the near-ground layer.
Comparing Tab. 1 with Tab. 6, we see that at the middle troposphere of
600—500 hPa, the anomalies in the Plateau and Indian regions have mainly
the opposite symbols, while at the layers of 300—30 hPa, the anomalies
are again mainly of same symbols (see Tab. 6). However, the symbol of
the anomaly of the stralosphere at the 30 hPa is not same as for the
troposphere.

VI. THE RELATIONSHIP BETWEEN THE HEATING EFFECT OF
PLATEAU AND THE PLATEAU SUMMER MONSOON

The Plateau summer monsoon is a wind system formed by a seasonal
change. This change is due to the heat difference between the Plateau

Tab. 6 The height anomaly symbols for the 500—30 hPa
levels during May-August, 1979.

		500 hPa	300 hPa	200 hPa	100 hPa	30 hPa
May	Plateau	+	+	-	-(+)*	-
	India	-	-	-	-	-
June	Plateau	+	+	-	-	+
	India	-	-	-	-(+)*	+
July	Plateau	+	-	-	-	+
	India	-	-	-	-	+
August	Plateau	+	-	-	-	+

(* see Tab. 4).

Tab. 7 The total atmospheric heating anomaly symbols
during May-August, 1979.

	May	June	July	August
Plateau	-	-	+	-
North and middle India	-	-	+	0

topography and the surrounding free atmosphere at the same height. The
intensity, extent and duration of this wind system are essentially
determined by the intensity of the heating effect of the Plateau. Using
the computed anomalies of total atmospheric heating over the Plateau
and India, we were able to construct Tab. 7[6] and find that the heating
effect on the Plateau was rather weak during the whole summer season
in 1979. Only in July did positive heating anomaly appear in a large
area. In addition, during the summer season of 1979, the total atmospheric static energy in the Plateau and south Asian regions was also
weak. In the summer of 1979 these conditions resulted in such climatic
characteristics as lower temperature on the ground layer, positive
pressure anomaly, frequent movements of cold air and a weak monsoon low
in the Plateau region.

CONCLUSION

The 1979 summer monsoon was weak in intensity, small in scope, and late in onset. The precipitation was less than in other years, leading to drought conditions. All these characteristics had much to do with the lower intensity of the atmospheric heat source, the frequent cold and dry weather in early summer and the strong advection of warm and dry air in midsummer over the south Asian regions (including the Qinghai-Xizang Plateau).

REFERENCES

[1] Ergebnisse des Synoptischen Dienstes im Jahre 1979, Meteorologische Abhandlungen, Band 23/Heft 5—8, 1979.

[2] Yuen Fumao, Gao Youxi. Some climatic characteristics of the Qinghai-Xizang Plateau monsoon and the Indian monsoon in 1979, Collected Papers of the Qinghai-Xizang Plateau Meteorological Science Experiment (1) Science Press, 1984 (in Chinese).

[3] Summer MONEX field phase report, FGGE operations report, Vol, 8, A 2. 3—11 3—16, 1981.

[4] R. Scherhag et al., Klimatologische Karten der Nordhemisphare, Meteorolongische Abhandlungen, Band 100/Heft 1, 1969.

[5] Tägliche Höhenkarten der 30-mbar-Fläche sowie monatliche mittelkarten für das Jahr 1979, Meteorologische Abhendlungen, Band 25/Heft 1—4, 1979.

[6] Yao Lanchang, et al., Studies on the mean atmospheric heat sources over Tibetan Plateau and its surrounding regions in summer, Collected Papers of Qinghai-Xizang Plateau Meteorological Science Experiment (1), Sciences Press, 1984 (in Chinese)

SEASONAL TRANSITION OF ATMOSPHERIC CIRCULATION
IN EARLY SUMMER,1979 AND THE ROLE
OF THE QINGHAI-XIZANG PLATEAU

Guo Qiyun

Institute of Geography, Chinese Academy of Sciences

One of the most important problems concerning meteorological experiments on the Qinghai-Xizang Plateau is the seasonal transition of
atmospheric circulation in early June, a subject this paper will address.
The focus of our discussion is on the seasonal transition of atmospheric
circulation in different geographic regions , and the onset process
of the summer monsoon in early June. The results of the analysis
indicate that three phases can be generally distinguished for the period
of circulation transition in early summer, 1979. The role of the
Qinghai-Xizang Plateau in this seasonal transition of atmospheric
circulation is discussed at the end of this paper.

I. THE CHARACTERISTICS OF ATMOSPHERIC CIRCULATION
IN THE TRANSITION SEASON--EARLY JUNE 1979

It is well known that the variation of zonal wind in the middle and
upper troposphere and the adjustment of long and ultra-long waves are
two important elements in the seasonal transition of atmospheric
circulation. Each of these elements is analysed in this section.

1. The Variations of Subtropical Westerlies in the Upper
   Troposphere

The variations of zonal wind are studied by using the wind data
set on a 5° × 5° grid within the area 0—180°E, 35°S—35°N, at 1000, 700,
500, 200 and 100 hPa levels. The updated wind observations over the
Qinghai-Xizang Plateau have also been included in the study. It is

found that not only in the Northern Hemisphere but also in the Southern Hemisphere the zonal wind experiences considerable modification in the early or middle part of June.

Fig. 1 shows the variations of pentad mean subtropical westerlies at 200 hPa for different meridians in both the Northern and Southern Hemispheres from May to July. In the Northern Hemisphere, the subtropical westerlies were in general weakened within this time interval. Along most of the meridians an abrupt weakening was observed by the end of May or in early June. If the absolute wind speed lower than 20m/s is used as a criterion of seasonal transition, the date of transition was found in the sixth pentad of May in the east Atlantic (0°E), in the first pentad of June in west Asia (60°E), in the third pentad near the Plateau (90°E) and in the fourth pentad over Asia (120°E). Generally the date of transition is delayed from the west to east.

At the same time, the intensity of subtropical westerlies was changed too in the Southern Hemisphere, though the seasonal transition (from summer to winter status) was not so clear as in the Northern Hemisphere because the non-seasonal variations in the Southern Hemisphere were stronger[1]. However, a moderate strengthening of westerlies could still be found between the third and fourth pentads of June in the Indian ocean (60°E). The strong westerlies predominated until the sixth pentad of July. This characteristic transition was obscured along other meridians, nevertheless, even here the trend toward increasing westerlies is still to some extent notable.

2. The Variations of Long and Ultra-long Waves

The zonal harmonic analysis was carried out for the pentad heights along the 30° latitude circle (representing the subtropical zone) and the 50° latitude circle (representing the temperate zone) in both the Northern and Southern Hemispheres at 1000, 500 and 100 hPa levels from May to July. The relative amplitudes and the trough positions of the first five harmonics were studied. The date of the general circulation adjustment was determined for each parameter according to its variation The dates of transition for most parameters in both hemispheres are concentrated in the first and second pentads of June.

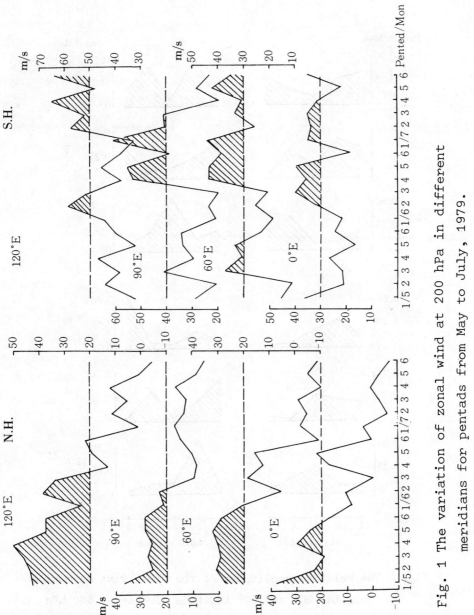

Fig. 1 The variation of zonal wind at 200 hPa in different meridians for pentads from May to July, 1979.

Fig. 2 gives the relative amplitudes of the first five harmonics along 30° and 50° latitude circles at 500 hPa. Only three pentads

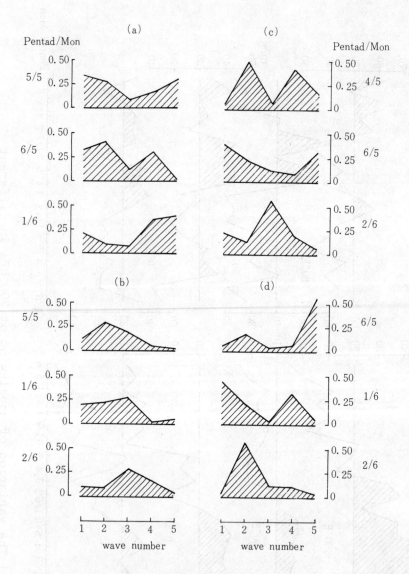

Fig. 2 The relative amplitudes of the first five harmonics along the 30° and 50° latitude circles at 500 hPa for pentad.

before and after the transition are shown here. It can be seen that the variation took place in the first pentad of June for 50°N (Fig. 2a). Before the first pentad of June the first and second harmonics predominated, after that they were weakened as the fourth and fifth harmonics became prominent. The date of adjustment falls in the second pentad of June for 30°N. This is later than for 50°N. This kind of variation took place in conjunction with the weakening of the first and second harmonics and the strengthening of the third and fourth (Fig. 2b). During that period the trough along the east coast of Asia gave way to the ridge at 500 hPa.

So in temperate latitudes the adjustment of planetary waves occurred earlier than in subtropical latitudes. The adjustment started in Europe, and then continued in a down-stream direction. The main trough along the east coast of Asia became weaker and retreated to the west. And one pentad later, the planetary wave system started to be modified in the subtropical latitudes. The third and fourth harmonics intensified and the ridge of subtropical high in the west Pacific extended to the east coast of Asia.

The date of the general circulation adjustment was the second of June in the Southern Hemisphere. The relative amplitudes of the third harmonic were small before adjustment, but after adjustment the third harmonic predominated for 50°S (Fig. 2c). Meanwhile, for 30°S (Fig. 2d). after adjustment the fourth and fifth harmonics became weaker and the second harmonic increased. It is readily apparent that the characteristics of planetary waves in the transitional period mentioned above are closely related to the seasonal transformation of general atmospheric circulation[2].

II. THE ESTABLISHMENT OF SUMMER MONSOON
CIRCULATION SYSTEMS IN ASIA

The abrupt set-in of the summer monsoon in the Asian continent during the transition period of atmospheric circulation in the Northern Hemisphere in early summer is a prominent meteorological phenomenon. During the transition period in June 1979, all synoptic components of the summer monsoon system experienced obvious modification. First of all, the Indian hot low became suddenly deeper in early June (not shown). About ten days later, in the third pentad of June, the high at 100 hPa suddenly moved to the west about 10 degress longitude between two successive pentads, while at the same time the Easterly at the south of the high also intensified (Fig. 3a), and the ITCZ along the 70°E profile shifted to the north from south of 5°N to 10°N[3].Accompanyin

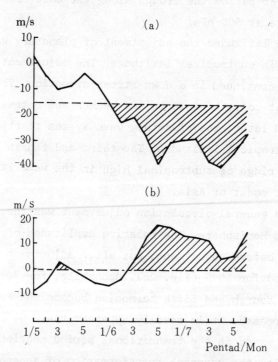

Fig. 3 The variations of zonal wind at 100hPa in 15°N,
80°E(a) and at 700 hPa in 10°N, 75°E(b) for
pentads from May to July.

the above mentioned series of circulation adjustments, the SW monsoon arrived at the southern end of Indian peninsula in third pentad of June (Fig. 3b). Then, as the Indian hot low deepened further, the sea-level pressure difference over the Indian peninsula between 10°–30°N decreased more considerably. At the same time the center of the high at the100 hPa level in the upper troposphere moved up to the Qinghai-Xizang

46

Plateau (not shown). With the intensification of the cold high from South Africa to Australia, the pressure situation became favorable to the enhancing of the meridional flux. As a result, the cross-equatorial air current in the lower troposphere was established definitely. During this interval the ITCZ and SW monsoons moved north with greater speed to 20°N over the Indian peninsula, while the subtropical high in the West Pacific jumped north with its ridge extending to the East Asian continent. In the end the summer monsoon predominated along the lower and middle reaches of Changjiang River, and the Mei-yu spell began (Fig. 4 a, b).

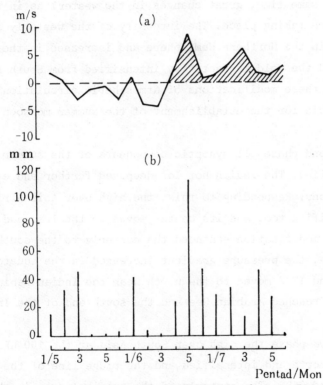

Fig. 4 The variations in meridional wind at 700 hPa in 30°N, 115°E (a) and mean rainfall in the middle and lower reaches of Changjiang River (b) for the pentads from May to July.

III. THE THREE PHASES IN THE SEASONAL TRANSITION OF ATMOSPHERIC CIRCULATION IN EARLY SUMMER

Three phases can be generally distinguished in the whole process of circulation transition in the early summer of 1979. In the first phase the long and ultra-long waves experienced dramatic modification in tropical and temperate zones in both hemispheres. It is interesting to note that this happened almost simultaneously but with some variance in the Northern Hemisphere, viz. the adjustment of circulation in the temperate zone preceding that in the subtropical zone by one pentad. At the same time, great changes in the westerlies in the upper troposphere were taking place. The intensity of the westerly at 200 hPa decreased in the Northern Hemisphere and increased in the Southern Hemisphere, and the cold air activity intensified from South Africa to Australia. All these modifications of atmospheric circulation provided a favorable basis for the establishment of the summer monsoon system in Asia.

In the second phase all synoptic components of the summer monsoon system intensified. The Indian hot low deepened further and expanded in all directions. Responding to this, the high over Asia-Africa at 100 hPa intensified too, and its center moved to the Iran and Xizang Platéaus. This modification enhanced the easterly to the south of the high. Meanwhile, the pressure gradient increased in the equatorial Indian Ocean and ITCZ moved to the north near the Indian peninsula. The SW monsoon reached Arabian Sea and the south end of the Indian peninsula.

In the third phase the high over South Asia at the 100 hPa level moved further north and intensified and the ridge line of the high pressure crossed 25°N. The easterly of the tropical zone in the upper troposphere increased in intensity. The ridge line of the subtropical high in the west Pacific at the 500 hPa level jumped to the north of 20°N. The cross-equatorial air current in the lower troposphere established itself as a result of the increase in sea level pressure in the Mascarene and Australia highs in the Southern Hemisphere. The southwest monsoon developed further, and moved to the north of 15°N over the

48

Indian peninsula. As the south-east monsoon predominated along the
lower and middle reaches of the Changjiang River, the Mei-yu spell
began. The characteristics of summer circulation became definitely
established.

IV. A DISCUSSION OF THE ROLE OF THE QINGHAI-XIZANG
PLATEAU IN THE SEASONAL TRANSITION
OF ATMOSPHERIC CIRCULATION

The role of the Qinghai-Xizang Plateau has been studied at from a
variety of different angles[4], so discrepancy is common in the
literature of the field. A synopsis can be constructed on the basis of
circulation variations in 1979.

1) It has already been suggested that the Plateau plays an important role in the seasonal transition of atmospheric circulation in
the Northern Hemisphere from spring to early summer. The regulation of
atmospheric circulation in temperate latitudes over Asia is caused by
the north-ward migration of a subtropical jet stream. But in 1979 the
regulation of the planetary waves took place about half a month before
subtropical jet stream began its migration. Thus this explanation seems
problematic.

2) The variation among the absolute amplitudes of the first three
harmonics at the 100 hPa level along 30°N shows that the second and
third harmonics began to increase earlier than the first harmonic
(Fig. 5). It is well know that the second and third harmonics are
linked to the establishment of troughs over the Atlantic and the
Pacific, and of ridges over Africa and Mexico, while the third is
related to the development of the Qinghai-Xizang high. Therefore one
cannot regard the development of the Qinghai-Xizang high as a preliminary stage which sets the adjustment of atmospheric circulation in
motion. The synoptic process in 1979 shows that the Qinghai-Xizang
high was strengthened only after other lower latitude systemssin the
upper troposphere had developed.

3) Concerning the seasonal variation of westerlies in the upper
troposphere south of the Plateau, it is generally accepted that the
temperature increases earlier over the Plateau, which acts in summer

49

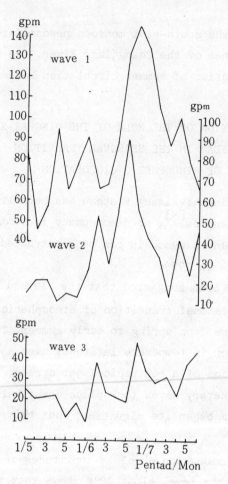

Fig. 5 The variations of absolute amplitudes of first
three harmonics at 100 hPa along 30°N for
pentads from May to July.

as a heat source, than it does in other places in the same latitude
zone. If this is true, the westerlies south of the Plateau must begin
to weahern earlier too. But the variation of westerlies in 1979 did
not support this theory. As seen from Fig. 1 the variation of westerlies
at the 200 hPa level was first found to Plateaus west and then propa-
gated from west to east. The variations in temperature supported this
finding. Data from the following stations is used to illustrate the
temperature variation:

i) Mangnai, Northern Plateau

ii) Shuanghu, Middle Plateau

iii) Zhongba, Southern Plateau

iv) Shiquanhe, Western Plateau

v) Qaindo, Eastern Plateau

vi) about 56°N, 80°E Northern Asia

vii) about 30°N, 45°E, Western Asia

viii) about 30°N, 135°E, Eastern Asia

It is indicated that in early June the temperature started increasing in the North of Asia (No. 6), then in the western part of the Plateau (No. 4) and finally in the west and east of Asia (No. 7 and No. 8). In the north and the middle of the Plateau the temperature started to increase in the middle of June. This feature is accords well with the variation in westerlies.

On the other hand, before the adjustment of general atmospheric circulation in June a significant modification of meridional temperature gradient takes place both in the Northern and Southern Hemispheres although the temperature gradient decreased in the Northern Hemisphere and increased in the Southern.The thicknesses between 1000–500 hPa levels characterize the mean temperature of the lower troposphere. The calculation of the difference between temperatures along 30°N (S) and 50°N (S) shows that there was significant variation in the meridional temperature gradient both in the Northern and Southern Hemispheres before the general adjustment of atmospheric circulation in June took place. It is known that planetary circulation is controlled by the intensity of the meridional temperature gradient. Therefore, the variation of the meridional temperature gradient plays a more important role than does the single temperature over the Plateau in the circulations adjustment in June.

REFERENCES

[1] Guo Qiyun, Acta Meteorological Sinica, Vol. 39, No. 3, 298—310, 1981.

[2] Meteorology of the Southern Hemisphere, Meteoral, Mongr. of the AM. Meteoral. SOC., 13, 1972.

[3] Sikka, D.R.,Results of summer MONEX field phase research (part B), No. 9, 87—95, 1980

[4] Guo Qiyun, Collected Papers of the Qinghai-Xizang Plateau Meteorology Science Experiment (1), 117—128, Science Press, 1984 (in Chinese).

SOME PHYSICAL CONCEPTS IN COMPLEX TERRAIN METEOROLOGY

DERIVED FROM THE U. S. DEPARTMENT OF ENERGY S ASCOT PROGRAM

William E. Clements

Atmospheric Sciences Group
Earth and Space Sciences Division
Los Alamos National Laboratory
Los Alamos, New Mexico, USA 87545

1. INTRODUCTION

In 1978 the United States Department of Energy (DOE) initiated a program directed specifically at atmospheric studies in complex terrain (ASCOT). The two broad objectives of the ASCOT program are:

- To improve fundamental knowledge of atmospheric transport and dispersion processes in complex terrain.
- Building on this improvement in the understanding of the physics to provide a methodology for performing air quality assessments.

These objectives are approached through an integrated program consisting of atmospheric physics theory, computer modeling, laboratory modeling, and field experimentation. The ASCOT team is composed of scientists from DOE-supported research laboratories, other federal laboratories, and universities. The accomplishments of the ASCOT program are the result of the combined efforts of many dedicated individuals. A complete list of participants is given by Gudiksen and Dickerson (1984). The initial emphases of the research are on nocturnal drainage winds, the effect terrain has on them, and their interactions with external flows. This paper describes some physical concepts of nocturnal drainage wind derived mainly from the results of the meteorological measurements portion of the field experiments.

53

Detailed descriptions of any part of the ASCOT program is not the intent of this overview paper. A comprehensive summary of all phases of the ASCOT program is given by Dickerson and Gudiksen (1984). Interested readers are encouraged to seek out this and all other references given in this paper for further information.

2. THE FIELD SITES

ASCOT has conducted intensive field studies at the five locations shown in Fig. 1. The topography of each of these sites is very different. The general characteristics of each site will be briefly described below in the order of their relative complexity. In each case references are given to more detailed descriptions of the sites and the experiments conducted.

RATTLESNAKE HILL: Located in the southeastern corner of the state of Washington, Rattlesnake Hill approximates a fairly uniform two-dimensional slope. It slopes from southwest at 595 m MSL to northeast at 366 m MSL for a distance of over 900 m. In the area of the hill where measurements were made there are two distinct angles to the slope. From the summit to about 495 m MSL the slope is 20° and below this point it is 8°. The average slope from the summit to the bottom is 16°. (Doran and Horst, 1983a,b; Horst and Doran, 1981, 1982)

PAJARITO MOUNTAIN: In 1979 and 1983 slope experiments were conducted on a fairly uniform ski run on Pajarito Mountain in the north central New Mexico. The ski run slopes from the south at 3090 m MSL to the north at 2830 m MSL for a distance of 850 m with an average slope angle of 18°. It is 50 m wide and is bordered on each side by dense stands of aspen and spruce approximately 15 m high. This site

54

approximates a uniform two-dimensional slope. (Clements and Nappo, 1983)

CORRAL GULCH: Corral Gulch, located in northwestern Colorado, was the site of field studies in 1980 and 1982. The experimental site is a shallow valley 15 km long. It drains roughly from the west at 2940 m MSL to the east at 1950 m MSL with an average slope of about 4°. At the location where measurements were made it is 60 m deep, 600 m wide, and has a local slope of only 1°. The sidewalls are steep (20 to 30°) and cut regularly by small tributaries. (Clements et al., 1981; Barr and Clements, 1981; Barr et al., 1983)

BRUSH CREEK: Brush Creek, also located in northwestern Colorado, is a 600 m deep valley with sidewalls that average 30° and at some places are much steeper. The valley floor drains for 22 km from the northwest at 2440 m MSL to the southeast at 1380 m MSL. The average slope is 2°. At the mouth of the valley the floor is about 1 km wide and near the headwaters it is very narrow. The sidewalls are cut regularly by sizable tributaries. Experiments were conducted here in 1982 and another set is planned for the fall of 1984. (Gudiksen, 1984)

GEYSERS: The study site in the Geysers area of northern California is a complex basin. It consists of a number of significant valleys and many small tributaries. The major drainage in the basin is from the northwest to the southeast ending in a restricted outflow region. Ridges composing the rim of the basin range in altitude from about 980 m MSL to over 1400 m MSL. The altitude of the most restricted portion of the outflow region is 390 m MSL. The basin can be thought of as circular with a diameter of 6 to 7 km. ASCOT conducted field experiments in this basin in 1979 and 1980. In 1981 the experimental site was moved from the Geysers Basin to Big Sulfur Creek, which is just over the western ridges of the basin. In this valley, which slopes down from the southeast to the northwest, cooling tower plume experiments were conducted. (Gudiksen, 1980, 1982, 1983; Dickerson, 1980)

These five experimental sites fall into three general topographic classifications

- Isolated simple slopes

  Rattlesnake Hill
  Pajarito Mountain

- Complex valleys

  Corral Gulch
  Brush Creek

- Complex basins

  Geysers

A complex valley is one with irregular sidewalls cut by tributaries. The isolated simple slope sites are used to study "simple drainage flows", which are the fundamental elements of the more complicated flows. The complex valleys and basins provide the setting for studying the effects of the merging of drainage winds.

3. THE FIELD STUDIES

The field studies at the above sites generally consist of intensive experimental periods of three to four weeks . During these times as many nocturnal periods with weather conditions conducive to good drainage winds are investigated as possible. This ranges from 3 to 6 nights depending on conditions. The period of the nocturnal studies usually runs from presunset to postsunrise in order to include the transitions into and out of the drainage wind regime. The intensive periods are usually complemented by a more modest long-term data collection. This is required to determine seasonal trends and the representativeness of the intensive studies.

During the intensive experiments two general types of studies are used to investigate the nocturnal drainage wind and its interaction with terrain and external flows. These are meteorological measurements and atmospheric tracer studies. The general types of meteorological instrumentation used in a particular intensive field experiment might include any or all of the following:

- Instrumented towers
- Optical anemometers
- Mobile meteorological sensors
- Free balloon-borne sensors
- Tethered balloon-borne sensors
- Monostatic acoustic sounders
- Bistatic doppler acoustic sounders
- Optically tracked neutrally buoyant balloons
- Radar tracked tetroons

Atmospheric tracer systems that have been applied in intensive studies are:

- BALLOONS:

    Neutrally buoyant balloons
    Tetroons

    Detection systems:

    Optical theodolites
    Radar

- PARTICLES:

    Smoke
    Fluorescent particles

Detection systems:

    Integrated surface samplers
    Sequential surface samplers
    Ground based lidar
    Airborne lidar

● GASES:

    Naturally occurring radon-222
    Sulfur hexafluoride
    Two perfluorocarbons (PMCH, C_7F_{14}; PDCH, C_8F_{16})
    Two heavy methanes ($^{13}CD_4$; $^{12}CD_4$)

Detection systems:

    Integrated surface samplers
    Sequential surface samplers
    Balloon-borne samplers
    Instrumented aircraft

Again the reader is referred to Dickerson and Gudiksen (1984) for further details of the instrumentation used in and the experimental design of each experiment. This reference will lead to even more detailed references. In this paper, I will concentrate on some of the results obtained from the meteorological measurements effort.

4. CONDITIONS FOR GOOD DRAINAGE WINDS

In this section some conditions for "good" drainage winds are discussed. The term "good" used in connection with drainage winds indicates that there is a minimum effect on the drainage winds by ambient meteorological conditions. In these cases the drainage wind is relatively stable and its characteristics are well defined.

On a clear night the absence of insolation allows the earth's surface to cool by long wave radiation. This surface cooling in turn attempts to cool the air near the ground. If this process is successful, a buoyancy deficit is created in the layer of air near the

58

ground with respect to the ambient air above. If this situation occurs on a sloping surface, a katabatic force is created and the layer will begin to move or "drain" downslope. This flow is retarded by frictional forces and adiabatic warming so that at some point a more-or-less steady state is reached. The resulting flow is referred to as the "nocturnal drainage wind" or, more often, simply the "drainage wind".

Once establised, the drainage wind is usually thought of as being decoupled from the ambient wind above. This is essentially correct if the ambient conditions are constant. However, the development and characteristics of a particular drainage wind depend strongly on the ambient conditions in which it was formed. Futhermore, changes in the ambient conditions usually will modify the drainage wind. For instance, as will be discussed later, intrusions of strong upper level winds during the night may completely destroy it. Also, a change in cloud cover will alter the radiative balance at the surface and, hence, change the characteristics of the drainage layer. I have observed nights that begin with a heavy cloud cover and no drainage wind. Later on the cloud cover will pass and the drainage wind will start soon afterwards.

Data from a number of ASCOT studies indicate some general conditions for good drainage winds to occur. First, radiative cooling at the ground must occur. This can be prevented or impeded, as discussed above, by cloud cover. Next, the layer of air near the ground must cool to produce a buoyancy deficit and, hence a downslope or katabatic force. Strong ambient winds at the surface can advect away any cooling of the surface air. Also, high humidity will impede the cooling of the layer compared to relatively dry air.

On simple isolated slopes a surface inversion of a few C° in a few tens of meters or less is usually sufficient to initiate drainage winds. These are subject to interruption by moderate ambient winds due to lack of shelter. In more complex situations, such as in valleys and basins, where the drainage wind is the result of the convergence of a number of drainage flows, the surface inversion may be a few C° in a few hundred meters. In the latter cases it is not at all clear that the observed inversions provide the total driving force for the the flow or that it is simply being advected by the flow. A bulk slope Richardson number, Ri_B, proved to be a good indicator of drainage wind on a slope in the Geysers area. Horst and Doran (based on some earlier work; Horst and Doran, 1981) define the bulk slope Richardson number as

$$Ri_B = \frac{\Delta T}{U^2} \cdot \frac{gh}{T} \tag{1}$$

where ΔT is the strength of the surface inversion of depth h, U and T are the wind speed and absolute temperature of the ambient air (at ridge top in the case of the Geysers study), and g is the gravitational acceleration.

Figure 2 shows the fraction of hours with good, fair and no drainage winds. These results, which at present have only been evaluated at the Geysers site, show that no drainage winds occur for $Ri_B < 0.15$, good slope flow never occurred for $Ri_B < 0.2$, and most of the good slope flow was observed for $Ri_B > 0.7$. Horst and Doran (see Chapter 3: Physical Concepts in Dickerson and Gudiksen, 1984) point out that Ri_B is a diagnostic rather than a prognostic indicator, because of its dependence on the strength and depth of the surface inversion.

In summary, good drainage flows occur on nights when there are few to no clouds, the ambient winds are light to calm, and the air is relatively dry. Drainage winds of varying quality occur in a wide variety of ambient conditions and sometimes under surprisingly severe situations. A great many factors are involved in the development of drainage winds in valleys and basins where they are sheltered. We are just beginning to understand some of these more complicated scenarios.

5. SOME GENERALIZED DRAINAGE WIND CHARACTERISTICS

The results of all the field studies show some characteristics of good drainage winds that appear to be independent of the complexity of the terrain. These features are useful in distinguishing drainage winds from synoptic winds that are channeled down a slope.

Before describing these characterisitics it is necessary to say something about the "depth" of a drainage wind. As discussed earlier established drainage flows are to a large extent decoupled from the ambient air above them. The distance from the ground up to the transition from drainage flow to ambient flow is a general definition for the depth. Although this sounds simple enough, the determination of the transition height is sometimes difficult and ambiguous. Changes in the vertical structure of temperature, wind direction, wind speed, and turbulent mixing are all used in attempts to define the depth of drainage winds. Horst and Doran (1981), King (1981), and Wolfsberg and Clements (1983) discuss the relative merits of these methods. It will suffice for the following discussions to accept that drainage winds exist over some finite depth near the ground.

Figure 3 shows the generalized vertical profiles of temperature, wind direction, and wind speed in and above a drainage wind layer of depth h. The following generalizations about these profiles can be made:

TEMPERATURE: The temperature increases from the surface upwards, approaching an isothermal state at the top of the layer. The temperature profile above the layer is most often isothermal.

WIND DIRECTION: The wind direction is downslope or downvalley in the drainage layer. Except in the frustrating cases when the ambient wind direction is downslope, there is a rather abrupt change in wind direction near the top of the layer. The wind direction immediatley above the drainage layer may be that of the synoptic wind or of an intermediate wind layer. The region between the top of the drainage layer and the synoptic wind is called the transition layer. The transition layer may be very simple as shown in Fig. 3. or extremely complex.

WIND SPEED: The wind speed profile within the drainage layer is characterized by a low-level jet with a wind speed minimum near the top of the layer. The maximum wind speed in the layer occurs

at about half the layer depth or less. The wind
speed above the layer can be zero when there are no
ambient winds, increase to the ambient wind speed,
or show secondary jets in the case of layered flow.

In addition to these features the turbulent structure in the
drainage layer is characteristically different from the ambient wind.
Clements and Nappo (1983) observed the range of the variation in wind
direction to change from about 60 degrees prior to the development of a
drainage wind on a simple slope to less than 10 degrees after drainage
began. There is limited data on the vertical structure of turbulence
in drainage winds. However it is expected to be quite different from
that in the stable boundary layer over flat terrain. This is because
of the different veritcal shear structure in the two cases. The
vertical variation of turbulent energy production in slope flow is
complex; however, the observed turbulence structure is generally
consistent with local shear production of turbulence. (See Physical
Concepts Chapter in Dickerson and Gudiksen, 1984).

Although the generalized characteristics given above are found in
most drainage flows, the values of the parameters associated with them
vary widely from site to site. In addition, as mentioned at the
outset, the depth of a drainage wind may take on different values
depending on which parameter is used for the determination. In the
next section specific examples of these generalizations will be given
and the discrepancies will be obvious.

6. SPECIFIC EXAMPLES OF DRAINAGE WIND CHARACTERISTICS

In this section examples of vertical profile data in drainage flows at each of the experimental sites are given and discussed. These examples have been chosen from good drainage flow situations; that is those cases wherein the drainage flow is steady and there is a minimum of influence from the ambient winds. The ASCOT data contain many examples that are not as clear cut as the ones presented here, which serve to remind us of how much we still have to learn about these phenomena.

RATTLESNAKE HILL: Figure 4 shows the vertical profiles of potential temperature and downslope wind speed at three towers on Rattlesnake Hill (a simple slope) from Doran and Horst (1983a). The towers are located at 193, 422, and 898 m along the slope (16° average) from the crest. The depth of the inversion and the maximum downslope wind speed both increase with downslope distance. These profiles also indicate that the depth of the drainage layer is increasing with downslope distance.

PAJARITO MOUNTAIN: Another example of drainage layer growth on a simple slope comes from data collected on Pajarito Mountain. In this data both temporal and spatial growth of the drainage layer can be seen. Figures 5 and 6 show hourly average profiles of wind speed, wind direction, and temperature for three hours at two locations on the slope. This data was collected from two towers located at 285 and 820 m along the 18° slope from the crest. At the upper tower in the three hours shown the inversion strength increases with time. The depth of the downslope wind direction increases from very shallow to

64

about 7 m in the three hour period. The maximum wind speed in the layer increases from the first to the second hour and then remains constant. At the lower tower the same type of trends are seen, but over greater depths. The final depth of the drainage wind in the last hour is in excess of 21 meters, which is the height of the highest instrument. Notice that the velocity maximum at both locations is considerably less than half the depth of the drainage flow.

CORRAL GULCH: Moving on to the first of our complex valleys, Corral Gulch, deeper drainage flows are found. Vertical profile data, including the mixing ratio, at this site obtained with a tethersonde are shown in Fig. 7. Here the top of the inversion, the point of wind direction shift, and the wind speed minimum all occur at about 200 m above the ground. During the four nights studied at this site, the depth of the drainage layer varied between 100 and 200 m. The wind speed maximum occurs at or slightly below the midpoint of the layer and averages between 2.0 and 2.5 m/s. The maximum shown in Fig. 7 is somewhat higher than average. There is a discontinuity in the mixing ratio somewhat lower than the top of the inversion. The fact that the mixing ratio is less in the drainage layer than above it is opposite to what is found at other sites and still remains somewhat of a mystery.

A much more subtle but persistent feature of the nocturnal flow is observed in this valley, which is not apparent in Fig. 7 (Barr and Clements, 1981). This is a weak down-valley windspeed maximum of about 1 m/s at roughly 10 m above the ground. This rather weak sublayer is easily obscured by stronger winds above.

A third layer with a wind speed maximum of 10 to 20 m/s at 500 m above the ground is described by Barr and Clements (1981). This layer is detected in pibal-minisonde profiles taken on open terrain about 7 km west of and 300 m above the tethersonde site. On two nights the flow in this layer is from the east when there is no evidence in the synoptic data for an easterly wind in the region. This layer appears to be due to a local topographic effect.

BRUSH CREEK: Figure 8 is vertical profile data taken in Brush Creek. Here potential temperature is plotted in place of the normal temperature. The wind speed profile shows the drainage layer jet with a well defined minimum at about 350 m. This is a little over half the depth of the valley. The maximum wind speed in the drainage layer is 7-8 m/s at about one-third the depth. This is typical of the Brush Creek drainage wind in summer. The only good indicator of the dainage depth in this case is the wind speed minimum above the jet.

Measurements made in Brush Creek in 1982 (Gudiksen, 1984) show an increase in the depth above the ground of the drainage layer with down-valley distance. There is some evidence that the top of the drainage layer is horizontal. This is consistent with the findings of Manins and Sawford (1979) who point out the analogy with a flooded internal jump.

GEYSERS: The last example is that of the drainage wind in the outflow region of the Geysers Basin. The vertical profile data, shown in Fig. 9, are somewhat atypical in the sense that this was not a particularly good drainage night. However, in spite of the lack of good drainage throughout the basin, there is a well defined drainage in the outflow region. This is most likely the result of very shallow,

66

perhaps below most instrument levels, drainage flows converging from throughout the basin. The outflow drainage layer depth is a little over 100 m. There is a well defined temperature inversion, low-level jet, and wind direction discontinuity. The mixing ratio is higher in the drainage layer than above it, as would be expected of a trapping process like the stable drainage layer. The more interesting feature of this data is the 100 m or so southerly jet above the northwest drainage flow. This is overlain by northwest to north winds above 300 m. The southerly jet could possibly be a return flow to the basin. This is a good example of the type of layered flow that can exists in complex topography.

Throughout the basin the slope and valley drainage flows grow with downslope distance from ridgetop to the "pooling region", which is discussed in section 8. Upon approaching the pooling region, which is growing in depth with time, the growth of the drainage winds decrease and they soon become part of the pooling flow.

For the isolated simple slope cases a relationship for the depth of the drainage layer, h, is given by Briggs (1981) as

$$h = 0.038 \cdot s \cdot (\sin \beta)^{2/3} \tag{2}$$

In equation (2) s is the distance along the slope from the crest and β is the slope angle. Data from Rattlesnake Hill and Pajarito Mountain generally verify this relationship.

Only a few examples of good drainage flow data have been given. The data from the ASCOT studies contain an enormous amount of information about nocturnal drainage winds and other related phenomena. Additional analyses have been and are continuing in parallel with the

67

ongoing field studies.

7. INTERACTIONS WITH EXTERNAL FLOWS

Locally driven wind fields, such as dainage winds, seldom occur in nature as isolated features. There is almost always some interaction with the ambient meteorological conditions. At the Geysers there are three dominant larger-scale wind fields that affect the drainage flow. These are: winds from migratory synoptic disturbances, seasonal sea breezes, and descending upper-level easterly winds.

At Geysers, the wintertime migratory synoptic features have an overriding influence on the drainage winds. In summer, when migratory storms occur less often, drainage winds are more persistent. In July and August good drainage winds occur on over 50 percent of the nights, while in January and February, they are present on less than 10 percent of the nights.

Very subtle features of the synoptic meteorological structure, such as the direction of a weak gradient flow, can influence the local drainage wind system. An easterly component over the Geysers area produces warming on the middle to upper slopes, destroys the drainage on the middle slopes, and gradually erodes the drainage flows at lower elevations. Figure 10 shows the gradual intrusion of northeasterly winds into the basin. This observed behavior is consistent with the systematic change in the wavelength of a nonlinear lee wave. If the wavelength is close to that of the lateral dimensions of the hill, the flow will tend to follow the topography. If the wavelength is significantly different, separation will occur. A relationship developed by Hunt et al. (1978) for hills of moderate slope and for a narrow range of wavelengths, which was tested with the Geysers data, was found to give a general indication of when intrusions of upper level winds occur. The condition for for the ambient wind to follow the terrain is given by

$$2L \leq \Lambda < 5L \qquad (3)$$

where Λ is the wavelength and L is the half width of the hill at half-height. The wavelength is given by

$$\Lambda = \frac{2\pi U}{N} \qquad (4)$$

where U is the ambient wind speed, N is the Brunt-Vaisala frequency,

$$N = \left[\frac{g}{\theta} \cdot \frac{\partial \theta}{\partial z} \right]^{1/2}, \qquad (5)$$

and θ is the potential temperature. The Geysers geometry is more complex than that considered by Hunt et al. (1978) so equation (3) offers only a broad guideline.

The invasion of Geysers by marine air is related to the coastal sub-tropical high pressure area and inland thermal low pressure area. A diurnal oscillation in the marine air is partly due to a similar oscillation in the coastal-inland horizontal pressure gradient ($P_{coast} - P_{inland}$). An increase in this gradient coincides with a decrease in the marine air influx and the beginning of the nocturnal drainage regime. The magnitude of the sea breeze influence is linked to synoptic scale structure.

8. LAYERING AND POOLING AT GEYSERS

In the Geysers there are some interesting nocturnal flow

phenomenon resulting from two topographic features of the basin. These are intersecting valleys within the basin and the restricted outflow region.

The nocturnal outflow from the Geysers Basin is mostly the result of the confluence of drainage flows from four major valleys. The flow in each valley is composed of drainage winds from its headwaters near ridgetop and inflows from its sidewalls and tributaries. Each of the valleys has its own drainage flow characteristics defined by its physical structure and orientation within the basin. Measurements made near the confluence of two or more valleys may vary unexpectedly due to an oscillating influence from each valley (Coulter, 1981). Because the flows from the different valleys may have different density structures, stratification may occur. This can produce layered flow in the convergence zones of the valleys and subsequently in the outflow region. Acoustic sounder records located in these areas show many instances of multilayered structure in the lowest 500 m. A different type of layering in the outflow region is shown in the data of Fig. 9. Here the layer above the drainage flow is almost in the opposite direction. This may be a case of a return flow to the basin above the drainage layer. The interaction of these stratified layers with the outer flow often generates gravity waves and instabilities at the layer interfaces. These contribute to the transition layer dynamics.

The outflow region of the Geysers is less than 1 km wide. The basin dimensions are 6 to 7 km. On nights with good drainage flow throughout the basin drainage air is produced faster than it can flow out of the basin. This results in a pooling region upstream of the

70

basin exit. The horizontal and vertical extent of the pooling region is a function of the strength of the drainage winds throughout the basin. This convergence in the pooling region manifests itself in the form of a vertical velocity reduction. That is, the negative vertical velocity associated with the drainage winds is reduced in magnitude and may become positive in the pooling region. Assuming incompressible flow, the change in the vertical velocity with height can be related to the horizontal convergence through the continuity equation

$$\frac{\partial u}{\partial x} + \frac{\partial v}{\partial y} + \frac{\partial w}{\partial z} = 0 \tag{6}$$

The depth through which the convergence occurs is also required.

The depth of the pooling region and a convergence parameter were estimated for the 1980 Geysers experiments from potential temperature patenrs derived from tethersonde profiles and doppler acoustic profile data. The convergence parameter is defined as

$$\text{Convergence Parameter} = \frac{\partial w}{\partial z} \tag{7}$$

The depth of the pooling region ranges from 120 to 250 m. The shallower depths occur, as expected, on nights with the weakest drainage flow. Values of the convergence parameter are between 0.0 and 0.001 s^{-1}; the smaller values occurring on nights with weak drainage winds. The pooling of the drainage air near the outflow of the basin is an interesting phenomenon that needs further investigation.

9. RADON AS AN INDICATOR OF DRAINAGE WINDS AT GEYSERS

Naturally occurring radon 222 was measured continuously 1 m above the ground in the outflow of the Geysers Basin during the 1980 study (Clements and Wilkening, 1981). Radon 222, an inert radioactive gas with a half-life of 3.8 days, can be considered in this application to be exhaled uniformly at a constant rate from the earth's surface throughout the basin. As cool slope winds move along the terrain and into the valley, the air masses involved accumulate radon through the night until morning instabilities mix it to greater depths. The diurnal trend of the near-surface radon in the outflow of the basin reflects the intergrated drainage flow conditions throughout the night, the morning and evening transition periods, daytime vertical mixing, and the influence of strong synoptic winds on the drainage flow.

On nights when good drainage flow occurred the diurnal trend in radon concentration reveals a characteristic pattern shown in Fig. 11. Low radon concentrations exists during the day followed by a sharp increase beginning just before sunset and continuing for about 2 hours. This period was typically followed by a gradual increase during the night. A sharp drop is observed after sunrise when surface heating and the resulting vertical mixing occur.

Departures for this standard pattern observed on nights when the external flow had a greater influence include: (1) almost no build up at night over daytime levels, (2) delayed build up occurring during the evening hours when unsettled conditions prevail, and (3) radon concentration decreases during some periods due to destabilization events or intrusion of the ambient winds.

10. SUMMARY

I have tried to give an overview of the United States Department of Energy's ASCOT program and touch upon some of the physical concepts derived mostly from the experimental phase of the reseach. Equally essential and interesting research is being conducted in the areas of atmospheric physics theory, computer modeling, laboratory experiments, and field tracer experiments. It is unfortunate that due to the diversity of the ASCOT program these other phases of the research could not be discussed in this paper. I encourage the interested reader to seek out the overview references given.

Since its beginning in 1978, the ASCOT program has strived to improve our understanding of the fundamental physical processes of complex terrain meteorology. It is hoped that this will provide a better basis for making air quality assessments in areas of nonuniform topography. Several terrain settings have been investigated in order to build a broad base of knowledge that at some point, hopefully, can begin to be generalized in a consistent theory. These studies have made it clear that the problem of transport and diffusion in complex terrain is complicated and that interactions on all scales are important. Only through the cooperation of all those concerned with complex terrain meteorology will we ever hope to reach a definitive understanding of these important problems.

ACKNOWLEDGMENTS

The ASCOT program is a team effort and every participant shares in any credit for its success. For this paper I am particularly indebted to Sumner Barr, Rich Coulter, Chris Doran, Paul Gudiksen, Tom Horst, Carmen Nappo, Monte Orgill, Bill Porch, and Marvin Wilkening. These individuals made major contributions to the sections of the Physical Concepts chapter of the ASCOT Technical Progress Report for 1979 through 1983 on which most of this paper is based. In fact, in places they will undoubtedly recognize close approximations to their own words. I greatly appreciate the support of this effort by Marv Dickerson (ASCOT Scientific Director) and Dave Ballantine (ASCOT

Program Manager, DOE/OHER). I also thank John Archuleta, Don Hoard, and Sue Noel for helping with the illustrations. This work was performed under the auspices of the United States Department of Energy.

REFERENCES

Barr, S. and W. E. Clements, 1981: Nocturnal wind characteristics in high terrain of the Piceance Basin. Preprints Second Conference on Mountain Meteorology, November 9-12, 1981, Steamboat Springs, Colorado, American Meteorological Society, 325-330.

Barr, S., T. G. Kyle, W. E. Clements, and W. Sedlacek, 1983: Plume dispersion in a nocturnal drainage wind. Atmos. Environ., 17, 1423-1429.

Clements, W. E. and M. Wilkening, 1981: Dirunal radon-222 concentrations in the outflow of a complex basin. Preprints Second Conference on Mountain Meteorology, November 9-12, 1981, Steamboat Springs, Colorado, American Meteorology Society, 212-217.

Clements, W. E. and C. J. Nappo, 1983, Observations of a drainage wind event on a high-altitude simple slope. J. Cli. and Appl. Meteor., 22, 331-335.

Clements, W. E., S. Goff, J. A. Archuleta, and S. Barr, 1981: Experimental design and data of the August 1980 Corral Gulch nocturnal wind experiment, Piceance Basin, northwestern Colorado. Report LA-8895-MS, Los Alamos National Laboratory, Los Alamos, New Mexico, 197 pp.

Coulter, R. L., 1981: Circulation characteristics during ASCOT 1980. Preprints Second Conference on Mountain Meteorology, November 9-12, Steamboat Springs, Colorado, American Meteorological Society,

316-320.

Dickerson, M. H. (editor), 1980: A collection of papers based on drainage wind studies in the Geysers area of northern California: Part I. Report: ASCOT-80-7, Lawrence Livermore National Laboratory, Livermore, California.

Dickerson, M. H. and P. H. Gudiksen (editors), 1984: Atmospheric studies in complex terrain, technical progress report, FY-1979 through FY-1983. Report: ASCOT-84-1, Lawrence Livermore National Laboratory, Livermore, California.

Doran, J. C. and T. W. Horst, 1983a: Observations of drainage winds on a simple slope. Preprints Sixth Symposium on Turbulence and Diffusion, March 22-25, 1983, Boston, American Meteorological Society, 166-168.

Doran, J. C. and T. W. Horst, 1983b: Observations and models of simple nocturnal slope flows. J. Atmos. Sci., 40, 708-717.

Gudiksen, P. H. and M. H. Dickerson (editors), 1984: Executive summary: atmospheric studies in complex terrain, technical progress report, FY-1979 through Fy-1983. Report: ASCOT-84-2, Lawrence Livermore National Laboratory, Livermore, California.

Gudiksen, P. H. (editor), 1980: ASCOT data from the 1979 field measurement program in Anderson Creek, California. Report: ASCOT-80-9, Lawrence Livermore National Laboratory, Livermore, California.

Gudiksen, P. H. (editor), 1982: ASCOT data from the 1981 cooling tower plume experiments in the Geysers geothermal area. Report: ASCOT-82-4, Vols. I and II, Lawrence Livermore National Laboratory, Livermore, California.

Gudiksen, P. H. (editor), 1983: ASCOT data from the 1980 field measurement program in the Anderson Creek Valley, California. Report: ASCOT-83-1, Vols. I, II, and III, Lawrence Livermore National Laboratory, Livermore, California.

Gudiksen, P. H. (editor), 1984: ASCOT data from the 1982 field measurement program in western colorado. Report: ASCOT (in press), Lawrence Livermore National Laboratory, Livermore, California.

Horst T. W. and J. C. Doran, 1981: Observations of the structure and development of nocturnal slope winds. Preprints Second Conference on Mountain Meteorology, November 9-12, 1981, Steamboat Springs, Colorado, American Meteorological Society, 201-205.

Horst T. W. and J. C. Doran, 1982: Simple nocturnal slope flow data from the Rattlesnake Mountain site. Report: PNL-4406/ASCOT-82-5, Pacific Northwest Laboratory, Richland, Washington, 98 pp.

Hunt, J. C. R., W. H. Snyder, and R. E. Lawson, Jr., 1978: Flow structure and turbulent diffusion around a three-dimensional hill. Environmental Protection Agency report: EPA-600/4-78-041, 84 pp.

King, C. W., 1981: Nocturnal drainage flow studies in complex terrain using an acoustic sounder and tethersonde. Preprints Second Conference on Mountain Meteorology, November 9-12, 1981, Steamboat Springs, Colorado, American Meteorological Society, 193-200.

Manins, P. C. and B. L. Sawford, 1979: A model of katabatic winds. J. Atmos. Sci., 36, 619-630.

Wolfsberg, D. and W. E. Clements, 1983: On determining depths of drainage winds. Preprints Sixth Symposium on Turbulence and Diffusion, March 22-25, 1983, Boston, American Meteorological Society, 121-122.

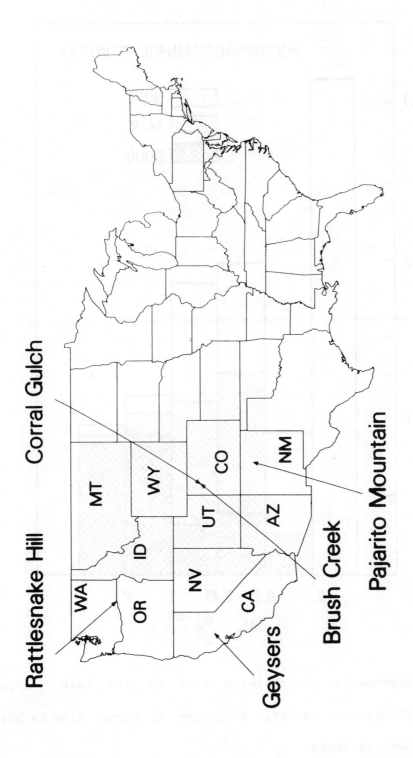

Fig. 1. Locations of ASCOT experimental sites.

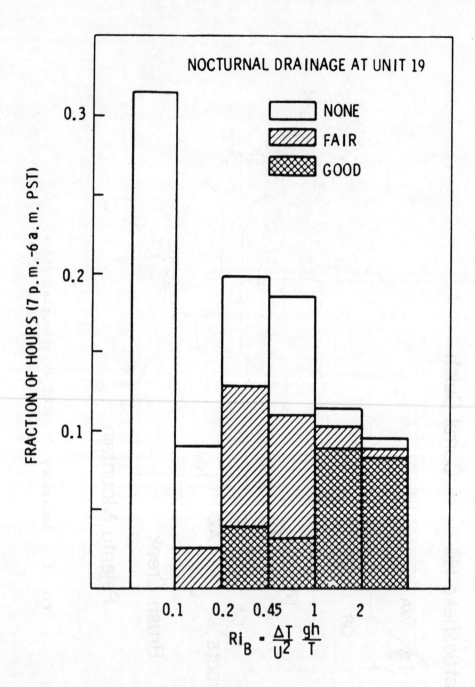

Fig. 2. Dependence of drainage wind on the bulk slope Richardson number, Ri_B, for a slope site in the Geysers Basin.

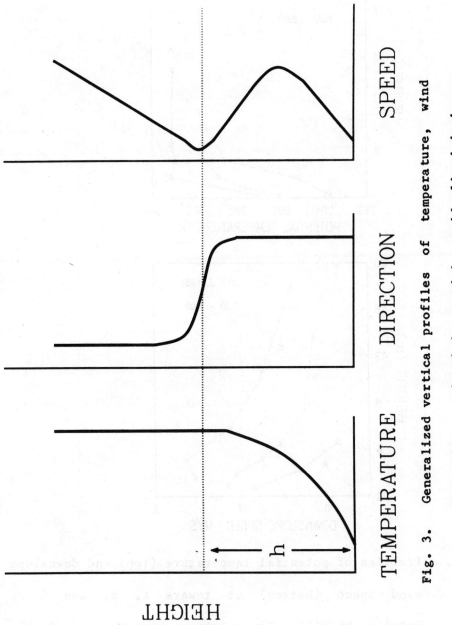

TEMPERATURE   DIRECTION   SPEED

Fig. 3. Generalized vertical profiles of temperature, wind
         direction, and wind speed in an idealized drainage
         wind of depth h.

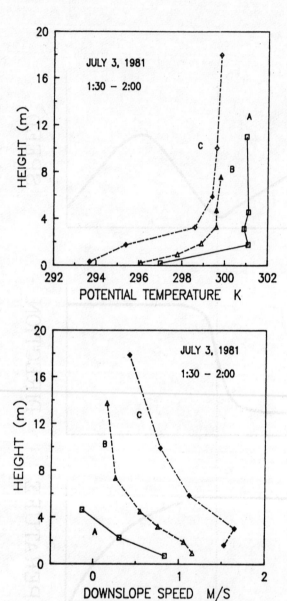

Fig. 4. Profiles of potential temperature (top) and downslope wind speed (bottom) at towers A, B, and C on Rattlesnake Hill. Tower A is nearest the top of the ridge and C is nearest the bottom. From Doran and Horst (1983).

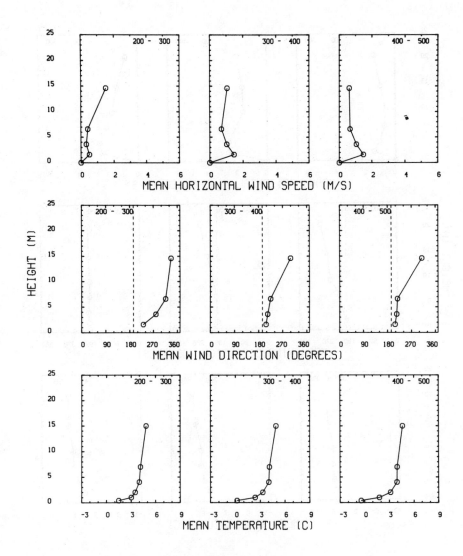

STATION 2; JULIAN DAY 310
RADIATION SUNSET 1445 HOURS
RADIATION SUNRISE 930 HOURS   TOWER2   02/23.

Fig. 5. Hourly average profiles of wind speed, wind direction, and temperature at the upper tower on Pajarito Mountain. Times shown in the upper right hand corner of each plot are mountain standard time on November 6, 1983.

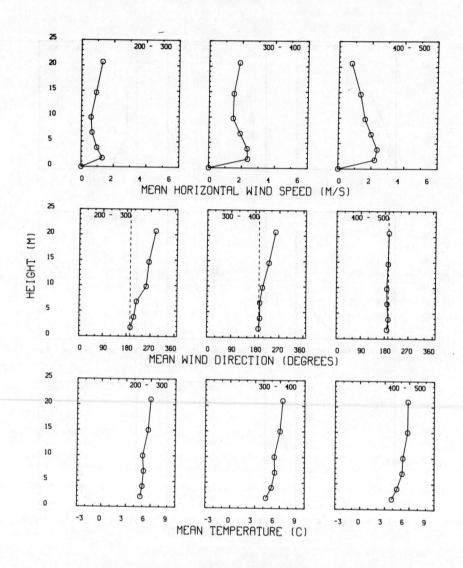

STATION 1; JULIAN DAY 310
RADIATION SUNSET 1445 HOURS
RADIATION SUNRISE 930 HOURS   TOWER1   02/23.

Fig. 6. Hourly average profiles of wind speed, wind
direction, and temperature at the lower tower on
Pajarito Mountain. Times shown in the upper right
hand corner of each plot are mountain standard time
on November 6, 1983.

CORRAL GULCH CO   TETHERSONDE FLT 29   DATE: 800810   0007 TO 0018 MST

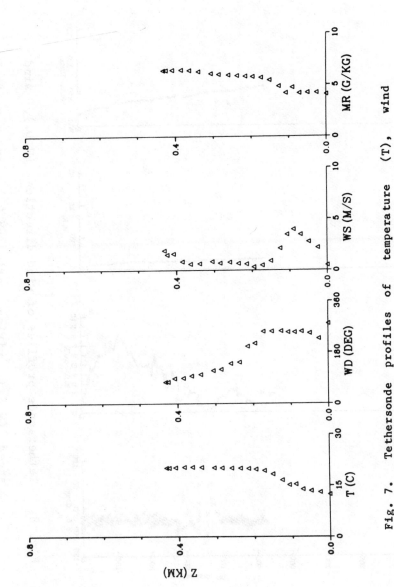

Fig. 7. Tethersonde profiles of temperature (T), wind
direction (WD), wind speed (WS), and mixing ratio
(MR) in Corral Gulch.

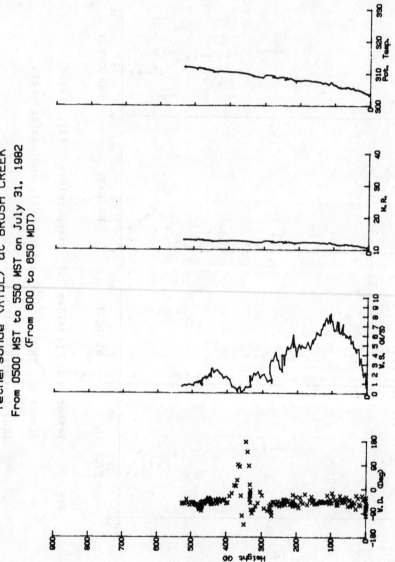

Tethersonde (ATDL) at BRUSH CREEK
From 0500 MST to 550 MST on July 31, 1982
(From 600 to 650 MDT)

Fig. 8. Tethersonde profiles of wind direction (W.D.), wind
speed (W.S.), mixing ratio (M.R.), and potential
temperature in Brush Creek.

84

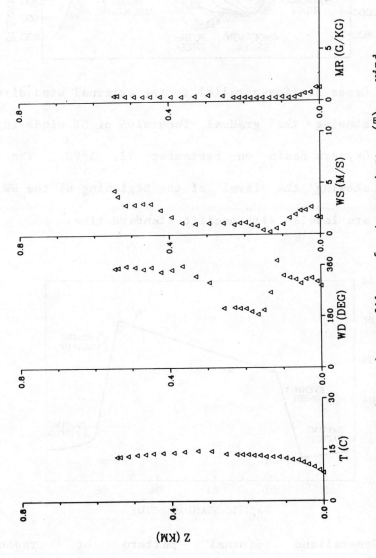

Fig. 9. Tethersonde profiles of temperature (T), wind direction (WD), wind speed (WS), and mixing ratio (MR) in the outflow region of the Geysers Basin.

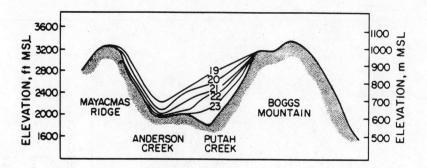

Fig. 10.   Cross section parallel to the external wind direction
           showing   the   gradual   incursion of NE winds into the
           Geysers Basin   on   September   22,   1980.   The   lines
           showing   the   level   of the beginning of the NW winds
           are labeled with pacific standard time.

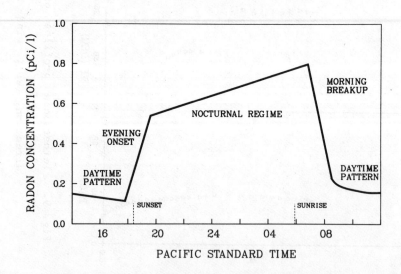

Fig. 11.   Generalized   diurnal   pattern   of   radon 222
           concentration   in the Geysers outflow region for good
           nocturnal drainage   conditions.   From   Clements   and
           Wilkening (1981).

Session 2: Observation and Analysis (2)

A GENERAL SURVEY OF SURFACE RADIATION AND HEAT BALANCE
INVESTIGATION OVER THE QINGHAI-XIZANG PLATEAU

Weng Duming

Nanjing Institute of Meteorology

Pan Shouwen

Nanjing University

Shen Zhibao

Lanzhou Institute of Plateau Atmospheric Physics,

Chinese Academy of Sciences

The protuberance of the Qinghai-Xizang Plateau plays an important role in the formation of the general circulation and thus affects the weather and climate not only over China but over Asia and even the Northern Hemisphere as a whole. Meteorologists have attached great impatance to the Plateau, and have especially focused upon the manner in which its surface heats the air. This latter concern has becomes one of the central issues for investigation. During the first QXPMEX from May to August, 1979, a group of scientific workers were sent to the inner part of the Plateau charged with making regular fixed-site field observations in an attempt to determine the surface heat sources. These worker detained a large quantity of data. They have provided a basis for examining surface radiation balance (net radiation) and heat balance together with seasonal variations.

I. A GENERAL SURVEY OF THE FIELD WORK

Field experiments were carried out from May to August, 1979, six observation posts being established so as to includs different natural landscapes as shown in Table 1. A control site was established during this period at Nanjing (located on the plain) to facilitade comparison. with the Plateau heating. In addition, control observations were made for a short period at Gunga in the vicinity of Lhasa. Spectrophotometric observations of direct solar radiation were also performed at Golmud with filters JB_2, CB_2, HB_{11} and HB_{16}. Some mobile stations carried out control measurements of surface albedo values.

Table 1 Geographical locations, types of ground surface and the operational period of the observing posts.

observing post	longitude ($^\circ$E)	latitude ($^\circ$N)	elevation above sea level (m)	type of ground surface	operational period
Shiquanhe	80°05'	32°30'	4278	waste grassland	May 17-Aug.31
Shuanghu	89°00'	32°38'	4920	grassland	May 21-Aug.31
Nagqu	92°03'	31°29'	4507	grassland	May 22-Aug.31
Lhasa	91°03'	29°40'	2633	grassland	May 6-Aug.31
Nyingchi	94°28'	29°34'	3000	wet grassland	May 11-Aug.31
Golmud I	94°54'	36°54'	2808	salinized land	May 15-July 6
Golmud II	94°48'	36°24'	2841	Gobi	June 28-Aug.31

Table 2 indicates the elements observed and the instruments used. Measurements were taken every two hours from 0400 to 2000 local mean solar time and only once during the night at 0000. To ensure the accuracy of the data all instruments were compared against standard instruments prior to and after field uses to check for the required stability and performance. Thousands of observed values were accumulated during this investigation.

II. DATA, ATLAS AND RESULTS

Some of the principal results of this scientific investigation were the compiling of an atlas showing decadal and monthly charts for all components of surface net radiation and heat balance over the Plateau during this period, and the publication of two volumes of net radiation and vertical gradient observational data. The atlas contains 192 charts for 12 elements such as components of radiation and heat

balances, sunshine duration, etc. All these charts were constructed on
the basis of the observed data from the experimental sites and permanent measurement stations. Calculations were done by empirical and semi-
empirical methods with the aid of conventional weather data from the
national meteorological network.

Since the charts in the atlas reflect the specific distribution
of decadal and monthly totals of various components for a particular
year, the difficulties encountered in their construction are much
greater than for mean charts constructed from multi-year records. Therefore, many schemes have been tried and their succes weighed before
the final design of a set of calculation methods suitable to the
Plateau. These methods include:

1) For the decadal average of daily totals of global radiation
the formula

$$Q = S_0(a + bS_1) \tag{1}$$

is used, where Q is the required average total; S_0, the average daily
total of astronomical radiation for the decade; S_1, the average sunshine duration in percentage; and a and b, empirical coefficients.
In order to improve accuracy, calculation was made, separately, for
seven regions. Table 3 indicates the regions and their respective coefficients. The calculation errors have been found to be 6.0% and 4.5%
for the decadal and monthly totals, respectively.

2) Albedo values on a 2°×2° longitude-latitude grid within the
Plateau region have been determined by means of information on soil,
vegetation and landforms found in "The Atlas of Physical Geography of
the People's Republic of China" and in the comprehensive physicogeo-
graphical survey of the Plateau; pictures from the earth resources
satellites of the PRC were also referred to. With seasonal variations
accounted for. Relative calculation errors are found to be within
10%[1].

3) For effective radiation F the empirical formula

$$F = \delta[\sigma T_0^4 - \sigma T^4(A + Bn + Ce)] \tag{2}$$

Table 2 The elements observed and instruments used.

element		instrument used
direct solar radiation	S	Japanese-built pyrheliometers Model MS-52
diffuse sky radiation	D	Japanese-built pyranometers Model MS-42 with sunshade Model MB-11
global radiation	Q	Japanese-built albedometers Model MR-21
surface reflective radiation	R_K	
upward terrestrial long wave radiation	U	U.S.-built precision infrared radiometers Model PIR
downward atmopsheric long wave radiation	G	
surface radiation balance	B	Japanese-built net radiometers Model CN-11
air temperatures and	T	
humidities at 0.2,0.5, 1.5 and 2.0 m above ground	e	home-made Assmann psychrometers
surface temperature including the maximum and minimum	T_0	home-made surface thermometers, including maximum and minimum thermometers
soil temperatures for depths 5,10,15,20,40,80 and 160 cm		home-made straight-and bent-stem soil and remotely-measuring thermometers
soil heat flux	H	Japanese-built heat-flow plates Model CN-81
wind speeds at 0.5,1.0, 2.0,3.0 and 5.0m above ground		Japanese-built anemo micro pairs
precipitation		home-made raingauges
evaporation		home-made small-sized evaporation pans
sunshine		home-made sunshine recorders

is employed[2], where F is the effective radiation; T_0 and T are the
decadal average of ground surface temperature and air temperature;
n, the decadal mean cloudiness; e, the mean vapor pressure in hPa; δ,
the relative radiation coefficient (taken as 0.95); σ, Stefan-Boltzmann's
constant; and A, B and C empirical coefficients. Calculation was done
separately for three regions with the respective coefficients as
indicated in Table 4. It should be noted that, since F gives the average
flux density of effective radiation, it should be multiplied by 14400
(if expressed in min.) to get the decadal total. The average relative
errors when applying the formula are found to be 6.3 and 4.5% for
decadal and monthly totals respectively.

Table 3 Values of a and b employed in Formula (1) for
different regions.

region	representive station	correlation coefficient	a	b
Tarim Basin	Kashi, Hotan, Ruo-qiang, Dunhuang	0.935	0.247	0.461
Hexi Corridor	Minqin	0.896	0.166	0.425
Central & western sectors of the plateau	Shiquanhe, Shuanghu Golmud and Nagqu	0.860	0.277	0.588
Southeastern sector of the Plateau & the Yarlung Zangbo River Valley	Lhasa and Nyingchi	0.937	0.314	0.527
Eastern sector of the Plateau	Qamdo, Yushu and Xining	0.915	0.223	0.555
The Plateau's eastern brim & Sichuan Basin	Lanzhou, Aba, Wei-ning, Chengdu and Mianyang	0.890	0.190	0.518
The Yunnan & Guizhou Plateau	Kunming and Tengchong	0.776	0.231	0.464

4) To calculate global radiation, surface albedo and effective radiation having been found, net radiation B can be obtained by the following expression:

$$B = Q(1-A) - F \qquad\qquad (3)$$

where A is surface albedo. Tests show that the average relative errors for the decadal and monthly totals are 9.7% and 4.9%, respectively.

Table 4 Empirical coefficients A, B and C in formula (2) for different regions.

region	representive station	A	B	C
The core of the Plateau & its surrounding arid regions	Golmud, Nagqu, Shuanghu and Shiquanhe	0.548	0.164	0.022
The eastern & southeastern portions of the Plateau	Lhasà and Nyingchi	0.631	0.200	0.0084
The plain land east of the Plateau	Nanjing	0.838	0.063	0.0021

5) For the sensible heat exchange P used in the calculation of heat balance components, we have

$$P = \rho C_P\, C_D\, V_{10}\, (T_0 - T), \qquad\qquad (4)$$

where ρ is the surface air density over the observing site, C_p specific heat under constant pressure (assumed to be $1\ \mathrm{J.g^{-1}.\,deg^{-1}}$), C_D the drag coefficient, V_{10} wind speed at a height of 10 m($\mathrm{m\cdot s^{-1}}$), and T_0 and T ground surface and air temperatures, respectively. A decadal total would be obtained when multiplied by 14400. C_D is determined empirically in an inverse operation[3]. For regions higher than 2800 m above sea level, the empirical expression is

$$C_D = 0.00112 + 0.10/V_{10} \qquad (5)$$

for the peripheral regions lower than 2800 m, an interpolation formula

$$C_D = 0.00112 + 0.10/V_{10} - 0.00362 \ (b-720)/280 \qquad (6)$$

is used, where b is the station pressure (hPa).

6) For heat of evaporation LE (latent heat), a remainder method is employed, i.e.,

$$LE = B - P - H \qquad (7)$$

where H is the heat exchange in soil, B and H having been specified above. The Bowen ratio method is also used independently to test the results[3]. It has been found that the results obtained by either method, are very close and their correlation coefficient reaches 0.89, the average absolute difference being 3.2 $W \cdot m^{-2}$.

7) For the heat exchange in soil[4] we employ

$$H = 9.8 + 8.9\Delta\theta_{5-20} \qquad (8)$$

where $\Delta\theta_{5-20}$ is the temperature difference between 5 and 20 cm in depth.

8) The heat-source intensity is defined as a characteristic quantity necessary for heating the air by the surface through transport of sensible and latent heat and is given by B-H.

Tested by the closed heat balance equation, the heat balance charts have calculation errors of less than 15% in most cases with an average of 10%. This satisfies the basic requirement of accuracy. Charts in the atlas reveal clearly the characteristic features of the surface radiation fields and surface warming patterns in summer months over the Plateau.

This field experiment and the construction of the atlas of net radiation and heat balance provide valuable information for investigation of the influence of net radiation and heat balance over the Plateau as well as their seasonal variations. In recent years a number of outstanding papers have been completed through the team-

work of the research group and further research work is under way.
The papers completed cover a wide range of subjects involving the
various aspects of surface radiation and heat balance, climatological
computation methods of various elements, estimation utilizing satellite
cloud pictures for data gaps, and the application of surface heat-
source data to other branches of studies.

III. PRINCIPAL FEATURES OF THE SURFACE RADIATION
AND HEAT BALANCE FIELDS OVER THE PLATEAU

Based on the atlas for the period of May-August, 1979, some basic
characteristic features can be generalized for the radiation and heat
balances fields over the Plateau.

Owing to its great elevation, the main part of the Plateau
manifested itself as a high center of global radiation throughout the
summer of 1979. The total global radiation exceeded 240 $W \cdot m^{-2}$ for each
of the summer months and amounted to more than 320 $W \cdot m^{-2}$ in west Xizang,
one of the regions with the most intense global radiation on earth[5].
Thus, the monthly total of net radiation was correspondingly large,
general more than 96 $W \cdot m^{-2}$ in the month of May prior to the onset of
the rainy season. The high center was found to be located in the
central part of the Plateau at around 30°N, 90°E with a maximum of
over 128 $W \cdot m^{-2}$. After the rainy season set in, however, the decrease
of albedo and effective radiation due to the high frequency of nocturnal
rain led to an obvious increase in net radiation in July with the high
center over the Yarlung Zangbo River Valley in Southern Xizang with
a maximum of over 128 $W \cdot m^{-2}$. The charateristic seasonal change of the
radiation balance distribution was also clearly shown on the daily
total curve at Lhasa[6] (Fig. 1). It can be clearly seen that a
discontinuity exists at the onset of the rainy season on the curve
for radiation balance (Fig. 1b), the daily total jumping from roughly
100 in the dry season to about 150 $W \cdot m^{-2}$ after June 17, when the
rainy period started. This increase well matched the sharp fall in
effective radiation (Fig. 1a) and, therefore, was associated with the
diurnal change in cloudiness owing to the nocturnal rain.

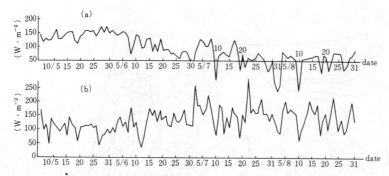

Fig. 1 The day to day change in the daily total of effective
radiation (a) and net radiation (b) at Lhasa for May-
Aug., 1979.

The characteristics of the surface heating pattern were very
similar to those of the distribution of net radiation. Calculation
indicates a correlation coefficient of more than 0.99 between the
monthly total of heat and that of net radiation at these observing
posts. From May to August 1979 the Plateau as a whole appeared to be
a rather intense huge heat-source. As shown in Fig. 2, the distribution
pattern of heat-sources indicated that a ridge extended toward the
northeast from the central and southern region with a maximum of 135
$W \cdot m^{-2}$ for July and that a lower-value band stretched from the Sichuan
Basin and the Qaidam Basin to the southern brim of the Tarim[7].
Although the intensity of heat sources varied from season to season,
the distribution pattern remained much the same. The radiation balance
near the surface and, hence, the heat-sources, intensified as the wet
season (after June 17) set in to replace the dry period (in May). If
the change in strength is represented in terms of the difference between
heat amounts for July and for May, an increase of heat is seen over all
the parts of the Plateau, most appreciably in the Yarlung Zangbo River
except for some places in the eastern sector of the Plateau.

A preliminary analysis indicates that between the heat-source distribution and the contours and isotherms of the 600 hPa surface there
exists such a relationship: the high-value center or band of the surface heat-sources generally correspond to the low pressure and high
temperature areas at 600 hPa. This indicates that the heating field
has considerable influence upon the pressure pattern.

The general characteristic of the heating of the air by the

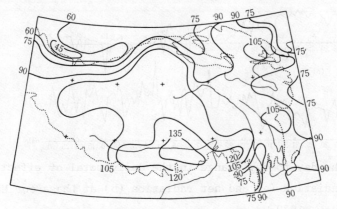

Fig. 2 The distribution of the total amount of heat by the
surface sources ($W \cdot m^{-2}$) over the Plateau for July, 1979.

surface is that, in the dry season, except for the moist southeastern
region of the Plateau and the Sichuan Basin together with the areas
northwest of it, warming process in the form of sensible heat
predominates. It contributes more than 50% of the heating as a whole,
more than 70% west of 90°E, and over 90% in the Tarim Basin and the
Qaidam Basin. In the wet season, the contribution from sensible heat
generally decreases to less than 50% in the eastern half of the Plateau
and even to 20-30% in the Yarlung Zangbo River and the Sichuan Basin,
with only the barren areas of the western and northern borders of the
Plateau being still dominated by sensible heat.

Figs 3 and 4 show the changes of sensible and latent heat over the

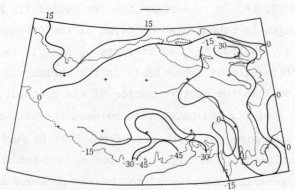

Fig. 3 The difference in surface sensible heat totals($W \cdot m^{-2}$)
May-July , 1979.

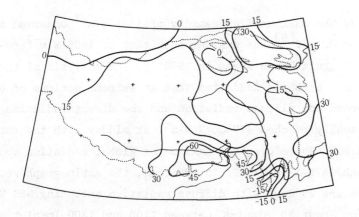

Fig. 4 The difference in surface latent heat totals ($W \cdot m^{-2}$) May-July, 1979.

Plateau from dry (May) to wet season (July). These changes are agree well with the foregoing analysis. Fig. 3 indicates a general decrease of sensible heat over the Plateau in the rainy period and Fig. 4 shows the considerable increase of latent heat, the greatest change occurring in the Yarlung Zangbo River of southern Xizang. This signifies that it is this region which is most strongly influenced by the transition from the dry to wet regime.

The air drag coefficient for each observing station was obtained from the heat balance data. Over the main part of the Plateau the value for C_D is of the order of 4×10^{-3} to 5×10^{-3}. Its general characteristic of distribution is that the value is higher in the east than in the west and also higher in the valleys than over the level regions, with a maximum of 9×10^{-3}. In addition, the C_D values also vary with the march of seasons, increasing somewhat from May to July and August.

IV. SOME OBSERVED FACTS OF THE PLATEAU RADIATION AND HEAT REGIME

1) As observations show, it is quite common for the summer solar global radiation flux density to exceed the solar constant (1382 $W \cdot m^{-2}$) Such a situation was observed 15 times by the six experiment stations. This situation was caused by a favorable combination of direct solar radiation and diffuse sky radiation under certain cloud and sky condi-

tions (Table 5). The maximum flux density of 1521 $W \cdot m^{-2}$ observed at
Lhasa at 1200 on July 2[8] was smaller than that of 1605 $W \cdot m^{-2}$ recorded in the Tanggula Range areas in 1976[9]. The analysis of the
actinograph traces at Lhasa indicates that an extreme maximum of 698
$W \cdot m^{-2}$ was recorded for diffuse radiation and the direct radiation
flux density usually reached 977–1047 $W \cdot m^{-2}$ at midday with the sun's
disk free of cloud. Therefore, the maximum of global radiation could
presumably reach 1675–1745 $W \cdot m^{-2}$. In addition, the actinograph records
show that there are 20 days with diffuse radiation exceeding 488 $W \cdot m^{-2}$
each lasting for about 30 minutes between 1100 and 1300 local time.
Hence, during this period of time, should the sun's rays reach the
ground through the breaks in the cloud cover, a global radiation flux

Table 5 Global radiation values higher than the solar
constant recorded by the observing stations

observing station	global radiation($W \cdot m^{-2}$)	operational time	cloudiness	cloud form
Lhasa	1521	1200,July 2	10/6	Cu,Ac
	1486	1200,June 21	10/5	Ac,Sc
	1389	1200,Aug. 15	8/2	Ac,Cu
Shuanghu	1486	1200,June 24	10/10	Cb,Cu,Fc
	1424	1200,June 27	8/8	Sc,Cu,Fc
	1458	1200,July 6	7/6	Cu,Cb,Ac
	1430	1200,July 23	6/6	Cb,Cu,Fc
	1389	1200,July 29	8/8	Cu,Fc
Nagqu	1444	1200,May 24	4/4	Cu
	1389	1200,May 26	8/7	Cu,Ci
	1396	1200,June 20	6/6	Cu
	1507	1200,June 30	8/8	Cu,Sc
Shiquanhe	1465	1200,June 13	10/0	Ac,Cu
Nyingchi	1472	1200,June 2	10/10	Sc,Cu
Golmud	1389	1200,May 16	4/4	Cu

higher than the solar constant would be observed. Undoubtedly, this is
one of the significant properties of the radiation climate of the
Plateau.

2) In connection with the high global radiation, high values of
net radiation were observed. The extreme maximum value of net
radiation flux of 1116 $W \cdot m^{-2}$ recorded at Lhasa occurred simultaneously
with the maximum value of global radiation flux of 1521 $W \cdot m^{-2}$. As far
as we know, this is the highest of all the values ever recorded on the
ground. To verify the result the simultaneously observed values of
all components are given in Table 6. Since they are actually observed
values, small discrepancy in the total is understandable. Besides,
six high values of net radiation balance exceeding 907 $W \cdot m^{-2}$ have been
recorded at other stations[8]. All these are far greater than the
maximum value in meteorological literature (768 $W \cdot m^{-2}$) recorded for
the Pamirs during July-August, 1956[10].

Table 6 Observational data of the net radiation
components ($W \cdot m^{-2}$) at 1200, July 2, 1979.

component	S'	D	Q	R_K	U	G	B
flux density	1.39	0.81	2.18	0.40	0.75	0.58	1.60

S' is the horizontal direct radiation and the other
letters denote the same as in Table 2.

New maxima of effective radiation have also been observed in our
experiment. The previous maximum of 307 $W \cdot m^{-2}$ indicated in the relevant
literature was observed in the Mid-Asian deserts[11]. And during our
investigation 10 records equal to or exceeding this value have been
registered at Nagqu, Shiquanhe and Lhasa, and at Nagqu at 1200, May
26 an extreme of 349 $W \cdot m^{-2}$ was reported.

3) The authors analyzed the transparency of the atmosphere over
the Plateau and its periphery and constructed a chart for the transparency coefficient P_2 distribution[12] shown in Fig. 5. This is the
first attempt to reveal the dependence of atmospheric transparency on
elevation. The P_2 value may reach 0.76 or even higher under the

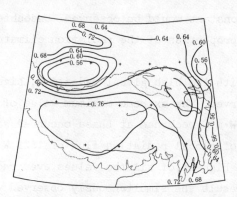

Fig. 5 The distribution of the atmosphere transparency
coefficient P_2(reduced to atmospheric absolute
mass) over the Plateau for August, 1979.

condition of an absolute atmospheric mass over the part of the
Plateau and it may be less than 0.56 in the Sichuan Basin and the
Tarim. This is quite a sharp constrast. Both diurnal and seasonal
changes in P_2 were found to be smaller on the Plateau than on the
plains.

In addition, more delicate observations of diminution of solar
radiation by dust and vapor were made with glass filters at Golmud
with some interesting results[13].

4) As indicated by a preliminary study, there exists a regularity
in the variation of all net radiation components with elevation under
clear skies. This can be accurately expressed by a set of empirical
equations.

Direct and global radiations increase with elevation exponentially
while diffuse radiation decreases negative-exponentially. These equations are:

$$Q = Q_0 \exp[a_Q(1-b/b_0)]$$

$$S' = S_0' \exp[a_s(1-b/b_0)]$$

$$D = D_0 \exp[a_D(b/b_0-1)]$$

where S_0', Q_0 and D_0 stand for direct, global and diffuse radiation at
sea level a_Q, a_s and a_D are empirical coefficients, and b and b_0
represent station pressure and pressure reduced to sea level.

For effective radiation and radiation from the atmosphere we have

102

$$F = F_\infty[1 - \frac{d}{H+d+1}]$$

and

$$G = G_0 \, e^{-\alpha H}$$

These show satisfactorily the variation of the components with station elevation. In these relations, F represents effective radiation at the station, G and G_0 are back radiation from the atmosphere observed at station level and at sea level respectively, H is the station elevation and F_∞, d and α are empirical coefficients, F_∞ being characteristic of escaping diffuse radiation at the upper limit of the atmosphere.

Some work has been done[14] to compare the radiation data obtained during the first QXPMEX with those of the MONEX of the FGGE and it is found that the value of back radiation from the atmosphere G satisfies the following empirical expression whether over the Plateau, the Indian Subcontinent or the Indian Ocean:

$$G = \varepsilon_0 \, \sigma T_a^4[1+(0.200-0.005 \, e)n]$$

where

$$\varepsilon_0 = 0.649 + 0.313 \, \lg\frac{b}{b_0} + 0.217 \, \lg e$$

in which T_a is air temperature in the screen (K), e surface vapor pressure (hPa), n average cloudiness, b and b_0 station and sea-level pressures respectively, and σ Stefan-Boltzmann's constant.

5) The authors have systematically analyzed the influence of cloud upon net radiation components. Transmission coefficients for direct radiation under various types of clouds have been computed statistically and computation has been done of values of diffuse, effective and back radiation and net radiation flux density under overcast skies of different types of clouds. In addition, analyses have been made to construct curves of daily amounts of net radiation components against amounts of cloudiness (Fig. 6) and similar effect has been found of cloudiness on the flux density observed at fixed time.

6) The observed albedo for the main part of the Plateau had a value of 16—18% for May-August, 1979. Its monthly variation was as-

103

sociated with the ground condition before and after the rainy season.
In the wet period vegetation began to flourish as the soil moisture
increased, resulting in a great diminution of albedo. Fig. 7 indicates
how albedo diminished with soil moisture at Lhasa. It is found that a
better relation existed between albedo and decadal or monthly rainfall
at some stations (e.g. at Nagqu). Using the results of this experiment,
albedo observed over the Plateau is found to be about 10% lower than
was estimated before.

7) Rather complete and typical curves of diurnal variations of
various radiation components under clear skies have been constructed
with the observed data for the stations. Actually observed diurnal
variation curves of heat flux between the ground surface and various
depths have also been established. Through the vertical gradient
observations, a better understanding has been attained of the

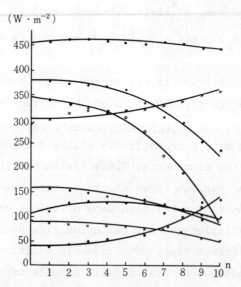

Fig. 6 The distribution of daily amounts of net radiation
components with different types of cloudiness
(based on data averaged for May-June, 1979 for
Lhasa).

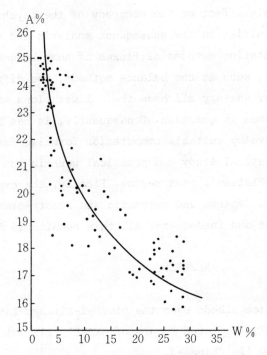

Fig. 7 The relationship between daily mean albedo and
soil wetness for May-August, 1979 at Lhasa.

physics of the near surface layer of the Plateau, including profiles
of various meteorological elements, order of magnitude of turbulent
exchange coefficients, stratification stability and surface roughness
parameters, etc. Various methods for computing heat balance components
now in use have been tested and compared in our investigation.

V. CONCLUDING REMARKS

The Plateau Heat Investigation under the First QXPMEX has
essentially attained its aims. Yet some defects do remain and need
to be overcome in future investigations. First, the experiment was
carried out only for the summer months, the results do not reflect
the annual heat regime of the region. The heat- and cold-sources in
other seasons of the year, especially in winter, need further investigation. Second, the dusty arid air in the surface layer during the dry

season had considerable effect on the accuracy of the psychrometer, giving rise to difficulties in the subsequent analysis and processing of data. Third, computation methods of fluxes of heat balance components now available, such as the balance methods, the diffusion methods and methods of analogy all have their limitations when applied to the high Plateau area in question. Consequently, it has become a central problem to develop suitable computation formulae for the Plateau through theoretical study and practical work in the continuation of study of the Plateau's heat regime. Finally, the experimenting stations seem to be too sparse and automatic and remote-sensing equipment insufficient and inadequate. All this remains to be improved.

REFERENCES

[1] Xie Xianqun, Surface albedo over the Qinghai-Xizang Plateau from May to August, 1979, Colleted Papers of QXPMEX (2), Science Press, 17—23, 1984 (in Chinese).

[2] Weng Duming, et al., Calculative method of total ten days and monthly effective radiation over Qinghai-Xizang Plateau, Ibid, 12—16, 1984. (in Chinese).

[3] Chen Wanlong and Weng duming, Preliminary study on the calculative method for the total ten days value of sensible heat and latent heat, Ibid, 35—45, 1984 (in Chinese).

[4] Xu Zhaosheng and Ma Yutang, The calculation of measurements and the method of popularization for soil heat flux on the Qinghai-Xizang Plateau, Ibid, 24—34, 1984 (in Chinese).

[5] Pan Shouwen, The characteristics of the components of the radiation balance during QXPMEX in 1979, Papers of Beijing International Symposium on Qinghai-Xizang Plateau and Mountain Meteorology, 1984 (in Chinese).

[6] Weng Duming, et al., Study on the characteristics of the solar radiation in the Lhasa River Valley, Collected Papers of QXPMEX (1), Science Press, 70—94, 1984 (in Chinese).

[7] Weng Duming, An analysis of the characteristic features of the surface heat source and heat balance over the Qinghai-Xizang

Plateau from May to August, 1979, Proceedings of International
Symposium on the Qinghai-Xizang (Tibet) Plateau and Mountain
Meteorology, March 20–24, 1984, Beijing, China.

[8] Xie Xianqun, Some characteristical value of the surface radiation
     field on the Qinghai-Xizang Plateau in summer, Kexue Tongbao, 7,
     425–429, 1983 (in Chinese).

[9] Lu Longhua and Dai Jujiaxi, The global and net radiation of the
     Tanggula region, Kexue Tongbao, 24 400–404, 1979 (in Chinese).

[10] E.A. Lopukhin, Trudy MGO, 13, 218–223, 1957 (in Russian).

[11] K. Ya. Kondratiev, Actinometry, Gidrometeoizdat, Leningrad,
     508,593,1965 (in Russian).

[12] Weng Duming, et al., The characteristic analysis on atmospheric
     transparency over the Qinghai-Xizang Plateau in summer, Bulletin
     of the Nanjing Meteorological Institute, 2, 1983 (in Chinese).

[13] Xiang Yueqin and Zhou Yunhua, Attenuation of solar radiation by
     desert aerosols and precipitable water in the atmosphere in the
     Qaidam Basin,Papers of Beijing International Symposium on Qinghai-
     Xizang Plateau and Mountain Meteorology, 1984 (in Chinese).

[14] Zhou Yunhua, The empirical calculation method of long-wave
     radiation surface long-wave radiation over the Qinghai-Xizang
     Plateau (to be published).

MOUNTAIN AIR RESOURCE MANAGEMENT

Douglas G. Fox[1] and Fred D. White[2]

1. INTRODUCTION

Mountains are special areas, generally rich in natural resources:
minerals, timber, unique and valuable flora and fauna, and aesthetics. They
contain the headwaters of major river systems that provide water to surrounding
populations. Often they represent special values as recreation centers. For
all these reasons, people desire to live in the mountains. Many mountain
locations are only now being developed because of modern transportation and
communications. Unless the resources are managed wisely, much of the value of
the mountains will be lost.

Wise use of resources requires scientific management. Renewable resource
(timber, wildlife, water) management involves gaining an understanding of the
dynamics of the resource (such as growth of trees), sufficient to allow
prediction of the natural state, and the effects of man's manipulations of the
natural state.

The atmosphere is not an unlimited resource. Although the atmosphere has
some capacity to clean itself, concentration of CO_2, for example, has increased
steadily since measurements began. Atmospheric deposition, particularly of
sulfur and nitrogen acids, has proven to be a significant threat to elements
of ecosystems far removed from population centers and industrial source
regions (Hutchinson and Havas, 1980). Concentrations of pollutants in small
mountain valley towns resulting from space heating (wood or coal stoves) and
strong ground-based inversions can exceed those measured in large urban

---

1.Douglas G. Fox is Chief Meteorologist, Rocky Mountain Forest and Range
Experiment Station, Fort Collins, Colorado, U.S.A.

2.Fred D. White is Chairman of the Steering Committee for the American
Meteorological Society/Environmental Protection Agency cooperative agreement
on air quality modeling.

complexes (Murphy, et al., 1981). Recognizing these conditions, the Congress of the United States has passed an increasingly strong set of laws providing policy and regulatory authority for managing air resources.

Air resource management is based on a concept of using the inherent dispersive capacity of the atmosphere to reduce the magnitude of ground-level concentrations. It assumes that a predictive capacity exists in the form of a relationship between pollution sources and the concentration at a distant ground-level location (receptor). These source receptor relationships are called air quality models. Using these relationships, a manager can identify the best locations for new sources and provide fair control of existing sources.

In the United States, each individual State has the authority to manage air resources within its territory, but the national government maintains responsibility for interstate pollution problems and for the protection of air-quality-related values of federally managed lands. Thus State regulatory agencies and Federal land managers are the air resource managers. The vast majority of federal land is located in mountainous topography requiring highly complex source receptor relationships for mountain air resource management.

This paper discusses air quality models. In Section 2 their use and their accuracy for simple topography are discussed. Section 3 reviews current knowledge about modeling in complex terrain. Section 4 presents one method and selected models suggested for mountain air resource management. The final section describes some efforts to evaluate these models.

2. AIR RESOURCE MANAGEMENT AND REGULATORY MODELS

An air quality model consists of theoretical and empirical equations that relates the emission from a source or sources to the ground-level concentration resulting at a distant receptor. The U.S. Clean Air Act provides for two uses of air quality models. First, the Act establishes clean air goals, setting standards for the ambient concentration of any pollutants that can adversely affect human health or welfare. Locations that do not meet these standards must develop plans to reduce emissions sufficient to reach these standards. Models are the primary tools used to develop these plans.

Second, the Act seeks to maintain the air quality in locations where it is now clean. This is accomplished through a permit process for new sources. In order to obtain a permit, the developer of a new source must use control technology to minimize its emissions and use modeling to show that its downwind concentrations do not exceed a small numerical increment. The numerical increment value depends on the pollutant, the averaging time over which concentration is estimated, and the classification of the location. Class I areas have the smallest numerical increment, and also require that air-quality-related values of the area will not be degraded. About 160 separate areas, each greater than 5000 ha in size, some much greater, and most located in high mountain areas, carry this Class I designation. Important decisions are based, almost exclusively, on model projections of likely impact because of the U.S. air quality laws. As a result, there has been strong interest in the quality of the models used for air regulatory purposes.

Selection and evaluation of air quality models

In 1978, the United States Environmental Protection Agency, recognizing the growing concern about models, issued a guideline on air quality models (EPA, 1978). Virtually all the models listed in the guideline are based on the hypothesis that concentrations can be described by a gaussian distribution in both the vertical and horizontal. The gaussian rate of spread is evaluated from empirical functions dependent on atmospheric stability. A similar empirically based formulation of momentum and buoyancy exchange between plume and environment is implemented in the models to determine the height of plume rise. These models are collectively termed gaussian plume models, and have been used almost exclusively for the management of air resources through the regulatory process.

The EPA modeling guideline generated controversy, in part because many felt that the scientific community was capable of producing something better. The EPA and the American Meteorological Society (AMS) therefore developed a cooperative agreement to improve the science of regulatory air quality modeling. The authors, along with a number of other air quality scientists, have been involved with this effort since 1979. Among the activities this cooperation has fostered include:

> Development of a model evaluation protocol (AMS, 1981)
>
> Identification of a set of performance measures for evaluating air quality models (Fox, 1981)
>
> Peer scientific review of air quality models (Smith, 1984)
>
> Identification of the sources of uncertainty in model analysis (Fox, 1984)
>
> Review of models for, and research in, complex terrain (Egan, 1984)
>
> Development of improved applied dispersion models (Weil, 1984)

Accuracy of regulatory models

This effort, and similar ones conducted by the Electric Power Research Institute (EPRI) (Hilst, 1978), have led to increased scientific understanding of how well regulatory air quality models work. Figure 2.1 (Bowne, et al., 1983), from the comprehensive EPRI study of plume model validity, illustrates that the models show very little accuracy when compared with observations. This figure shows the highest hourly averaged and normalized concentration of

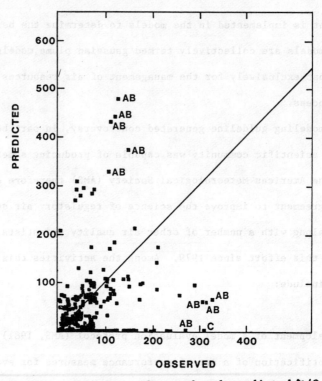

Figure 2.1 Comparison of highest observed and predicted χ/Q values for

1-hour SF_6 concentration averages, CEQM Ib.   (Bowne et al., 1983).

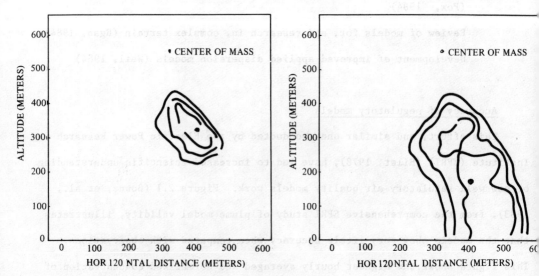

Figure 2.2   Lidar data showing fumigation of a plume in 6 minutes

(Bowne et al., 1983).

a tracer released from the stack of a large coal-burning electric generating station located in flat terrain and measured at a large network of points. These measurements are compared against predictions from one of the guideline models, using meteorological data collected at a tower on the site. A number of high concentrations were observed where the modeling predicted little or none. The AB notation signifies that the atmosphere was unstable during the time of the observation. These convective conditions give rise to the highest measured and predicted concentrations because, during these conditions, the plume can be physically transported to the ground by the vertical motion field within the boundary layer (called fumigation). Figure 2.2 (Bowne, et al., 1983), from the same study, shows a lidar observation of this fumigation process. These convective conditions in flat terrain represent the highest concentrations measured, and appear to be the least well modeled by the gaussian models.

Figure 2.3 (Landagen, et al., 1982) illustrates another method of comparing observations and predictions based on frequency distribution of maximums. There is no pairing of observations with the model predictions for the same time or location within the monitoring network. These data are from a different, large coal-fired power plant located in fairly flat terrain. These data are typical of the type of data collected by regulatory agencies. The results here show that the peak or top end of the frequency distribution of observations is quite well simulated by the model. Similar results have been found for many different evaluations, including the EPRI study mentioned above. Below the top few observations the model does not reproduce the observation frequency distribution. This data suggests that the use of models is acceptable for regulatory purposes only when the top of the frequency distribution is used.

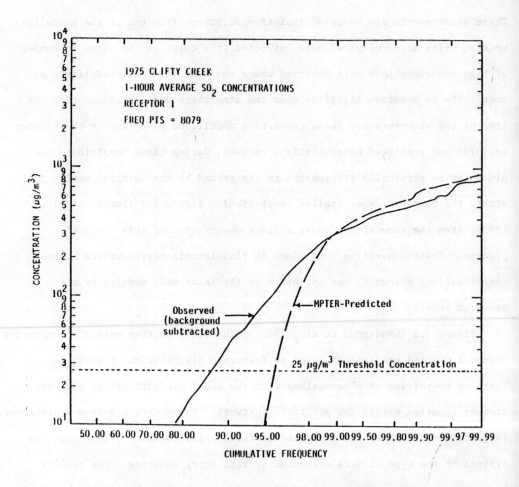

Figure 2.3 Cumulative frequency distribution of observed and model
(EPA-MPTER) predicted concentrations (Londergan et al., 1982)

Model uncertainties

At a recent workshop addressing uncertainty about specific air quality model predictions, attendees concluded that, at the present time, there really is no acceptable way to quantify the uncertainty of a model prediction (Fox, 1984). Differences between observations and predictions will always be large. The significant questions to model users are first, how much of this difference can be attributed to model inadequacies, and secondly, how small can this difference be made?

There are four sources of difference between observations and predictions: 1) measurement error in the concentration sampling, 2) systematic errors in the model physics and its parameterization, 3) uncertainty in model input parameters, and 4) natural variability caused by atmospheric motions. The first two are easily dealt with. The third and fourth give rise to major problems. The atmosphere is rarely in a steady state. Observations of the atmospheric state when averaged over a time period therefore are uncertain (have a mean and a variance). Atmospheric motion is defined by field variables which exhibit, especially on length and time scales of interest to air resource managers, small variations compared with the measurement scale. The representativeness of any single measurement within such a field depends very much on the nature of the field, increasing the uncertainty. This uncertainty resulting from the complexity of space and time atmospheric patterns may be the primary uncertainty in this system. The fourth source of difference between observation and prediction is the natural variability of atmospheric motions. The atmosphere is turbulent. Even if an experiment was performed repeatedly, such that the set of input variables was held fixed, the measured concentrations would not be the same. Rather, they would approach an ensemble mean value as a limit over a number of repeats (realizations) of the same experiment. In the real atmosphere, which is neither stationary nor homogeneous, each observation represents an individual realization. Even when these observations are averaged over long time there remains uncertainty due to the fact that ensemble and time averaging are not the same.

Thus one should expect differences between observations and predictions, and realize that they will not be small even if a perfect model were developed.

3. AIR QUALITY MODELING IN MOUNTAIN TERRAIN

The above discussion has dealt with models used for air resource management under idealized circumstances. In mountain areas, the problem is considerably different. Rugged terrain introduces significant constraints on air pollution movement that we cannot quantify at present. The influence of topography on the flow field is especially obvious near the ground and often even far removed from the surface. A recent workshop conducted under the AMS/EPA cooperative agreement focused on dispersion over complex topography (Egan, 1984). Information presented at the workshop resulted largely from two major field efforts conducted by the EPA and the Department of Energy. The EPA program focused initially on idealized topography (small, round hill, two-dimensional ridge) and stable stratification. The DOE program, ASCOT (Atmospheric Studies in Complex Terrain) dealt with night-time drainage flow in a valley (Dickerson, 1980). Both of these programs are presented elsewhere in this conference, so we won't describe them further.

Windward effects

The AMS/EPA workshop addressed phenomena considered important for air quality regulatory applications. These included windward terrain interaction with elevated plumes, lee side effects, valley situations, and long distance transport under stable conditions. Modeling windward interaction of plumes

116

with terrain has been controversial for some time. The air flow that impinges on a mountain is distorted and deflected, causing mean field variations in the winds as well as increases in the turbulence levels. At issue is whether these mean field variations are the result of the plume being carried into, around, or over a barrier. What happens in simple situations depends upon the upwind Froude number, Fr=U/Nh, where:

U = wind speed of the approach flow

$$N = \left(- \frac{g}{\rho} \frac{\partial p}{\partial z}\right)^{\frac{1}{2}} = \text{Brunt-Vaissala frequency}$$

h = height of the terrain barrier.

When the flow is strongly stable, 0<Fr<1, a critical height appears to exist below which the flow remains virtually horizontal, going around isolated barriers, and above which the flow goes over the barrier. This height can be estimated by considering the level at which an air parcel acquires sufficient kinetic energy to overcome the stability (Sheppard, 1956). A plume located near this critical height will impinge on the barrier. The ground-level concentration at this point is approximately what would be experienced at the plume centerline in the absence of a topographic barrier. This phenomenon has been observed for small hills (order 100 m, Strimaitis, et al., 1983). Recent results from a more substantial mountain (order 600 m) suggest that the critical height exists, but more complex flow patterns are generally observed (Furman and Wooldridge, 1984). The patterns are not simply determined by the Froude number, but depend on often subtle variations in the approach wind and temperature profiles. Under neutral, unstable, or slightly stable conditions, the flow is likely to go over the topography with some divergence causing a component of flow around the feature. Simple flow models based on potential flow or a conservation of mass approach appear to describe these flow regimes.

Lee side effects

Lee side effects can also lead to high ground-level concentrations. In neutral flow, both separation and recirculation phenomena can cause this concentration. The wake structure behind topography is complex and often includes significant spacial structure. Simple models such as discussed in section 3 will simulate a slowing down of the wind, but no recirculation.

Valley effects

Valleys can trap pollution due to combinations of a strong inversion and physical blocking of winds by the side walls. However, there is also, in general, some surface flow along these valley side walls. Valleys exhibit upslope flows in afternoons and drainage flows at night. Recent studies indicate that the various drainage patterns are often unique to the particular valley. The workshop chose to classify valleys as shallow, deep and draining, and closed. Shallow valleys are those where the plume can be considered to stay above the valley walls. This situation can be simulated with models that alter the plume path through a coefficient adjusting its height relative to the topography and as a function of the stability present. A number of the EPA guideline models can do this. Deep draining valleys can be modeled by recognizing the presence of the surface flows and embedding a diffusion model within them. Finally, the closed valley can be modeled using a box model approach. The key parameter here is the depth of the inversion, which represen the lid of the box.

Future research

Scientists participating in the workshop on complex topography made recommendations about research needed to improve models used for the management

118

of air resources. The scientists agreed with the current emphasis on field data collection and recommended its continuation. In particular, the use of physically simple terrain in which to conduct experiments was recognized as appropriate for learning something general about an otherwise non-general situation. Combinations of field work, physical model simulation (such as wind tunnels, water channels), and theoretical work were considered to be most desirable. The scientists reiterated that precise prediction of concentration at particular space and time coordinates was unrealistic. Accurate prediction of a mean value seems possible only when a time average which is long compared to the turbulence time scales of the flow is desired.

Meteorological measurements used for input to models in complex terrain are very important. No single measurement point can be representative of an entire flow field. Thus, measurements should be made at as many locations as practical. Measurements of a lagrangian nature, such as balloon tracking, should also be used. In this regard, use of meteorological models to generate a wind distribution was considered an appropriate step so long as the model could be relied upon to produce acceptable flow field data at reasonable cost. Such a combination of flow field and dispersion models was also suggested for the long range transport problem. Finally, a number of research areas were recommended for further study.

4.   <u>AN AIR RESOURCE MODELING SYSTEM FOR APPLICATIONS IN MOUNTAINS</u>

Although the models discussed here do not meet all the requirements identified in the previous section, they go a step in that direction. Some considerations of their validity are addressed in the last section.

The Topographic Air Pollution Analysis System (TAPAS) is a computer system designed to allow non-scientists an opportunity to apply the complex air quality models needed to manage air resources in mountains. TAPAS was developed by the USDA Forest Service and the USDI Bureau of Land Management for application in the western United States (Fox, et al, 1984). An overview of the major components of TAPAS is shown in figure 4.1. TAPAS is easy to apply. Many of the components allow menu-driven interaction with the computer: the user requests a particular model, and the computer calls up directions and instructions to make its use simple. Results can be displayed graphically and scaled to overlay any size map. This feature allows a directly useful result for users who require map-based information. Furthermore, TAPAS is a growing system with new models replacing old ones as they become available.

A particularly valuable TAPAS component, the terrain data management module, provides digital elevation data at grid points every 30 seconds of latitude and longitude for the entire United States. The user need only input the latitude and longitude of the corners of his analysis area, and the computer will generate the terrain grid. This terrain grid can in turn be input to any other TAPAS model which may need it. For example, the two- and three-dimensional wind models and the puff dispersion model all need terrain elevations as boundary conditions.

The two-dimensional wind model is a diagnostic simulation of the depth-averaged flow over topography. The model is based on solving the continuity equation for the layer, assuming that the flow is everywhere parallel with the terrain at the terrain surface and that vertical velocity is zero at the top of the layer. The divergence is calculated in a terrain-following coordinate system, and then related to velocity components by inverting the stream function. Friction and thermal forces are added as small correction terms (Fosberg, et al., 1976). The solution is driven by a background flow assumed to exist above the terrain-influenced layer. The background flow can be determined in a number of ways: from direct observations, from synoptic

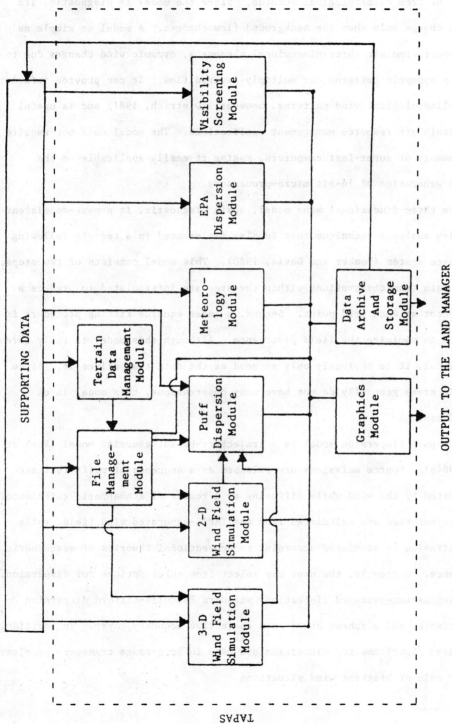

OUTPUT TO THE LAND MANAGER

Figure 4.1   Overview of TAPAS.

charts, or from climatological records. Since the model is diagnostic, its results change only when the background flow changes. A model as simple as this cannot simulate three-dimensional phenomena, dynamic wind changes due to evolving synoptic patterns, or multiply-layered flow. It can provide information about climatological wind patterns, however, (Dietrich, 1981) and is useful for certain air resource management applications. The model does not require large memory or super-fast computers, making it easily applicable on the current generation of 16-bit micro-processors.

The three-dimensional wind model, again diagnostic, is a mass-consistent objective analysis technique that is also implemented in a terrain-following coordinate system (Bunker and Davis, 1980). This model consists of two steps. First, data from observations within the area are interpolated to produce a wind vector at each grid point. Second, a least squares fitting procedure is employed to minimize the field divergence. Although this model is fully three dimensional, it is obviously only as good as the data that drives it. Since mountain areas generally do not have many observations, this model is of limited use.

The puff dispersion model is a trajectory-based, gaussian model (Ross et al., 1984a). Source emissions are released as a sequence of puffs that are transported by the wind while diffusing as a result of atmospheric turbulence. Puff trajectories are calculated from the TAPAS-generated wind field, while puff diffusion is simulated according to conventional theories of atmospheric turbulence. Currently, the user may select from three options for dispersion, including sector-averaged dispersion, standard Pasquill-Gifford dispersion coefficients, and a scheme based on a convective boundary layer. In addition, time-based algorithms for dispersion are used in long-range transport problems and for calm or stagnant wind situations.

The sector-averaged dispersion option is based on the extensive studies of model evaluation for complex terrain conducted by the EPA (Lavery, et al., 1982). These studies suggest that the best agreement with observation for "straight-line" gaussian model predictions occurs when sector averaged dispersion is employed. However, it is important to note that, since sector averaging is used in part to include the additional effects of plume meander in complex terrain, one should use care in applying such coefficients. The puff model can directly incorporate these effects separately by using a time-varying or strongly space-varying wind field. Finally, the "PPSP" code as developed by Weil and Brower (1982), which incorporates some of the modern concepts of the convective boundary layer and similarity scaling, is also available as an option.

Line, area, and volume source distributions can be simulated by the use of virtual point sources. Removal of pollutant by decay or dry deposition is incorporated in the model via a linear depletion of mass from the puffs. Plume rise is calculated according to standard methods such as recommended in guideline models by the EPA (Pierce and Turner, 1980). The model also allows inclusion of a terrain adjustment factor, which raises or lowers puff height over topography according to stability considerations (Lavery, et al., 1982).

TAPAS has been applied to estimate wind flows, pollution potential, and pollutant concentrations for a variety of air resource management applications, including industrial and recreational development and forest fire smoke management. Analyses are typically applied with a grid spacing of 0.5 to several kilometers, covering areas from 100 to 100,000 km^2.

Results from a recent study to evaluate potential air quality effects of oil shale and other synthetic fuel development in western Colorado (Dietrich et al., 1983a, 1983b) are provided as an example of TAPAS analytical and

123

graphic output capability. The sampling of figures provided here represents typical results.

Figure 4.2 is a base map of the area considered. Two of the 18 industrial sources considered in the analysis are located on the base map along with nearby sensitive class I areas. (The U.S. Clean Air Act provides special protection to such areas. TAPAS was initially developed to analyze impacts for this regulation.) The area is approximately 240 km E-W by 220 km N-S, and is characterized by grid points at 2-minute intervals (81 grid points E-W by 61 grid points N-S). The trajectories of puffs from two oil shale processing facilities for a steady west wind condition are illustrated in Figure 4.3. These sources had plume rises of approximately 300 m. Figure 4.4 provides the ground-level SO_2 impact from the two sources, for one of several high-production scenarios considered.

Although this system has been used for applications such as the above, it still requires more evaluation. In fact, the only applications that can be considered scientifically acceptable at present are those where the uncertainti and inadequacies are recognized and the results treated accordingly. In the foregoing example, assumptions were made that could be generally accepted by all parties to the analysis as being conservative. This allowed the results to be interpreted as representing an upper bound on air quality. In addition, an uncertainty was placed on results based essentially on the likelihood of achieving the fixed (in time) wind pattern assumed. TAPAS is appropriate for general screening analysis of this sort.

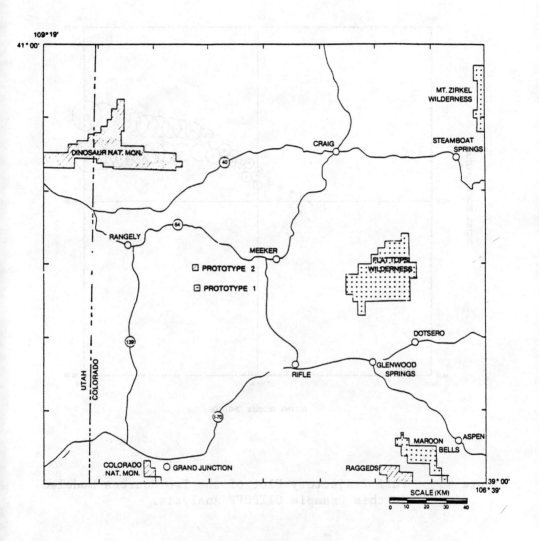

Figure 4.2   Base Map of the Analysis Area Considered
in the Example CITPUFF Analysis.

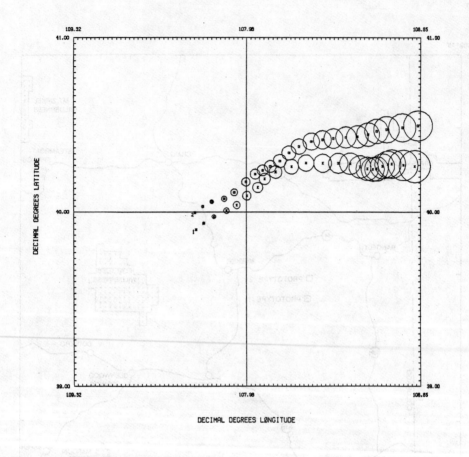

Figure 4.3   Puff Trajectory Plot of the Two Sources Modeled
in this Example CITPUFF Analysis.

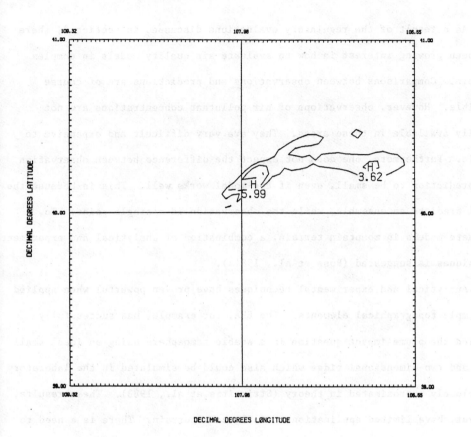

Figure 4.4 SO$_2$ Ground-Level Pollutant Concentrations ($\mu g/m^3$) Resulting from this Example CITPUFF Analysis.

5. EVALUATING AIR QUALITY MODELS FOR MOUNTAIN APPLICATIONS

As a result of the regulatory evaluations discussed in section 2, there has been growing interest in how to evaluate air quality models in complex terrain. Comparisons between observations and predictions are of course possible. However, observations of air pollutant concentrations are not readily available in these areas. They are very difficult and expensive to obtain. Furthermore, one does not expect the difference between observation and prediction to be small, even if the model works well. This is because the model predicts an ensemble, while the observation is a single point. To evaluate models in mountain terrain, a combination of analytical and experimental techniques is suggested (Ross et al., 1984a).

Analytical and experimental techniques have proven powerful when applied to simple topographical elements. The EPA, for example, has successfully studied the plume impact question in a stable atmosphere using an ideal small hill and two-dimensional ridge which also could be simulated in the laboratory and closely approximated in theory (Strimaitis et al., 1983). These results, however, have limited application to full-scale terrain. There is a need to evaluate more realistic situations in the same manner.

Wind measurements

One such study uses a conically-shaped mountain (Fig. 5.1) in north central New Mexico. Mount San Antonio is approximately 600m high (2700m base to 3,325 m top) about 8 km in diameter, and separated at least 25 km from the nearest topographic features (McCutchan et al., 1982).

The wind field on Mount San Antonio has been measured using nine remote automated surface weather stations. These stations are solar powered and use the Geostationary Orbiting Earth Satellite (GOES) for communicating hourly

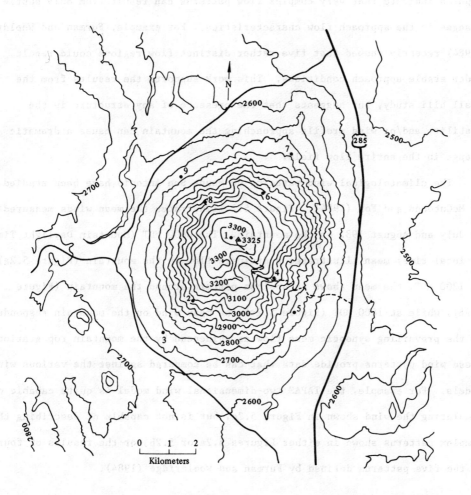

Figure 5.1   Location of nine remote automatic weather stations (RAWS) on
San Antonio Mountain. Terrain contours in meters, contour
interval 50 meters.

values of wind, temperatures, humidity, and other parameters. Preliminary
studies indicate that very complex flow patterns can result from only subtle
changes in the approach flow characteristics. For example, Furman and Wooldridge
(1984) recently showed that five rather distinct flow regions could result
under stable approach conditions. This work supports the results from the
small hill study, but suggests that the presence of any structure in the
stability and/or wind profile approaching the mountain can cause a dramatic
change in the entire flow field.

The climatological wind patterns on Mount San Antonio have been studied
by McCutchan and Fox (1984). Figure 5.2 illustrates the mean winds measured
in July and August 1982 on the mountain. The 0600 MDT (Mountain Daylight Time
or local time) mean illustrates a drainage flow on the mountain (Figure 5.2a).
At 1200 MDT, the mean shows an upslope pattern around the mountain (Figure
5.2b), while at 1800 MDT (Figure 5.2c) the mean wind on the mountain responds
to the prevailing synoptic mean flow as evidenced by the mountain top station.
These wind patterns provide data that can be compared against the various wind
models. For example, the TAPAS two-dimensional wind model is quite capable of
simulating the wind shown in Figure 5.2c, but is not capable of describing the
complex patterns shown in either Figures 5.2a or 5.2b, or the results of four
of the five patterns defined by Furman and Wooldridge (1984).

Wind modeling

Analytical solutions for flow around a simple hemisphere have been
compared with results from the TAPAS three-dimensional wind model by Ross
et al. (1984a). They show how critically important the location of observations
are to the solution. Figure 5.3 shows the three-dimensional wind model
results for flow past a hemisphere when a linear profile is input only at
point A. Figure 5.4 a shows the result when the liner profile is provided

130

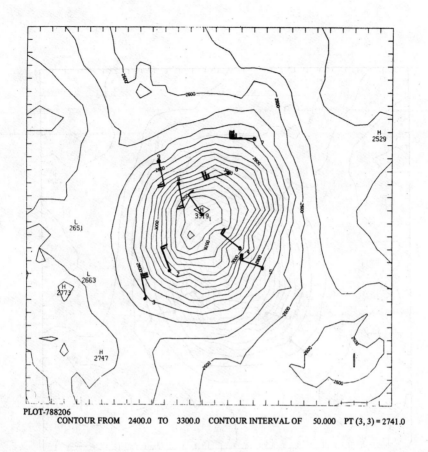

PLOT-788206
CONTOUR FROM   2400.0 TO 3300.0   CONTOUR INTERVAL OF   50.000   PT (3, 3) = 2741.0

Figure 5.2

Resultant wind speeds and directions for the mean u- and v-wind components for 10 days in July and August 1982 at nine stations on San Antonio Mountain at: (a) 0600 MDT, (b) 1200 MDT, and (c) 1800 MDT. Full barb 1 m/s, and half-barb 0.5 m/s. Terrain contours in meters, contour interval 50 meters.

(a)   0600 MDT

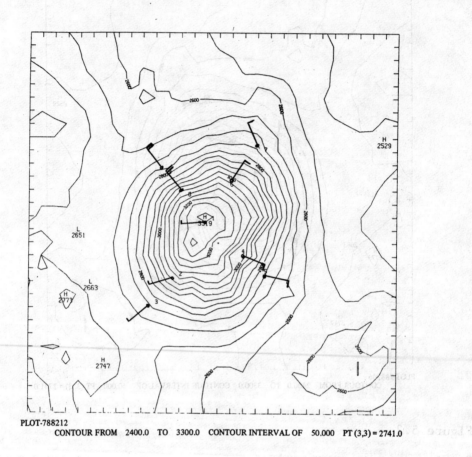

PLOT-788212

CONTOUR FROM 2400.0 TO 3300.0 CONTOUR INTERVAL OF 50.000 PT (3,3) = 2741.0

Figure 5.2   (b) 1200 MDT

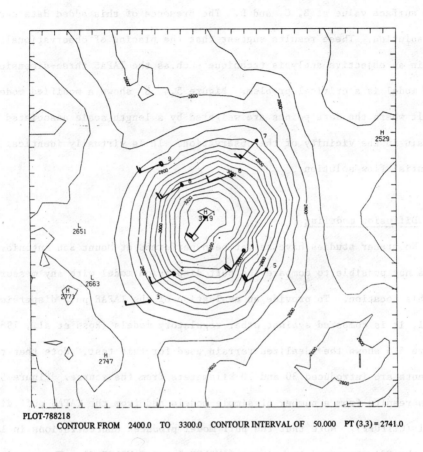

PLOT-788218

CONTOUR FROM 2400.0 TO 3300.0 CONTOUR INTERVAL OF 50.000 PT (3,3) = 2741.0

Figure 5.2   (c) 1800 MDT

133

at the point A and the actual potential flow analytical solution is provided as a surface value at B, C, and D. The presence of this added data deteriorates the solution. These results suggest that the placing of observational data within an objective analysis technique such as the TAPAS three-dimensional wind model is a critical problem. Figure 5.4 b shows a modified model result where the data points are weighted by a length scale associated with terrain in the vicinity of the observation. It is virtually identical to the potential flow solution.

Diffusion modeling

No tracer studies have as yet been performed at Mount San Antonio. Thus, it is not possible to compare the puff dispersion model with any measurements at this location. To provide an evaluation of the TAPAS puff dispersion model, it is compared against other regulatory models (Ross et al., 1984b). Figure 5.5 shows the idealized terrain used for this test. Note that topographi elements are introduced 30 and 70 kilometers from the source. Figure 5.6 shows results from a number of simple models and from the TAPAS puff dispersion model (CITPUFF). Note that the puff model produces concentrations in line with the EPA suggested techniques COMPLEX I and COMPLEX II. These models have been subject to considerable testing against tracer data for sources within a kilometer or two from topographic elements. More work is needed in the evaluation of these models before one model can be selected over another.

6. CONCLUSIONS

We have reviewed the current state of regulatory modeling for mountain air resources management, and illustrated the accuracy of these models in flat

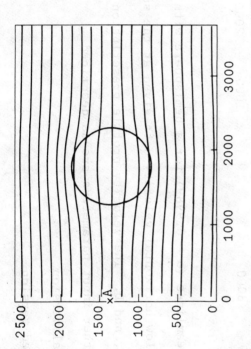

Figure 5.3

Surface streamlines for a hemisphere from TAPAS three-dimensional
wind model. A uniform 10 m/s velocity is specified at A.

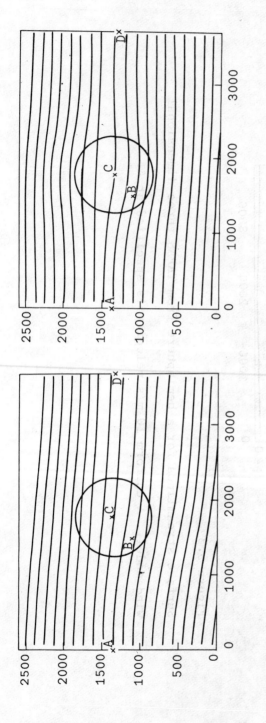

Figure 5.4   Surface streamlines ($\sigma = 0.9989$) for a hemisphere with additional

potential flow horizontal velocity data given at the surface

points B, C, and D. The velocity profile prescription at location

A and the other input parameters are as described

(a) original version

(b) modified version

136

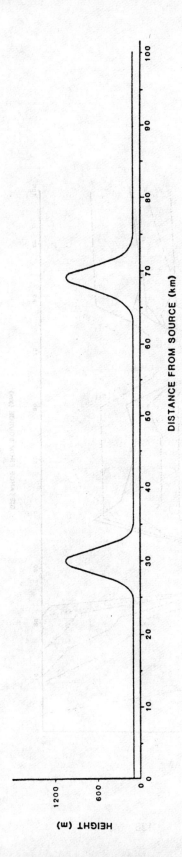

Figure 5.5   Cross Section of Mounded Terrain Data for EPA COMPLEX/CITPUFF Comparisons.

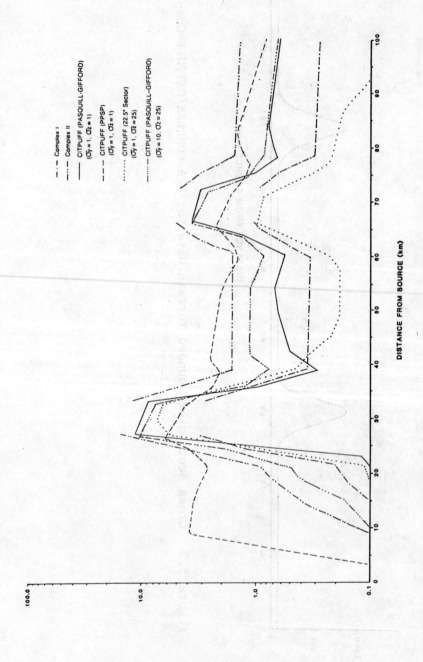

Figure 5.6 Mounded Terrain Comparison of COMPLEX I, COMPLEX II and CITPUFF. (Plume Height = 300, Emissions = 100 g/s, Wind--West at 4 m/s measured at 10m, Stability = E, CITPUFF runs vary by the definition of initial σy and σz at the source and the dispersion scheme in CITPUFF.

terrain applications. For complex mountain topography, the Topographic Air
Pollution Analysis System (TAPAS), consisting of two- and three-dimensional
wind simulation models along with a gaussian puff dispersion technique, has
unique capabilities. A field study to gather data for the validation of these
techniques was described. Preliminary results illustrate a considerably more
complex picture than had been found with more limited studies of smaller
terrain features. Although the models have not been evaluated against these
data, as yet, results of comparing them against simple analytical solutions
have indicated areas of needed improvement.

REFERENCES

American Meteorological Society. 1981. Air Quality Modeling and the Clean Air Act. A report prepared under cooperative agreement with the Environmental Protection Agency. Amer. Meteor. Soc., Boston, Mass. 41 p.

Bowne, N.E., et al. 1983. Overview, Results and Conclusions for the EPRI Plume Model Validation and Development Project: Plains Sites. EPRI-EA-3074, Project 1616-1 EPRI. Palo Alto, Calif. 230 p.

Bunker, S.S., and C.G. Davis. 1980 Mass-consistent wind fields - July 22, Geysers area. In ASCOT Rep. UCID-18884, Lawrence Livermore Laboratory, Livermore, Calif.

Dickerson, M.H., Ed. 1980. A collection of papers based on drainage wind studies in the Geysers area of northern California. U.S. DOE Report ASCOT-80-7 UCID 18884, Lawrence Livermore National Laboratory, Livermore, Calif. 248 p.

Dietrich, D.L. 1981. Wind climatology in complex terrain. Ph.D. Thesis, 149 p. Colorado State University, Dept. of Earth Resources, Fort Collins, Colo.

Dietrich, D.L., D.G. Fox, D.G. Ross, and W.E. Marlatt. 1983a. Technical Report - Air Quality Impact Assessment for the Supplemental Environmental Impact Statement in the Prototype Oil Shale Leasing Program.

Dietrich, D.L., D.G. Fox, D.G. Ross, and W.E. Marlatt. 1983b. Technical Report - Air Quality Impact Assessment for the Environmental Impact Statement in the Federal Oil Shale Management Program. Bureau of Land Management, Denver Service Center. 285 p.

Egan, B.A. 1984. Dispersion in Complex Terrain. A report of a workshop held at Keystone, Colo., May 1983. Amer. Meteor. Soc., Boston, Mass. (To appear)

EPA. 1978. Guidelines on Air Quality Models. EPA-450/2-78-027 OAQPS Guidelin Services. 84 p.

Fosberg, M.A., W.E. Marlatt, and L. Krupnak. 1976. Estimating air flow patterns over complex terrain. USDA Forest Service Research Paper RM-162, 16 p. Rocky Mountain Forest and Range Experiment Station, Fort Collins, Colo.

Fox, D.G. 1981. Judging Air Quality Performance. Bull. Am. Meteor. Soc. 62, 599-609.

Fox, D.G. 1984. Uncertainty in Air Quality Modeling. Bull. Am. Meteorol. Soc., 65, 27-36.

Fox, D.G., D.L. Dietrich, and J.E. Childs. 1984. Overview of the Topographic Air Pollution Analysis System (TAPAS): User's/Programmer's Guide. General Technical Report, Rocky Mountain Forest and Range Experiment Station, Fort Collins, Colo. 60 p. (To appear)

Furman, R.W., and G.L. Wooldridge. 1984. Observations of surface flow over a three-dimensional hill. Submitted to Quart. J. Royal. Meteorol. Soc.

Hilst, G.R. 1978. Plume Model Validation EA-917-SY, October 1978, Electric Power Research Institute, Palo Alto, Calif. 59 p.

Hutchinson, J.C. and M. Havas, Ed. 1980. Effects of Acid Precipitation on Terrestrial Ecosystems. Plenum Press, New York. 654 p.

Lavery, T.F., A. Bass, D.G. Strimaitis, B. Venkatram, B.A. Gren, P.J. Drivas, and B.A. Egan. 1982. EPA Complex Terrain Model Development First Milestone; Report 1981. EPA-600/3-82-036, USEPA. Research Triangle Park, N.C. 327 p.

Londergan, R.J. et al. 1982. Evaluation of rural air quality models. TRC Environmental Consultants, Inc. East Hartford, Conn. EPA-450/4-83-003, October 1982, Research Triangle Park, N.C. 300 p.

McCutchan M.H., Fox, D.G. and R.W. Furman. 1982. San Antonio Mountain Experiment (SAMEX). Bull. Amer. Meteor. Soc. 63(10), 1123-1131.

McCutchan, M.H. and D.G. Fox. 1984. The Effect of Elevation and Aspect in Wind, Temperature and Humidity. (To appear)

Murphy, D.J., R.M. Buchan and D.G. Fox. 1981. Ambient particulate and Benzo (2) Pyrene concentrations from residential wood combustion in a mountain resort community. In Proc. 1981 Intl. Conf. on Residential Solid Fuels. June 1981, Portland, Oreg. Oregon Graduate Center, Beaverton, Oreg. p. 495-585.

Pierce, T.D. and D.B. Turner. 1980. USEPA Guide for MPTER. Pub. No. EPA-600/8/80-16. USEPA, Research Triangle Park, N.C. 187 p.

Ross, D.G., D.G. Fox, D.L. Dietrich, J.E. Childs and W.E. Marlatt. 1984a. CITPUFF: A Gaussian puff model for estimating pollutant concentration on complex terrain. Research Paper Rocky Mountain Forest and Range Experiment Station, Fort Collins, Colo. (To appear)

Ross, D.G., D.G. Fox and M.C. Thompson. 1984b. On the Application and Evaluation of a Complex Terrain Air Quality Modeling System. In Proc. 8th Intnl. Clean Air Conference. Melbourne, Australia, May 1984.

Sheppard, P.A. 1956. Airflow over mountains. Q.J. Roy. Meter. Soc. 82. 528-529.

Smith, M.E. 1984. Review of the Attributes and Performance of Ten Rural Diffusion Models. Bull. Amer. Meteor. Soc., 65. (To appear)

Strimaitis, D.G., A. Venkatram, B.R. Greene, S.R. Hanne, S. Heisler, T.F. Lavery, A. Bass, and B.A. Egan. 1983. EPA Complex Terrain Model Development Program: Second Milestone Report - 1982. EPA-600/3-83-015, USEPA, Research Triangle Park, N.C. 375 p.

Weil, J.C. and R.B. Brower. 1982. The Maryland PPSP dispersion model for tall stacks, Environmental Centre. Martin Marietta Corporation Report PPSP-MP-36 to Maryland Department of Natural Resources.

Weil, J. 1984. Updating Applied Diffusion Models. A report of a workshop held at Clearwater Beach, Fla., February 1983. Amer. Meteor. Soc., Boston, Mass. (To appear)

THE CHARACTERISTICS OF RADIATION BALANCE
COMPONENTS DURING QXPMEX IN 1979

Pan Shouwen

Department of Meteorology, Nanjing University

I. INTRODUCTION

Many different studies about synoptic meteorology and climatology
pointed out that as a massive prominence the Qinghai-Xizang Plateau
have an important influence on the weather and climate of China, as
well as on the general circulation over Asia and the Northern Hemi-
sphere.In order to study the heating of the atmosphere by the Qinghai-
Xizang Plateau ground and the distribution features of the surface
heat source, it is necessary to understand the energy budget over
the plateau. Thus, during the Qinghai-Xizang Plateau Meteorological
Science Experiment (QXPMEX) from May to August 1979 in the same period
as FEEG, the six fixed-point stations carried out exploratory observations. The radiation measurement near the ground is one of the
scientific explorations mentioned above. These six exploring stations
were set up not only with an eye toward their geographic distribution,
but also toward the features of the underlying surface, the natural
landscape belt and climatic conditions over the Plateau. A control
station was established in Nanjing on the eastern plain of China. The
positions of the stations and the date of observations are given in
Table 1, and the observational items and instruments used are shown in
Table 2. Measurements were taken every two hours from 0400 h to 2000
h in the local mean time and only once at 0000 h in the nighttime.

Table 1 Conditions at the exploring stations and the dates of observations during QXPMEX 1979.

Station	Long. (E)	Lat. (N)	Alt. (m)	Surface	Date
Shiquanhe	80°05'	32°30'	4278	desert steppe	May 17— Aug. 31
Shuanghu	89°00'	32°38'	4920	"	May 21— Aug.31
Nagqu	92°03'	31°29'	4920	grass-land	May 22— Aug. 31
Lhasa	91°08'	29°42'	3658	"	May 6— Aug. 31
Nyingchi	94°28'	29°34'	3000	wet grass	May 11— Aug. 31
Golmud	94°54'	36°54'	2808	saline soil	May 15— July 6
Golmud	94°48'	36°24'	2841	gobi	June 28— Aug. 31

II. DATA AND METHOD

Through the observational data of solar and atmospheric radiation derived from the six exploring stations mentioned above,and through the records of 125 meteorological stations spread over the Qinghai-Xizang Plateau and its surrounding regions during QXPMEX, scientists became familier with the principal characteristics and the geographic distribution of radiation balance components in the region. They were also able to compile 112 maps of geographic distribution for May-August 1979, including those for global radiation (Q), albedo (A), effective radiation (F) and net radiation (B), absorption radiation by underlying surface Q (1-A), sunshine duration (s) and relative sunshine (s_1). The calculation methods used are as follows:

1. Global Radiation

The mean daily total amount of global radiation Q in $W.m^{-2}$ is given by

$$Q = Q_0 (a+bs_1) \tag{1}$$

where Q_0 is incident solar radiation at the top of the atmosphere for the daily sum in $W.m^{-2}$, s_1 is the relative sunshine in the form of a

143

percentage, a and b are two empirical coefficients depending on locations. The goal was to estimate global solar radiation in period of dekad average, and of monthly average, keeping in mind certain statistical considerations about the relationship between the global radiation and such climatological parameters as sunshine duration. For this reason, we divide the Plateau into seven subregions, of which empirical coefficients a and b are obtained seperately. According to equation (1), the mean relative error of calculated values is measured to within 10%. In fact, the percentage error is 6.0% for the dekad average, and 4.5% for the monthly average.

It must be pointed out that the mean relative error is not larger than 10%, if Q_0 is the global radiation under ideal atmosphere instead of astronomical radiation. That is to say, the percentage error is 6.6% for dekad and 4.0% monthly.

2. Albedo

The Plateau is divided into several subregions, the area of each of which is $2° \times 2°$ on the latitude-longitude grid. Using these latitude-longitude parameters we can consult maps of soil, plant cover, and topography as well as photomaps from the resource survey satellite over the Plateau. Having done this, it is possible to obtain the figure for the average albedo of different subregions according to the percentage of their surface pattern. Thus, the average relative error of calculated albedo is within 10%.

3. Effective Radiation

The mean daily total amount of effective radiation F in $W \cdot m^{-2}$ is given by equation (2):

$$F = \varepsilon(\sigma \overline{T}^4 - \sigma \overline{T}_0^{\ 4}\ (a' + b'\overline{n} + c'\overline{e}))$$ (2)

where ε is surface emissivity ($\varepsilon=0.95$), σ is the Stefan-Boltzmann constant, $\overline{T}$ and $\overline{T}_0$ are the average temperatures of the air and the surface in dekad, $\overline{n}$ is the mean total cloud cover, $\overline{e}$ is the mean water-vapour pressure (hPa), a', b', c', are empirical coefficients dependent on location. The Qinghai-Xizang Plateau is divided into three subregions according to the varying degrees of moisture content in the atmosphere. Coefficients are estimated separately. According to equation (2), the

Table 2 Observational items and instruments.

Term	Symbol	Unit	Instruments
Direct solar radiation	S, S'	$W \cdot m^{-2}$	Pyrheliometer model MS-52 made in Japan
Sky radiation	D	"	Pyranometer with shadow band stand model in Japan
Global radiation	Q	"	1. Q=S'+D 2. Albedometer model RM-21 made in Japan
Reflected global radiation	R_k	"	Albedometer
Surface albedo	A	%	According to $A=R_k/Q$
Upward terrestrial radiation	U	$W \cdot m^{-2}$	Epply pyrgeometer made in America
Atmospheric ciounter radiation	G	"	"
Effective radiation	F	"	According to F=U-G
Radiation balance	B	"	1. $B=Q-R_k-F$ 2. Net radiameter model CN-11 made in Japan

mean relative error of calculated values is measured to be 6.3% for dekad and 4.5% monthly.

4. Radiation Balance

According to the above results, the radiation balance is given as follows:

$$B = Q (1-A) - F \tag{3}$$

The mean relative error of calculated value is calculated to be 9.7% for dekad and 4.9% monthly.

III. THE PRINCIPAL CHARACTERISTICS OF COMPONENTS
OF THE RADIATION BALANCE

Through analysis and studies of the observational data at the six stations, we are able to uncover some interesting facts and phenomena.

1. Global Radiation

The monthly sum of global radiation over the main part of the Qinghai-Xizang Plateau is larger than 225 $W \cdot m^{-2}$, particularly in the western region of the Plateau where the sum exceeds 300 $W \cdot m^{-2}$. This is one of the regions where the global radiation is at a maximum for the whole Earth. This is represented in Table 3. It is clear from Table 3 that the global radiation over the Plateau is far greater than over the plain regions in the same latitude, as for example, in Nanjing. For convenience sake, if we regard the average of three months in Nanjing as 100%, referring to Table 3, the percentage of different stations over the Plateau then will be:

Shiquanhe	Shuanghu	Nagqu	Lhasa	Nyingchi	Golmud
159%	141%	118%	129%	109%	134%

Table 3 A comparison of global radiation at different stations during QXPMEX ($W \cdot m^{-2}$).

Stations	June	July	August
Shiquanhe	355	303	314
Shuanghu	307	264	290
Nagqu	267	248	210
Lhasa	291	275	227
Nyingchi	250	207	212
Golmud	282	276	261
Nanjing	222	186	204

It is apparent that a rapid increase of direct solar radiation occurs with altitude over the Plateau, and that the direct radiation is much stronger than the sky radiation on the Plateau, with the monthly sum of the direct radiation greater than diffuse radiation except in Nyingchi station. Generally, the direct radiation is about 55-78% of the total radiation.

It is worth noticing that the measured values of global radiation surpassed the solar constant at noon on some days at the six exploring stations (Table 4). This phenomenon has been measured by some foreign investigators, such as Takasu (1953) on Mount Shirouma (2720) in Japan and in the upper Ötz valley at a height of 1940 m. The absolute maxima of global radiation from May to July are found between 1535 and 1570 $W \cdot m^{-2}$ (Turner, 1958). Because of special geographic and meteorological conditions, the phenomena have been observed repeatedly over the Qinghai-Xizang Plateau.

2. Surface Albedo

The monthly average of albedo on the Plateau from May to August is about 16—18%, depending on weather and surface conditions, except in the Qaidam Basin and Tarim Basin where albedo increases to 30% under desert conditions. This is shown in Table 5, where the monthly average of albedo over the Plateau is presented.

The variation of albedo on the Plateau is related to the wetness of the underlying surface. Albedo before and after the arrival of the rainy season shows an evident change. The relationship between albedo and the precipitation for the warm season over the Plateau is shown in Fig. 1. The curves in Fig. 1 display the tendency change during QXPMEX.

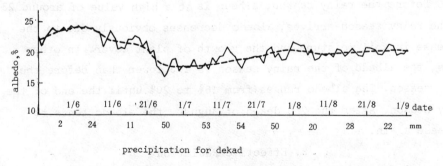

Fig. 1 The daily variation of the average albedo
from May to August at Nagqu.

Table 4 Records of global radiation flux (S'+D)
greater than the solar constant (1382)
at noon ($W \cdot m^{-2}$).

stations	Q	D	sun's disc	cloudiness and cloud forms	time
Shiquanhe	1514	468	θ	[10]/0 Ac Cu	VI.13
Shuanghu	1486	468	θ,θ²	[10] / [10] Cb Cu Fc	VI.24
	1424	586	θ,θ°	8/8 Sc Cu Fc	VI.27
	1458	544	θ	7/6 Cu Ac Fc	VII.6
	1430	433	θ²,θ°	6/6 Cb Cu Fc	VII.23
	1389	342	θ²,θ	8/8 Cu Fc	VII.29
Nagqu	1444	349	θ	4/4 Cu	V.24
	1389	279	θ	8/7 Cu Ci	V.26
	1396	307	θ	6/6 Cu	VI.20
	1507	440	θ	8/8 Cu Sc	VI.30
	1424	377	θ	8/7 Sc Ci	VII.2
Lhasa	1486	516	θ	[10]/5 Ac Sc	VI.21
	1535	565	θ	[10]/6 Cu Ac	VII.2
	1389	321	θ	8/2 Ac Cu	VIII.15
Nyingchi	1472	398	θ	[10] / [10] Sc Cu	VI.1
Golmud	1389	321	θ	4/4 Cu	V.16

The rainy season in 1979 begins at the Nagqu station on the 8th of
June. Before the rainy season, albedo is at a high value of around 23—25%.
As the rainy season arrives, albedo decreases obviously due to the
increase in soil moisture and the growth of plant cover. In other
words, the albedo of the rainy season is 1/3 lower than before the
rainy season. The albedo ranges from 16% to 20% until the end of the
rainy season. After the 2nd dekad of August, the albedo tends to
increase again.

3. Effective Radiation

The level of monthly effective radiation over the Plateau is two
or three times as large as in the plain regions of East China (Table 6)

Table 5 The mean albedo for each month and dekad
over the Qinghai-Xizang Plateau (%).

Time	Shi-quanhe	Shuang-hu	Nagqu	Lhasa	Nying-chi	Glomud
May	-	-	-	-	-	-
1st dekad	-	-	-	-	-	-
2nd dekad	-	-	-	24	18	-
3rd dekad	25	23	23	24	19	31
June	26	18	21	22	17	28
1st dekad	26	19	23	24	20	30
2nd dekad	26	19	22	22	17	24
3rd dekad	27	17	19	21	14	28
July	24	16	19	19	16	20
1st dekad	24	16	18	19	15	18
2nd dekad	23	16	19	20	16	20
3rd dekad	26	15	20	18	16	21
August	27	17	20	17	16	20
1st dekad	26	17	20	16	16	20
2nd dekad	26	17	20	17	16	19
3rd dekad	29	18	20	18	16	21

As far as the monthly average of the warm season is concerned, the effective radiation shows a tendency to increase its strength with height. For example, the effective radiation at the Nyingchi station is 1.25 times larger than that at Nanjing, at Lhasa 1.8 times, at Nagqu 1.9 times, at Golmud 2.6 times, at Shuanghu 2.7 times and at Shiquanhe 3.1 times.

4. Radiation Balance

Based on measured values of global radiation and surface albedo and of effective radiation near the ground, the net radiation balance is given by equation (4). Owing to the high value of global and effective radiation, the net radiation balance over the Plateau is not larger than on the neighboring regions, and the difference between the Plateau and the plain of East China is not evident. This is shown in Table 7.

Table 6 The effective radiation during QXPMEX
$(W \cdot m^{-2})$.

Stations	June	July	August
Shiquanhe	141	122	137
Shuanghu	120	101	126
Nagqu	96	75	74
Lhasa	106	71	57
Nyingchi	67	47	50
Golmud	110	113	110
Nanjing	56	33	39

Table 7 A comparison of the net radiation balance at
seven stations during QXPMEX
$(W \cdot m^{-2})$.

Stations	June	July	August
Shiquanhe	122	110	93
Shuanghu	133	122	114
Nagqu	115	126	95
Lhasa	122	152	131
Nyingchi	141	128	128
Golmud	93	108	99
Nanjing	122	117	125

IV. THE GEOGRAPHIC DISTRIBUTION FEATURES
OF THE RADIATION BALANCE

During QXPMEX a higher value of the radiation balance was maintained over the whole Plateau. In the main part of the Plateau the
value was more than 90 $W \cdot m^{-2}$. The high value center was situated around
90°E and 30°N and the center value was greater than 120 $W \cdot m^{-2}$. The
geographic distribution of radiation balance in May and July are given
in Fig. 2 and Fig. 3.

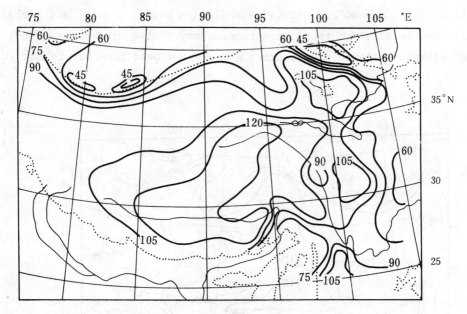

Fig. 2 The distribution of the radiation balance for the
earth's surface over the Plateau in May 1979
$(W.m^{-2})$.

We know from Fig. 2 that over the main part of the Plateau the
higher value of the radiation balance in May to August is closely
related to the high value for global radiation. When the rainy season
comes to the Plateau, albedo and the effective radiation decrease even
further than does global radiation. As a result, the radiation balance
in the rainy season rapidly increases. This is how the high-value center
over the main part of the Plateau is maintained during QXPMEX.

In the southeastern region of the Plateau under the influence of
the Bengal wet current, there is greater moisture content and more
cloudy weather. Although albedo and effective radiation are less,
global radiation still decreases obviously due to the lower transparency
of the atmosphere and higher cloud cover. The radiation balance is also
far less than that over the main part of the Plateau and becomes a low
value belt.

In the northeastern region (the Qaidam Basin) and the northwestern
region (the south site of the Tarim Basin) of the Plateau under the

arid climate, albedo and effective radiation become greater and global
radiation less due to the lower transparency of the atmosphere. Thus,
the radiation balance also becomes a low value belt from May to August.

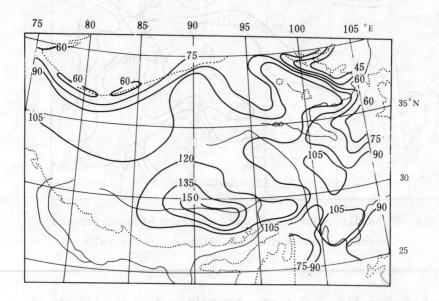

Fig. 3 The distribution of the radiation balance for the
earth's surface over the Plateau in July 1979
$(W \cdot m^{-2})$.

In spite of the fact that the radiation balance may vary in
intensity due to the transition of seasons over the Plateau, the
distribuiton pattern remains the same. Fig. 2 and 3 show the
influence of the transition of seasons upon the intensity and the
distribution of radiation balance between May and July. The map for May
represents the distribution in the dry season and the map for July
shows it in the rainy season. The intensity of the radiation balance
in July is about 25% larger than that in May. Meanwhile, the position
of the high value center moves southward to the river-valley region in
south Xizang. Where the low value albedo dut to the wetness of the surface and the high value global radiation due to the high frequency of
night rain in rainy season occur simultaneously, the radiation balance

increases evidently. For instance, the global radiation of Lhasa in rainy season decreases in intensity by 17% over the dry season, but effective radiation amounts to 51% in addition to the decrease of albedo. As a result, the radiation balance increases by 30% or so.

It is clear from Fig. 4 that the transition of seasons on the Plateau augments the radiation balance. In comparison with the dry season, the radiation balance in the rainy season increases by 15—30 $W.m^{-2}$, particularly in the river valley region of the southern Plateau.

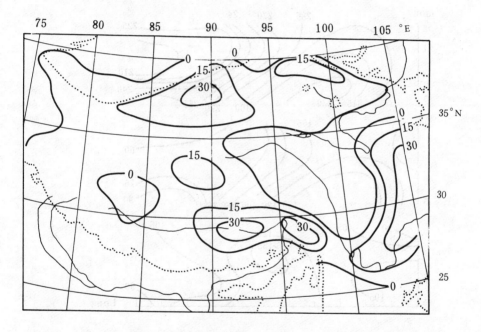

Fig. 4 The distribution of the difference between radiation balance in July and in May
$(W.m^{-2})$.

In the western and northwestern part of the Plateau, the influence of the transition of seasons on the radiation balance is not evident due to less precipitation. In the southeastern part, because the variation in precipitation with the transition of season is not evident, the variation of radiation balance with the transition of season is not significant either.

Fig. 5 summarizes the variation of components of the radiation balance with the transition of season. Fig. 5 is shown as a time cross-section along the latitude of 31°N. It is clear from Fig. 5 that the highest inter-monthly variation of components of radiation balance is represented in the regions from 88°E to 92°E. It is closely related to the change in weather-climatic conditions, when the transition of seasons over the Plateau occurs.

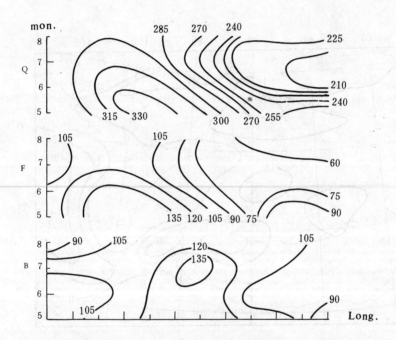

Fig. 5 The time cross-section of global radiation Q, effective
radiation F, and radiation balance B along the latitude
of 31°N (W·m^{-2}).

V. CONCLUSIONS

Based on the studies mentioned above we are able to come to the following conclusions:

1) Because of special geographic and climatic conditions, the total sum of the global radiation over the Qinghai-Xizang Plateau is

more than 300–330 $W \cdot m^{-2}$. Thus the level of global radiation in this region is among the highest in the world, and undoubtedly, this is one of the most significant facts to remember about radiaiton climate over the Plateau.

2) In the main part of the Plateau, the monthly sum of the radiation balance exceeds 120–150 $W \cdot m^{-2}$. This is the same level for those regions in the Northern Hemisphere where the net radiation is the highest.

3) During QXPMEX the measured values of the global radiation flux density surpassed the solar constant at noon on some days. They were measured at six exploring stations with a maximum value of 1535 $W \cdot m^{-2}$.

4) The average for surface albedo on the Plateau during QXPMEX was about 16–18%, with the variation due to weather and surface conditions. The albedo over desert regions increased to 30% It decreases evidently with the arrival of the rainy season over the Plateau.

5) The geographic distribution pattern of the radiation balance and its variation with the seasonal transition over the Plateau is similar to that of the global radiation of the Plateau. A high value center is maintained over the whole Plateau from beginning to end.

6) The radiation balance augments with the seasonal transition. The high value center of the radiation balance is located in the river-valley of south Xizang where the low value albedo due to the surface wetness and the high value global radiation due to the high frequency of night rain in rainy season occur simultaneously.

REFERENCES

[1] Collected Papers of the Qinghai-Xizang Plateau Meteorological Science Experiment (1), Science Press, pp. 1–60, 70–103, 1984.
[2] Collected Papers of the Qinghai-Xizang Plateau Meteorological Science Experiment (2), Science Press, pp. 1–23, 75–84, 1984.
[3] Lu Longhua and Dai Jiaxi, Global and net radiation in Tanggula Mountains, Science Report, 24, pp. 400–404, 1979.
[4] Deacon, E.L., Physical Processes near the surface of the earth,

general climatology, 2. world Survey of Climatology, vol. 2,
edited by H. Flohn, pp. 39—109, 1969.

[5] Timanov Skaya, R.G. and Faraponova, G.P., Measurement of the
radiation heat influx in the atmospheric ground layer, Izv. Akad.
Nauk. S.S.S.R., Ser. Atmos. Ocean. Phys., 3, pp. 742—746, 1967.

[6] Maps of radiation balance and heat balance during QXPMEX, Meteorology Press, 1984.

ATTENUATION OF SOLAR RADIATION BY DESERT AEROSOLS
AND PRECIPITABLE WATER IN THE ATMOSPHERE
IN THE QAIDAM BASIN

Xiang Yueqin and Zhou Yunhua

Institute of Geography, Chinese Academy of Sciences

The Golmud region is located in the central southern part of the Qaidam Basin. Its average elevation is between 2700 and 2800 meters. The major parts of its surface are desert and gobi.

A total of 162 sets of solar spectral radiation measurements were made using an AT-50 Yanishefsgi Actinometer mounted with wideband cut-off filters CB3, HB11 and HB16 relative to wavelengths 523, 630 and 700mμ respectively from May to August, 1979.

The load of desert aerosols and the attenuation of solar radiation by aerosols and precipitable water along the northeast boundar of the Qinghai-Xizang Plateau were analyzed using these measurements. This information provided a background for constructing a picture of the thermal and radiation budget over the Plateau.

I. METHOD FOR ESTIMATING ATMOSPHERIC TURBIDITY PARAMETERS

1. Estimation of Ångström Turbidity Parameters β and α

The irradiance of solar radiation, S_λ (a,d), passing through the atmosphere may be described in the ultraviolet and visible bands where water vapor has negligible absorption, by

$$S_\lambda(a,d) = \frac{1}{F} \int_0^\lambda I_{0\lambda} \, e^{-(\sigma_{a\lambda}+\sigma_{0_3\lambda}+\sigma_{d\lambda})M_h} \, d\lambda \tag{1}$$

where $I_{0\lambda}$ is the extra-terrestrial spectral irradiance at the wavelength λ, and $\sigma_{a\lambda}, \sigma_{0_3\lambda}$ and $\sigma_{d\lambda}$ are the optical depths for air molecular scattering, absorption of ozone, and aerosol extinction at λ, respectively. M_h is the relative air mass.

According to Ångström's empirical equation

$$\sigma_{d\lambda} = \beta \lambda^{-\alpha} \qquad (2)$$

where β and α are parameters, β is defined as the Ångström turbidity coefficient which may characterize the load of aerosols in the vertical column of air above an observing site and α is defined as a wave length exponent which may characterize the size distribution of aerosols in the vertical column of air. Therefore the load of aerosols may be described by the combination of β and α. First based on the measurements of S_λ (a,d) in the ultraviolet and visible bands $\sigma_{d\lambda}$ can be obtained by reasonable simplification and numerical integration from Equation (1) in which the values of $\sigma_{a\lambda}, \sigma_{0_3\lambda}$ and $I_{0\lambda}$ can be found in handbooks and M_h can be found from the elevation of the sun at the time of observation. Secondly, β and α may be estimated by placing $\sigma_{d\lambda}$ into Equation (2).

2. Estimation of Linke Turbidity Factor T, Dry Turbidity Factor D and Wet Turbidity Factor W

By definition, the Linke turbidity factor T may be expressed numerically as a ratio of attenuation of solar radiation by the actual atmosphere to the attenuation by air molecular scattering. T=1 if only molecular scattering is considered. T>1 for attenuation by the actual atmosphere.

The attenuation of solar radiation is caused mainly by aerosols, precipitable water and air molecules in the atmosphere. So that T can be expressed as

$$T = 1 + D + W \qquad (3)$$

where D and W are defined as the dry turbidity and wet turbidity factors, respectively. T and D can be derived from Equations (4) and (5)

$$T = p(m)\ln \frac{S_0}{FS} \qquad (4)$$

$$1 + D = p(m)\ln \frac{S_0}{FS(a,d)} \qquad (5)$$

where p(m) is the Linke turbidity which depends on the absolute air mass, m. S_0 is a solar constant and S is the irradiance of solar radiation on the ground. F is a factor for correcting the sun-earth distance. S(a,d) is the irradiance incident on the ground when the atmosphere is dry. Then

$$S(a,d) = \frac{1}{F} \int_0^\infty I_{0\lambda} \, e^{-(\sigma_{a\lambda} + \sigma_{0_3}\lambda + \beta\lambda^{-\alpha})M_h} \, d\lambda \qquad (6)$$

where each term on the right side of Equation (6) is the same as in Eq. (1). Based on the known α and β S(a,d) may be estimated by integrating Eq. (6) numerically. T and D can be calculated from measurements of S and S(a,d) using Eq. (4) and (5) for which p(m), F and S_0 can be found in handbooks. Then W can be derived by placing T and D into Eq. (3). Thus the attenuation of solar radiation by aerosols and precipitable water combined with that by moleculars in the air may be described by T, D and W.

II. THE LOAD OF AEROSOLS OVER THE GOLMUD REGION
DURING SPRING AND SUMMER

The values of β and α in Table 1 show that the monthly average load of atmospheric aerosols decreases steadily from May to August while the size of aerosols reached the maximum in June, the minimum in August. In general the load of aerosols in the air was higher and The size was larger in spring than that in summer.

Figure 1 shows that the daily averaged load of aerosols fluctuates greatly in spring and summer. Analysis of the synoptic situation suggested two causes for the increasing load of aerosols. The first was that a large quantity of desert dust was lifted into the air by the strong wind caused by the frequent invasion of cold air before the middle of June. From the observations it is that strong anticyclones occurred three times from 22 May to 3 June, 1979. During this period the instantaneous wind speed was occasionally 18–$24 \text{m} \cdot \text{s}^{-1}$, causing intense sandstorms. Secondly, high concentration of aerosols was ob-

Table 1 The monthly average value of atmospheric
turbidity over the Golmud Region, 1979.

	The latter part of May	June	July	August
α	1.1	0.5	1.4	1.7
β	0.200	0.161	0.102	0.060
T	4.76	4.26	3.83	3.44
D	2.15	2.01	1.36	0.84
W	1.62	1.24	1.46	1.60

served in the atmosphere in the presence of strong thermal convection
associated with a shear in the upper atmosphere in summer. When the
convection was stronger the wind speed was often more than $10m \cdot s^{-1}$.
The daily fluctuation of the load of aerosols from 30 July to 4 August
was caused by this action. Generally the strong wind was much more
important in initiating and maintaining the load os aerosols in the
atmosphere than in the thermal convection.

Based on data of wind speeds of less than $10m \cdot s^{-1}$ an empirical
equation for β and α was obtained with the result that $\beta = 0.161e^{-0.88\alpha}$,
the correlation coefficient being 0.88. But this equation is not true
for the data collected under the conditions of strong wind. It was
shown that the size distribution of the aerosols in the atmosphere was
changed because of a large quantity of desert dust of larger size
transported into the air above the region by strong wind.

III. ATTENUATION OF SOLAR RADIATION BY AEROSOLS
AND PRECIPITABLE WATER

The value of D and W in Table 1 shows that the attenuation of
solar radiation by aerosols decreased gradually whereas that by
precipitable water was relatively unchanged from May to August. In
May and June attenuation of solar radiation by aerosols was twice
greater than that by molecules, and attenuation by precipitable water
was about one and half times that due to molecules. In contrast the

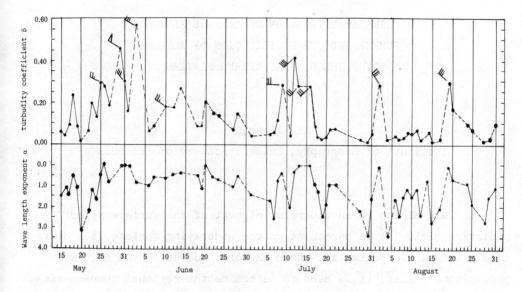

Fig. 1. The daily average values of Ångström atmospheric
turbidity parameters β and α over the Golmud region
from May to August, 1979.

attenuation of solar radiation by precipitable water in August was
the same as in May, but attenuation by aerosols was less. The attenuation by aerosols was the same as that by precipitable water in
July. It is shown that the change of atmospheric transparency depended
mainly on the load of aerosols, especially during the windy season.
The effect of precipitable water on atmospheric transparency should
not be ignored. However, variation in the precipitable water content
was small even in a rainy season because of the dry climate in the
region.

Normally, only the effect of molecular scattering on attenuation
of solar radiation is taken into accout because of the high elevation
and the usually clean air over the Qinghai-Xizang Plateau. But the
above analysis shows that the attenuation by aerosols and precipitable
water should be considered as a factor at least for the norther margin
of the Plateau during spring and summer.

A CLIMATIC ESTIMATION OF SURFACE ALBEDO IN
SUMMER AND THE DISTRIBUTION OF ALBEDO
OVER THE QINGHAI-XIZANG PLATEAU

Xie Xianqun

Institute of Geography, Chinese Academy of Sciences

Surface albedo is an important element of the surface solar
radiation field. It is dependent on the underlying surface, sun elevation, atmospheric regime and other factors.Because of the high and
precipitous relief, it is hard to determine the regional characteristics of surface albedo distribution over the Plateau based on the
routine solar observation of a few stations of this area. But using
the data obtained during the QXPMEX from May to August of 1979, an
integrated method is employed here to estimate the albedo, that is
to calculate the regional albedo weighted by different grids of latitude
and longitude from the correlation between the albedo and moisture or
precipitation, and then to weight the value for the monthly and ten
day albedo.

CALCULATION OF REGIONAL AVERAGE ALBEDO

1. Determination of the Underlying Surfaces and Their Distribution
   The pedological, botanical and geomorphological maps in the
Physiographic Atlas of China were adopted as the basic maps of our
analysis. With reference to the earth's resource satellite images,
Tirous-N satellite cloud maps from May to August of 1979, and other
information from integrative surveys over the Plateau by field investigators[1-3], a working map has been made to examine the features
of the underlying surface of the Plateau. Within the area between
75°—105°E and 24°—40°N, a grid net is divided every 2° both in

longitude and latitude. The type of underlying surface is evaluated for each grid. Let the total area be 10 and the fraction of various surfaces be n_1, n_2,..., n_i, then $n_1+n_2+...+n_i=10$.

2. Estimation of Average Albedo in Various Surfaces

The average albedo of various surfaces is determined on the basis of:

1) Albedo observation at six observation points over the Plateau from May to August, 1979 (see Table 1).

2) Albedo observation at standard stations of the Plateau and surrounding area.

3) Albedo observation over different surfaces features and altitudes in 1966 and 1968 Scientific Investigation at Mount Qomolangma[4].

4) Albedo observation as reported in various literature[5—9].

Table 1 The surface albedo in six standard radiation stations.

Stations / Surface / Date	Shuanghu	Nagqu	Shiquanhe	Lhasa	Nyingchi	Golmud*
	Desert grassland cold desert	Meadow grass-land	Desert grass-land	Mountain grassland Valley	Forest grass-land	Desert solonetz gobi
May	0.20	0.22	0.28	0.24	0.19	0.31
June	0.18	0.21	0.27	0.22	0.17	0.27
July	0.16	0.19	0.25	0.19	0.16	0.20
August	0.18	0.20	0.28	0.17	0.16	0.20

* Observations in Golmud are obtained from desert solonetz during May to June and gobi during July to August.

Knowing the average albedo and the area fraction of various surfaces, the average albedo of each grid can be calculated by the weighted average formula:

$$A = (n_1A_1+n_2A_2+...+n_iA_i)/10$$

where A is the average albedo of each grid.

3. Calculation of Actual Average Albedo in the Plateau

The surface albedo depends on the soil moisture and decreases exponentially with it[10]. There have not been any soil moisture observations over the large areas in the Plateau, and hence we have to use precipitation observations for albedo correction instead of soil moisture. According to our records in six observation points, monthly albedo decreases exponentially with the increase of monthly precipitation. The relationship is shown in Fig. 1 and the correlation coefficient is 0.878. From the curve in Fig. 1 we can find out how

Table 2 The mean relative error and correlation coefficient
between the average albedo in the grid and measured albedo
in May to August, 1979 on the Qinghai-Xizang Plateau*.

Date	δ	γ	Date	δ	γ	Date	δ	γ
May	0.088	0.916	Second ten days	0.084	0.902	Last ten days	0.082	0.895
June	0.098	0.895	Last ten days	0.110	0.897	First ten days of Aug.	0.072	0.891
July	0.096	0.891	First ten days of July	0.101	0.878	Second ten days	0.084	0.891
August	0.081	0.890	Second ten days	0.081	0.890	Last ten days	0.083	0.894
First ten days of June	0.094	0.888						

* Mean relative error $\delta = (\dfrac{|A_{grid} - A_{measured}|}{A_{measured}})$ average of 21 stations

much albedo will change in response to the fluctuation in precipitation, for instance, <5mm, 5—10mm, 11—15mm, 16—25mm, 26—50mm, 51—75mm, 76—100mm, >100mm etc.. From precipitation maps of May to August on the Plateau[11], the fluctuation in precipitation is estimated and the correction of albedo for each grid obtained. Then the monthly ten days average albedo of each grid can be found.

Measurements of albedo in 21 stations over the Plateau were taken as a benchmark for the comparison, and they indicate, that the mean relative error of the estimated values is within 10% (Table 2).

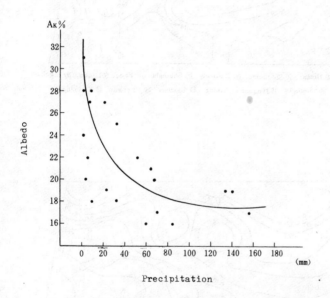

Fig. 1 The relationship between albedo and precipitation of
six stations on the Plateau from May to August 1979.

THE GEOGRAPHICAL CHARACTERISTICS OF SURFACE
ALBEDO OVER THE PLATEAU

A series of maps of monthly and ten days averages of albedo over the Plateau were compiled on the basis of calculation by the method described above. Some characteristics of albedo over the Plateau can be delineated as follows (Fig. 2-Fig. 5).

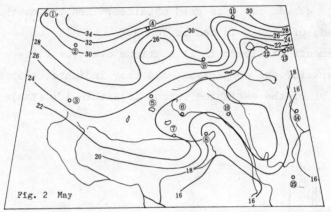

Fig. 2 May

①Kashi ② Hotan ③ Shiquanhe ④ Ruoqiang ⑤ Shuanghu ⑥ Nagqu ⑦Lhasa⑧ Nyingchi
⑨ Golmud ⑩ Changdu ⑪Jiuquan⑫ Xining ⑬ Lanzhou ⑭ Chengdu ⑮ Kunming

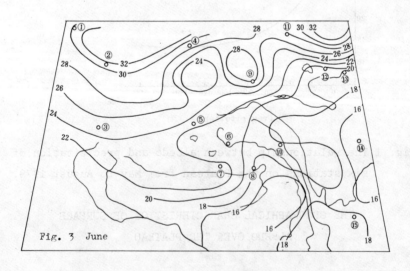

Fig. 3 June

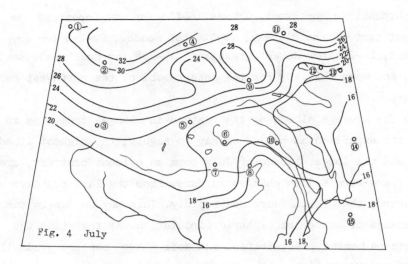

Fig. 4 July

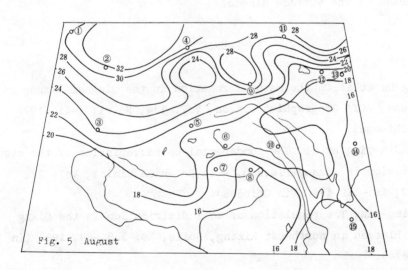

Fig. 5 August

Fig. 2-Fig. 5 The distribution of monthly average albedo
over the Qinghai-Xizang Plateau, May-August
1979.

1) The monthly and ten day averages of albedo over the Plateau appear in a zonal distribution corresponding to the physiographical zonal distribution of the Plateau. The albedo gradually increases from southeast to northwest. It was 16—19% for wet forest of the southeast part of the Plateau, 19—22% for meadow, 22—24% for dry grassland, 24—26% for desert grassland, 26—28% for desert and semi-desert, and more than 30% for the desert-gobi of the northwest part of the Plateau.

2) The average albedo of the Plateau is changes from time to time. During the experimental period of May to August, the highest albedo of May and the lowest albedo of July form an evident contrast, even though there were sudden changes in June. From the last ten days of June onward, the albedo decreases rapidly. This may be due to the rapid change of summer atmospheric circulation. At that time the rainy season begins,precipitation and soil water content gradually increase and vegetation on the Plateau grows luxuriantly, resulting in the decrease of the surface albedo.

REFERENCES

[1] Zheng Du et al,On the natural zonation in the Qinghai-Xizang Plateau, ACTA Geographica Sinica, Vol. 134, No. 1, 1—11, 1979 (in Chinese).

[2] Wang Jin-ting et al, The fundamental characteristics of the steppe vegetation in Xizang Plateau, ACTA Botanica Sinica, Vol. 22, No. 2, 161—169 1980 (in Chinese).

[3] Li Ming-sen, The regulation of soil distribution in the Qiang Tang Plateau in northwest Xizang, Soils, No. 3 81—84, 1980 (in Chinese).

[4] The report of the Scientific Investigation of Mount Qomolangma in 1966—1968, Meteorology and Solar radiation, Science Press. 1975.

[5] Chen Jiansui, The distribution of surface albedo and its varia-tion in China, Scientia Geographica Sinica, 30, 85—93, 1964 (in (Chinese).

[6] E.C. Kung, R.A. Bryson and D.H. Lenshow, Study of a continental surface albedo on the basis of flight measurement and structure of the earth's surface over North America, Monthly Weather Review, Vol. 92. No. 12, 1964.

[7] R.G. Barry and Chambers, A preliminany map of summer albedo over England and Wales, Quarterly J. Roy. Met. Soc. 92, No. 394, 1966.

[8] V.L.Gaevski, On the role of albedo in the formation of surface radiation state, Trudy MGO, 39, 1953 (in Russian).

[9] V.V. Mukhenbert, The surface albedo of the global continents, Trudy MGO, 193, 1967 (in Russian).

[10] K. Ya. Kondratiev, Actinometry, Gidrometeoizdat, Leningrad, 1965 (in Russian).

[11] Ye Fangde, Zhou Yenhua, et al., Studies of the Qinghai-Xizang Plateau summer precipitation by the use of satellite cloud picture, Plateau Meteorology, Vol. 2, No. 2, 26—35. 1983 (in Chinese).

Mountain Climate Data for Long-Term Ecological Research

R. G. Barry
Cooperative Institute for Research in Environmental Sciences
and Department of Geography
University of Colorado

Introduction

Ecological research in mountain regions raises important questions concerning the best treatment of incomplete field data from remote sites, of data representativeness, and of the selection of suitable climatic indices for ecosystem studies. These problems of basic and applied climate research in mountain areas are addressed in general terms and illustrated by reference to a project in the Colorado Rocky Mountains.

Under a recently-established pilot program of the U.S. National Science Foundation, research on long-term trends in natural ecosystems is being conducted in a range of different biomes. The University of Colorado Long-Term Ecological Research (LTER) program is focusing on possible effects of environmental disturbance on alpine and subalpine ecosystems on Niwot Ridge ($\sim$40°03'N, 105°35'W) in the Colorado Front Range (Figure 1). In support of these studies, mountain climate data collected in the area since 1952, principally at sites named Niwot Ridge, 3743m, and Como, 3018m, (Barry, 1973; Losleben, Personal Communication, 1982) are being reanalyzed and new air quality, snow depth and hydrological measurements begun. Limited statistical analysis of the temperature data has been made earlier by Joseph (1973).

Reliability and Representativeness of Basic Observations

In most mountain regions there are few homogeneous records, even for the common climatic elements. Extreme winds and blowing snow in winter make precipitation data from above timberline unreliable unless special screened gauges are installed, although forest clearings provide adequate sites below timberline. Temperature records tend to be fragmented when severe winter weather prevents regular servicing of alpine stations. In the absence of nearby alpine stations that can provide estimated data for incomplete records, mean temperatures and total precipitation for months with up to 25 percent of missing values can be estimated by substituting the long-term daily averages for the missing days in those months.

On Niwot Ridge, a snow-fence shielding was installed around the precipitation gauge at 3743 m in late 1964 (Barry, 1973), resulting in a 60 percent increase in the annual totals. Only post-1964 data are used in this study. For monthly precipitation totals, 15 percent of the record at Niwot Ridge and 4 percent at Como have been adjusted by substituting the daily mean for missing data.

An alternative approach would be to use the data from a neighboring station as an estimate (Tabony, 1983, for example). However, comparisons of monthly mean temperatures at Niwot Ridge and Berthoud Pass (3448m) some 30 km to the southwest and located just west of the Continental Divide, show minimum correlations of ∨ 0.5 in July and November (just significant at the 5 percent level) and in other months around 0.7-0.8, significant at better than the 1 percent level (see Figure 2). Hence these stations typically share only 50-60 percent of the variance of the monthly mean temperatures and, at worst, 25 percent.

A similar analysis of the trends of mean June-August temperatures at Niwot Ridge station, Como (3018 m) at the south-eastern end of Niwot Ridge, and Berthoud Pass (Figure 3) also reveals low correlations over time in spite of the visual similarity of the deviations in several of the extreme seasons. These results highlight the uncertainty of the spatial representativeness of mountain station data.

Analysis of subsets of the 1952-82 averages at Como indicates that mean monthly temperatures in winter can be estimated within about 1.0-1.5°C of the 30-year average with only a 5-year record length and within 0.7-1.0°C for a

10-year record, but the determination of the return period of extreme values, which are important for plant studies, requires at least a 20-year record (Barry, 1973).

Average precipitation totals are less stable. For the Colorado data the use of median rather than mean values has been found preferable in characterizing the annual precipitation regime (Figure 4). Barry (1973) suggested, on the basis of mean monthly precipitation amounts, that the High Plains regime with a spring maximum resulting from easterly "upslope" storms (Barry et al., 1981; Diaz et al., 1982), prevailed up to at least the elevation of Como (3018 m). Niwot Ridge was considered to have a "west slope" regime with a winter maximum attributable to Pacific storms. However, the dispersion diagrams in Figure 4 illustrate that mean values in winter and spring at both stations will be biased upward by some extreme months. This effect was exacerbated by the shorter record previously available. Figure 4 shows that the regime at the Niwot Ridge station is transitional in character between Como (7 km to the east) and Berthoud Pass (30 km to the southwest). There is a clear March maximum at Niwot Ridge (March-April at Como), but the pattern in early winter and the occurrence of several individually snowy winter and spring months is more like the regime at Berthoud.

The spatial representativeness of data is commonly determined 'after the fact,' since most mountain station sites are chosen on the basis of logistical considerations. A recommended solution would be to construct maps of the potential solar radiation and surface energy balance using now available computer programs with gridded terrain data (Williams et al., 1972; Barry, 1981) and to locate stations at sites with pre-selected radiation or energy balance regimes. Statistical or physical models of precipitation variation with altitude (see Barry, 1981, pp. 184-199), or of temperature variability with relief (Hess, et al., 1976) might also be useful in considering station siting. Data were

collected at different topoclimatic sites on Niwot Ridge in 1952-53 (Marr, 1967)
and intermittently since, but the measurements are too limited in scope and
duration to be of much assistance in estimating the topoclimatic variability.
Wind exposure is a special problem in this area. Mines Peak (3,808m) near
Berthoud Pass has similar average speeds in winter (Judson, 1977) to Niwot Ridge,
but short-term observations at a knoll site (3608m), 1 km east of the Niwot Ridge
station, show considerably more extreme winter gust conditions.

Climatic Indices for Ecosystem Studies

A major problem in ecosystem studies is the selection of suitable climatic
indices to assess thresholds for critical hydrological and geomorphic events and
for ecological processes. Conventional climatic statistics may have little
relevance in this context and several derived indices have therefore been
prepared.

Hydrologic and geomorphic events are frequently related to precipitation
intensity. For Niwot Ridge, about 43 percent of summer days have measureable
precipitation (with a mean of 5 mm/day) but the wettest 13 percent of
precipitation days contribute about 45 percent of total summer precipitation; this
fraction is comparable with other areas of western Colorado.

For plant life, daily frequencies of mean and minimum temperatures below
freezing during the growing season are of considerable significance. Analysis
of the daily temperatures on Niwot Ridge shows that there is almost no frost-free
season (Figure 5). Even in July, night air frost occurs on almost every day about
once in 10 to 20 years. Sharp seasonal transitions occur about 20 June and 10
September when the frequency of minima $\leq 0°C$ crosses the 50 percent frequency
level.

An integrated measure of summer warmth is provided by cumulative deviations
from the average daily mean temperature (Myers and Pitelka, 1979). Sums of these

173

deviations for July and for the three summer months have correlations with the July daily mean and summer daily mean temperatures, of 0.47 and 0.98, respectively, but display considerably greater interannual variability than the mean temperatures. Phenophase data on alpine meadow plants (May and Webber, 1982) show a two-week advance in summer 1974 growth compared with 1973, for example, and the corresponding cumulative temperature deviations for June through July at Niwot Ridge totalled +53°C in 1974 and -31°C in 1973. In contrast, the monthly mean temperature departures were only +1.9°C in June and +0.2°C in July, 1974, and 0.0°C in June and -0.7°C in July, 1973. Figure 6 illustrates plots of cumulative daily mean temperature deviations at Niwot Ridge for a near-average summer (1973), the coldest in the 25-year record (1982) and the second warmest (1963). Similar trends were observed at Como, but with less pronouced cold in 1982. It is interesting that few of the corresponding plots for the other years of record display patterns where strong positive/negative deviations early in the summer were reversed later. This contrasts with the findings of Myers and Pitelka (1979) for Barrow, Alaska, and suggests that in the Colorado Front Range there is less tendency for the late summer to reverse an anomaly pattern established earlier in the season.

An hypothesis that can be examined in the context of ecological research is Billings' (1974) statement that "alpine environments are characterized by short, cold, unpredictable growing seaons." The brevity and degree of cold have been amply confirmed, but two approaches are used here to assess summer temperature variability at Niwot Ridge.

The first assessment uses statistical quality control analysis (Moroney, 1956) to examine the variability between pentads (5-day intervals). The mean and range of daily temperatures grouped in pentads have been determined for July 1970-79. The six sets of pentad values for each July are treated as independent samples of July average temperature. The estimated population standard deviation

can be estimated from $\hat{s} = \bar{w}/d$ where $\bar{w}$ = the mean standard range and d = 2.326 for

samples of 5 days (Moroney, 1956, p. 155). For July 1970-79, $\hat{s}$ = 1.62. For a

normal distribution, 95 percent of the samples should lie within $\pm 2\ \hat{s}$ of the mean;

i.e. 8.1 $\pm$ 3.2°C, or 4.9 to 11.3°C. In fact, 8 percent of the sample means lie

outside this range and if July 1980-82 is also analyzed, 12 percent of pentads are

outside the range, compared with the 5 percent expectation. Moreover, 4 percent

lie outside the $\pm 3\ \hat{s}$, compared with only 0.3 percent in a normal distribution.

These results imply a greater than expected variability in temperatures on a

pentad time scale.

Another approach for assessing inter-diurnal variability makes use of the

autocorrelation of daily mean temperatures. Autocorrelations for lag-1 and lag-2

have been determined first for the average mean daily temperatures at Niwot Ridge

in July; the values are 0.53 and 0.03, respectively. Table 1 summarizes the

autocorrelations for each July 1970-1982; lag-2 and lag-3 values are included

wherever the lag-1 value exceeds that for the mean daily temperatures. In order

to compare differences between correlation coefficients, based on small samples

(<30 pairs), R.A. Fisher's Z transformation must be used (Weatherburn, 1962, p.

200). The 95 percent limits of the lag-1 autocorrelation for mean daily

temperatures are 0.20 and 0.75. The value for 1971 is outside this range, but

this could arise by chance one year in twenty. However, intercomparing individual

summers shows significant differences between 1970 (or 1974, or 1980) and 1971 or

1973; also between 1971, on the one hand, and 1972, 1975, 1977 and 1978, on the

other hand. The lag-2 correlation in 1971 and 1973 also differ significantly from

the value for the daily mean temperatures, but this is not true of the other lag-2

correlations in Table 1. These results indicate first, that the daily

temperatures can be characterized basically by a first-order Markov process, since there is generally no dependence beyond one day. Second, there are inter-annual differences in the short-term (inter-diurnal) dependence and, therefore, the 'predictability' of daily temperatures.

Table 1. Autocorrelations of daily mean temperature in July at Niwot Ridge (3743m)

	Lag-1†	Lag-2+	Lag-3*
Average daily means, 1953-82	0.53	0.03	---
1970	0.34		
1971	0.82	0.62	0.51
1972	0.45		
1973	0.75	0.56	0.47
1974	0.26		
1975	0.46		
1976	0.61	0.30	
1977	0.42		
1978	0.42		
1979	0.61	0.37	
1980	0.37		
1981	0.52		
1982	0.73	0.46	

† The 5% and 1% points are 0.36 and 0.46, respectively.
+ The 5% and 1% points are 0.37 and 0.47, respectively.
* The 5% and 1% points are 0.37 and 0.48, respectively.

Both sets of statistical analyses lend a measure of support to the description of Billings, cited above, concerning the unpredictable nature of the growing season in the alpine of Colorado.

Long-Term Considerations

In order to make long-term ecosystem assessments, it is essential to know whether the system is in equilibrium with present climate and its resilience to possible changes. More severe climatic conditions may tend to reduce the number of timberline trees but not significantly lower the treeline (Ives and Hansen-Bristow, 1983). Glacier fluctuations provide a more complex climatic "barometer" since they take place at undetermined lags. Consequently, the magnitude of climatic changes in the mountains can only be estimated.

Evidence in the Colorado Rockies suggests that the present timberline may have been established at least a thousand years ago during a more favorable climatic regime and withstood subsequent deteriorations in conditions (Ives and Hansen-Bristow, 1983). Cirque glaciers in the area have undergone major retreat since the mid-nineteenth century. The Arapahoe Glacier, the largest in Colorado, 5 km southwest of Niwot Ridge weather station has been observed since 1900. Photographic evidence (Figure 7) indicates a frontal retreat of 80-230 m and considerable thinning (Waldrop, 1964) although it has remained fairly stable since about 1960. Climatic records for the Rocky Mountain area indicate warmer springs and summers and drier winters in 1920s-40s than around the turn of the century (Bradley, 1980; Barry et al., 1981). Spring and fall in the 1960s and 1970s have become cooler, but summers showed little change. The 1962-82 data for Como, Niwot, and Berthoud (Figure 3) also show no recent trend in summer temperature.

Summary

This study has examined various facets of mountain climatology in the Colorado Rocky Mountains using two neighboring stations with 25-30 years of data. The uncertainty of spatial representativeness is highlighted by low inter-station correlations of monthly mean temperature. This problem consequently eliminates most standard procedures for interpolating estimates for missing records.

Appropriate climatic indices are required for ecological applications and several examples are provided. Measures of integrated summer warmth suggest that summers at Niwot Ridge tend not to display reversals of strong anomalies established in the first part of the season. However, high inter-diurnal and inter-pentad variability of temperatures lends support to the view expressed by Billings (1974) that the growing season in the alpine environment is short, cold and unpredictable. Average daily mean temperatures in July at Niwot Ridge resemble a first-order Markov process although in two individual years during 1970-82, the daily means in July showed significant autocorrelation for up to a 3-day lag.

The records indicate no recent trends in summer temperature in this part of the Colorado Rocky Mountains although other evidence suggests significant climatic fluctuations have occurred during the twentieth century as well as on longer time scales.

Acknowledgements

This study was supported in part by the University of Colorado's Long Term Ecological Research Project, under the National Science Foundation, Division of Biological Sciences grant DEB 80-12095. Participation in the International Symposium in Beijing was made possible by support from the National Science Foundation, Division of Atmospheric Sciences, the Cooperative Institute for Research in Environmental Sciences and the CULTER Project.

References

Barry, R.G., 1973. A climatological transect on the east slope of the Front Range, Colorado. Arct. Alp. Res., 5, 89–110.

Barry, R.G., 1981. Mountain Weather and Climate, Methuen, New York. 313 pp.

Barry, R.G., Kiladis, G. and Bradley, R.S., 1981. Synoptic climatology of the western United States in relation to climatic fluctuations during the twentieth century. J. Climatol., 1: 97–113.

Billings, W.D., 1974. Adaptations and origins of alpine plants. Arct. Alp. Res. 6: 129–42.

Bradley, R.S., 1980. Secular fluctuations of temperature in the Rocky Mountain states and a comparison with precipitation fluctuations. Mon. Wea. Rev., 108: 873–85.

Diaz, H.F., Barry, R.G., and Kiladis, G., 1982. Climatic characteristics of Pike's Peak, Colorado (1874–1888) and comparisons with other Colorado stations. Mountain Res. Development 2: 359–71.

Hess, M., Niedweidz, T. and Obrebska-Starkel, B., 1976. The method of characterizing the climate of the mountains and uplands in the macro-, meso- and microscale (exemplified by Southern Poland). Zesz. Nauk., Univ. Jagiellon, Prace Geog. 43: 83–102.

Ives, J.D. and Hansen-Bristow, K.J., 1983. Stability and instability of natural and modified upper timberline landscapes in the Colorado Rocky Mountains, U.S.A. Mountain Res. Development 3: 149–55.

Joseph, E., 1973. Spectral analysis of daily maximum and minimum temperature series on the east slope of the Colorado Front Range. Mon. Wea. Rev. 101: 505–09.

Judson, A., 1977. Climatological data from the Berthoud Pass area, Colorado. U.S.D.A. Forest Service, General Tech. Rept. RM-42, Rocky Mt. For. and Range Exp. Stn., Fort Collins, Colorado, 94 pp.

Losleben, M.V., 1983. Personal communication.

Marr, J.W., 1967. Data on mountain environments. I. Front Range, Colorado, sixteen sites, 1952-1953. Univ. of Colorado, Boulder, Ser. Biol., No. 27, 110 pp.

May, D.E. and Webber, P.J., 1982. Ecological studies in the Colorado alpine. Occas. Pap. No. 37, Inst. Arct. Alp. Res., Univ. of Colorado, Boulder: 35-62.

Moroney, M.J., 1956. Facts from Figures, Penguin Books, Harmondsworth, Middlesex, Third Edn., 472 pp.

Myers, J.P. and Pitelka, F.A., 1979. Variations in summer temperature patterns near Barrow, Alaska: Analysis and ecological interpretation. Arct. Alp. Res., 11: 131-44.

Tabony, R.C., 1983. The estimation of missing climatological data. J. Climatol., 3: 297-314.

Waldrop, H.A., 1964. The Arapahoe Glacier: a sixty-year record. Univ. of Colorado Studies, Series in Geology, no. 3, Univ. of Colorado Press, Boulder, Colorado, 37 pp.

Weatherburn, C.E., 1962. A First Course in Mathematical Statistics, Cambridge University Press, 277 pp.

Williams, L.D., Barry, R.G. and Andrews, J.T., 1972. Application of computed global radiation for areas of high relief. J. Appl. Met., 11: 526-33.

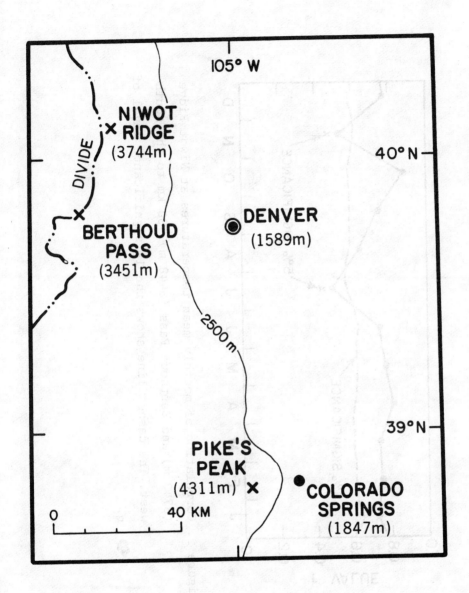

Figure 1. Location map.

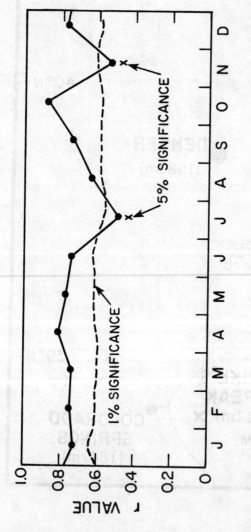

Figure 2. Correlation of monthly mean temperatures at Niwot Ridge (3743 m) and Berthoud Pass (3448 m), 30 km to the southwest. The dashed line shows the 1% significance level of r.

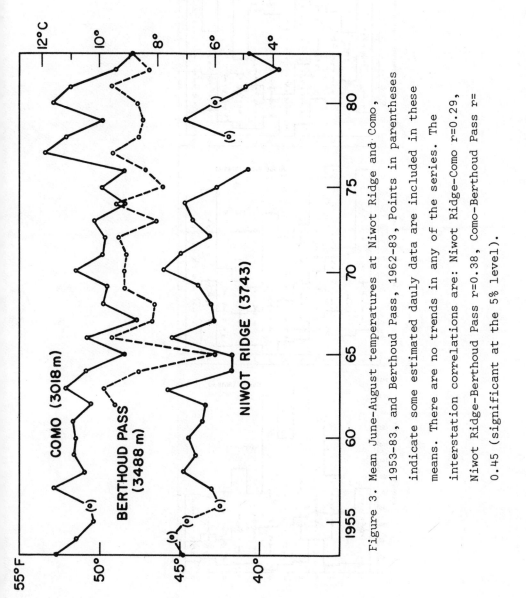

Figure 3. Mean June-August temperatures at Niwot Ridge and Como, 1953-83, and Berthoud Pass, 1962-83, Points in parentheses indicate some estimated daily data are included in these means. There are no trends in any of the series. The interstation correlations are: Niwot Ridge-Como r=0.29, Niwot Ridge-Berthoud Pass r=0.38, Como-Berthoud Pass r= 0.45 (significant at the 5% level).

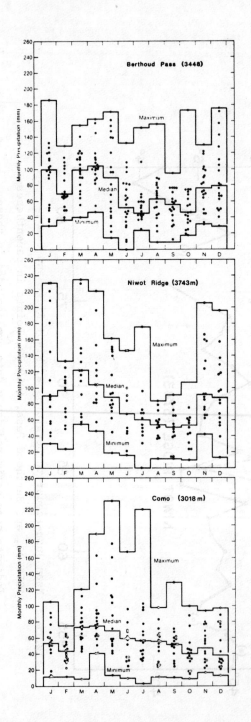

Figure 4. Monthly precipitation at Berthoud Pass (1963-82) Niwot Ridge and Como, 1965-82, Some monthly data are missing for the last two stations Maximum, median and minimum monthly values are indicated.

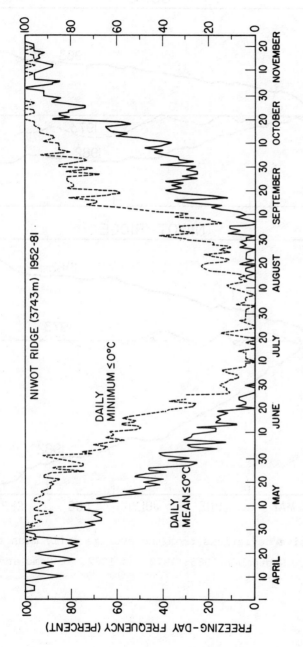

Figure 5. The frequency of freezing temperatures at Niwot Ridge (3743 m), 1952-81, for daily minimum and daily mean temperature.

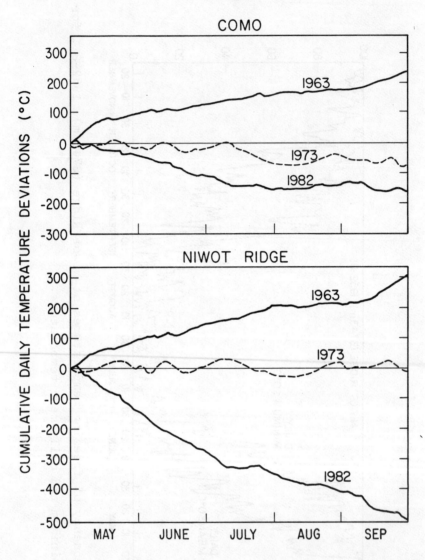

Figure 6. Cumulative deviations from the average daily mean temperature, 1 May–30 September 1963, 1972 and 1982, at Como and Niwot Ridge

Figure 7. The Arapahoe Glacier, Boulder County, Colorado, in late summer 1954 (top) and 1910 (bottom) [courtesy WDC-A for Glaciology].

BASIC PHYSICAL CHARACTERISTICS OF THE ATMOSPHERE
NEAR THE GROUND OVER THE QINGHAI-XIZANG PLATEAU

Ma Yutang* and Xu Zhaosheng
Institute of Geography, Chinese Academy of Sciences

I. INTRODUCTION

There have been many studies on the Qinghai-Xizang Plateau mainly concerned with synoptic and climatological analysis[1,2]. Due to a lack of opportunity for field experimentation, it is difficult to do research on the physics of the atmosphere near the ground. However, during May to August 1979, the QXPMEX was carried out over the Plateau. The data detained from this study has made it possible for us how to study this subject.

Six radiation stations have been adopted in this study. In order to compare the radiation in different sorts of land, five of the stations are in the Plateau, and one is in the east plain of China. The geographical coordinates and the surface pressure of these stations are outlined in Table 1.

II. THE BASIC CHARACTERISTICS OF STABILITY

Turbulent motion is a basic characteristic of air near the ground. There have been various parameters for stability, namely ε given by Laihatman, m by Budyko and $\frac{L}{Z}$ by Monin-Obuhov. All these parameters show the effects of stability on turbulance. But the most direct and simple calculating parameter is the Richardson number Ri. The stability through Ri will be investigated as follows:

---

* Present address-Meteorological Office of Weifang, Shandong Province.

Table 1 Geographical features of radiation stations

Station		Lat.(N)	Long.(E)	Altitude(m)	Surface Pressure(hPa)
Plateau	Golmud	36°25'	95°54'	2808	720
	Lhasa	29°40'	91°03'	3633	650
	Shiquanhe	32°30'	80°05'	4278	600
	Nagqe	31°29'	92°03'	4507	590
	Shuanghu	32°38'	89°00'	4920	560
Plain	Nanjing	32°31'	118°48'	25	1000

$$Ri = \frac{g}{T} \frac{\frac{\partial \theta}{\partial z}}{(\frac{\partial u}{\partial z})^2} \tag{1}$$

where g is the acceleration of gravity, T is the air temperature in K, θ is potential temperature, u is wind speed and Z, the height. The gradient of temperature and wind speed can be determined by the following equation[3,4]:

$$\frac{\partial \theta}{\partial Z} = \frac{\theta_2 - \theta_1}{Z \ln\frac{z_2}{z_1}} \; ; \qquad \frac{\partial u}{\partial Z} = \frac{u_2 - u_1}{Z \ln\frac{z_2}{z_1}} \tag{2}$$

Where $z = \sqrt{z_1 z_2}$, θ_2, θ_1, u_2 and u_1 are the potential temperature and wind speed at the heights z_2, z_1 respectively. And Ri can be presented by Equation (3)

$$Ri = \frac{gZ}{T} \frac{\Delta\theta}{(\Delta u)^2} \ln \frac{z_2}{z_1} \tag{3}$$

where $\Delta\theta = \theta_2 - \theta_1$ and $\Delta u = u_2 - u_1$.

Then Ri is calculated by Eq. 3 for all observational data of stations in Tab. 1. For the sake of easier calculation, $\Delta\theta$ and Δu are taken to be the differences between the height of 2 and 0.5 meter, and $Z = \sqrt{z_1 z_2} = 1$m. Thus the calculated Ri is referred to as the height of 1 meter.

Fig. 1 shows the averaged diurnal changes of Ri from May to August. It is obvious that the Ri during the warm season over the Plateau is higher than that over the east plain. It means that a very strong unstable condition develops during the day and a strong stable condition at night. At noon, Ri over the Plateau is about 10 times greater than that over the Plain.

Ri in the northeastern plain of China[5] is also used here to enable comparison with Ri in Nanjing, and this comparison shows that Ri in Nanjing can be taken as representative of that over the East Plain.

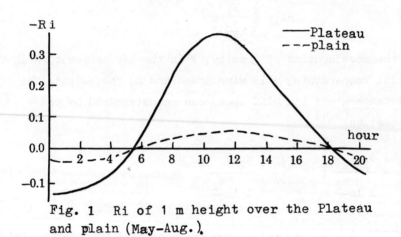

Fig. 1   Ri of 1 m height over the Plateau and plain (May-Aug.).

III. THE BASIC CHARACTERISTICS OF Δθ AND Δu

In order to discover the factors affecting the Ri over the Plateau Δθ and Δu are analyzed for the Plateau and the Plain. Fig. 2 shows there is no significant difference between the values of Δθ over the Plateau and over the Plain. Fig. 3 shows that Δu over the Plateau is much smaller than that over the Plain. On the average, Δu over the Plateau is about 1/2 or 1/3 of that over the Plain.

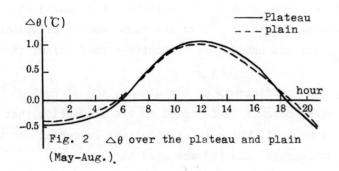

Fig. 2   △θ over the plateau and plain
(May-Aug.)

The smallness of Δu over the Plateau shows there must be some factors weakening the turbulence. The vertical shear in wind velocity (Δu) near the ground is an expression of air viscosity, which is connected with the air density. The small air density over the Plateau, for the viscosity is quite small there, makes a weak shear of the wind velocity. As a result, there is only a small Δu over the Plateau.

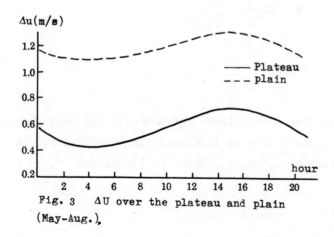

Fig. 3   ΔU over the plateau and plain
(May-Aug.)

IV. THE CHARACTERISTICS OF EXCHANGE

Both stability and Δu over the Plateau affect the characteristics of turbulence. It is believed that the turbulence coefficients of moisture and heat are almost equal[3], so the turbulence coefficients of momentum and heat are discussed here without the coefficient of moisture.

In calculating the coefficients of turbulence, the diffusion method is not used here, because it is based on the assumption that the vertical layer is neutral, which is particularly in curred for the Plateau. Instead equations (4) and (5) are used here.

$$\phi_m = \frac{KZ}{V_*} \frac{\partial u}{\partial z} \tag{4}$$

$$\phi_h = \frac{KZ}{T_*} \frac{\partial \theta}{\partial z} \tag{5}$$

$T_* = -\frac{p}{c_p \rho} / v_*$, V_* is dynamic velocity, and K, the Carmen constant. It may be achieved as follows:

$$K = \frac{1}{\phi_m^2} \frac{K^2 \, \Delta u}{\ln \frac{z_2}{z_1}} \tag{6}$$

$$K_T = \frac{1}{\phi_m \phi_h} \frac{K^2 \, \Delta u}{\ln \frac{z_2}{z_1}} \tag{7}$$

K, K_T are turbulence coefficients for momentum and heat exchange respectively. Now the main problem in calculating coefficients is deciding how to determine ϕ_m and ϕ_h. Based on literature[6], ϕ_m and ϕ_h may be shown as:

$$\phi_m = \begin{cases} 1 + \beta_m Z/L & (Z/L \geq 0) \\ (1 - \gamma_m Z/L)^{-\frac{1}{4}} & (Z/L \leq 0) \end{cases} \tag{8}$$

$$\phi_h = \begin{cases} 1 + \beta_h Z/L & (Z/L \geq 0) \\ (1 - \gamma_h Z/L)^{-\frac{1}{2}} & (Z/L \leq 0) \end{cases} \tag{9}$$

L is the Monin-Obuhov Length, parameters in these equations are determined experimentally. Tab. 2 shows the various results from different authors[6].

Table 2 Some statistics by different authors

Author	γ_m	β_m	γ_h	β_h	$\phi_h(0)$	K
Businger etal (1971)	15	4.7	9	6.4	0.74	0.35
Poulson (1970) Badgley (1972)	16	7	16	7	1	0.40
Webb (1970)	18	5.2	9	5.2	1	0.41
Dyer and Hicks	16	-	16	-	1	0.40

$$\frac{1}{\phi_m \phi_h} = - \frac{\ln\frac{z_2}{z_1}}{c_p \rho \, K^2 \Delta u \Delta \theta} \cdot P \tag{10}$$

C_p is the specific heat at constant pressure, ρ is the air density, P is the turbulence heat flux. If the values of Δu, $\Delta \theta$ and P can be estimated, $1/\phi_m \phi_h$ will be determined by Equation (10). In Fig. 4. Our experimental values are distributed in the area covering curves (1), (2), (4) and (5). The values actually conform to those from different authors.

Then K and K_T are calculated for 6 stations using all observational data. Δu and $\Delta \theta$ are taken between 0.5—2m, so they can represent the value at 1m height.

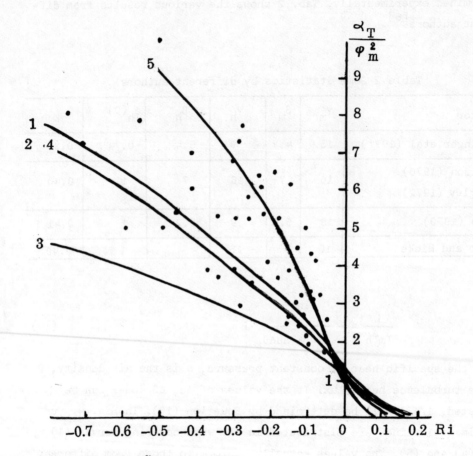

Fig. 4 $\dfrac{\alpha_T}{\varphi_m^2}$ observations and former results.

•observations, 1 from Businger, 2 from
paulson. Badgley, 3 from Webb, 4 from
Dyer. Hicks, 5 from legotina-Orlenko.

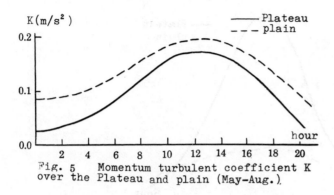

Fig. 5   Momentum turbulent coefficient K
over the Plateau and plain (May–Aug.)

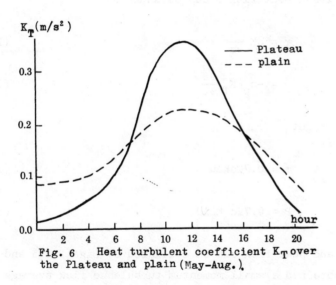

Fig. 6   Heat turbulent coefficient K_T over
the Plateau and plain (May–Aug.)

Figs 5,6 and 7 demonstrate that there are definite differences between the Plateau and the Plain. The momentum turbulence coefficient is smaller over the Plateau than over the Plain during day or night. But the heat turbulence coefficient over the Plateau is larger than that over the Plain in daylight (unstable), Though it is smaller over the Plateau during night (stable).

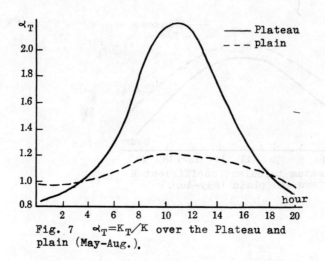

Fig. 7 $\alpha_T = K_T/K$ over the Plateau and plain (May-Aug.).

Momentum flux τ and heat flux P can be shown as:

$$\tau = \rho K \frac{\partial u}{\partial z} \qquad (11)$$

$$P = -c_p \rho K_T \frac{\partial \theta}{\partial z} \qquad (12)$$

substitute by (2), we get

$$\tau = 0.72 \rho K \Delta u \qquad (13)$$

$$P = 0.72 c_p K_T \Delta \theta \qquad (14)$$

Then momentum and heat flux can be evaluated by Figs 2,3,5 and 6. By this method we obtained a small momentum turbulence flux average over the Plateau only 1/4 — 1/7 of that over the Plain. Because there is no evident difference of $\Delta \theta$ between Plateau and Plain, the heat flux will depend mainly on K_T and ρ. During the daytime, K_T is large over the Plateau, and is 60% larger than over the plain at noon. But ρ over the Plateau is only 60% of that over the Plain, so the heat flux in the near ground layer over the Plateau is equal to or smaller than that over the Plain. The daily total depends on the value during

196

daytime, so it is easy to estimate that the total turbulence heat flux over the Plateau will approximate that over the Plain. Since the air mass over the Plateau is 50% of the Plain, for the same amount of turbulence heat flux, the effective heating over the Plateau will be double that over the Plain.

V. THE BASIC CHARACTERISTICS OF TURBULENT ENERGY BUDGET

We have discussed Ri, $\Delta\theta$, Δu, K, K_T, and τ, P over both the Plateau and the Plain, which reflect the physical features of the near ground layer. Now we will discuss the components of the turbulent energy budget.

$$\frac{\Phi E}{\Phi t} = K(\frac{\partial u}{\partial z})^2 - K_T \frac{g}{T} \frac{\Delta\theta}{\partial z} - D \qquad (15)$$

Where E is turbulence momentum per unit air mass, $K (\frac{\partial u}{\partial z})^2$ is the rate of momentum energy converted to turbulent energy under the effect of friction (generally expressed by Tr), $K_T\frac{g}{T}\frac{\partial\theta}{\partial z}$ is the consuming energy from anti-Archimed to turbulent energy expressed by A and D is the turbulent energy lost by converting to heat. From Equations (2), (6 and (7) Tr and A can be estimated. It is believed that the development of turbulence gets its inflexion point in the early morning, at noon and in the evening. $\frac{\partial F}{\partial T}$ may be seen as zero, then we get

$$D = T_r - A \qquad (16)$$

The results are given in Fig. 8. With the exception of A during the daytime, all the components over the Plateau are larger than over the Plain. The converting rate from momentum energy to turbulence over the Plateau is 20—30% of that over the Plain in daytime, and 5—10% during night, with almost the same amounts for P. This may be due to the value of Trbeing larger than that of than A.

That the absolute A over the Plateau is larger than that over the Plain is the second feature. This means that over the Plateau the

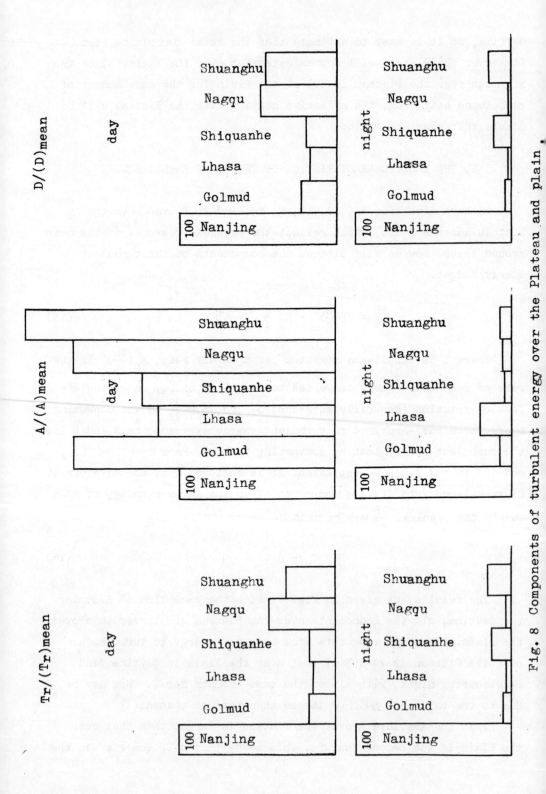

Fig. 8 Components of turbulent energy over the Plateau and plain.

contribution from the heating factor is larger than it is over the Plain. On average, it is 1.4:1 during the daytime. But during night, A over the Plateau is 10% of that over the Plain.

VI. CONCLUSIONS

1) The primary feature of the atmosphere near the ground over the Plateau is that the diurnal change of stability is quite large. In comparison with the East Plain, the air over the Plateau is very unstable in daytime and quite stable during night. So the turbulence over the Plateau tends to deviate often from the neutral state.

2) The main factor affecting stability is the significant decrease of the vertical gradient of wind velocity (Δu). The average value of $\Delta u_{2.0-0.5}$ over the Plateau is only 1/3–1/2 of that over the Plain, being closely associated with the small air density over the Plateau. The small viscosity over the Plateau weakens the vertical shear of the wind velocity.

3) In general the turbulence momentum exchange coefficient K over the Plateau is 80% of that over the Plain. The heat turbulence coefficient K_T over the Plateau is 1.6 times greater than that over the Plain. Small momentum turbulence fluxes over the Plateau appear only 1.4–1/7 as often as over the Plain. The same amount of turbulence heat flux results in more heating over the Plateau than over the Plain because of the small air density over the Plateau.

4) The difference in the components of turbulent energy flux balance near the ground between the Plateau and Plain is discussed. It is shown that there is a low converting rate of momentum to the energy of turbulence over the Plateau, on the average only 20–30% of the value of the Plain during daytime and 5–10% at night. But the turbulence energy coming from heat buoyance is higher over the Plateau, about 40% greater than that over the Plain. So the effect of the thermal factor is larger over the Plateau and the effect of the dynamic factors is smaller in comparison with that over the Plain.

REFERENCES

[1] Yang Chien-chu, Dao Shih-yen, Yeh Tu-cheng, Koo Chen-chao, Xizang Plateau Meteorology, Science press, 1960 (in Chinese).

[2] Yeh Tu-cheng, Kao Yu-hsi, Meteorology of the Qinghai-Xizang Plateau, Science press, 1979 (in Chinese).

[3] Laihatman D.L., Atmospheric physics in boundary layer, Gidro-meteoizdat, 1970 (in Russian).

[4] Legotina, S.I. and Orlenko L.R., The calculation of the turbulent fluxes of heat and moisture used by data of gradient measurements, Trudy MGO, 402, 29—39, 1978 (in Russian).

[5] Ma Yu-tang, Yao Wen-quan, Xu Zhao-sheng, Some Microclimatic Effects of Reclamation, ACTA METEOROLOGICA SINICA, Vol. 40, No. 3, 1982 (in Chinese).

[6] Busch, N.E., On the mechanics of atmospheric turbulence, Workshop on Micrometeorology, Boston, 1973, 1—65.

AN ANALYSIS OF THE CHARACTERISTIC FEATURES OF THE SURFACE HEAT SOURCE AND HEAT BALANCE OVER THE QINGHAI-XIZANG PLATEAU FROM MAY TO AUGUST,1979

Weng Duming

Nanjing Institute of Meteorology

I. INTRODUCTION

The surface heating of the atmosphere over the Plateau has been a main subject of interest in Qinghai-Xizang meteorology, attracting the attention of meteorologists all over the world. Several attempts of field observation on net radiation and heat balance have made by Chinese meteorologists, yielding some significant results[1—4]. Since these attempts were made at isolated places, they could not give a full picture of the effect as a surface heat-source of the Plateau, which makes up about 1/4 of the Chinese mainland. During the first QXPMEX from May to August, 1979, systematic field observations were made at six stations in the hinterland of the Plateau, providing invaluable data for further investigation of the heat-source problem.

For a variety of reasons, most of the climatological methods available for computation of surface heat-balance components are empirical in nature, based on limited observed data mostly from stations on low lying plains[5,6]. For the Plateau and its surroundings, therefore, a formula for computing heat-balance components suitable for elevations ranging from sea level to a height of 5000 meters has to be developed and applied to the Plateau. Maps showing distributions of decadal and monthly totals of heat-balance components for the QXPMEX can then be drawn to exhibit the basic properties and their seasonal variations of the surface heating over the Plateau.

Thanks to the efforts of a study group, an atlas of surface net radiation and heat balance over the region for the period of May-Aug., 1979 has been completed[7]. This article is a preliminary analysis of the surface heat source and the heat-balance regime over the Plateau, applying information on heat balance from the atlas. New maps and charts are drawn in addition to those adopted from the atlas.

II. COMPUTATION OF HEAT-BALANCE COMPONENTS WE HAVE THE SURFACE HEAT BALANCE EQUATION

$$B = P + LE + H \qquad (1)$$

where B is the net radiation, P the sensible heat transfer, LE the latent heat transfer (heat of evaporation) and H the heat exchange in soil. Eq. (1) may be written as

$$B - H = P + LE \qquad (2)$$

which indicates whether the surface acts as a heat source or a cold source, a heat source when (B-H)>0 and a cold one when (B-H)<0.

During the QXPMEX, B and H were directly observed by net radiometers and heat-flow plates. Sensible and latent heat fluxes may be determined by the heat-balance method (for super-adiabatic atmosphere) or with the aid of the Monin-Kazansky diagram (for stable atmosphere), using the data obtained through vertical gradient observations. In this way, the decadal and monthly totals of the observed values of the heat-balance components during the experiment are computed.

The climatological method of computation of the totals of radiation balance (net radiation) components is dealt with in another paper[8]. Only the computation procedure for heat-balance components is briefly described in the following.

(1) For computing the mean daily total of sensible heat P, we have

$$P = \rho C_P \, C_D \, \bar{V}_{10} (\bar{T}_0 - \bar{T}) \qquad (3)$$

where ρ and C_P are air density and specific heat at constant pressure respectively, C_D is the drag coefficient, $\bar{V}_{10}$ the average wind speed

202

at 10 m above the ground, and $\bar{T}_0$ and $\bar{T}$ mean surface-and air-temperatures. To find the decadal total, the result will have to be multiplied by 14400, the number of minutes in a decade.

Evidently, correct determination of C_D is essential in sensible heat computation. Strictly speaking, C_D is a function of atmospheric instability. For instantaneous flux, we have found an empirical relationship between C_D and $(T_0-T)/V_1^2$. In climatological computation, however, since mean values of a decade or a month are to be computed, the effect of atmospheric instability is smoothed out to a great extent thus, a simplified formula may be used. For stations with elevations above 2800 m*, we have empirically[9]

$$C_D = 0.00112 + \frac{0.010}{\bar{V}_{10}} \qquad (4)$$

For stations with elevations below 2800 m, linear interpolations may be performed with the aid of Eq. (4) and the observed results at Nanjing, a station for control comparison in East China, to get

$$C_D = 0.00112 + \frac{0.010}{\bar{V}_{10}} - 0.00362 \frac{(B'-720)}{280} \qquad (5)$$

in which B' is the air pressure (hPa) at the station involved. Climatological computation of sensible heat can thus be performed. Judging by the test of the results and by the comparison of the computed values with the radiation balance ratio, the computation seems to be quite reasonable.

(2) For computation of latent heat, the remainder method is used. Also employed for verification is the Bowen ratio method, i.e.,

$$LE = \frac{B-H}{1+\beta} \qquad (6)$$

where β is the Bowen ratio. For computing decadal totals, it is empirically found that

---

* In the six plateau stations, the lowest one, Golmud, has an elevation of 2800 m.

$$\beta = 4.83 \left[\frac{\bar{T}_0 - \bar{T}}{e_0}\right]^{1.014}$$

where

$$\bar{e}_0 = \bar{e}/10^{1.34 \ln (B'/B_0')}$$

in which $\bar{e}$ and $\bar{e}_0$ are respectively mean water vapor pressure at station level and that reduced to sea level, B' and B_0' are the corresponding atmospheric pressures. Results obtained by the remainder and the Bowen ratio methods are in general agreement, the correlation coefficient between them amounting to 0.89 and the mean absolute error being 3 $W \cdot m^{-2}$.

(3) Computation of heat exchange in soil H. With the simple assumption that heat flux in soil is proportional to the gradient of soil temperatures, Xu and others[10] establish an empirical relation between the mean decadal heat exchange in soil and difference of soil temperatures of 5 and 20 cm below the ground surface, $\Delta\theta_{5-20}$, with a correlation coefficient as high as 0.98.

This relation can be written as

$$H = 9.8 + 8.9\Delta\theta_{5-20}$$

Test computations show that it gives satisfactory results.

III. ANALYSIS OF HEAT SOURCE AT THE PLATEAU SURFACE

Computation by Eq. (2) shows the obvious effect of the surface of the Plateau serving as a heat source.

May is dry season for the Plateau. During that period, the whole Plateau serves as a strong heat source centered in the central and southern portion with intensity exceeding 105 $W \cdot m^{-2}$, a ridge of high intensity extending from the center to the northeastern brim. Around the Qinghai Lake is a small high center of 90 $W \cdot m^{-2}$. Over the Sichuan Basin is located a low area of less than 60 $W \cdot m^{-2}$, which tends to extend to the northwest and southwest to join another low at the south side of the Qaidam Basin and Tarim Basin to the northwrth

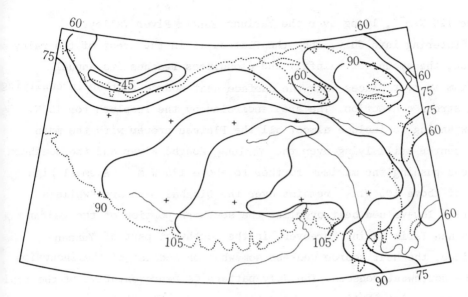

Fig. 1 Distribution of surface heat sources in May
$(W \cdot m^{-2})$.

(Fig. 1). This pattern of distribution of the surface heat sources
is associated with the atmospheric circulation as well as the topography
of the Plateau. Examination of distribution of rainfall and cloudiness
for the same period (charts omitted) shows that surface heat source
is notably intensified over the Plateau proper due to lack of rainfall
and cloudiness and thus abundance of solar radiation. Over the Sichuan
Basin, where rainfall increases with the onset of the rainy season,
and over the southern part of the Tarim, where the westerly troughs
and the windward slopes gives rise to large amount of cloudiness,
the heat sources are relatively weak.

The general pattern of heat-source distribution in June is similar
to that of May. However, with the increased intensity of solar radiation in June, the surface heat sources intensity strengthen instead
of weakening, because the approaching rainy season does not block the
daytime solar radiation very seriously, though it is spreading from
east to west with a high frequency of nocturnal rain which lowers the
surface albedo and the effective ground radiation. This is a particular
but typical phenomenon over the Plateau region with nocturnal rainfall.
In June, the high heat-source center may strengthen to a value of

above 120 $W \cdot m^{-2}$, lying over the Yarlung Zangbo River Valley.

Entering into July, when the Plateau is in the crest of the rainy season, the amount of rainfall and cloudiness reaches its annual maximum in most places with the surface heat source further intensifying with strong convection. The area encircled by the isopleth for $105W.m^{-2}$ is expanding to include almost all the Plateau proper with the main high center still lying over the Yarlung Zongbo River and the southern Plateau pushing the maximum further to above 135 $W.m^{-2}$. A small high area of above 105 $W.m^{-2}$ remains over the Qinghai Lake area while a relative low-valued belt overlies the southern portion of the Qaidam Basin and Tarim Basin. Moreover, in the northwest part of Yunnan Province, the heat source weakens somewhat on account of the incursion of the southwest monsoon. The July pattern is representative of the typi heat-source distribution of the wet season over the Plateau, whose intense surface heating may play an important role in the development of both the summer monsoon and the nocturnal rain there.

The heat-source distribution of August is similar to that of July except for a slight decrease in intensity due to weakening solar radiation and lesser amount of rainfall.

In order to exhibit the seasonal variation of the surface heating, a map is constructed showing the distribution of difference in heat-source intensity between July and May (Fig. 2). It is seen that, after the onset of the rainy season, except for the eastern part, the sur-face heat-sources intensify all over the Plateau, particularly over the Yarlung Zangbo River in southern Xizang and over the Sichuan Basin and Qaidam Basin. Of course, the mechanism of development of these heat-sources may be different at different localities, for instance, the intensification of heat-sources over the Yarlung Zangbo River and the Sichuan Basin may be chiefly due to the nocturnal rain. (The scattered local summer drought in the Sichuan Basin may play a signific-ant role.) And for the Qaidam Basin, or more broadly speaking, for the wh area of the Plateau north of 35°N, the intense heating is associated with a general increase of solar radiation at higher latitudes. A region showing negative values of difference, though small, occurs mainly between 30°N and 35°N in the eastern part of the Plateau. This

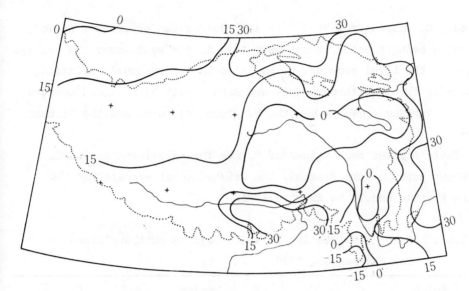

Fig. 2 Distribution of difference of heat-source values
between July and May ($W \cdot m^{-2}$).

covers the same area as the region of excessive rainfall, for two
reasons: (i) overcast skies predominate and (ii) similarly, the negative areas are over the Hengduan Mountains (which run in a north-south
direction, distinctive in China's topography) and NW Yunnan, due to the
increased rainfall and cloudiness.

A preliminary study of the summer intensity of the surface heat-
sources and the heights and temperatures of the 600 hPa surface shows
that they are distinctively related. The general rule is that the
heat-source highs (zones of high thermal values) are usually coinci-
dent with the low and warm centers on that isobaric surface. This
clearly shows the significant influence of surface heating on the
pressure systems.

IV. DRAG COEFFICIENT C_D AND SENSIBLE HEAT TRANSPORT

As we know, the heating of the atmosphere by the ground surface
is caused by sensible and latent transport. The drag coefficient C_D
and the two types of heat transport have been computed by using Eqs.
(3) and (5). The results suggest that, over the main part of the

Plateau, C_D ranges from 3×10^{-3} to 10^{-2} during the experiment period, its value being greater in the east than in the west, over the valleys than over the level grounds. C_D may also have a seasonal variation, increasing from May through July and August with high-valued centers over the Yarlung Zangbo River, the Hengduan Mountains and the Sichuan Basin.

Table 1 gives mean values of C_D for the experiment period in different parts of the Plateau. The weighed areal average for the entire Plateau should be about $4 - 5 \times 10^{-3}$.

Table 1 Mean C_D values in different regions of the Plateau for May-August, 1979.

region	C_D	region	C_D
Hengduan Mountains	$6.5 \quad 10^{-3}$	central Plateau	$4.9 \quad 10^{-3}$
eastern Plateau	$5.4 \quad 10^{-3}$	southern Plateau	$5.8 \quad 10^{-3}$
Sichuan Basin & hilly countries of Northwest Sichuan Province	$6.1 \quad 10^{-3}$	western & part of central Plateau	$3.8 \quad 10^{-3}$
northeastern Plateau	$4.0 \quad 10^{-3}$	Tarim Basin	$4.0 \quad 10^{-3}$

The distribution pattern of sensible heat transport is again quite stable. Take May for example. As shown in Fig. 3, a high-valued ridge extends northeastwards from the West Plateau along a narrow belt over the north slope of the Himalayas with a maximum of over 90 $W \cdot m^{-2}$. The values are smaller over the SE Plateau, the Sichuan Basin and the windward slope of the southern brim of the Tarim Basin on account of increased rainfall and cloudiness, with a minimum over the Sichuan Basin of less than 15 $W \cdot m^{-2}$, the values over the Hengduan Mountains areas shadowed by the rain being relatively greater. With the onset of the rainy season following June, the low-valued belt of sensible heat expands while the high-valued zone weakens, contracts and confines itself to the West Plateau. However, the general pattern

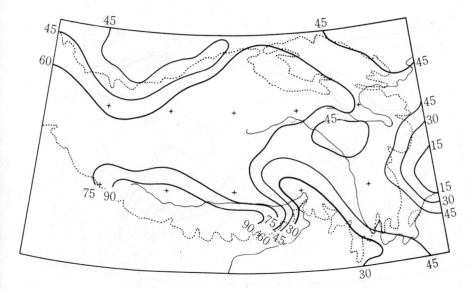

Fig. 3 The distribution of surface sensible heat totals
for May, 1979 ($W \cdot m^{-2}$).

of distribution remains unchanged. An interesting fact is that our
results are in complete agreement with those obtained from an analysis
of multiannual mean difference of ground and air temperatures[11].

To further examine the effect of the approaching seasons on the
variation of the surface sensible heat field, Fig. 4 is a map showing
the distribution of sensible heat differences between July and May.
It clearly shows that the sensible heat totals diminish over the entire Plateau in the wet month (July) in comparison with the dry month
(May), the biggest drop occurring again in the Yarlung Zangbo River
Valley, a drop of over 45 $W \cdot m^{-2}$. The eastern part and the northern rim
of the Plateau are the only places with a slight increase.

It is of significance to investigate the contribution of sensible
heat to the surface heat-sources, which may be expressed in terms of
the percentage ratio of sensible heat to the heat source. Fig. 5
depicts this contribution in both dry and wet months. As shown in
Fig. 5a, sensible heat contributes more than 50% of the heating of
the atmosphere by the plateau surface in the dry month except for the
southeastern and northeastern rims and the hilly areas northwest of
Sichuan, with the vast territory west of 90°E exceeding 70% and the
Qaidam Basin and Tarim Basin even amounting to 90%. Upon entering into the

209

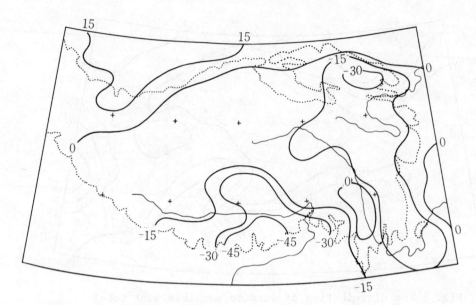

Fig. 4 The distribution of sensible heat differences between
July and May, 1979 $(W \cdot m^{-2})$.

wet season (Fig. 5b), the contribution by the sensible heat diminishes
on the whole and the 50% isoline shifts westward, implying that in the
region east of this isoline latent heat exceeds sensible heat in the
contribution and becomes the major factor in the heating of the atmosphere by the ground surface. The contribution by sensible heat over
the Yarlung Zangbo River and the Sichuan Basin even reduces to less
than 20—30%. Only in the west, especially the desert area near the
border, does the sensible heat stay to play a major role. Figure 6
gives the difference of contributions by sensible heat in July and in
May, which depicts the transition of heat-source contribution from
sensible to latent heat. The seasonal variation of the heating process
by the ground surface must have a significant influence on the weather
and climate of the Plateau.

V. LATENT HEAT TRANSPORT

Latent heat transport to the air concentrates mainly in the East
Plateau and during the rainy season. In the dry month (May) on account
of the dry ground surface, the amount of latent heat transport is

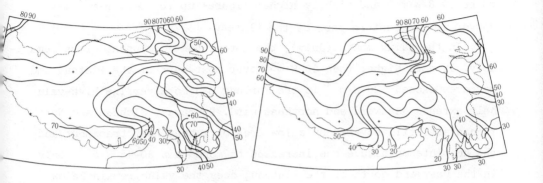

Fig. 5 Contribution of sensible heat to the heat-sources(%)
       (a) May, 1979; (b) July, 1979.

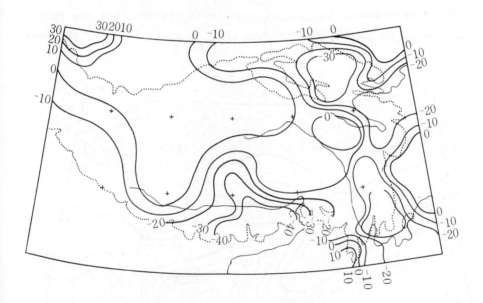

Fig. 6 Difference of contributions by sensible heat in July
       and in May.

generally small, being less than 30 $W \cdot m^{-2}$ and even less in the Hengduan Mountains areas shadowed by rain. Over the Tarim Basin and Qaidam

Basin and along the Corridor west of the Huang He River, it may drop to below 8 $W \cdot m^{-2}$ and sightly higher figures up to 30—60 $W \cdot m^{-2}$ may appear in the eastern part of the Plateau.

In the rainy month (July), the latent heat component grows remarkably. It gets higher than 30 $W \cdot m^{-2}$ over most part of the Plateau and may even exceed 120 $W \cdot m^{-2}$ in the Yarlung Zangbo River. The high-value belt extends eastward and northeastward. Southwest Sichuan and North Yunnan are still covered by a low area of latent heat transport and the absolute magnitude has increased to more than 30—45 $W \cdot m^{-2}$. Only in the west and north of the Plateau, does the value remain below 8 $W \cdot m^{-2}$.

In August, latent heat at the surface diminishes because of the decrease in rainfall as well as the weakening of the heat-sources, though the distribution pattern remains much the same as in July.

The alteration of the latent heat transport from the dry to the wet season is in good agreement with that of sensible heat. Wherever

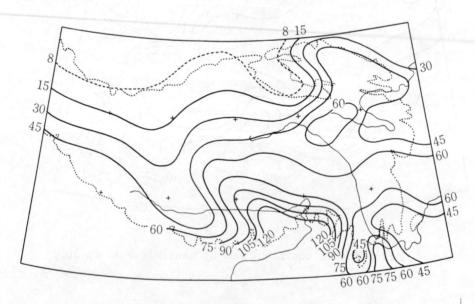

Fig. 7 Distribution of latent heat totals on the surface, July, 1979 ($W \cdot m^{-2}$).

there is an increase of latent heat, there is a corresponding decrease

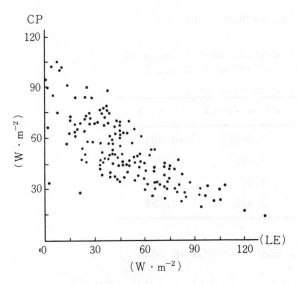

Fig. 8 Correlation of decadal and monthly totals of
sensible and latent heat.

of sensible heat. By plotting the scatter diagram of the decadal and
monthly totals of sensible and latent heat, their negative correlation
can be clearly seen (Fig. 8). The divergence of the dots on the diagram
is caused mainly by the diversity of the local heat-sources, that is,
the difference in geographic latitude, underlying surface and conditions of sky of the observing stations.

The contribution of sensible heat to the surface heat-sources
depicted by Figs. 7 and 8 may be used to determine the contribution
of latent heat through the interaction of sensible and latent heat (By
Eq. (2), the contribution by sensible heat plus that by latent heat
should be 100%).

Finally, decade-to-decade variation of latent heat totals of
Lhasa, Shiquanhe, Xining and Chengdu have been analyzed, indicating
that there is a marked increase in decadal totals from the dry to the
wet period. The characteristic curves of decadal heat-source totals
of all the stations run analogously while they run in an opposite
direction to the curves of sensible heat totals (Figure omitted).
By computing the correlation coefficients between the decadal latent
heat totals LE, the heat-source totals (B-H) and the sensible heat

totals P of these stations, we have Table 2.

Table 2 Correlation coefficients of
LE vs.(B-H) and LE vs. P
for four stations.

station	LE vs.(B-H)	LE vs.P
Lhasa	0.960	-0.931
Shiquanhe	0.866	-0.776
Xining	0.709	-0.862
Chengdu	0.826	-0.356

VI. HEAT EXCHANGE IN SOIL

During the QXPMEX, the soil heat-exchange fluxes are all found positive over the Plateau, though their magnitudes are less than 15 $W \cdot m^{-2}$, not exceeding, in most cases, 10% of the net radiation totals. Their distribution makes little difference from month to month. The basic pattern is that a high-value belt lies over the northern portion of the Plateau extending east-by-south, the entire humid southeast and the southern borders are covered by a low-value zone.

VII. CONCLUDING REMARKS

Through a preliminary analysis of the characteristic features of the surface heat-sources and the components of heat balance over the Qinghai-Xizang Plateau, the principal results are as follows:

1) The characteristics of distribution of the surface heat-sources and their seasonal variation have been revealed. It is of particular interest to note that marked intensification of the surface heat-sources occurs over almost the entire Plateau with the transition from the dry to wet season, especially over the Yarlung Zangbo River. And the cause of the intensification is found to be the moist ground surface and the high nocturnal rain frequency in summer. The feedback mechanism of nocturnal rain to the surface heat-source should be of significance in the formation of the summer weather and climate there.

2) Sensible heat transport is found to be the principal process for heating the atmosphere by the surface during the dry period, especially over the western portion of the Plateau. The contribution by sensible heat amounts to over 70% over the desert area at the northern rim and even runs as high as 90%. Upon entering into the wet period, contribution by sensible heat diminishes and that of latent heat increases markedly. The latter may exceed 50% for the entire eastern half of the Plateau and may attain a high percentage of 70—80% over the Yarlung Zangbo River Valley and the adjacent areas.

3) Through the analysis of records from 4 representative stations, decadal totals of heat-sources, sensible heat and latent heat and their decade-to-decade variation curves have been plotted, which could reflect their continuous seasonal changes. The heat-source curve and that of latent heat run in good agreement, the sensibleheat curve having a negative correlation.

4) The surface drag coefficient C_D over the Plateau is found to be small in the west and large in the east, ranging from 3×10^{-3} to 10^{-2}, a suitable average may be taken as $4-5\times10^{-3}$.

REFERENCES

[1] Zeng Qunzhu and Kou Youguan, The heat balance during the Rongbu Glacier melting period, Reports of scientific exploration of the Mount Qomolangma area (1966—1968) — Glacier and Geomorphy, Science Press, 1975 (in Chinese).

[2] Lu Longhua and Dai Jiaxi, Thermal state of the Tanggula Range region, Kexue Tongbao 25, 505—509, 1980.

[3] Zeng Qunzhu and Xie Yingqen, On the thermal effect of Qinghai-Xizang Plateau from the surface radiation balance and heat balance, Kexue Tongbao, 25, 552—554, 1980 (in Chinese).

[4] Weng Duming, The aspects of heat balance at the Kunlun Mountains Spur, Bulletin of the Nanjing Meteorological Institute, Supplement to No. 1, 20—27, 1979 (in Chinese).

[5] Gao Auedong and Lu Yurong, Surface radiation and heat balance, Science Press, 1982 (in Chinese).

[6] Budyko, M.I., Climate and life, Gidrometeoizdat, Leningrad, 1971 (in Russian).

[7] Pan Shouwen, The characteristics of components of the radiation balance components during QXPMEX in 1979, compiled in this volume.

[8] Atlas of surface radiation and heat balance field over the Qinghai-Xizang Plateau, May-August, 1979, Meteorological Press, 1—196, 1984 (in Chinese).

[9] Chen Wanlung and Weng Duming, Preliminary study on the calculative method for the total ten days value of sensible heat and latent heat, Collected papers of QXPMEX (2), Science Press, 35—45, 1984 (in Chinese).

[10] Xu Zhaosheng and Ma Yutang, The calculation of measurements and the method of popularization for soil heat flux on the Qinghai-Xizang Plateau, Ibid, 24—34, 1984 (in Chinese).

[11] Ye Duzheng and Gao Youxi, Meteorology of the Qinghai-Xizang Plateau, Science Press, 1—278, 1979 (in Chinese).

Session 3: Modeling and Theory (1)

Impact of orography on global-scale weather forecasting

S. Tibaldi and L. Bengtsson

European Centre for Medium Rauge Weather Forecast

The importance of the corrugations of the lower boundary of the
Earth's atmosphere in causing both weather and climate to depart from
axial symmetry has long been recognized (e.g. Charney and Eliassen,
1949; Bolin, 1950). General Circulation Models (GCMs), if they are to
represent realistic mean atmospheric states and departures from them,
must have an adequate representation of this effect.

There have been several attempts to assess the role played by
orography in determining the climate of a GCM, and , with a number of
assumptions, of the real atmosphere. Some of the more notable attempts
are those by Manabe and Terpstra (1974) a decade ago and, much more
recently, by Held (1983).

Tibaldi and Buzzi (1983) and Ji and Tibaldi (1983), also recently,
made an attempt to investigate the role played by orography, as opposed
to the one played by land-sea contrast, on the onset and maintenance
of observed blocking episodes. They concluded that, while diabatic
effects might be important in successfully modelling the deep cyclogeneses episodes that are usually associated with the onset of Euro-
Atlantic blocks, the role of the global orography is dominating in
maintaining the quasi-stationarity of the planetary ultra-long waves,
and that this stationarity is crucial to maintain the blocking feature
and to prevent its dispersion. This is consistent with the different
synoptic characteristics of blocking occurrence in the Southern
Hemisphere (where topography is likely to be of lesser importance),
where the average duration of a blocking episode seems to be several
days shorter than in the Northern Hemisphere, the amplitude of the

winter stationary waves is lower and the mid-latitude tropospheric
mean westerlies are stronger (van Loon et al., 1973).

Unfortunately, however, the legitimacy of applying to the real
atmosphere conclusions drawn on the above mentioned GCM simulations is
based on the assumption that large scale orographic effects are adequately represented in those numerical models used to simulate the
atmosphere. Recently, Wallace et al, (1983) have put forward the
hypothesis that a considerable proportion of the so-called Systematic
Errors characteristic of most GCMs (and of the ECMWF one in particular)
can be explained in terms of under-estimated orographic "forcing".

A natural way to investigate the deficiencies shown by those GCMs
that are routinely used to forecast the weather on a given time range
(a week to ten days, say) is to compare the ensemble mean of forecast
fields to the corresponding ensemble mean of observed (analysed) fields.
The difference field between such observed and forecast values is usually referred to as the model's Systematic Error (SE). Conversely,
GCMs used to simulate climate are often evaluated on the basis of how
well, during suitably long integrations, they reproduce some time-mean
fields, compared to observed atmospheric (time mean) quantities. Their
mean error is therefore referred to as Climatic Error (CE).

It is easy to see that, if one extends in time the range of the
integrations of a forecasting model to periods of the same order of
magnitude of the relaxation time of the model to its own climate (e.g.
one to two months), the SE tends towards the CE. This is why very often
the existence of SEs is also referred to as the problem of the model's
Climate Drift.

Fig. 1 shows the Climate Drift of the ECMWF Global Grid-Point
model during the winter 1980-1981 as it progressively changes from the
day 1 forecast ensemble to the day 10 forecast ensemble. The figure
shows the SE field superimposed on the observed mean flow for the same
period, to evidentiate that the horizontal structures of the error
evolve from a situation in which they are strongly related to the
underlying orography (day 1) to a situation in which they are negatively
correlated with the mean flow (day 10). This suggests an inadequate

representation, in the model's atmosphere, of those effects that are
mostly responsible for the maintenance of the ultra-long planetary
waves (e.g. orography). From Fig. 1 it is also evident that there is
progressively more area covered by negative error than there is by
positive error. This effect being stronger at higher levels than at
low levels (not shown), implies that the model's troposphere progressively cools. Negative error centres, moreover, occupy on average more
northerly positions, while positive error centres lie on more southerly
positions, indicating (in the geostrophic assumption) that an error in
the mid-latitude mean westerlies also develops, with the model's atmosphere becoming, on average, too westerly after 10 days. A further
characteristic of the SEs is their equivalent barotropic character,
with their amplitudes only changing with height and the phase of their
large scale minima and maxima remaining more or less fixed.

Wallace et al. (1983) suggested that, if an under-representation
of the orographic effect was partially responsible for the lack of
forcing present in the ECMWF Global model, an enhanced orography would
have improved both the climate of the model and the ensemble mean forecast.

Fig. 2 shows the enhanced "envelope" orography used in their set
of forecast experiments. This was constructed adding, to the grid-square
mean terrain height, twice the sub-grid standard deviation, as evaluated from a very high resolution global terrain height dataset originally produced by the U.S. Navy.

On the basis of their encouraging results, a more extensive set
of experiments was later performed at ECMWF on data covering the entire
January 1981 period, notably the period during which the SE of the
ECMWF operational model had been at its peak (Tibaldi, 1984). Fig. 3
documents the decreasing of the 500 hPa geopotential heights SE achieved by the use of such envelope orography. It is evident that all major
characteristics of the SE have been positively affected. The bias has
been reduced (cooling), and the mean westerly wind error has also been
reduced, together with the evident improvement in the amplitude of this
quasi-stationary planetary waves.

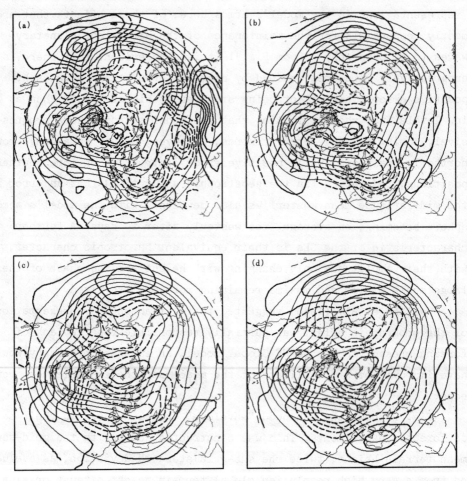

Fig. 1 Ensemble mean forecast error fields for ECMWF operational forecasts of 500 hPa height for the 100 day period 1 December 1980 - 10 March 1981, inclusive. (a) Day 1 forecasts, contour intervals 5 m; (b) Day 4 forecasts, contour interval 16 m; (c) Day 7 forecasts, contour interval 30 m; (d) Day 10 forecasts, contour interval 30 m. Background field (lighter contours) is the mean 500 hpa height field based on ECMWF operational analyses for the same period, contour interval 80 m. Negative contours are dashed. (From Wallace, Tibaldi and Simmons, 1983).

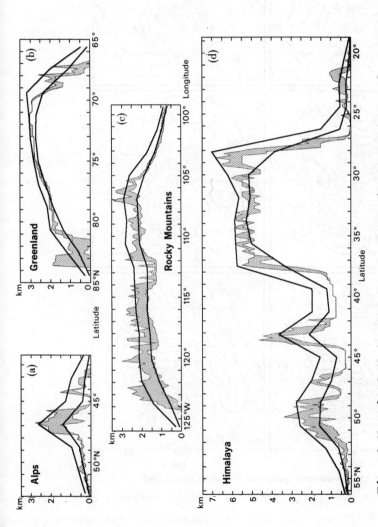

Fig. 2 Vertical N-S or E-W cross-sections through selected mountain ranges showing envelope orography (upper curve) and average orography (lower curve). Shading indicates the range of terrain heights between the maximum and minimum value in the U.S. Navy high resolution grid. (a) N-S along 11.25°E, (b) N-S along 33.75°W, (c) E-W along 41.25°N and (d) N-S along 85-25°E. (From Wallace, Tibaldi and Simmons, 1983).

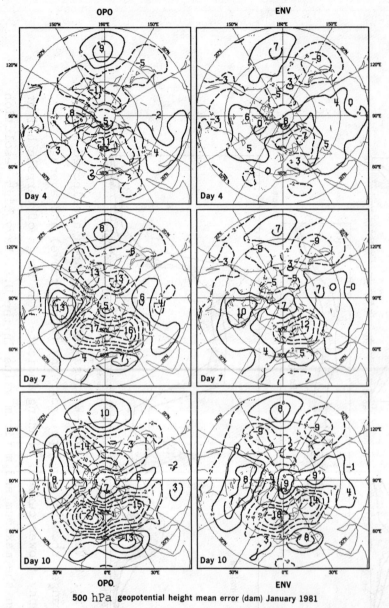

500 hPa geopotential height mean error (dam) January 1981

Fig. 3 Ensemble 500 hPa geopotential height forecast error fields
for the 31 cases of January 1981 with the operational average-type
orography (OPO, left) and the envelope orography (ENV, right). From
top to bottom, Day 4, Day 7 and Day 10. Dashed contours represent
negative SE, full contours positive SE; isolines drawn every 4 dams.
(From Tibaldi, 1984).

In parallel to these experiments, other, more detailed, studies of the local effects of orgraphy have been carried out (e.g. Dell'Osso and Tibaldi, 1983; Radinovic, 1984; Wu and Chen, 1984). From these studies a few considerations emerge: firstly the amount of "envelope" enhancement needed at higher resolutions is considerably smaller than expected.

Secondly, there is conflicting evidence on the benefit of the use of the envelope orography on comparatively smaller scales (e.g. synoptic and sub-synoptic scales) meteorological phenomena. Modelling of cyclogenesis in the lee of the Alps clearly benefits from an enhanced local orography, while successful representation of a mesoscale vortex on the eastern flank of the Tibetan plateau seems to be favoured by realistic (rather than enhanced) local orography. Moreover, the comparatively poor performance of an alternative "valley-filled" global orography casts some doubts on the rationale at the base of the envelope orography justification presented in Wallace et al.(1983).

What so far emerges from this composite picture is that the influence that orography exerts on all scales of atmospheric motion, from the planetary to the local mesoscale, is still comparatively poorly represented in numerical general circulation models. Lack of understanding of the detailed dynamics of this influence lies at the base of our current problems in representing satisfactorily the interactions between the earth's atmosphere and its lower boundary.

References

Bolin, B. 1950, On the influence of the earth's orography on the general
    character of the westerlies. Tellus, 2, 184-195.
Charney, J.G. and Eliassen, A. 1949, A numerical method for predicting
    the perturbations of the middle latitude westerlies. Tellus, 38-
    38-54.
Dell'Osso, L. and Tibaldi, S. 1982, Some preliminary modelling results
    on an ALPEX case of lee cyclogenesis. ALPEX Preliminary Scientific
    Results-GARP-ALPEX Report No. 7, WMO, Geneva, 3-19.

Held, I. 1983, Stationary and quasi-stationary eddies in the extratropical troposphere: Theory. In large scale dynamical processes in
the atmosphere, Ed. B.J. Hoskins and R.P. Pearce, Academic Press.

Ji, L.R. and Tibaldi, S. 1983, Numerical Simulations of a case of blocking: the effects of orography and land-sea contrast. Mon. Wea.
Rev. III, 2068-2086.

Manabe, S. and Terpstra, T.B. 1974, The effects of mountains on the
general circulation of the atmosphere as identified by numerical
experiments. J. Atmos. Sci., 31, 3-42.

Radinovic, D. 1983, A Dynamic approach to orography representation in
numerical weather forecast models - Unpublished Manuscript
available from ECMWF.

Tibaldi, S. 1984, Systematic Error of the ECMWF Grid-Point Forecast
Model and their relationship with orographic forcing: some more
extensive results. Rivista di Meteorologia Aeronautica. In print.

Tibaldi, S. and Buzzi, A.1983, Effects of orography on Mediterranean
lee Cyclogenesis and its relationship to European Blocking. Tellus, 35A, 583-602.

Van Loon, H., Jenne, R.L. and Labitzke, K. 1973, Zonal harmonic standing
waves. J. Geophy. Res., 78, 4463-4471.

Wallace, J.M., Tibaldi, S. 1983, Reduction of systematic forecast erros
in the ECMWF model through the introduction of an envelope
orography. Quart. J.R. Met. Soc., 109, 683-717.

Wu Q.X. and Chen, S.J. 1984, The effect of mechanical forcing on the
formation of a mesoscale vortex. To be submitted for publication.

INTERPRETATION OF SATELLITE OBSERVATIONS OVER MOUNTAINOUS AREAS

Hans-Jürgen Bolle

Institut für Meteorologie und Geophysik der Univ.Innsbruck

ABSTRACT

Satellite measurements made over mountains impose a number of interpretation problems. Some of them will be addressed in this paper.

One problem is the visual discrimination of clouds against snow and the establishment of reliable cloud climatologies over mountains in connection with the International Satellite Cloud Climatology Project.

Of considerable interest is the effect of mountains as elevated heat sources on the atmosphere. Surface temperature determinations are therefore an important application of satellite measurements.

Because of the generation of orographic clouds the use of clouds as tracers for wind vector determinations has to be carefully considered in the presence of mountains.

Another interesting application is the construction of snow maps by means of radiometric microwave measurements undisturbed by clouds.

Examples of the evaluation of satellite measurements by application of interactive digital data processing methods will be presented.

THE EFFECTS OF THE QINGHAI-XIZANG PLATEAU ON THE MEAN GENERAL CIRCULATION IN EAST ASIA IN SUMMER

Wang Qianqian   Wang Anyu

Lanzhou Institute of Plateau Atmospheric Physics,

Chinese Academy of Sciences

Li Xuefong and Li Shuren

Department of Geology and Geography, Lanzhou University

In order to investigate the effects of large-scale topography, and heating sources and sinks on the mean general circulation in East Asia in summer, we use a 5-layer primitive equation numerical model (Fig.1) developed originally by Kuo and Qian[1] to make four kinds of exper-

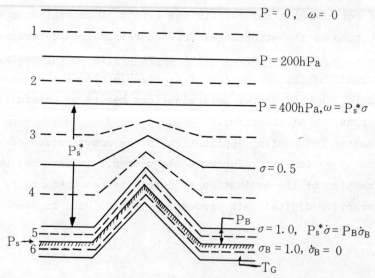

Fig.1 Sketch of the vertical structure of the numerical model.

iments. The topography used in the model is close to the real one (Fig.2). The heating sources and sinks of the model are the stationary ones computed by Yiao et al.[2] with the actual data of July, 1979.

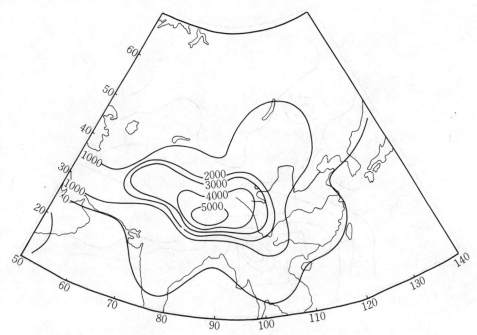

Fig. 2 Smoothed topography used in the model for the Qinghai-
Xizang Plateau area, in meters.

The initial geopotential heights used in this investigation are taken
from the mean July zonal average. The model covers the area 40°-150°E
and 10°-65°N with a horizontal grid size of 5 degrees in both the
latitudinal and the longitudinal directions. The time step is 15 min.
and all the integrations are carried out to 5 days.

To match the model grid size we smoothed the Yao's monthly mean
heat source distributions and vertically distributed the heating rates
to each layer of the model (Fig.3).

We have done 4 experiments of simulation. The first one (designated
by HN) has no diabatic heating but includes orography and the second
(by HH) includes. Both diabatic and orographic factors. Because in
summer the two most important heat sources are located in the Qinghai-
Xizang Plateau and Bay of Bengal regions in East Asia, we designed the
third and the fourth experiments based on the HH experiment. The third
(by PN) has no heating in the Plateau, while the fourth (by MN) has no

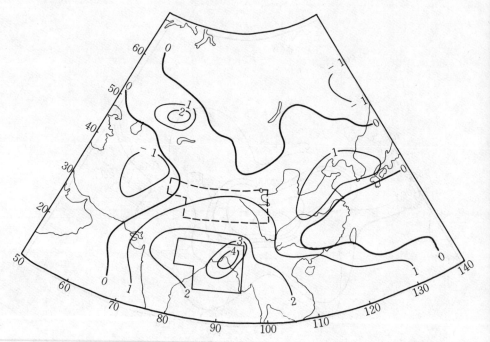

Fig. 3 Ideal July heating fields (Solid line area is the Bay of
Bengal, the area with dashed lines is the Plateau),
in °C/ d^{-1}.

heating in the Bay of Bengal region. The main results of the experiments
are stated in the following sections.

I. THE HORIZONTAL TEMPERATURE AND GEOPOTENTIAL HEIGHT FIELDS

By comparing the results obtained from the HN experiment with
those from the HH experiment we can reach the following conclusions:

1). In the lower and middle troposphere the westerlies influenced
by the dynamic-orographic effects tend to flow around rather than to
cross over the Qinghai-Xizang Plateau. The disturbances produced by the
rounding flows are found to be generally more obvious in the mid-
latitudes (north of 40°N) than in lower latitudes and are not clear in
the upper troposphere.

2) Figs.4 (a,b) and 5 (a,b) are the simulated geopotential height and temperature fields at 100 hPa and the 500 hPa, respectively. By comparing these fields we can see that the heat sources produced by

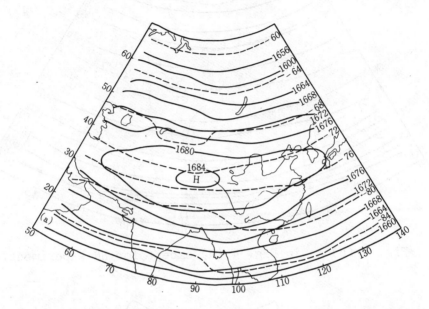

Fig. 4 (a) HH five-day simulated 100 hPa geopotential height and temperature.

the Plateau and the latent heat source to the south of the Plateau make great contributions 1) to the formation of the strong anticyclone located at the Plateau in the upper troposphere and lower stratosphere, 2) to the deepening of the low trough located at India in the lower troposphere, and 3) to the break of the subtropical high belt at the 500 hPa.

II. THE HORIZONTAL FLOW FIELD

Fig. 6 (a,b) illustrates simulated v-components at the P_7^* level of the HH and the HN experiments, By comparing these figures we can see that in the results of the HH experiment, the strong southerly winds originating from the ocean are well simulated. The seasonal pre-

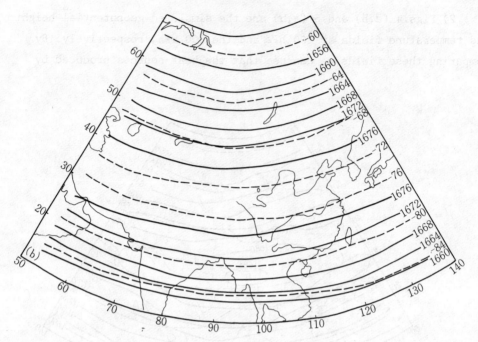

Fig. 4 (b) As in Fig. 4 (a) except for HN experiment.

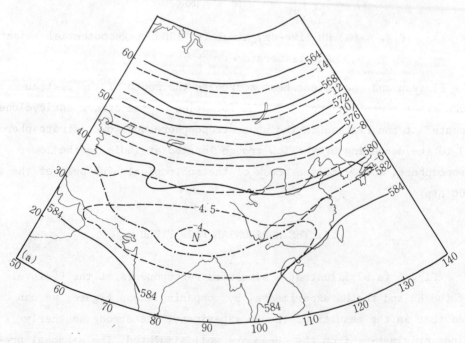

Fig. 5 (a) As in Fig. 4 (a) except for The 500 hPa level.

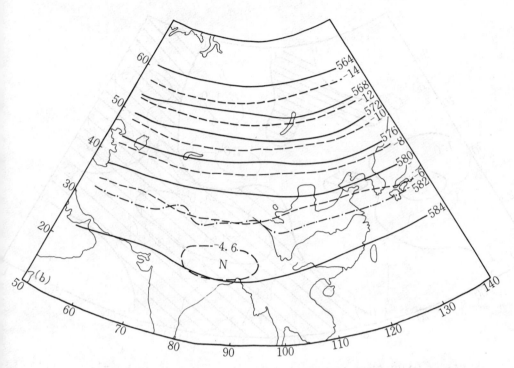

Fig. 5 (b) As in Fig. 4 (b) except for the 500 hPa level.

cipitation is closely linked to the southerly winds. In the results
of the HN experiment there are also southerly winds to the east and
south of the Plateau, but they are weaker, and not linked to ocean.

III. "MONSOON CIRCULATION TUBE"

Based on analysis of the data, Yin Sue-xin et al[3]. Pointed out
that in summer there exists a "monsoon circulation tube" in East Asia.
In the experimental results of the HH experiment the monsoon circulation tube has been well simulated in the meridional vertical circulations at every 5° longitude. However the simulated tube is longer than
the tube defined by Yin. In the simulated area (40°E−150°E) there are
two vertical circulation tubes. One exists to the west of 80°E and is
a Hadley cell tube to the south of the Plateau (Fig. 7 a). The other
is located to the east of 80°E and is like a Ferrel cell tube (so called "monsoon tube"). The second one can be subdivided into two sections

233

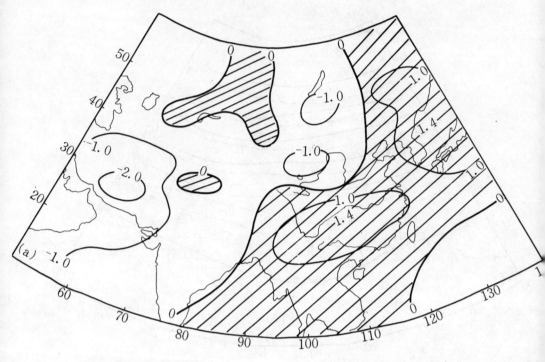

Fig. 6 (a) HH five-day simulated v-component (unit:m•s^{-1}) at the P_7^* . Hatching indicates the southerly winds.

The first section is located to the south of 40°N from 80°E to 105°E. It appears first at 300 hPa at 80°E, then starts to stretch extensively downward and eastward. At about 90°E it expands vertically to the whole troposphere (Fig. 7 b). In addition, to the west of 95°E this section is composed of a small monsoon tube over the Qinghai-Xizang Plateau and a larger one over India. At 95°E the above mentioned two tubes merge into one. Then it further develops; at 100°E its ascent branch can reach 40°N (Fig. 7 c). At 105°E the tube suddenly shrinks (Fig.7d). From about 110°E it becomes strong again, however it only appears in the upper troposphere and its ascent branch is near 35°N while its descent branch is near the ridge line of the subtropical high of the western Pacific (Fig. 7 e).

In the individual meridional circulation profiles obtained from experiment HN the monsoon circulation tubes have not been found, and at the south of 40°N there is a Hadley cell tube only (Fig. 8).

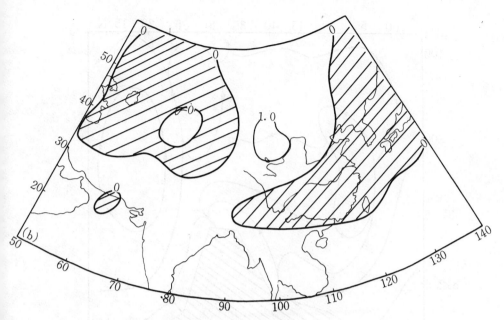

Fig. 6 (b) As in Fig. 6 (a) except for HN.

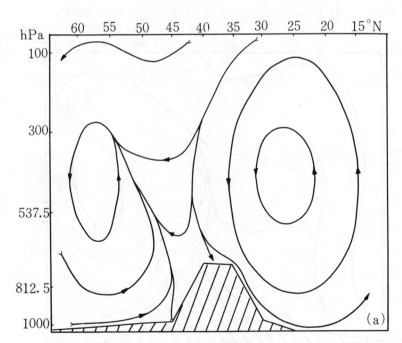

Fig. 7 (a) HH Simulated N-S vertical circulations along 70°E
with vertical velocity magnified by 400 times.

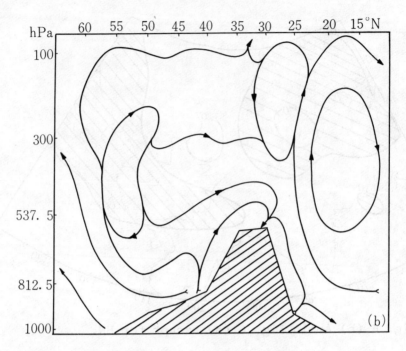

Fig. 7 (b) As in Fig. 7 (a) except for 90°E.

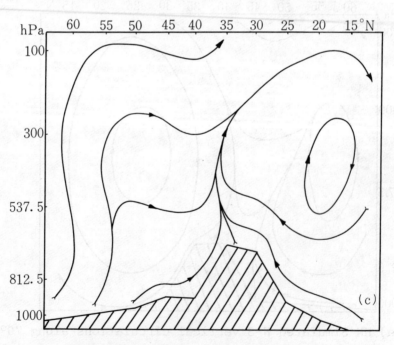

Fig. 7 (c) As in Fig. 7 (a) except for 100°E.

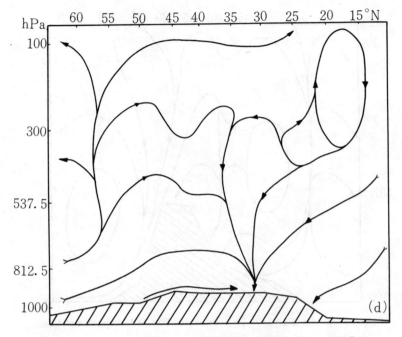

Fig. 7 (d) As in Fig. 7 (a) except for 105°E.

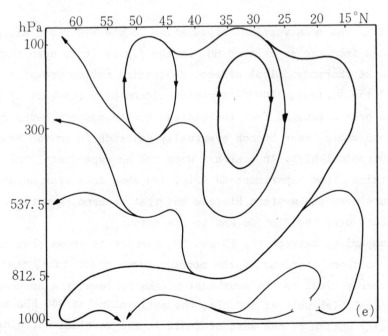

Fig. 7 (e) As in Fig. 7 (a) except for 120°E.

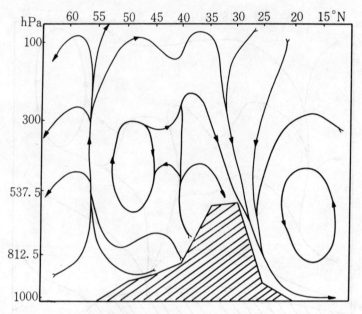

Fig. 8 As in Fig. 7 (b) except for HN experiment.

IV. THE MEAN VERTICAL CIRCULATION IN THE W-E PLANE

Fig. 9 is the mean vertical circulation in the W-E Plane along 35°N obtained from the experiment HH. In the figure it is seen that because of the thermodynamical effect, descending motion prevails over the west of the Plateau, ascending motion occurs over the east of 80°E whose upper branch extends from the Plateau to the eastern boundary of the model and whose lower branch eventually descends in to the Western Pacific subtropical high. In contrast with the HH experiment, the results obtained from experiment HN (Fig. 10) show that weak ascending motion occurs over the western Plateau and that descending motion dominates both over the Plateau and to its east.

The comparison between the PN and MN experiments shows that the middle and southern sections of the monsoon circulation tube, the strong southerly winds to the southeast of the Plateau, the break of the subtropical high belt at 500 hPa, the anticyclone at 300 hPa and the descending motion to the west of India in the vertical circulation profile along 25°N are all more closely related to the heat source

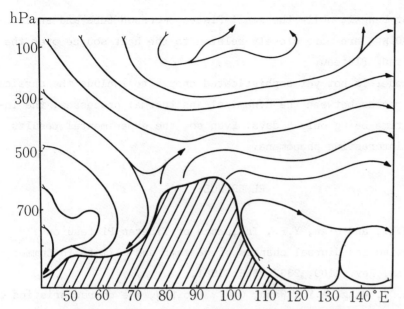

Fig. 9 HH simulated W-E vertical circulations along 35°N
with vertical velocity magnified 700 times.

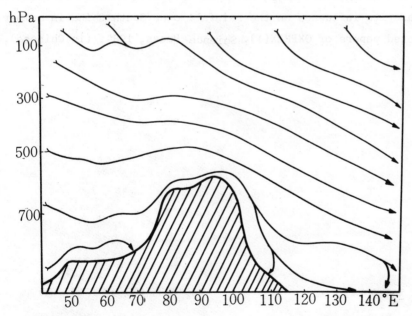

Fig. 10 As in Fig.9 except for the HN experiment.

over Bay of Bengal, while the small Plateau monsoon tube and anticyclone at 100 hPa are more closely related to the heat source over the
Qinghai-Xizang Plateau.

Our model is not yet sophisticated enough to include the physical
feedback process between the dynamical and thermal processes, the integration time being only 5 days. Even so, the experimental results
show many interesting phenomena.

REFERENCES

[1] Kuo, H.L., and Qian, Y.F., Influence of Tibetan Plateau on
    cumulative and diurnal changes of weather and climate in Summer,
    Mon. Wea. Rev., 109, 2337-2356, 1981.
[2] Yao, L.C., Wang, A.Y., Wang, Q.Q., Luo,S.W, The characteristics of
    the monthly heating field distribution in the Qinghai-Xizang
    Plateau and its Surrounding region in Summer of 1979, Collected
    Papers of QXPMEX(1), 1984, Science Press (in Chinese).
[3] Yin,S.X., Liang,H.M., Liou, M.Zh. et al., Asia Monsoon circulation
    tube and its influence on the rainfall over South China in summer,
    collected papers of QXPMEX(1), Science Press, 1984, (in Chinese).

The Influences of Orography on Large-scale Atmospheric Flow Simulated by a General Circulation Model

Ngar-Cheung Lau
Geophysical Fluid Dynamics Laboratory Princeton University

1. Introduction

The role of orography on the atmospheric general circulation has been a scientific problem of considerable interest to meteorologists during the past few decades. Different facets of this issue have been examined by numerous investigators using model tools of varying degree of complexity. Among the most sophisticated models used for this purpose is the fully nonlinear general circulation models (GCMs) incorporating a wide array of dynamical and physical processes. To date, the most detailed analyses of orographic influences on the circulation features appearing in a grid-point GCM are probably those reported by Manabe and Terpstra (1974) and Kasahara et al. (1973), who examined simulations of relatively limited durations (say, one or two seasons). The rationale for reexamining this problem in the present study is three-fold:

- With the development of a new generation of GCMs based on the spectral formulation, as well as the availability of advanced computing devices, it is now feasible to perform model integrations of very long durations (say, 10 years or more). The history tapes for such extended experiments ensure that the sensitivity of the model atmosphere to orography may be determined with a high degree of statistical significance.

- With the increased use of spectral models in long-range numerical weather prediction and in sensitivity experiments, it is of

interest to document in some detail the response of such models
to orography.

• With the introduction of more incisive theoretical tools as well as
  diagnostic techniques, the past decade has witnessed considerable
  progress in our understanding of various large-scale circulation
  features. By analyzing the GCM experiments in light of these
  modern results, one should gain a deeper insight into the nature of
  orographic influences on the atmosphere.

The experimental design for the present study is essentially analogous to that for the precious GCM investigations. A pair of 15-year
integrations have been conducted. In one experiment, hereafter referred
to as 'Experiment M', the mountain complexes are incorporated in the
lower boundary. In the other experiment, hereafter referred to as
'Experiment NM', the lower boundary is uniformly flat. In both experiments M and NM, the present-day distribution of continent and ocean is
prescribed.

The detailed formulation of the GCM examined in this study has
been described in detail by Gordon and Stern (1982). The spectral model
uses a rhomboidal truncation at wavenumber 15, and vertical variations
are represented at 9 sigma levels. Various essential physical processes, such as those associated with the hydrologic cycle, radiation
and subgrid-scale transfers, are incorporated in the model design.

The effects of orography on the model atmosphere are inferred by
contrasting selected circulation statistics from the two model runs. In
this report, we shall describe the influences of mountain complexes on
the stationary flow field at various altitudes in the troposphere, and

on the behavior of synoptic scale wave disturbances. The seasonal dependence of these influences will also be discussed.

2. Orographic effects on the stationary flow field

2.1 Streamfunction and velocity potential at 200 hPa

The vectorial difference between the long-term averaged horizontal wind field at 200 hPa for Experiment M and the corresponding field for Experiment NM is computed, and the rotational and divergent components of the different field thus obtained are then depicted by the patterns of streamfunction and velocity potential, respectively. The latter patterns are shown in Fig. 1 for Northern Hemisphere winter [panels (a) and (b)] and summer [panels (c) and (d)].

2.1.1 Winter

The streamfunction response to orography in the winter season (Fig. 1a) is dominated by two well-defined wavetrains. One such group of perturbations extend southeastward from the Sea of Okhotsk to the subtropical Pacific, with vortices labelled as L_1, H_1 and L_2 in Fig. 1a. The other group of waves span southeastward from the Gulf of Alaska towards the Caribbean Sea, with vortices H_2, L_3 and H_3. The individual vorticity centers exhibit a pronounced southwest to northeast tilt, thus implying a preferential equatorward energy propagation. The response shown in Fig. 1a is remarkably similar to the steady state solutions for the linear barotropic vorticity equation forced by realistic topography. The latter solutions have been discussed in an extended review by Held (1983, Figs. 6.7 to 6.9), who further demonstrated that the wavetrain consisting of L_1, H_1 and L_2 is essentially forced by the Tibetan Plateau; whereas the wavetrain consisting of H_2, L_3 and H_3 is mainly a

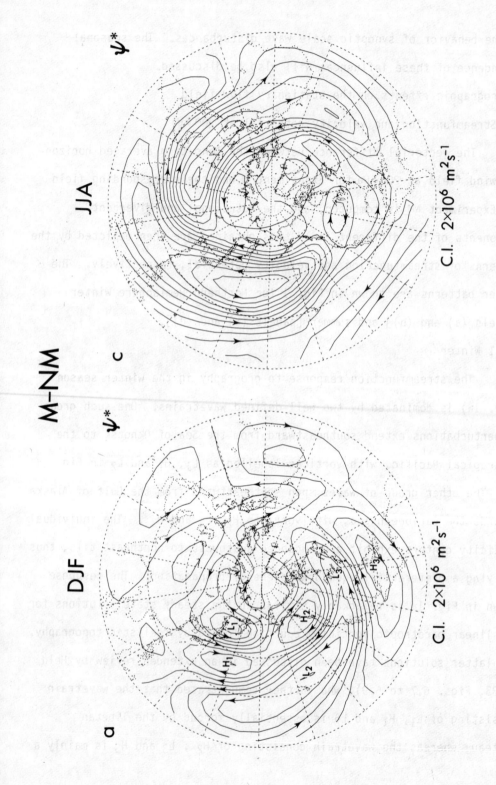

ψ^*

M−NM

JJA

c

C.I. 2×10^6 m^{2}s^{-1}

ψ^*

DJF

a

C.I. 2×10^6 m^{2}s^{-1}

H$_1$ L$_1$ H$_2$ L$_3$ H$_3$

L$_2$

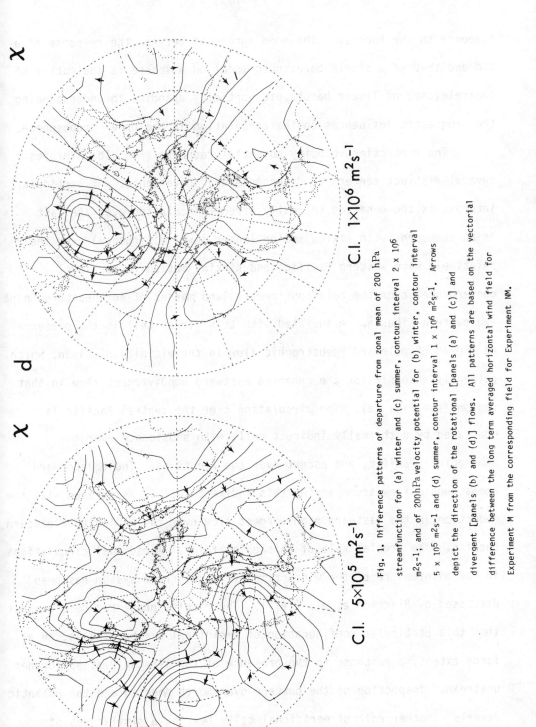

Fig. 1. Difference patterns of departure from zonal mean of 200 hPa streamfunction for (a) winter and (c) summer, contour interval 2×10^6 m²s⁻¹; and of 200 hPa velocity potential for (b) winter, contour interval 5×10^5 m²s⁻¹ and (d) summer, contour interval 1×10^6 m²s⁻¹. Arrows depict the direction of the rotational [panels (a) and (c)] and divergent [panels (b) and (d)] flows. All patterns are based on the vectorial difference between the long term averaged horizontal wind field for Experiment M from the corresponding field for Experiment NM.

response to the Rockies. The good agreement between the response of a GCM and that of a simple barotropic model on a sphere is indicative of the relevance of linear barotropic vorticity dyanmics in understanding the orographic influences on the circulation in the upper troposphere.

The wintertime velocity potential response (Fig. 1b) exhibits several distinct centers of divergence and convergence. Of particular interest is the enhanced thermally direct circulation over the East Asian seaboard, with rising motion (upper level divergence) over the relatively warm western Pacific, and sinking motion (upper level convergence) over the cold continental land masses of northeastern China and eastern Siberia. Associated with this direct cell is the presence at 200 hPa of poleward ageostrophic flow in the vicinity of Japan, which evidently accounts for the enhanced eastward nondivergent flow in that region (see Fig. 1a). The circulation over the central Pacific is dominated by a thermally indirect cell, with subsidence over the subtropical Pacific, and ascent over the Aleutians. The equatorward ageostrophic flow accompanying this indirect cell is associated with the reduction in the eastward nondivergent flow over the midlatitude eastern Pacific. The crucial role of this pair of local meridional circulations in the maintenance of the wintertime Asian jetstream system has been discussed by Blackmon et al. (1977). The results presented here suggest that this particular configuration of meridional overturning is, to a large extent, a response to the presence of the Tibetan Plateau further upstream. Inspection of the pattern over North America and the Atlantic reveals another pair of meridional cells bearing a similar set of relationships with the North American jetstream system. Also notable in the streamfunction and velocity potential patterns is a tendency for the centers of divergence (convergence) in Fig. 1b to be located east (west) of the cyclonic vortex centers in Fig. 1a, thus indicating a dominant counterbalance between advection of relative vorticity and the divergence term in the vorticity budget.

2.1.2 Summer

The spatial scale of the streamfunction response in summer (Fig. 1c) is considerably larger than that of its counterpart in winter (Fig. 1a). The summertime pattern is characterized by anticyclonic centers over the Eurasian and North American land masses, and cyclonic centers over the two ocean basins in the temperate latitudes. The features in Fig. 1d are dominated by rising motion over the central portion of Asia and North America, and sinking motion over the midlatitude Pacific and Atlantic. Comparison between Figs. 2c and 2d indicates that the centers of divergence (convergence) are located west (east) of the cyclonic centers. Hence, contrary to the situation in winter, the divergence term in the summertime vorticity budget is primarily balanced by advection of planetary vorticity.

2.2 Circulation near the surface

In Fig. 2 are shown the distribution of the long-term averaged horizontal vector wind field at the model sigma level closest to the surface (i.e., $\sigma = p/p* = 0.99$, where p is the pressure at the selected sigma level and p* is the pressure at the earth's surface). The maps presented here portray the patterns for Experiments M and NM, as well as the difference between these two model climatologies (M-NM), for January [panels (a), (b) and (c)] and July [panels (d), (e) and (f)].

247

SURFACE FLOW

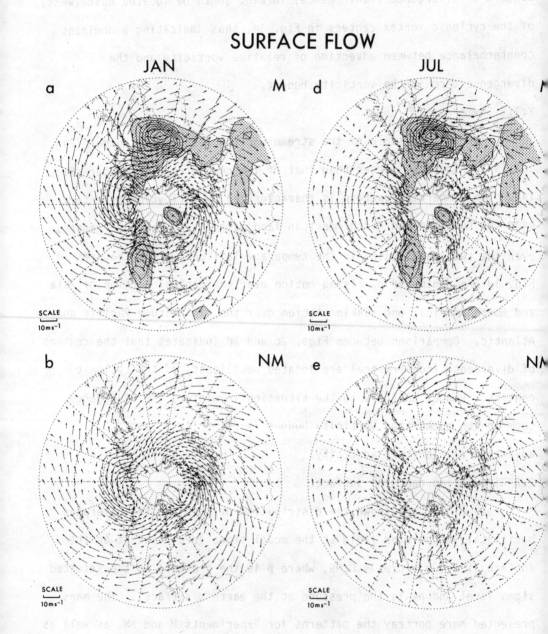

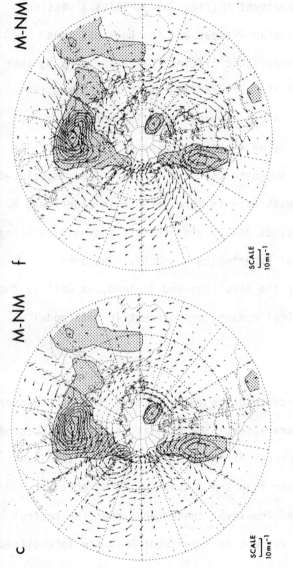

Fig. 2. Horizontal wind field at the lowest model sigma level ($\sigma = 0.99$) for Experiment M in (a) January and (d) July, for Experiment NM in (b) January and (e) July, and for the difference between Experiments M and NM in (c) January and (f) July. Contours depicting the model orography are plotted at an interval of 500 m.

249

2.2.1 January

The differences between Experiments M and NM in the surface flow field are most notable in the vicinity of the mountain complexes. The wintertime surface wind in Experiment M (Fig. 2a) exhibits a distinct tendency to flow around the Tibetan Plateau and the Rockies, thus resulting in relatively weak orographic lifting. In Experiment M, there is a noticeable retardation of the surface flow as it approaches the windward side of the mountain ranges; whereas the flow pattern on the leeward side is characterized by an enhanced northerly component. The difference pattern in Fig. 2c suggests that the offshore flow associated with the winter monsoon over east Asia is more intense in Experiment M.

Comparison between Figs. 2a and 2b also reveals many similarities in the surface climatologies of Experiments M and NM. The semi-permanent cyclone centers over the Aleutians and Iceland, as well as the anticyclones over the subtropical oceans, are present in both model runs.

2.2.2 July

In summer, the surface circulation in Experiment M (Fig. 2d) over the Tibetan Plateau acquires characteristics similar to those associated with heat lows. The wind component directed up the slope of the local terrain is considerably stronger than the wintertime flow, thus resulting in much more enhanced low-level convergence over the central portion of the Plateau. Also evident in Experiment M is the intensified southwesterly summer monsoon over the Indian Subcontinent and the Arabian Sea. Along the west coast of North America, the Rockies act as an effective barrier to the prevailing midlatitude westerlies. The anticyclones over the Pacific and Atlantic ocean basins are present in both Experiments M and NM.

2.3 Vertical structure

The vertical structure of the stationary waves in Experiments M and NM is compared in Fig. 3, which shows the longitude-pressure cross-sections of the departure from zonal symmetry of the geopotential height field along the 51.75°N latitude circle in winter [panels (a), (b) and (c)], and along the 42.75°N latitude circle in summer [panels (d), (e) and (f)]. The stationary waves appearing in Experiment NM [panels (b) and (e)] may be interpreted as the response to diabatic heating processes associated with continent-ocean contrast alone; whereas the patterns for Experiment M [panels (a) and (d)] portray the total response to both diabatic heating and orography.

2.3.1 Winter

The pattern for Experiment M (Fig. 3a) bears a strong resemblance to the vertical structure of the observed stationary waves. The characteristic westward tilt with increasing height of the trough and ridge axes, as well as the high and low pressure centers near the surface, are most evident in Experiment NM (Fig. 3b); whereas the pattern for M-NM (Fig. 3c) exhibits a much more barotropic structure, with only weak circulation features in the lower troposphere. Comparison between the patterns for Experiments M and NM suggests that the forcing by diabatic processes alone accounts for about one-half of the amplitude of the total response, and that the stationary waves appearing in both

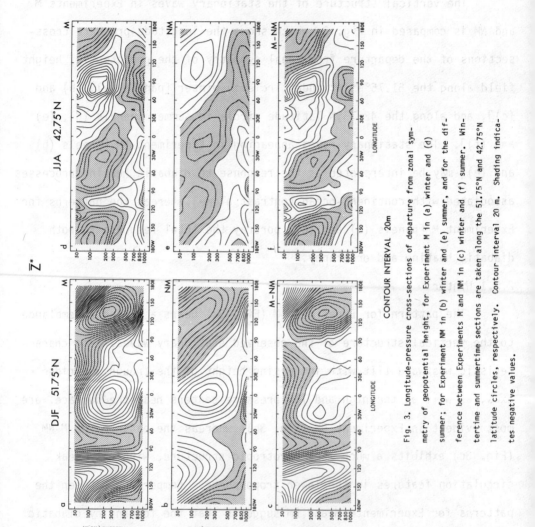

Fig. 3. Longitude-pressure cross-sections of departure from zonal symmetry of geopotential height, for Experiment M in (a) winter and (d) summer; for Experiment NM in (b) winter and (e) summer, and for the difference between Experiments M and NM in (c) winter and (f) summer. Wintertime and summertime sections are taken along the 51.75°N and 42.75°N latitude circles, respectively. Contour interval 20 m. Shading indicates negative values.

experiments tend to be in phase.

2.3.2 Summer

In the pattern for Experiment M (Fig. 3d), which also agrees quite well with observations, perturbations of a given polarity tend to be located directly above perturbations of the opposite polarity. The response to diabatic processes in summer (Experiment NM, Fig. 3e) is almost exactly out-of-phase with the corresponding response in winter (Fig. 3b). The diabatic forcing is seen to contribute significantly to the total forcing. However, contrary to the wintertime situation, there is less evidence of an in-phase relationship between the summertime upper tropospheric response in Experiments M and NM. Along certain longitudinal sectors, the perturbations at the tropopause level in Experiment M actually tend to be in quadrature with those in Experiment NM.

3. Orographic influences on the behavior of wintertime cyclone scale
    disturbances

The notable differences between the stationary response in Experiments M and NM, as described in the previous section, may have significant implications on the properties of transient disturbances on various time scales. In this section we shall contrast the behavior of wintertime wave disturbances with time scales between 2.5 and 6 days in the two model runs. In order to isolate those fluctuations with this particular range of time scales, the daily model data have been processed through a band pass filter identical to that described in Blackmon and Lau (1980).

3.1 Geographical location of the principal cyclone tracks

253

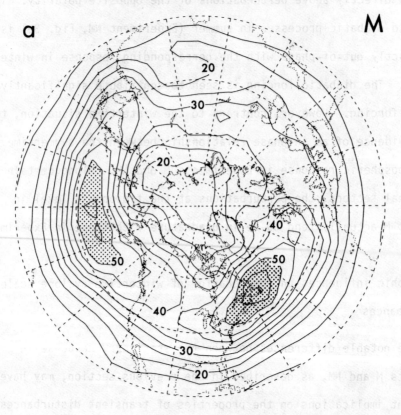

DJF
BAND PASS 500hPa $(\overline{Z'^2})^{1/2}$

a

M

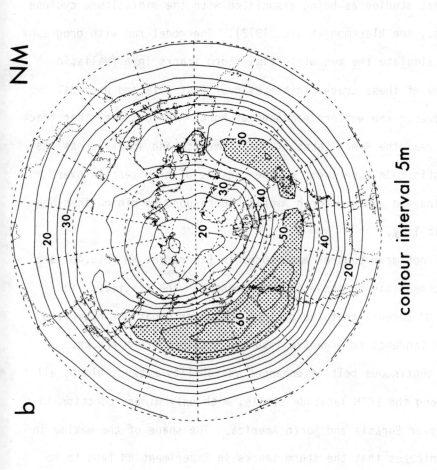

NM

contour interval 5m

Fig. 4. Root-mean-squares of the band pass filtered 500hPa height in winter, for (a) Experiment M and (b) Experiment NM. Contour interval 5 m. Shading indicates values exceeding 50 m.

In Fig. 4 are shown the distributions of the root-mean-squares of band pass filtered 500 mb height in winter, for (a) Experiment M and (b) Experiment NM. The pattern for Experiment M exhibits two elongated maxima which are distinctly separated from each other. These regions of enhanced synoptic scale activity have been interpreted in previous observational studies as being associated with the midlatitude cyclone tracks (e.g., see Blackmon et al., 1977). The model run with orography is seen to simulate the two wintertime storm tracks in a realistic manner. One of these tracks initiates over East Asia and extends eastward towards the western seaboard of North America; the other track originates near the Great Lakes of North America and spans across the North Atlantic. In Experiment M, the regions of high terrain over western China and western North America correspond to minima in synoptic scale variability.

The root-mean-squares amplitudes in Experiment NM (Fig.4b) are generally comparable to those in Experiment M. The statistics for Experiment NM suggest that the cyclone scale perturbations exhibit a much weaker tendency to organize into two distinct storm tracks. Instead, a continuous belt of enhanced variability extends almost all the way along the 50°N latitude circle, with only minor reduction in amplitude over Eurasia and North America. The shape of the maxima in Fig. 5b indicates that the storm tracks in Experiment NM tend to be zonally oriented; whereas the corresponding pattern for Experiment M implies preference for a southwest to northeast orientation over the oceans.

3.2 Horizontal structure and temporal evolution

The propagation characteristics and structural properties of the disturbances embedded in the wintertime cyclone tracks described above may be delineated by the distributions of time-lagged autocorrelation statistics, shown in Fig. 5 for (a) Experiment M and (b) Experiment NM. These plots, often referred to as teleconnection patterns, are constructed by mapping the correlation coefficients between the band pass filtered 500 hPa height at individual grid points and the corresponding values at a fixed reference point, which in this case is chosen to be the model grid point located at 42.75°N, 142.5°E. The various panels displayed here depict the patterns so obtained using different time lags (-3 days, -2 days, ... , +2 days, +3 days) between the time series at individual grid points and the corresponding series at the reference point. A similar technique has been employed by Blackmon et al. (1984) to diagnose the behavior of observed atmospehric disturbances.

The overall similarity of the patterns for Experiments M and NM suggests that the different stationary wave responses in these model runs do not alter the fundamental properties of the cyclone waves. It is seen that disturbances with synoptic time scales tend to be elongated in the meridional direction; their typical zonalwavelength (as inferred from the longitudinal separation between two successive extrema of the same polarity) is approximately 6000 km; and their typical phase speeds (as inferred from the time taken for two successive extrema of the same polarity to pass through a given geographical location) is approximately +17 m/s. Over much of Asian continental land mass, which lies upstream

257

BAND-PASS FILTERED (2.5 TO 6 DAYS)
500hPa HEIGHT DATA

M   NM

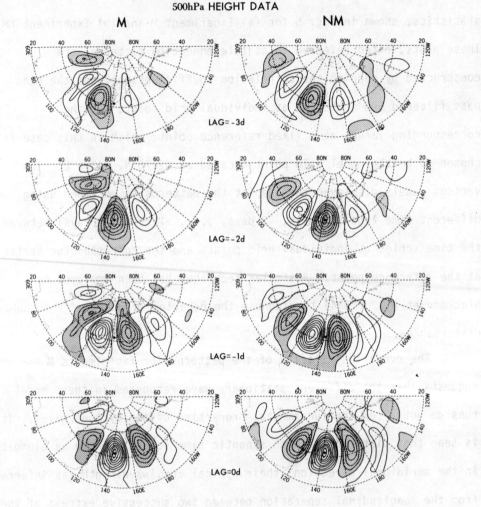

LAG= -3d

LAG= -2d

LAG= -1d

LAG= 0d

258

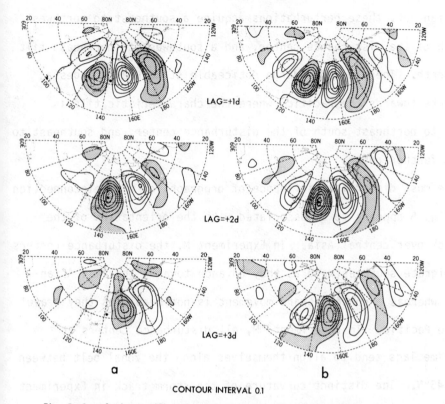

CONTOUR INTERVAL 0.1

Fig. 5. Correlation coefficients at various time lags between band pass
filtered 500hPa height in winter at the reference point (42.75°N, 142.5°
E) and the corresponding values at all grid points, for (a) Experiment M
and (b) Experiment NM. Contour interval 0.1. The zero contour is
omitted for sake of clarity. Shading indicates values smaller than -0.1.
Negative (positive) lags indicate that the teleconnection patterns have been
constructed with the fluctuations at the reference point lagging (leading) the
fluctuations at other grid points.

of the Asian jet, these perturbations acquire a southeast to northwest
tilt south of the disturbance center, and a southwest to northeast tilt
further north. These waves undergo noticeable structural changes as
they migrate towards the Pacific, where the characteristic tilt is
southwest to northeast south of the disturbance center, and southeast to
northwest further north.

The most discernible influence of orography on the teleconnection
maps in Fig. 5 appears to be associated with the orientation of the
storm track over central Asia. In Experiment M, the disturbance centers
tend to migrate southeastward as they advance towards the East Asian
seaboard; whereas a northeastward movement is noticeable as they travel
across the Pacific. In Experiment NM, all disturbance centers at
various time lags tend to align themselves along the zonal belt between
40°N and 45°N. The distinct curvature of the storm track in Experiment
M is apparently a consequence of the steering effect associated with the
intensified stationary trough in that integration (see Fig. 1a).

3.3 Vertical structure

In this subsection the vertical phase structure of the cyclone
waves is depicted by the coherence and phase between geopotential height
fluctuations at 1000 and 500 hPa over the same site. The phase and
coherence between each pair of time series may be determined by the
cross-spectral analysis procedure described in Lau (1979). The phase
difference at a given grid point is represented by the orientation of
the arrow at that grid point, an arrow pointing directly poleward
indicates zero phase difference, and the arrow rotates clockwise by 1°
for each degree of phase lag of the fluctuations at 500 hPa behind those

at 1000 hPa Hence an arrow pointly eastward indicates that the 500 hPa fluctuations lag the 1000 hPa fluctuations by one-quarter cycle; whereas an arrow pointing southward indicates an exact out-of-phase relationship, and so on. The coherence, which provides for a measure of the statistical significance of the phase difference, is represented by the length of each arrow, according to the scale shown at the lower left corner. Regions with coherence exceeding 0.5 are also depicted by light shading. In Fig. 6 are shown the phase and coherence patterns for the frequency band corresponding to a period of 4 days, and for (a) Experiment M and (b) Experiment NM.

It is seen at most midlatitude locations that the 500 hPa fluctuations tend to lag behind those near the surface. For eastward propagating waves, such a phase difference implies a westward tilt with increasing altitude. In Experiment M (Fig. 6a), the strongest vertical tilts (up to almost 90° phase lags) are detected over the western Pacific and eastern North America, which correspond to the western portion of the two oceanic storm tracks, shown in Fig. 4a. The phase differences are systematically reduced as the cyclones migrate eastward along these storm tracks, indicating a gradual transition from a baroclinic structure to a more barotropic behavior. The pattern over the Eurasian land mass suggests the occurrence of two preferred tracks as the disturbances approach the Tibetan Plateau. One of these tracks lies along the northern periphery of the Plateau, and is associated with baroclinic perturbations; whereas the disturbances along the other track, which curves around the southern edge of the Plateau, are distinctly barotropic.

DJF
COHERENCE BETWEEN 500 AND 1000hPa Z
a
M

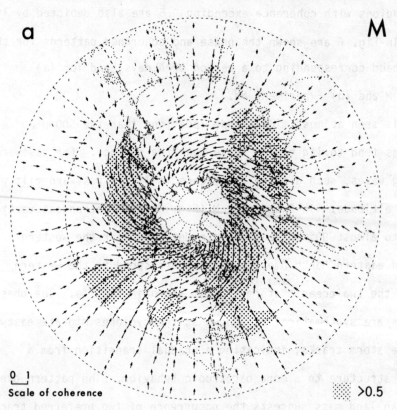

0 1
Scale of coherence

>0.5

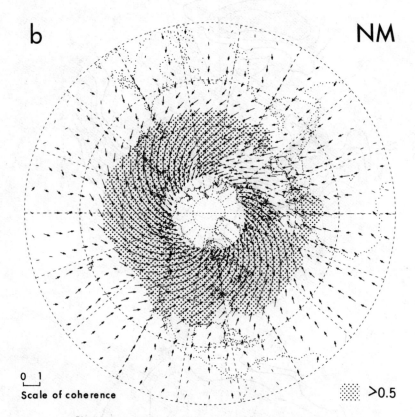

0 1
Scale of coherence

▦ >0.5

Fig. 6. Coherence and phase between the wintertime geopotential height fluctuations at 500hPa and those at 1000hPa as depicted respectively by the length and orientation of the individual arrows (see text for detailed explanation of the convention used), for (a) Experiment M and (b) Experiment NM. Shading indicates coherence exceeding 0.5. All results are based on spectral estimates for wave periods of 4 days.

VERTICALLY AVERAGED DIABATIC HEATING

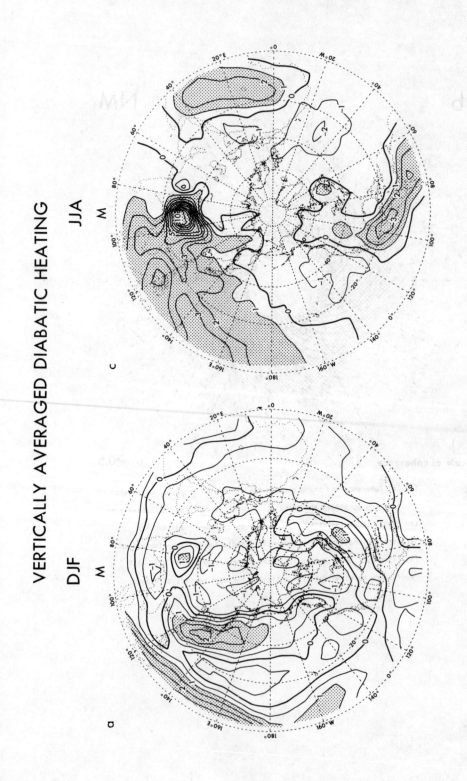

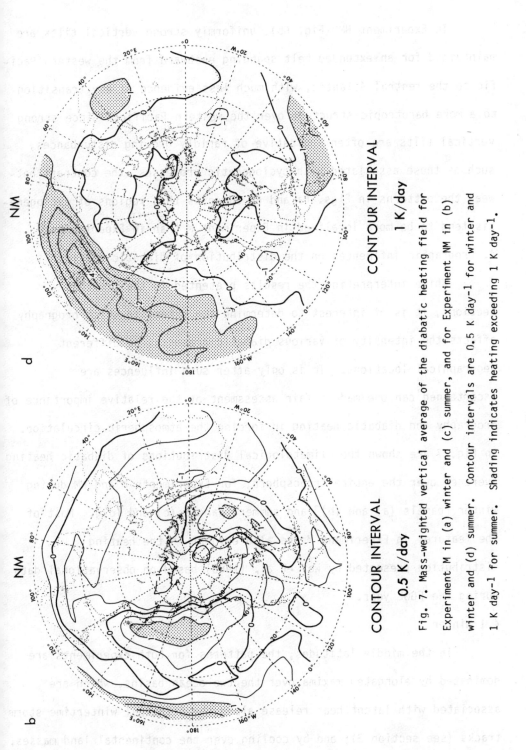

NM

CONTOUR INTERVAL
0.5 K/day

NM

CONTOUR INTERVAL
1 K/day

Fig. 7. Mass-weighted vertical average of the diabatic heating field for
Experiment M in (a) winter and (c) summer, and for Experiment NM in (b)
winter and (d) summer. Contour intervals are 0.5 K day⁻¹ for winter and
1 K day⁻¹ for summer. Shading indicates heating exceeding 1 K day⁻¹.

In Experiment NM (Fig. 6b), uniformly strong vertical tilts are maintained for an extended belt spanning eastward from the western Pacific to the central Atlantic, with much less evidence of any transition to a more barotropic structure over the western Pacific. Since strong vertical tilts are often indicative of rapidly growing disturbances, such as those associated with cyclogenetic processes, the contrast between the patterns in Figs. 6a and 6b suggests that regions of cyclogenesis tend to be more localized in Experiment M than in Experiment NM.

4. Orographic influences on the distribution of diabatic heating

While interpreting the results presented in the previous sections, it is of interest to determine the extent to which orography affects the intensity of various diabatic processes at different geographical locations. It is only after such influences are ascertained can one made a fair assessment of the relative importance of orography and diabatic heating in forcing the atmospheric circulation. In Fig. 7 are shown the climatological distributions of diabatic heating averaged over the entire troposphere, for Experiments M and NM during winter [panels (a) and (b)] and summer [panels (c) and (d)]. Most of the features in Experiment M are also evident in the heating distribution presented by Wei et al. (1983) based on observational data during the FGGE year.

4.1 Winter

In the middle latitudes, the patterns for both experiments are dominated by elongated maxima over the two ocean basins, which are associated with latent heat release along the principal wintertime storm tracks (see section 3); and by cooling over the continental land masses.

Intense latent heating also occurs over the equatorial Pacific, South America and Africa. With the exception in Experiment M of relatively stronger heating along the midlatitude Pacific cyclone track, and of the heating over the Tibetan Plateau, the patterns in Figs. 7a and 7b are qualitatively similar to each other.

4.2 Summer

The impact of orography on the diabatic heating patterns is much more notable in summer than in winter. The pattern for Experiment M (Fig. 7c) exhibits a relatively more distinct land-sea contrast, with heating over the eastern portion of the Tibetan Plateau, the eastern Rockies, Central America, the African continent equatorward of 10°N and the western Pacific. Cooling prevails over the eastern Pacific and much of the Atlantic. The extremely intense heating over the Tibetan Plateau is a consequence of latent heat release associated with excessive precipitation in the model simulation over that region. The heating pattern for Experiment NM (Fig. 7d) is dominated by an elongated positive maximum over the tropical western Pacific, and cooling over the eastern portion of the two ocean basins. The pattern over the major continental land masses is rather featureless.

5. Concluding Remarks

By contrasting the climatological statistics for two extended GCM integrations, one with and other without orography, we note that the large-scale mountain complexes have a demonstrable impact on various circulation features through a complicated chain of processes. In particular, it is seen that the amplitude and three-dimensional structure of the stationary waves are significantly altered in the presence of mountains. The forced, steady-state response in turn modifi

the behavior of rapidly moving transient waves through steering effects

and through providing preferred geographical sites for baroclinic

growth. The location of the storm tracks determines to a large extent

the pattern of latent heating in the middle latitudes, which in turn

play a major role in the forcing of the stationary waves. During the

summer season, orography is also seen to influence the heating pattern

through modulating the precipitation pattern associated with monsoon

circulations.

Recent forecast experiments also exhibit demonstrable sensitivity

of prediction skill to various representations of the earth's orography

in numerical models (Wallace et al., 1983). Hence an understanding of

the orographic influences on various dynamical and physical processes in

the atmosphere is not only of academic interest, but may find many

practical applications as well. Diagnosis of the complicated processes

involved by using both observation and simulated data should shed more

light on this important problem. Such studies should also identify

those aspects of the orographic forcing in the model which call for

further improvement.

References

Blackmon, M.L., and N-C. Lau, 1980: Regional characteristics of the
     Northern Hemisphere wintertime circulation: A comparison of the
     simulation of a GFDL general circulation model with observations.
     J. Atmos. Sci., 37, 497-514.

Blackmon, M.L., Y-H. Lee, J.M. Wallace and H-H. Hsu, 1984: Time variation of
     500 mb height fluctuations with long, intermediate, and short time scales
     as deduced from lag-correlation statistics. J. Atmos. Sci., 41, in press.

Blackmon, M.L., J.M. Wallace, N-C. Lau and S.L. Mullen, 1977: An observational
     study of the Northern Hemisphere wintertime circulation. J. Atmos. Sci.,
     34, 1040-1053.

Gordon, C.T., and W.F. Stern, 1982: A description of the GFDL global spectral
     model. Mon. Wea. Rev., 110, 625-644.

Held, I.M., 1983: Stationary and quasi-stationary eddies in the extratropical
     troposphere: theory. Large-scale Dynamical Processes in the Atmosphere,
     B.J. Hoskins and R.P. Pearce, editors. Academic Press. pp.127-168.

Kasahara, A., T. Sasamori and W.M. Washington, 1973: Simulation experiments
     with a 12-layer stratospheric global circulation model. I. Dynamical
     effect of the earth's orography and thermal influence of continentality.
     J. Atmos. Sci., 30, 1229-1251.

Lau, N-C., 1979: The structure and energetics of transient disturbances in
     the Northern Hemisphere wintertime circulation. J. Atmos. Sci., 36, 982-995.

Manabe, S., and T.B. Terpstra, 1974: The effects of mountains on the general
     circulation of the atmosphere as identified by numerical experiments.
     J. Atmos. Sci., 31, 3-42.

Wallace, J.M., S. Tibaldi and A.J. Simmons, 1983: Reduction of systematic
     forecast errors in the ECMWF model through the introduction of an
     envelope orography. Quart. J. Roy. Meteor. Soc., in press.

Wei, M-Y., D.R. Johnson and R.D. Townsend, 1983: Seasonal distributions of
     diabatic heating during the First GARP Global Experiment. Tellus, 35A,
     241-255.

THE THERMAL EFFECT OF THE QINGHAI-XIZANG PLATEAU ON THE FORMATION AND MAINTENANCE OF STATIONARY PLANETARY-SCALE CIRCULATION DURING SUMMER IN THE NORTHERN HEMISPHERE

Huang Ronghui

Institute of Atmospheric Physics,
Chinese Academy of Sciences

I. INTRODUCTION

Many computations showed that the Qinghai-Xizang Plateau is a strong heat source during summer in the northern hemisphere, and acts as a heat island in "the atmospheric ocean". Yeh and Gao computed the sensible heat flux and latent heat flux on the Qinghai-Xizang Plateau, based on long-term records of surface observations[1]. Their results show that there are large upward fluxes of sensible heat in the western Plateau from May to July and significant contributions to the latent heat release in the eastern Plateau from June to August. Ashe computed the planetary-scale distribution of heat sources during summer in the northern hemisphere from the climatic data[2]. The results also show that the greatest heat source during summer in the northern hemisphere is located on the Qinghai-Xizang Plateau. Very recently, Nitta, Luo and Yanai computed the distribution of heat source over the Qinghai-Xizang Plateau for the summer in 1979, respectively[3,4]. Their computed results show that the maximum mean heating rate is located in the 400—500 hPa layer over the Qinghai-Xizang Plateau.

The stationary planetary waves responding to the forcing induced by the heat source over the Qinghai-Xizang Plateau, and their propagations may play an important role in the formation and maintenance of stationary planetary-scale circulation over the northern hemisphere, especially, in the mean summer monsoon circulation over South Asia. Thus, it is necessary to investigate, theoretically and numerically, the propagations of stationary planetary waves responding to the forcing induced by the diabatic heating over the Qinghai-Xizang Plateau and the thermal effect of the Qinghai-Xizang Plateau on the formation and maintenance of stationary planetary-scale circulation during summer in the northern hemisphere.

II. STATIONARY PLANETARY WAVES PROPAGATIONS IN SUMMER

If we assume that there are no heat sources and viscosity in the atmosphere, and the Newtonian cooling effect is not been taken into consideration, the linearized vorticity and thermodynamic equation in a sperical coordinate system can be expressed as

$$(\frac{\partial}{\partial t} + \bar{U} \frac{\partial}{\cos\phi\partial\lambda})\zeta' + v' \frac{\partial}{a\partial\phi}(\bar{\zeta} + f) = f\frac{\partial\omega}{\partial p}, \tag{1}$$

$$(\frac{\partial}{\partial t} + \bar{U} \frac{\partial}{\cos\phi\partial\lambda})(\frac{\partial\phi'}{\partial p}) - 2\Omega_0 \sin\phi\frac{\partial\bar{U}}{\partial p}v' + \sigma\omega = 0, \tag{2}$$

respectively. Here, a is the radius of the earth, ϕ is the latitude, λ is the longitude. $\bar{U}$ is the zonal mean wind speed. v' is the meridional component of perturbation motion. ϕ' is the geopotential of perturbation. $\bar{\zeta}$ is the vertical component of relative vorticity of the basic state. $\sigma = - \frac{\partial\ln\theta}{\partial p}$ is the static stability. f is Coriolis parameter. ω the vertical velocity. Ω_0 is the rotation rate of the earth. The absolute vorticity advection term in equation (1) can be expressed as

$$v' \frac{\partial}{a\partial\phi}(\bar{\zeta}+f) = \frac{1}{a}[2(\Omega_0+\hat{\Omega})- \frac{\partial^2\hat{\Omega}}{\partial\phi^2} + 3\tan\frac{\partial\hat{\Omega}}{\partial\phi}]\cos\phi\times v', \tag{3}$$

where $\hat{\Omega}$ is defined as

$$\hat{\Omega} = \frac{\bar{U}}{a \cos\phi},$$

It is the angular speed of the basic flow.

Eliminating ω between (1) and (2), using Z-ordinate defined by $z = -H_0 \ln(\frac{p}{p_0})$ instead of p-ordinate. (p_0: reference pressure, H_0: representative scale height), introducing a new variable Ψ, i.e.,

$$\phi'(\lambda,\phi,z,t) = \text{Re} \sum_{k=1}^{K} \phi_k(\phi,z,t)e^{ik\lambda}, \qquad (4)$$

$$\Psi_k(\phi,z,t) = e^{-z/2H_0} \phi_k(\phi,z,t), \qquad (5)$$

and assuming that the atmosphere is nearly isothermal, then, the linearized potential vorticity equation in a spherical and Z coordinate may be written as follows

$$\frac{\sin\phi}{\cos\phi} \frac{\partial}{\partial\phi}(\frac{\cos\phi}{\sin\phi} \frac{\partial\Psi_k}{\partial\phi}) + 1^2\sin^2\phi \frac{\partial^2\Psi_k}{\partial z^2} + Q_k\Psi_k = 0, \qquad (6)$$

where $1 = 2\Omega_0 a/\tilde{N}$, $\tilde{N}$ is Brunt-Välsälä frequency. Q_k is called the refractive index squared for zonal wave number k, and can be expressed as

$$Q_k = Q_0 - \frac{k^2}{\cos^2\phi}, \qquad (7)$$

For stationary planetary waves, Q_0 is

$$Q_0 = [2(\Omega_0+\hat{\Omega}) - \frac{\partial^2\hat{\Omega}}{\partial\phi^2} + 3\tan\phi \frac{\partial\hat{\Omega}}{\partial\phi}$$

$$- 1^2\sin^2\phi(\frac{\partial^2\hat{\Omega}}{\partial z^2} - \frac{1}{H_0} \frac{\partial\hat{\Omega}}{\partial z})]/\hat{\Omega}-1^2\sin^2\phi \frac{1}{4H_0^2} . \qquad (8)$$

It may be considered as the refractive index squared for wave number 0. The characteristics of the distribution of Q_0 may play an important role in wave propagations.

We can see from equation (6) that if $Q_k>0$, the vertical and lateral propagations of stationary planetary waves are freely permitted. Vice versa, the propagations of stationary planetary waves are inhibited.

However, because of the vertical and meridional shears in the

272

realistic zonal mean wind and the sphericity of the earth, the ray of
wave propagations is refracted during propagating. We define the directions of the local group velocity of wave as the propagating ray of
planetary waves and assume that the angle between a propagating ray
and the horizontal direction is $\hat{\alpha}$. By using WKBJ method we have

$$\frac{d_g \hat{\alpha}}{dt} = \frac{1}{Q_k} i \; \vec{c}_g' \times \nabla Q_0 , \qquad\qquad (9)$$

where i is unit vector in λ direction, and

$$\frac{d_g}{dt} = \frac{\partial}{\partial t} + \vec{c}_g' \cdot \nabla$$

as the advective derivative moving with the group velocity of waves.
It can be concluded that the variations of stationary planetary wave
propagating ray are determined by Q_k and the graients of Q_0, which is
in connection with the basic state. Moreover, the propagating rays are
always refracted toward the direction of gradients of Q_0.

The distributions of the refractive index squared Q_0 and Q_k in
summer are computed from the zonal mean wind shown in Fig. 1 by equation (8), (7), and Q_1 are shown in Fig. 2 (dashed curves). In Fig. 2,
the schematic wave guides of stationary planetary wave for k = 1 are
denoted by arrows. Stationary planetary waves forced by a forcing
mechanism in summer cannot propagate into stratosphere. However, they
can propagate from low latitudes toward middle and high latitudes in
the troposphere and can propagate into the upper troposphere over the
subtropics.

III. THE PROPAGATIONS OF STATIONARY PLANETARY WAVES
FORCED BY IDEALIZED HEAT SOURCE OVER THE
QINGHAI-XIZANG PLATEAU

The propagations of stationary planetary waves responding to the
forcing induced by idealized heat sources over the Qinghai-Xizang
Plateau are simulated by using a quasi-geostrophic steady-state,

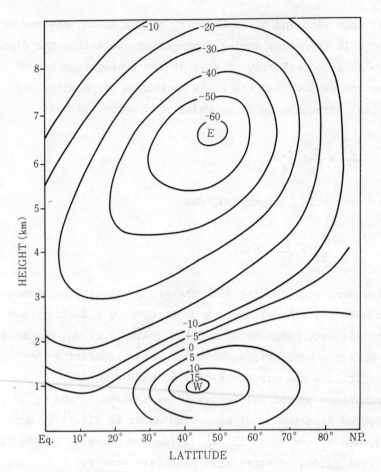

Fig. 1 The vertical distribution of the zonal mean wind in
summer (units in m·s).

34-level spherical coordinate model in which Rayleigh friction, the
effect of Newtonian cooling and horizontal thermal diffusivity are
included. The model equations and parameters are given in reference[5,6], respectively. In the derivation of the model equations, we
have included the nongeostrophic component to v' in the planetary
vorticity advection term in the model equations in order to get an
equation which a reasonable energy equation follows[7].

We assume that a vertical distribution of an idealized heat
source is over the Qinghai-Xizang Plateau.

274

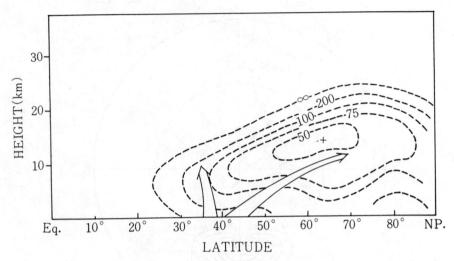

Fig. 2 The schematic wave guides for k =1. Dashed curves show
the refractive index squared Q_1 for wave number k = 1.

$$H_0(\lambda,\ \phi,p) = \hat{H}_0(\lambda,\ \phi)\exp(-(\frac{p-\bar{p}}{d})^2), \qquad (10)$$

where d = 300 hPa, p = 500 hPa. This means that the distribution of
diabatic heating has a maximum at 500 hPa, which is close to the observed distribution. Moreover, the horizontal variation of the idealized heat source $\hat{H}_0\ (\lambda,\ \phi)$ is given by

$$\hat{H}_0(\lambda,\phi)=\begin{cases}\hat{H}_0\Big(\sin\frac{\pi(\phi-\phi_1)}{(\phi_2-\phi_1)}\sin\frac{\pi(\lambda-\lambda_1)}{(\lambda_2-\lambda_1)}\Big)^2, & \lambda_1<\lambda<\lambda_2,\quad \phi_1<\phi<\phi_2,\\ 0 & \text{otherwise.}\end{cases}$$

where, $\frac{1}{c_p}\hat{H}_0$ =1.38 K·day^{-1}, λ_1=45°E, λ_2=1.35°E, ϕ_1=15°N,ϕ_2=45°N. Fig. 3
shows the horizontal distribution of this heat source. We can find
the center of this heat source located over Qinghai-Xizang Plateau.

We have computed the vertical distribution of amplitude and phase
of stationary planetary waves responding to the forcing induced by the
idealized heat source for wave number 1—3. Fig. 4 shows the distribution of amplitude and phase for wave number 1. In this figure, we can
find that the amplitude for k=1 has a maximum in the upper troposphere
near 30°N and has a secondary peak in the upper troposphere at high

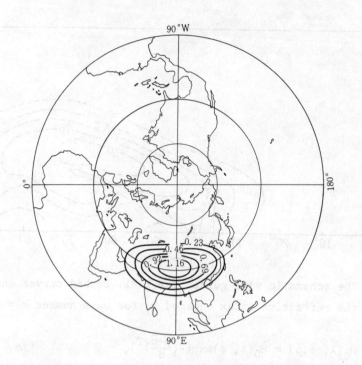

Fig. 3 The horizontal distribution of an idendized heat
source at 500hPa level (units in $K \cdot d^{-1}$).

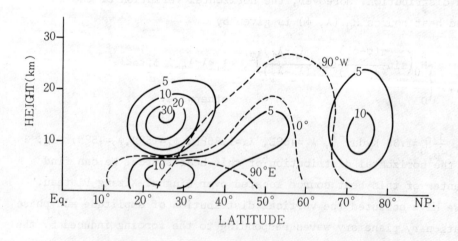

Fig. 4 The vertical distribution of amplitude (m) (solid curves)
and phase (dashed curves) responding to forcing by the
idealized heat source over the Qinghai-Xizang plateau for
K=1.

latitudes. The distribution of phase is westward with the increasing
latitude from the subtropics. Therefore, we may see that stationary
planetary waves propagate from the subtropics toward high latitudes.
This may verify the above conclusion obtained by the theoretical
analysis.

In the following, the stationary disturbance pattern, responding
to the forcing induced by this idealized heat source at constant
height levels has been computed. Figs. 5,6 show the computed disturbance pattern at 9 km and 6 km height level, respectively. In these

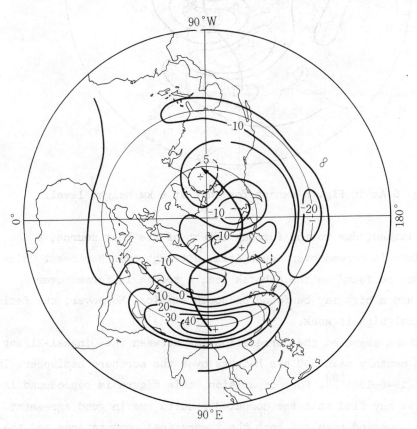

Fig. 5 The stationary disturbance patterns (m) forced by
the idealized heat source over the Qinghai-Xizang
Plateau at 9 km height level.

two figures, we can see that when a heat source is over the Qinghai-

277

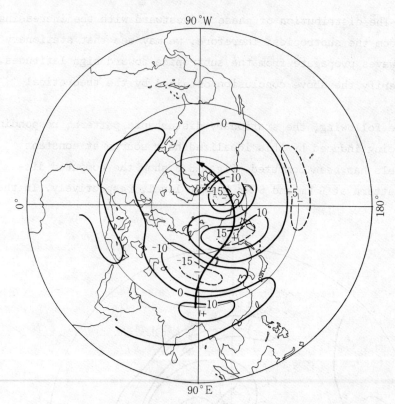

Fig. 6 As in Fig. 5 except for that at 6 km height level.

Xizang Plateau, due to the forcing effect of the heat source, a anticyclone is formed over South Asia, a trough over Northeast China, a high may be found on the Okhotsk Sea, a trough is formed over Alaska, and a high may be found on North America, Moreover, the Pacific subtropical high is weak.

Asakura computed the correlation map between the Qinghai-Xizang high and monthly mean 500 hPa heights over the northern hemisphere in July of 1946—1965[8]. For comparision, this figure is reproduced in Fig. 7. We may find that the computed results are in good agreement with the observed results. Both the theoretical computations and the observations show the following:

When the anticyclone over the Qinghai-Xizang Plateau develops, i) the anticyclone will cover South China and West Japan, entailing a hot summer in South China and West Japan; ii) a trough will be

278

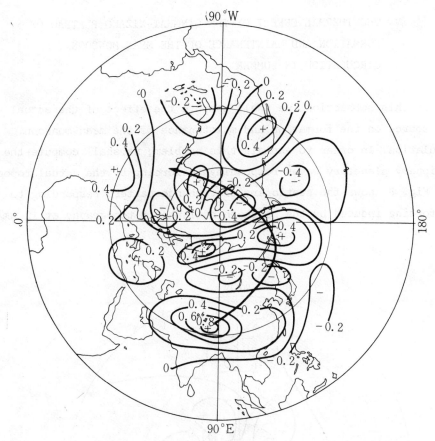

Fig. 7 Simultaneous correlation map between the Qinghai-Xizang
high and 500 hPa heights over the northern hemisphere
in July of 1946—1965 (After Asakura).

formed from Northeast China to North Japan, resulting in a cold summer
in there; iii) an anticyclone may be found over the Okhotsk Sea and
a trough is formed over Alaska; iv) the Pacific subtropical high
will be weak. Thus, we are able to conclude that the stationary
planetary waves responding to the forcing induced by the diabatic
heating over the Qinghai-Xizang Plateau can influence not only the
circulation over middle and high latitudes, but also the circulation
over North America.

IV. THE THERMAL EFFECT OF THE QINGHAI-XIZANG PLATEAU ON
    FORMATION AND MAINTENANCE OF THE MEAN MONSOON
    CIRCULATION IN SUMMER

In this subsection, we shall compute the effect of the actual
heat source on the formation and maintenance of the mean monsoon
circulation. In order to explain this problem, we shall compute the
stationary planetary waves responding to forcing by the actual topography. Fig. 8 shows the stationary disturbance pattern responding to
the forcing induced by the northern hemispheric topography at 12 km

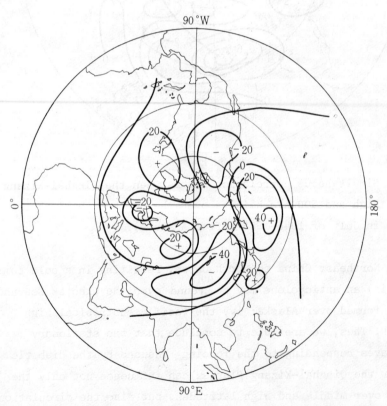

Fig. 8 The stationary disturbance patterns (m) responding to
        the forcing induced by the topography, at 12 km height
        level.

level by synthesizing wave components of k = 1–3. A remarkable feature
is that the disturbance is mainly confined to middle latitudes. The
disturbance pattern over Asia is a negative anomaly, while a major
positive anomaly is found over the Pacific Ocean. This disturbance
pattern differs appreciably from the mean summer monsoon circulation.

However, as mentioned in the introduction, according to the observed results, a strong heat source is found over the Qinghai-Xizang
plateau. Thus, we shall compute the stationary disturbance pattern
responding to the forcing induced by the actual topography and heat
sources at constant height. Fig. 9 shows the computed disturbance

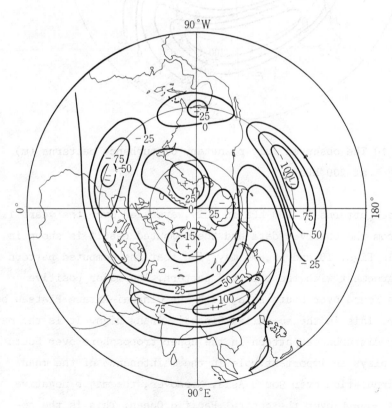

Fig. 9 As in Fig. 8 except for that forced by both the
topography and heat sources.

pattern at 12 km height. For comparison, we also calculated the

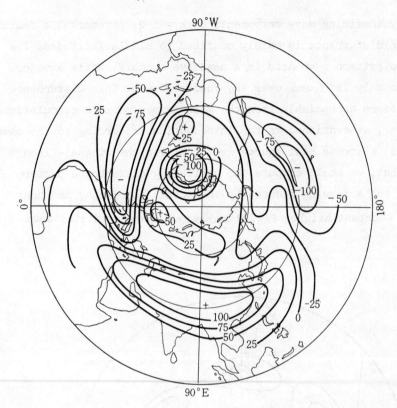

Fig. 10 The observed July planetary disturbance patterns (m)
at 200 hPa.

disturbance patterns at 200 hPa in July, averaged over the years 1972
to 1977 from the observed data and the computed result is shown in
Fig. 10. In Figs. 10 and 11, we may find that the computed pattern is
in good agreement with the observed pattern. The major positive
anomaly is found over South Asia, with the Qinghai-Xizang Plateau been
the center. This is the so-called Qinghai-Xizang high. It is the major
stationary disturbance pattern in the upper troposphere over South
Asia, and plays an important role in the maintenance of the mean
monsoon circulation over South Asia. Moreover, the major negative
anomaly is found over the central Pacific Ocean. This is the so-
called T.U.T.T.. Fig. 11 shows the computed disturbance pattern at
3 km height level. The major negative anomaly is found over South Asia
with the east of the Qinghai-Xizang Plateau being the center. The major

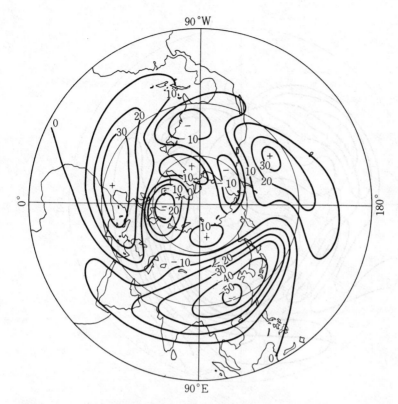

Fig. 11 As in Fig. 9 except for at 3 km height level.

positive anomaly is found at the center of the Pacific Ocean. This is the so-called Pacific subtropical high. This is the monsoon circulation in the lower troposphere over South Asia in summer.

Moreover, we have also computed the perturbance geostrophic wind speed from the computed disturbance pattern at 3 km height level. Fig. 12 shows the computed results. We find that the strong South-West wind belt passes through the Indian peninsula, Bay of Bengal and South China to South Japan. This strong South-West wind belt is found in the South-West monsoon region and plays an important role in the maintenance of the monsoon circulation.

V. CONCLUSIONS

In this paper, the three-dimensional propagations of the forced

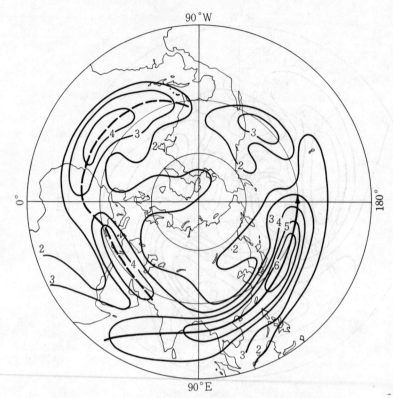

Fig. 12 The perturbance geostrophic wind speed (ms^{-1}), 3km
height level.

stationary planetary waves in a realistic summer current are discussed
using the refractive index squared of waves in a spherical coordinate
system. The theoretical analysis shows that the forced stationary
planetary waves can propagate from low latitude toward middle and
high latitudes in the troposphere and can propagate into the upper
subtropical troposphere.

In this paper, a steady-state, quasi-geostrophic, 34-level
spherical coordinate model with Rayleigh friction, Newtonian cooling
and the horizontal thermal diffusivity is used to investigate the
propagations of the stationary disturbance pattern responding to the
forcing induced by an idealized heat source over the Qinghai-Xizang
Plateau in summer. The computed results show that when the anticyclone
over the Qinghai-Xizang Plateau develops.

284

1) the anticyclone will cover South China and West Japan, entailing a hot summer in South China and West Japan.

2) a trough will be formed from Northeast China to North Japan, resulting in a cool summer in there.

3) an anticyclone may be found over the Okhotsk Sea and a trough is formed over Alaska.

4) the Pacific subtropical high will be weak. These results are in good agreement with the observed results.

The stationary disturbance patterns forced by the actual topography and heat sources are computed by using the above-mentioned model. The computed results show that the thermal effect of the Qinghai-Xizang Plateau may play an important role in the formation and maintenance of the stationary planetaryscale circulation, especially, in the formation and maintenance of the mean monsoon circulation over South Asia.

REFERENCES

[1] Ye Duzheng and. Gao Youxi et al., The Meteorology of the Qinghai-Xizang Plateau, Science Press, 278 pp, 1979 (in Chinese).
[2] Ashe, S., A nonlinear model of the time average axially asymmetric flow induced by topography and diabatic heating, J. Atmos. Sci., 36, 109–126, 1979.
[3] Nitta, T., Observational study of heat sources over the eastern Tibetan Plateau during the summer monsoon, J. Meteor. Soc., Japan, 61, 590–605, 1983.
[4] Luo, H., and M. Yanai, The large-scale circulation and heat sources over the Tibetan Plateau and surrounding areas during the early summer of 1979, Mon. Wea. Rev., 112 (to be publised), 1984.
[5] Huang Ronghui and K. Gambo, The response of a hemispheric multi-level model atmosphere to forcing by topography and stationary heat sources, Part I, J. Meteor. Soc. Japan, 60, 79–82, 1982.
[6] Huang Ronghui The characteristics of the forced stationary planetary wave propagations in summer in the northern hemisphere, Adv. Atmos. Sci., 1, No. 1, 1984.

[7] Huang Ronghui The wave-action conservation equation of stationary planetary waves and the wave guides of stationary planetary waves propagations shown by the wave action flux in a spherical atmosphere, Scientia Sinica, No. 4, 1984.

[8] Asakura, T., Dynamical climatology of atmospheric circulation over East Asia centered in Japan, Papers in Meteorology and Geophysics, 19, 1—68, 1968.

A MODEL INTERCOMPARISON FOR HOMOGENEOUS FLOW
OVER LARGE-SCALE OROGRAPHY

Joseph Egger

Meteorologisches Institute der Universität

München Munich, FRG

1. Introduction

The major mountain massifs on earth act as huge barriers to the
air flow. Correspondingly there have been intensive research efforts
to understand and simulate the flow over and around such obstacles.
For example, a bibliography of papers on orographic effects published
in 1980 (GARP) contains the titles of more than two hundred theoretical
papers on flow over mountains in rotating systems. The approaches
chosen differ widely. For example, there is a large group of papers
wherein the linear theory of mountain induced planetary waves is developed and extended. Within this group we find papers which rely on
the barotropic vorticity equation (e.g. Charney and Eliassen, 1949),
on the quasigeostrophic potential vorticity equation (Staff members,
1958) or on the primitive equations (e.g. Webster, 1972). There are
papers on the solution of the nonlinear barotropic voricity equation
(e.g. Stewart, 1948) and of the shallow water equations (e.g. Kasahara,
1966). The geostrophic momentum approximation has been invoked (Merkine and Kalnay-Rivas, 1976) and the full set of the primitive equations has been integrated as well (e.g. Huppert and Bryan, 1976).
Moreover general circulation models have been used to study the impact
of the large-scale mountain massifs on the atmosphere (e.g. Manabe
and Terpstra, 1974).

Given this enormous variety of approaches, it is somewhat surprising to see that only few papers have been devoted to an intercom-

parison of the various methods and an examination of their respective validity. Bannon (1980) integrated the viscid shallow water equations with a Newtonian forcing and compared the steadystate solution to those obtained for inertialess viscid flow, nonlinear inviscid quasi-geostrophic and shallow water flow. Vallis and Roads (1984) made a comparison of linear and nonlinear solutions for baroclinic channel flow. A rather weak resemblance of linear and nonlinear flow patterns has been found. Despite these efforts there remain many questions which have yet to be answered. For example, we do not know for what kind of flow and for what type of obstacles the linear theory breaks down. It is not clear which situations can be described reasonably well by the barotropic vorticity equation. In general we have to compare solutions which have been obtained with different methods. Verification poses also a difficult problem. In some cases results from laboratory experiments will be available for verification. Comparison of model results with observed flows in the atmosphere is notoriously difficult. A systematic approach to all these problems would be extremely ambitious. Here we restrict our attention to a rather simple flow situation. We consider homogeneous flow over and around steep orography on the f-plane. A numerical model is used to study this problem. The results provided by this model are accepted as reference flow patterns. Then we start to simplify the equations of the reference model. We consider first the shallow water flow for the same obstacle and compare the result to the reference flow patterns. Next we proceed to study the barotropic vorticity equation in a fully nonlinear form. In addition the linearized vorticity equation will be solved. Our choice of the basic equations has been motivated by the fact that the barotropic vorticity equation has been widely used in research on mountain flow problems. Our reference model, on the other hand, is based on the primitive equations and is three-dimensional. Therefore its basic equations are more general than the vorticity equation and the reference model includes a number of effects which are discarded in the barotropic vorticity equation. However, since stratification is excluded in the reference model the results obtained with this model should still be close enough to those obtained with the barotropic vorticity equation to make a comparison meaningful.

Since each case requires the running of four different models we cannot cover a wide range of mountain shapes and mean flow conditions. So we consider only few cases which bear a similarity to flow situations and obstacles found on the earth. The solutions of the shallow water equations and of the barotropic vorticity equation in nonlinear and/or linear form will be compared to those provided by the reference model for a mountain which resembles the Himalayas. This way we hope to find out about the validity of these equations for 'realistic' flow situations.

2. Basic equations

The flow situation we are going to examine is shown in Fig. 1. We consider a homogeneous inviscid fluid of depth $H = H_0 + \eta - h$ where H_0 is the mean depth, η the height deviation of the free surface and h the orographic profile. The basic equations are

$$\frac{\partial u}{\partial t} + u\frac{\partial u}{\partial x} + v\frac{\partial u}{\partial y} + w\frac{\partial u}{\partial z} - fv = -g\frac{\partial \eta}{\partial x}, \tag{2.1}$$

$$\frac{\partial v}{\partial t} + u\frac{\partial v}{\partial x} + v\frac{\partial v}{\partial y} + w\frac{\partial v}{\partial z} + fu = -g\frac{\partial \eta}{\partial y}, \tag{2.2}$$

$$\frac{\partial \eta}{\partial t} + u\frac{\partial \eta}{\partial x} + v\frac{\partial \eta}{\partial y} = w_t, \tag{2.3}$$

$$\frac{\partial u}{\partial x} + \frac{\partial v}{\partial y} + \frac{\partial w}{\partial z} = 0, \tag{2.4}$$

where (u,v,w) is the velocity and w_t is the vertical component of the velocity at the free surface. The Coriolis parameter is denoted by f and g is the acceleration due to gravity. As is well known the shallow water equations are derived from (2.1)—(2.4) by integrating over the depth of the fluid under the assumption that neither u nor v depend on z. With the vertical component of velocity at the bottom

$$w_b = u\frac{\partial \eta}{\partial x} + v\frac{\partial \eta}{\partial y}, \tag{2.5}$$

we have

$$\frac{\partial u}{\partial t} + u\frac{\partial u}{\partial x} + v\frac{\partial u}{\partial y} - fv = -g\frac{\partial \eta}{\partial x}, \qquad (2.6)$$

$$\frac{\partial v}{\partial t} + u\frac{\partial v}{\partial x} + v\frac{\partial v}{\partial y} + fu = -g\frac{\partial \eta}{\partial y}, \qquad (2.7)$$

and the equation of continuity

$$\frac{\partial H}{\partial t} + \frac{\partial}{\partial x}(uH) + \frac{\partial}{\partial y}(vH) = 0. \qquad (2.8)$$

From (2.6)—(2.8) we can proceed to the barotropic vorticity equation

$$\frac{\partial}{\partial t}(\nabla^2 - \lambda^2)\psi + J(\psi_1 \nabla^2_\psi + \frac{hf}{H}) = 0, \qquad (2.9)$$

where ψ is the stream function of the geostrophic wind

$$\psi = \eta g/f, \qquad (2.10)$$

and $\lambda^{-1} = (gH_0)^{1/2}/f$ is a radius of deformation. The derivation of (2.9) from (2.6)—(2.9) is presented, for example, in pedlosky (1979). Main assumptions are $h \ll H_0$ and the neglect of the ageostrophic advection of vorticity.

Finally we linearize (2.9) with respect to a zonal mean flow U_0 which does not depend on y. This yields

$$\frac{\partial}{\partial t}(\nabla^2 - \lambda^2)\psi' + u_0\frac{\partial}{\partial x}(\nabla^2\psi' + hf/H_0) = 0, \qquad (2.11)$$

where ψ' is the deviation of the stream function from that of the mean flow.

We have now four sets of equations at hand. The system (2.1)—(2.4) serves as a reference model for the mountain flow problem.

We have to compare the solutions obtained by aid of the shallow water model (2.6)—(2.8), and through (2.9) and (2.11) to that given by the reference model. That will give us some indications under what conditions these approximate equations are valid.

3. NUMERICAL MODELS

All models are designed for f-plane flow in a channel of length L and width D with solid walls as southern and northern boundaries.

We consider only cases with westerly mean flow so that the inflow at the western boundary must be prescribed. The eastern outflow boundary is open and we prescribe zero-gradient conditions there.

a) Homogeneous three-dimensional flow

We integrate (2.1)—(2.4) numerically using the C-grid (e.g., Mesinger and Arakawa, 1976). In this grid the velocity components u, v, w are located such that the fluxes through a mesh cube can be controlled rather easily (Fig. 1). The height is carried at the center of the mesh squares. In most runs we have three levels in the vertical. As horizontal grid increment we choose Dx = 300 km. It is straight-forward to write down the finite difference form of (2.1)—(2.4) and the resulting formulae will not be given. The upper boundary condition (2.3) is implemented by integrating the horizontal divergence over all levels of the model so that w_t is the vertical velocity at $z = H_0$ and not at $z = H_0 + \eta$. Moreover we approximate (2.3) by

$$\frac{\partial \eta}{\partial t} + u_t \frac{\partial \eta}{\partial x} + v_t \frac{\partial \eta}{\partial y} = w_t, \qquad (3.1)$$

where u_t, v_t are the velocity components at the uppermost computational level of the model and not at $z = H_0 + \eta$ (see Fig. 1). The leap-frog scheme is used in time.

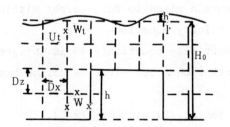

Fig. 1. Homogeneous fluid with a free surface and mountain
block with vertical walls. Grid of the reference
model (dashed) and location of u and w in the grid.
See text for further details.

We introduce steep orography using a blocking technique which has been proposed by Egger (1972) and refined, for example, by Mesinger (1982). Mountains are treated as obstacles with vertical walls which block all flow perpendicular to the wall. It is best to place such a wall in the grid such that the boundaries of the mountain coincide with the boundaries of mesh cubes. For example we have just to prescribe u = ∂u/∂t = 0 at the eastern and western wall of the mountain in Fig. 1 and w = ∂w/∂t = 0 on top of the block. Inside the mountain, there is no flow, of course.

We prescribe westerly flow $u = U_0$ at the inflow boundary, where we require v = 0. We have v = 0 at the walls to the north and south and $\partial u/\partial x = \dfrac{\partial v}{\partial x} = \partial \eta/\partial x = 0$ at the eastern outflow boundary. The initialization of the flow is done in a most straightforward manner. We prescribe $u = U_0$, v = 0 at all grid points of the flow domain. The height of the free surface is in geostrophic balance with the zonal flow initially.

b) Shallow water equations

Of course, little effort is required to reduce a model of homogeneous three-dimensional flow to a shallow water flow model. We can retain the grid structure in the horizontal and we just have to reduce the number of grid levels so as to have only one. On the other hand, we can no longer represent the mountain as a block. We prescribe the orographic profile h(x,y) used in the homogeneous flow model in (2.8). In particular the mountain rises to full height within grid distance.

c) Barotropic vorticity equation

The barotropic vorticity equation(2.9) is integrated in time using a pseudo-spectral technique. First we define basic spatial modes

$$F_{1n}^{m} = \cos(n\pi y/D)\,\sin((m+\tfrac{1}{2})\pi x/L), \qquad (3.2)$$

$$F_{2n}^{m} = \sin(n\pi y/D)\cos((m+\tfrac{1}{2})\pi x/L), \qquad (3.3)$$

$$F_{3n}^{m} = \sin(n\pi y/D)\sin((m+\tfrac{1}{2})\pi x/L), \qquad (3.4)$$

with mode indices n = 1,......N, m = 0,......M. The southern boundary of the channel is at y = 0. To allow for a mean flow we add a mode

$$F_0 = -y/L. \tag{3.5}$$

The stream function ψ is projected onto the third basic mode and on F_0:

$$\psi = \underset{m,n}{\Sigma} \, \psi_n^m F_{3n}^m + u_0 F_0 L, \tag{3.6}$$

with expansion coefficients ψ_n^m. This choice ensures that v vanishes at the walls of the domain. Moreover, we have $\psi = -U_0 y$ at the inflow boundary $x = 0$ and $\partial\psi/\partial x = 0$ at the outflow boundary $x = L$. The meridional velocity is projected onto the second mode

$$V = \underset{m,n}{\Sigma} \, v_n^m \, F_{2n}^m, \tag{3.7}$$

with $v_n^m = (m+\frac{1}{2})\pi\psi_n^m/L$. Of course, the meridional flow component does not necessarily vanish at the inflow boundary. Moreover we have

$$u = \underset{m,n}{\Sigma} \, u_n^m F_{1n}^m + u_0, \tag{3.8}$$

with $u_n^m = -n\pi\psi_n^m/D$. The zonal velocity u does not vanish at the boundary walls. The vorticity $\nabla^2\psi$ is projected onto the modes F_{3n}^m as is the orography. The spectral coefficients h_n^m of the orography are determined by carrying out a Fourier transform of the mountain profile prescribed in the reference model.

The Jacobian in (2.9) is evaluated as follows. Given ψ according to (3.6) at a time we compute v,u and the gradients of vorticity in the spectral domain according to (3.7), (3.8) and corresponding formulae for the gradient of vorticity. However, the multiplications like $u\frac{\partial}{\partial x} \nabla^2\psi$ as required to determine the Jacobian are carried out in a two-dimensional grid. To do this we define a grid which covers the channel and compute the values of $u\frac{\partial}{\partial x} \nabla^2\psi$ etc. at these grid points using the spectral representation of all these fields. The multiplications and all the other algebraic manipulations necessary are carried out at these grid points so that we have then the value of the Jacobian at these points. The Jacobian is next transformed back to the spectral domain using a Fourier transformation which is based on the modes F_{3n}^m. This way we obtain a spectral representation of the tendency $\frac{\partial}{\partial t}(\nabla^2-\lambda^2)$. The inversion of the Laplacian is an easy matter since the basic functions (3.2)—(3.4) are eigenfunctions of the Laplacian. So we arrive easily at a spectral representation of the tendency $\frac{\partial}{\partial t}\psi$ of the stream function which is based on the modes $F_{3,n}^m$, of course. Given $\frac{\partial\psi}{\partial t}$ we can step the model forward in time using a leap frog scheme. The mesh size of the computational grid is the same as that of the reference model.

d) Linear vorticity equation

It is, of course, no problem to integrate (2.11) in time when the nonlinear flow model is available. We just have to discard all nonlinear interaction terms. The integration could be done analytically. However, we prefer to use the code of the nonlinear model for reasons of consistency.

4. RESULTS

We choose an obstacle which is vaguely reminiscent of the Himalayas. A cubic elongated mountain of length B = 2700 km, width 2 W = 1500 km and height h_0= 2660 m is placed at the channel's axis (Fig. 2) The results obtained with the homogeneous flow model are shown first. This model is run with three levels in the vertical and a vertical spacing which equals the mountain height so that the mountain blocks the lowest layer of the fluid. We have weak westerly inflow U_0 = 4 ms^{-1}. With that choice of the parameter we have a Rossby number R = U_0/fW ~ 0.1 so that we expect to find quasigeostrophic flow. It is not immediately obvious which times of the flow evolution to choose for the intercomparison. Most often models are integrated in time till a quasiinsteady state is attained. On the other hand, we cannot expect to reach a truely steady state in inviscid flow. Moreover we hardly ever observe anything like a steady state in the atmosphere where the flow impinging at the mountains changes in time. We decided to terminate most runs after two days. As will be shown later on the important features of the flow appear to be established by then. Moreover frictional effects can be expected to be unimportant after two days.

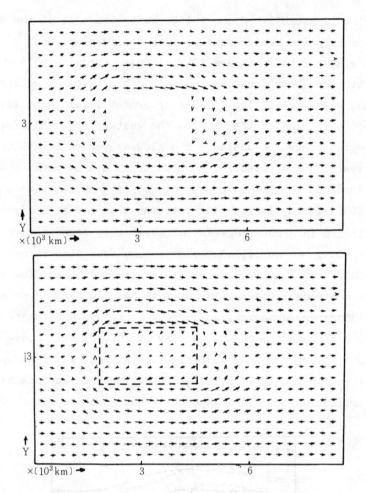

Fig. 2. Horizontal velocity after two days. Reference model
with three levels; $U_0=4ms^{-1}$. $H_0=8000m$, $h_0=2660m$.
The length of the arrows at the inflow boundary
corresponds to a speed of $4ms^{-1}$. a) level z=1330m;
b) level z=6670m; contour of the mountain dashed;
Dx=300 km; Dz=h_0.

In Fig. 2 we show the flow at the level $z = h_0/2$ and at the uppermost level $z = H_0-h_0/2$ after two days starting from westerly flow.
Below the crest height of the mountain the flow splits into two branches
in front of the obstacle. These branches have about equal intensity. In
the lee there is a kind of wake with weak winds but almost undisturbed

westerly flow is established about 1500 km to the east of the mountain.
This pattern is quite reminiscent of flow observations near the Himalayas in winter (Staff members 1957, 1958; Murakami, 1981) where
this splitting has been found up to the 500 hPa level as well as the
zone of weak winds in the lee. At the uppermost level there is still
a pronounced splitting of the flow to the west of the mountain. In the
interior region above the obstacle the flow is reflected upstream
before transversing the obstacle. Such patterns are commonly observed
in nearly inviscid low Rossby number flow. For example, Hide and
Ibbetson (1966) found similar flow patterns when towing a right cylinder
through a liquid in a rotating tank although they found also regions
of stagnant flow, the so called Taylor column (see also Baines and
Davies, 1980). Vaziri and Boyer, (1971) obtained similar flow patterns
by means of numerical integration. We must bear in mind, however, that
all these laboratory observations correspond to a steady state whereas
we consider a certain moment in the flow evolution. Well above the
Himalayas one observes strong westerly flow without any flow splitting
in contrast to our result (Murakami, 1981). The free surface has
height deviations as one would expect for quasigeostrophic flow. In

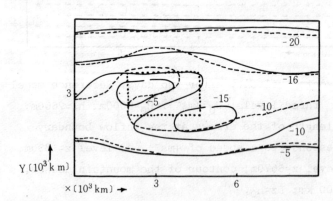

Fig. 3. Stream function $\psi(10^6 m^2 s^{-1}$; bold) as obtained from
the nonlinear barotropic vorticity equation and
geostrophic free surface stream function (dashed)
as obtained from the reference model after two days.
$U_0 = 4ms^{-1}$; $h_0 = 2660m$; $H_0 = 8000m$; contours of the mountain
dottes. M=14, N=14.

Fig. 3 we show the geostrophic stream function gη/f corresponding to this height field. There is a center of anticyclonic circulation to the northwest of the mountain and a 'trough' to the southeast. The perturbation stream function ψ' is mainly antisymmetric with respect to the channel's axis. A comparison with the results provided by the barotropic vorticity equation requires the evaluation of vertical averages. After all, the stream function ψ in (2.9) represents the

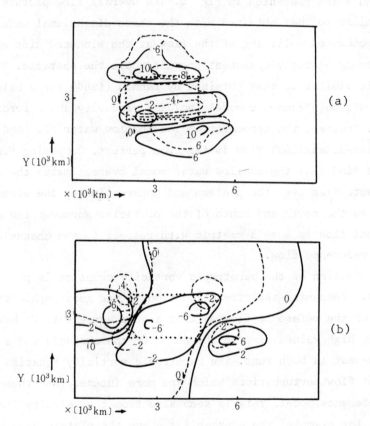

Fig. 4. Velocity components (ms^{-1}) as obtained from the non-linear barotropic vorticity equation (bold) and from the reference model (dashed) after two days. U_0=4ms^{-1}; h_0=2660 m; H_0=8000m; contours of the mountain dotted only the most prominent features are given. a) zonal wind u. b) meridional wind v. The reference velocities are vertical averages.

vertically averaged flow. In Fig. 4 we present u,v as averaged over the depth of the fluid. The zonal flow component has negative values above the mountain and intensified flow to the north and south of the obstacle. The northern branch is slightly more vigorous. The vertically averaged meridional flow is almost antisymmetric with respect to the channel's axis. Above the obstacle the mean flow is northerly.

Let us turn now to the solution of the shallow water equations. Both u and v are presented in Fig. 5. The overall flow picture is quite similar to that obtained with the three-dimensional model. We have a pronounced splitting of the flow at the windward side and winds with a strong easterly component prevail above the obstacle. This pattern is similar to that obtained by Bannon (1980) for a bell-shaped mountain although Bannon's model includes viscosity and a forcing of the flow. However, the agreement of the shallow water flow and the fully three-dimensional flow is far from perfect. Comparing Fig. 4 and Fig. 5 we find that the shallow water model overestimates the intensity of the easterlies over the plateau and underestimates the strength of the jets to the north and south of the obstacle. Moreover the shallow water zonal flow is more symmetric with respect to the channel's axis than the reference flow.

The solution of the barotropic vorticity equation is presented in Fig. 3. Comparing the stream function to the geostrophic stream function of the reference flow we find a good agreement. We have relatively high values of ψ to the north of the obstacle and a low to the south-east in both runs. the barotropic vorticity equation appears to predict flow perturbations which are more intense than those given by the reference model. This is seen also from the velocity fields (Fig. 4). For example, the northerlies above the plateau have now maximum values of 6 ms^{-1} whereas the reference model gives rather weak maximum intensities of 2 ms^{-1}. The barotropic vorticity equation produces rather strong jets at the northern and southern rim of the mountain. By and large, however, the barotropic vorticity equation is capable of handling the situation. This comes as a positive surprise After all one would have thought that ageostrophic motions are important near a mountain with so extremely steep slopes. Moreover, the

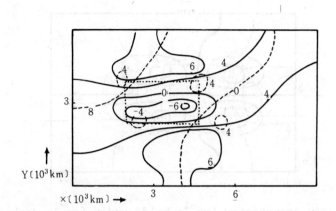

Fig. 5. Velocity components u(ms^{-1}; bold) and v (ms^{-1}; dashed)
as obtained from the shallow water equations after two
days. U_0=4ms^{-1}; h_0=2660 m; H_0=8000 m; contours of
mountain dotted; Dx=300 km.

barotropic vorticity equation (2.9) is derived under the condition
h_0<<H, a condition which is not satisfied in this case.

The linear solution is presented in Fig. 6. We have a v-field
which is exactly symmetric with respect to the channel's axis and so
is, of course, the stream function ψ' (not shown). On the windward
side of the mountain there is a ridge with anticyclonic circulation
and we have a trough in the lee. Since ψ' is symmetric, the zonal
velocity deviation $u-u_0$ must be antisymmetric. Of course there are
no jets at the northern and southern rim of the obstacle and the
strongest zonal winds are liked to the ridge and trough to the west and
to the east of the mountain , respectively. It is easy to understand
why ψ has to be symmetric. If the orographic profile h in (2.11) is
symmetric with respect to the channel's axis then $\frac{\partial}{\partial t}(\nabla^2-\lambda^2)\psi'$ is
symmetric too and therefore, ψ' is symmetric at all times if no
asymmetry is introduced at the boundaries. In the nonlinear run,
however, the Jacobian J (ψ,fh/H) creates asymmetry even if ψ is symmetric at t = 0.

It is rather dissappointing to see that the linear solution is
so far off the mark. In Fig. 3 we have a ridge to the northwest and a

299

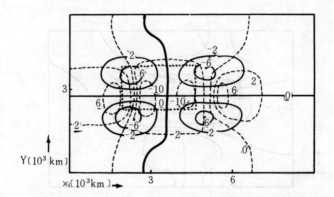

Fig. 6. Velocity components u (ms^{-1}; bold) and v (ms^{-1}; dashed)
as obtained from the linear barotropic vorticity equa-
tion after two days. U_0=4ms^{-1}; h_0=2660m; H_0=8000 m;
contours of mountain dotted.

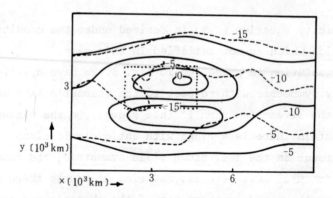

Fig. 7. The same as Fig. 3 but for h_0=4000m.

trough to the south-east of the obstacle whereas the linear theory
places a ridge to the west and a trough to the east of the mountain.
So there is not even a qualitative agreement of the linear and the
nonlinear solution of the barotropic vorticity equation.

It would be somewhat premature to draw even tentative conclusions
on the basis of one case only. Instead we have to perform such inter-
comparisons at least for some sections of the parameter space. Although
we do not want to change the horizontal dimensions of the obstacle
numerical experiments with a larger h_0=4000 m will be discussed. After

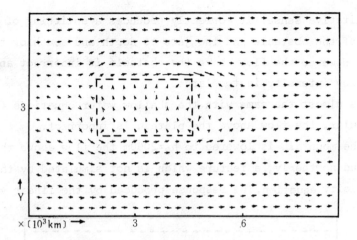

Fig. 8. Vertical average of the horizontal velocity after
two days. Reference model with four levels. $U_0=$
$4ms^{-1}$; $H_0=800$ m; $h_0=4000m$. The length of the
arrows at the inflow boundary corresponds to
$4ms^{-1}$. Contour of the mountain dashed. Dx=300 km;
Dz=2000 m.

all, the Tibetan plateau has a mean height of about 4000 m. Our choice
of a rather weak mean flow speed $U_0=4ms^{-1}$ has been motivated by the
fact that the mean surface winds at the Tibetan Plateau hardly exceed
5 ms^{-1} (Murakami, 1981). On the other hand, surface winds may be much
stronger under favorable synoptic conditions. So experiments with
$U_0=8$ ms^{-1} and $U_0=12ms^{-1}$ have been carried out.

In Fig. 7 we show the geostrophic stream function as obtained
from the reference model (dashed) and from the barotropic vorticity
equation for $U_0=4ms^{-1}$ and $h_0=4000m$. The reference model has been run
in a four-level version where the lowest two layers are blocked by
the mountain. It is surprising to see that the response of the reference
model appears to be weaker than for $h_0=2700$ m (Fig. 3). On the other
hand the barotropic vorticity equation predict an strong response
which is almost exactly antisymmetric with respect to the channel's
axis. In particular we have to expect strong easterlies above the
plateau whereas the reference model predicts weak southerly flow there
(Fig. 8). So the barotropic vorticity equation is correct only in a

most qualitative sense. It correctly produces a splitting of the flow
in front of the obstacle and two jets to north and south of the
obstacle. However the flow above the obstacle is incorrect and the
overall flow intensity is too large.

It is almost the same with the shallow water equations (Fig. 9).
In particular, the flow above the plateau is incorrect.

If the mountain is lowered to $h_0 = 2000m$ the flow above the plateau
is weak and northerly, a feature which is not simulated by the barotropic vorticity equation. The response produced by the linear vorticity

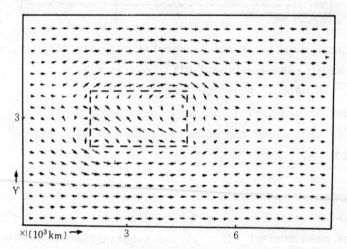

Fig. 9 Horiontal velocity after two days as obtained from the
shallow water equations. $U_0 = 4ms^{-1}$; $H_0 = 8000$ m;
$h_0 = 4000$ m. The tength of the arrows at the inflow
corresponds fo a speed of $4ms^{-1}$; $p_x = 300$ km.

equation is proportional to the mountain height and, therefore, essentially the same as shown in Fig. 6. Of course, this response is
unsatisfactory.

An increase of the zonal inflow to $U_0 = 12ms^{-1}$ does not change the
basic characteristics of the response. Fig. 10 shows the vertically
averaged winds as obtained from the reference model. We have again a
splitting of the flow on the windward side of the obstacle. There is
now a pronounced northerly flow component in the lee of the mountain
whereas an almost calm wake has been observed for $U_0 = 4ms^{-1}$. Above the

302

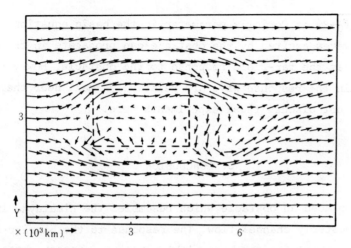

Fig. 10 Verical average of the horizontal velocity after two
days. $U_0 = 12$ ms^{-1}; $H_0 = 8000$ m; $h_0 = 2660$ m. contour
of the mountain dashed. The length of the arrows at
the inflow boundary corresponds to speed of 12 ms^{-1}.
Reference model; $D_x = 300$ km; $D_z = 2660$ m.

obstacle we have northeasterly flow just as in Fig. 2b. The free-
surface stream function shows the ridge-trough structure familiar from
Fig. 3. The shallow water equations appear to be capable of describing
the response quite well. The nonlinear barotropic vorticity equation
produces a response which clearly overestimates the flow intensities.
However, the positions of the ridge and trough are predicted correctly.
The linear vorticity equation fails also in this case since the
response is similar to that shown in Fig. 6.

A few additional experiments have been made to find out about
the long-term behaviour of the flows. The experiments for the case
$U_0=4$ms^{-1}, $h_0=2700$ m have been extended to six days. It turned out
that the basic flow characteristics remained unaltered, i.e. there
is still a splitting of the flow in front of the obstacle and a wake
with weak winds. However, the flow over the obstacle is much weaker
after six days and almost from the south. The barotropic vorticity
equation produces also little change during these 4 more days. However,
the flow above the obstacle remains strong and easterly in contradiction to the reference flow. Furthermore, β-plane experiments have been

303

made with $f=f_0+\beta\gamma$ as Coriolis parameter in the basic equation. With $\beta\neq0$, Rossby waves excited by the obstacle can propagate in the channel. This influences the far-field response to the obstacle. However, the differences in the response wave are relatively small near the obstacle and we shall not discuss these experiments here.

5. CONCLUSION

We have placed an obstacle which resembles the Himalayas into homogeneous westerly channel flow. The response to the barrier after two days was essentially the same within the range $2000\leq h_0\leq4000$m, $4\leq U_0\leq12$ ms^{-1}. The flow is splitting into two branches in front of the obstacle at all levels, the northern branch being more intense. The flow above the obstacle is generally weaker than the flow around the mountain. Its direction depends on the circumstances. Laboratory experiments tell us that the flow obtained with the reference model is essentially realistic although the aspect ratio of the obstacles used in these experiments is much larger than that of our mountain block. Moreover, there is no inviscid fluid in the laboratory, so that both cases are not strictly comparable.

Both the shallow water equations and the nonlinear barotropic vorticity equation are correct in simulating the flow splitting. Both equations tend to give too large flow velocities and they can hardly cope with the situation above the plateau. In particular, the batrotropi vorticity equation produces quite strong easterlies above the obstacle which are unrealistic. The solution of the linear vorticity equation is clearly unsatisfactory. According to this equation the perturbation stream function is symmetric with respect to the channel's axis. There is a ridge in front of the mountain and a trough in the lee. There is no flow splitting at all.

This failure of the linear theory is somewhat surprising. After all it has been claimed quite often and evidence has been presented that the linear barotropic vorticity equation can describe a surprisingly large part of the standing wave patterns observed in the atmosphere(e.g. Charney and Eliassen, 19; Held, 1983). Moreover, Egger (1977) and Ashe (1979) studied the impact of nonlinear terms in a planetary standing wave model. They found that the impact of orographically induced nonlinearities is minor. In addition, it is demonstrated rather easily that the solution ψ' of the steady state version of (2.11)

$$\nabla^2 \psi' + hf/H_0 = 0 \qquad (5.1)$$

is also a solution of the steady-state nonlinear barotropic vorticity equation. Obviously, the solution of (5.1) satisfies also the equation

$$J(\psi_1 \nabla^2 \psi + fh/H_0) = 0 \qquad (5.2)$$

Of course, the solution of (5.2) is also symmetric with respect to the channel's axis. With all that in mind one may wonder why the linear theory fails completely in our case. The answer appears to have been given by Johnson (1978; see also Huppert and Bryan, 1976). As the experiments are started, the vertical filaments of fluid initially above the obstacle move off downstream. Vortex lines are stretched and a cyclonic vortex is formed in the lee. Simultanously, compression of filaments moving over the obstacle means that an anticyclonic vortex is formed above the obstacle. Below a critical mean flow velocity, the interaction of this vortices is so strong that the cyclonic vortex is not washed away and a steady-state may result where the anticyclonic vortex stays above the obstacle and the cyclonic vortex is found to the south much the same way as in Fig. 3. Now the nonlinear solution (5.2), (5.1) appears to correspond to the situation where the cyclonic vortex has been washed away whereas the cyclonic vortex stays near the obstacles in all our experiments. According to Johnson (1978) the cyclonic vortex will stay near the obstacle if $HR_0/h_0 < 0.25$. This criterion is satisfied in all the runs considered here. The work of Egger (1977) and Ashe (1979) deals with an expansion of a linear solution like (5.1). It is not surprising, therefore, that they found only minor effects.

It is rather difficult to draw any conclusions with respect to the

atmospheric flow near the Himalayas. After all, the atmosphere is not
an homogeneous and inviscid fluid. The comparison with the observed
flow in winter (Murakami, 1981) showed that the reference model is
qualitatively correct at levels below the plateau height. However the
vertically integrated flow as shown in Fig. 3, is unrealistic, and so
is, of course, the response computed by aid of the shallow water equations and by the nonlinear vorticity equation. It can, therefore, not
be recommended to use the barotropic vorticity equation for a simulation of the vertically averaged flow near the Himalayas. While the
nonlinear barotropic vorticity equation captures at least the basic
characteristics of the flow at low elevations, the linear version of
this equation is completely useless. This has been anticipated by Murakami (1981) who demonstrated that the surface winds near the Tibetan
Plateau were not predominantly zonal so that the forcing term in (2.11)
is incorrect. Nevertheless this conclusion appears to be somewhat in
contrast to studies where the usefulness of the barotropic linear
barotropic vorticity equation has been demonstrated for the simulation
of planetary scale standing waves. However, it has to be borne in mind
that these studies (e.g. Held, 1983) concentrate on the far-field response to topographic forcing, i.e., one looks at the planetary propagation of waves generated at the mountains. In our paper we were interested in the flow close to the obstacle. Moreover, these linear
theories rely on the assumption of steady state whereas the flow evolution in our case is clearly influenced by the initial state. Nevertheless it remains somewhat surprising that an equation like the linear
barotropic vorticity equation which performed so poorly in our case
should be satisfactory when we look at the mountain problem on a global
scale.

REFERENCES

Ashe, S., 1977: A nonlinear model of the time-averaged axially asym-
    metric flow induced by topography and diabatic forcing. J. Atm.
      Sc. 36, 109—126.
Baines, P.G. and P. Davies, 1980: Laboratory studies of topographic

effects in rotating and/or stratified fluids. WMO/ICSU; GARP publ. Ser. No. 23; Orographic effects in planetary flows. 233–300.

Bannon, P., 1980: Rotating barotropic flow over finite isolated topography. J. Fl. Mech. 101, 281–306.

Charney, J. and Eliassen, A., 1949: A numerical method for predicting the perturbation of the midlatitude westerlies. Tellus, 1, 38–54.

Egger, J., 1972: Incorporation of steep mountains into numerical forecasting models. Tellus, 24, 324–335.

Egger, J., 1977: Nonlinear aspects of the theory of standing planetary waves. Contr. Atm. Phys. 49, 71–80.

GARP, 1980: Orographic effects in planetary flows. WMO/ICSU; GARP Publ. Ser. 23.

Held, I., 1983: Stationary and quasi-stationary eddies in the extra-tropical atmosphere: theory. In: Large-scale dynamical processes in the atmosphere. Hoskins and Pearce, eds. Academic Press, 127–168.

Huppert, H. and K. Bryan, 1976: Topographically generated eddies. Deep Sea Res., 23, 655–679.

Johnson, E.R., 1978: Trapped vortices in rotating flow. J. Fl. Mech. 86, 209–224.

Kasahara, A., 1966: The dynamical influence of orography on the large-scale motion of the atmosphere. J. Atm. Sc. 23, 259–271.

Manabe, S. and T.B. Terpstra, 1974: The effects of mountains on the general circulation of the atmosphere as identified by numerical experiments. J. Atm. Sc. 31, 3–42.

Merkine, L.O. and E. Kalnay-Rivas, 1976: Rotating stratified flow over finite isolated topography. J. Atm. Sc. 33, 908–922.

Mesinger, F. and A. Arakawa, 1976: Numerical methods used in atmos-pheric models. WMO/ICSU, GARP Publ. Ser. 17, 1–64.

Mesinger, F., 1982: Representation of mountains in numerical models; sigma system vs. the blocking method. ITAM'82, Deutscher Wetterdienst, 20–22.

Murakami, T., 1981: Orographic influence of the Tibetan Plateau on the Asiatic winter monsoon circulation. Part. I. Largescale aspects.

J. Met. Soc. Japan, 59, 40—65.

Pedlosky, J., 1979: Geophysical fluid dynamics. Springer; New-York-Heidelberg-Berlin; 86—93.

Staff members of academia sinica, 1957: On the general circulation over Eastern Asia (I). Tellus, 9. 432—446.

Staff members of academia sinica, 1958: On the general circulation over Eastern Asia (II), (III), Tellus 10,58—75, 299—312.

Stewart, H.J., 1948: A theory of the effects of obstacles on the waves in the westerkies. J. Met. 5, 236—243.

Vallis, G. and J. Roads, 1984: Resonant and turbulent effects in numerical models of flow over topography. Subm. to J. Atm. Sc.

Vaziri, A. and D. Boyer, 1971: Rotating flow over shallow topographies. J. Fl. Mech. 50, 79—95.

Webster, P. 1972: Response of tropical atmosphere to local, steady forcing. Mon. Weath. Rev. 100, 518—541.

TERRAIN AERODYNAMICS AND PLUME DISPERSION

A Perspective View Gained from Fluid Modeling Studies

William H. Snyder[1]

Meteorology and Assessment Division
Environmental Sciences Research Laboratory
Environmental Protection Agency

[1]On assignment from the National Oceanic and Atmospheric Administration, U.S. Department of Commerce.

309

1. INTRODUCTION

The ability to predict ground-level concentrations of air pollutants released from sources in or near complex terrain is required in order to determine the environmental impact from existing sources, to evaluate alternative new source locations and designs, and to estimate the effects of possible modifications to existing sources. Mathematical models that reliably predict concentrations when plumes are affected by complex terrain are not yet available. Field studies are very expensive and time consuming, and the results are not generally transferable to other sites. Previous wind-tunnel studies on dispersion of effluents from industrial plants located in complex terrain have usually been designed to answer specific questions such as the suitability of a particular plant location or the stack height necessary to avoid downwash; the results have not been generally applicable to other sites.

In the early wind-tunnel studies, care was taken to ensure a uniform wind profile and low turbulence intensity in the flow approaching the model. More recent studies have shown that the effects of shear and turbulence intensity (and scales) in the approach flow can strongly affect the basic flow structure over a model, e.g., the turbulence can move the locations of separation and reattachment points on the lee sides of hills. Hence, in the past 15 years, serious attempt have been made to simulate the atmospheric boundary layer, including stratification, and to conduct "generic studies", wherein idealized terrain features have been used to obtain basic physical understanding. This is contrasted with "engineering case studies", wherein very specific questions are answered with regard to particular installations.

This paper attempts to summarize the results of recent neutral (wind-tunnel) and stably-stratified (towing-tank) studies that were designed to obtain basic physical understanding of flow and diffusion in complex terrain.

These results provide guidance on locating sources in complex terrain as well as "rules of thumb" for estimating concentrations when a source is located in complex terrain. Most of the results and conclusions reported herein are based on studies conducted in the Meteorological Wind Tunnel (Snyder, 1979) and the Towing Tank (Thompson and Snyder, 1976) of the EPA. Each of these studies followed the guidelines of Snyder (1981), but the details may have differed considerably from one study to the next. Direct inter-comparisons of the results of these studies are made to obtain a perspective view, but the reader is cautioned to study the original reports in detail before attempting to apply the conclusions to the real atmosphere.

2. NEUTRAL-FLOW WIND-TUNNEL STUDIES

The effects of terrain on the flow can be demonstrated through the deflection of streamlines over and around ridges and hills. For example, the displacement of the mean streamlines determines how near to the surface the centerline of the plume will reach. The convergence and divergence of the streamlines in the directions normal to and, in the case of three-dimen-sional flows, parallel to the surface affect the plume width (see Hunt, Puttock and Snyder, 1979). If flow separation occurs, the size and shape of the recirculating cavity that ensues and the position of the source with respect to this cavity can be very important in determining subsequent plume behavior.

A simple way to evaluate the effects of terrain on concentration is to calculate a terrain amplification factor A which is defined as the ratio of the maximum surface concentration occurring in the presence of the ter-rain χ_{mx} to the maximum that would occur from a similar source in flat ter-rain χ_{mx}° (see Figure 1). This definition is useful only for elevated sources, of course, because for ground-level sources, the maximum surface concentration occurs at the source itself.

Using these ideas, the results of studies using two- and three-dimensional terrain features will be described.

2.1 Two-Dimensional Hills

Numerous studies have been conducted on two-dimensional terrain features: (a) a ramp with a slope of 14° followed by an elevated plateau (unpublished), (b) a bell-shaped hill with a maximum slope of 12° by Courtney and Arya (1980), (c) a steep triangular ridge with a slope of 68° (Arya and Shipman, 1981; Arya et al, 1981), but perhaps the most illustrative is (d) a series of smooth-shaped hills by Khurshudyan et al (1981).

In the latter study, three hill shapes were generated from a set of parametric equations. The aspect ratios (streamwise half-length/height) of the hills were 3, 5 and 8; maximum slopes were 26°, 16° and 10°, respectively. These hills are referred to by their aspect ratios, e.g., hill 5.

Streamlines derived from mean velocity measurements with hot-wire and pulsed-wire anemometers are shown in Figures 2, 3 and 4. No flow separation was observed on the lee sides of hills 8 or 5; however, the pulsed-wire measurements showed that, whereas the mean flow on the lee side of hill 5 was everywhere downstream, instantaneous flow reversals occurred frequently. Indeed, at one position, reverse-flow was observed up to 40% of the time. A large cavity clearly formed on the lee side of hill 3.

From a study of the streamline patterns of Figures 2 through 4 we may anticipate qualitative effects of the terrain on sources placed at various positions with respect to the three hills. For example, consider sources located at the upwind bases of the three hills. Because of the stronger convergence of streamlines (plume centerline closer to surface) upwind of hill 3 (as compared with hill 8) we may expect larger terrain amplification factors (A) there. Because of the divergence of the streamlines downwind of the hill centers, we may expect A's less than 1.0, and smaller values

312

for the steeper sloped hills. When sources are located at the downwind bases of the hills, we may expect quite different behavior because of the strong differences in streamline patterns. For hill 8 (10° slope), the streamline pattern appears nearly symmetrical fore and aft of the hill center, so that we might at first expect similar terrain amplification factors as found for upwind sources. However, because of the much larger turbulence intensities in the lee of the hill, considerably larger A's were observed there (to be shown later). For low sources located at the downwind base of hill 3 (26° slope), Figure 4 shows that it would be located within the recirculation cavity, and we may, in fact, expect to find the maximum surface concentration upwind of the source. For higher sources (i.e., for sources located on the separation-reattachment streamline), we may expect the plumes to impinge directly on the surface downwind, with very large terrain amplification factors resulting. Because of the strong convergence downwind of the base of hill 5 (16° slope), we may expect large A's for sources located there. These statements, however, are speculative, because we have not accounted for changes in turbulence intensity as effected by the terrain. As we shall see shortly, the statements are generally true, but a few surprises are in store.

Ground-level concentrations (glc's) were measured downwind of sources of various heights located at the upwind bases, the tops, and the downwind bases of the hills. Three sets of glc patterns are compared with those measured over flat terrain in Figures 5, 6 and 7. The source height H_s is half the hill height h in each of these cases, x_s/h is the distance downwind from the source, and χ is the nondimensional concentration, $\chi = CUh^2/Q$, where C is the concentration, U is the free-stream wind speed, and Q is the source flow rate. In Figure 5, the sources were located at the upwind bases of the hills, and the maximum glc's occurred at the tops of the hills in all cases. A's increased from 1.1 to 2.5 as the slope of the hill increased from

10° to 26°. The influence of the recirculating cavity for hill 3 is clearly evidenced by a sharp drop in concentration at the separation point and a secondary peak at the reattachment point.

Figure 6 shows that the amplification factors were less than unity for sources located at the hill tops. The effect of the recirculation cavity is again clearly evidenced for hill 3; effluent entered the cavity near the reattachment point and was transported up the lee slope to the separation point.

Large values of A were observed for all three hills when the sources were at the downwind base locations (Figure 7). For hill 3, the maximum glc was located upwind of the source because of the reverse flow within the cavity. A somewhat surprising result, however, is that the largest A occurred for hill 5, where the mean flow did not separate. As mentioned earlier, due to the intermittent separation, the mean velocity was extremely small at the downwind base (exemplified by the large spacing of streamlines near the downwind base, shown in Figure 3), and, therefore, the local turbulence intensities were extremely large.

Table 1 summarizes the observed terrain amplification factors for the various stacks at the three locations. A maximum A of 15 occurred with a stack of height one-fourth the hill height and located at the downwind base of hill 5. This was due to the very small mean transport but very rapid turbulent transport at that location. A's nearly as large occurred when the source was located near the separation-reattachment streamline, because in this case, the plume was advected directly toward the surface (reattachment point). Upwind sources resulted in terrain amplification factors in the range of 1.1 to 3, with the larger values being observed for the steeper hills. Finally, the hill top source location resulted in amplification factors less than unity, with smaller values being observed for steeper hills.

314

TABLE 1. TERRAIN AMPLIFICATION FACTORS FOR TWO-DIMENSIONAL HILLS

HILL	H_s/h	SOURCE LOCATION A		
		UPWIND	TOP	DOWNWIND
8	1/4	1.5	0.9	3.4
	1/2	1.1	0.6	3.0
	1	1.5	0.8	2.4
	1-1/2	1.2	0.8	1.7
5	1/4	2.0	0.5	15.0
	1/2	2.0	0.6	8.0
	1	1.7	0.9	5.6
	1-1/2	1.2	1.0	2.9
3	1/4	2.8	0.3	7.5
	1/2	2.5	0.7	6.4
	1	1.8	0.9	10.8
	1-1/2	1.9	0.9	7.8

2.2 Three-Dimensional Hills

Two studies have been conducted to determine the effects of the cross-wind aspect ratio of a hill on dispersion from nearby sources. Triangular ridges of different crosswind lengths were constructed by cutting a cone in half and inserting straight triangular sections between the two halves. A sketch of these hills is given and the aspect ratio is defined in Figure 8. The slope of the ridges was 1:2, or 26.5°. Four configurations were used: C2 was the basic cone, C4 and C6 had short and long straight sections, respectively, inserted between the two halves, and CX was a two-dimensional triangular ridge that extended completely across the test section of the wind tunnel.

Snyder and Britter (unpublished) investigated surface concentrations on the ridges resulting from upwind sources. Ground-level concentrations were measured downwind of stacks of height 0, 0.5 and 1 h, with the stacks located 3.7 h upwind of the hill centers. Results for the cone and the two-dimensional ridge are shown in Figures 9 and 10, respectively. For the ground-level source, downwind concentrations were reduced by the presence

315

of the hills due to the excess turbulence and divergence of the flow around the hills. For the elevated sources, the maximum glc's occurred at the crest or on the lee sides of the hills. The maximum values for the cone were 2 to 3 times those for the two-dimensional ridge and 2 to 4 times those in flat terrain.

Castro and Snyder (1982) extended this study by locating sources at various downwind positions. Flow separation was observed on the lee sides of these hills because of the steep lee slopes and the salient edge at the crests. The shapes of the recirculating cavities were measured with pulsed-wire anemometry. Loci of zero-longitudinal mean velocity are shown in Figure 11. Sources were placed downwind of these hills, with the locations being selected as predetermined fractions of the distances to the reattachment points, x_s/x_R, where x_R is the distance to the reattachment point (both measured with respect to the hill center).

Figure 12 provides a summary of the terrain amplification factors. The largest A's were observed when the effluent was emitted near the separation-reattachment streamline (e.g., $x_s/x_R \sim 0.5$ and $H_s/h \sim 1$). In general, A's increased with an increase in aspect ratio; the largest value of A (= 11) was observed for the two-dimensional ridge.

2.3 Summary for Neutral Flows

The cases discussed above are summarized in Table 2 by listing them in order of decreasing A. From the standpoint of a fixed stack height, it appears that the worst location for a source is just downwind of a two-dimensional ridge and the best is on top of a ridge.

Downwind sources generally result in larger glc's because of the excess turbulence generated by the hills and because the effluent is generally emitted into streamlines that are descending toward the surface. Maximum terrain

316

TABLE 2.   SUMMARY OF TERRAIN AMPLIFICATION FACTORS FOR NEUTRAL FLOW

SOURCE LOCATION	HILL TYPE	A
DOWNWIND	TWO-DIMENSIONAL	10-15
DOWNWIND	THREE-DIMENSIONAL	5-6
UPWIND	THREE-DIMENSIONAL	2-4
UPWIND	TWO-DIMENSIONAL	1-3
TOP	TWO-DIMENSIONAL	0.5-1

amplification factors are considerably larger downwind of two-dimensional hills than those downwind of three-dimensional hills. A probable cause of this effect is that, in three-dimensional flows, lateral and vertical turbulence intensities are enhanced by roughly equal factors, whereas in two-dimensional flows, the lateral turbulence intensities are not enhanced as much as are the vertical turbulence intensities (because of the two-dimensionality). Since the maximum glc depends upon the ratio σ_z/σ_y (Pasquill, 1974), we may expect the A's downwind of two-dimensional hills to be larger than those downwind of three-dimensional hills. Also, the sizes of the recirculating cavity regions of three-dimensional hills are generally much smaller than those of two-dimensional ridges.

With regard to upwind sources, terrain amplification factors are larger for three-dimensional hills because, in such flows, streamlines can impinge on the surface and/or approach the surface more closely than in two-dimensional flows (see Hunt and Snyder, 1980; Hunt, Puttock and Snyder, 1979; Egan, 1975).

3.   STABLY STRATIFIED TOWING-TANK STUDIES

Plume behavior in complex terrain is dramatically altered by the addition of stable stratification: plume growth is severely inhibited, possibly

reduced to zero growth (Britter et al, 1983); under strong enough stratification, the flow may have insufficient kinetic energy to overcome the
potential energy of the stratification to surmount the hill, so the plume
may be forced to impact on the front surface of the hill and travel round
the sides (Hunt and Snyder, 1980); yet, under appropriate conditions, the
plume may be rolled-up in a free-stream rotor above the hill top or possibly in an attached rotor on the lee side of the hill (Castro et al, 1983).
The dividing-streamline concept has been shown to be a useful indicator in
determining whether a plume will impact on the windward surface of a hill
or surmount the top (Snyder et al, 1984). The dividing-streamline height
is applicable to a wide range of hill shapes, density profile shapes, and
wind angles, and in strong shear flows as well.

3.1 Dividing-Streamline Height

The basic concept of the dividing-streamline height is that in strongly
stratified flows over three-dimensional hills, the flow is composed of two
layers. In a lower layer of essentially horizontal flow, the approach flow
has insufficient kinetic energy to overcome the potential energy of the
stratification to surmount the top. Plumes from upwind sources impinge
directly on the hill surface. In an upper layer, the approach flow does
have sufficient kinetic energy, so that plumes from upwind sources may pass
over the hill top. The depth of the lower layer is the dividing-streamline
height, H_D. It is primarily a function of the Froude number, $F = U/Nh$,
where U is the approach flow velocity, N is the Brunt-Vaisala frequency, and
h is the hill height.

The utility of the two-layer concept arises because the transport and
diffusion in each layer may be analyzed quite separately and independently
of one another using different but well-established techniques in each layer.
In the lower layer, for sufficiently strong stratification ($F \ll 1$), the flow

318

is approximately horizontal. Outside the wake, the velocity field at height $z < H_D$ is the same as two-dimensional flow about a cylinder. The shape of the cylinder is defined by the contour of the hill at height z. Given this mean flow field and the fact that stable stratification limits vertical diffusion, the calculation of dispersion from a point source at height H_S ($< H_D$) is the same as that from a line source near a cylinder (See Figure 13 and Hunt, Puttock and Snyder, 1979, and Weil et al, 1981).

In the upper layer, buoyancy and inertial forces control the flow as it passes over the hill. The simplest and easiest assumption to make in constructing a mathematical model (and the only one made to date) is that the dividing-streamline surface is perfectly flat, i.e., that the upper- and lower-layer flow regimes are separated by a horizontal plane. As discussed by Snyder and Hunt (1984), a plume released in the upper-layer flow (at H_S) upstream of a hill of height h may be treated like a release from a stack of height $H_S - H_D$ upstream of a hill of height $h - H_D = Fh$, i.e., as if a ground plane were inserted at height H_D. The stratification above H_D, of course, has important influences on the diffusion as well as the vertical convergence and horizontal divergence of streamlines, so that the flow structure over the hill of height Fh must be treated like that with $F = 1$, i.e., the Froude number of the flow over this smaller hill is unity (Bass et al, 1981; Hunt et al, 1984).

3.2 Hill-Surface Concentrations

Snyder et al (1980) and Snyder and Hunt (1984) have shown that when a plume embedded in the lower layer (i.e., released below H_D) impacts on a hill surface, the maximum glc's can be essentially equal to those observed at the plume centerline in the absence of the hill. Plume centerline concentrations for stable plumes are typically 3 to 4 orders-of-magnitude larger than maximum glc's from stable plumes over flat terrain. Hence, the

319

terrain amplification factor that was used for neutral conditions results in extremely large numbers that would be difficult to interpret, and a new factor A_s is defined for stable flows as the ratio of the maximum observed glc in the presence of the terrain to that observed at the plume centerline at the same downwind distance in the absence of the hill.

3.3 Flat Dividing-Streamline Surface

Snyder and Lawson (1983) conducted a series of towing-tank studies to examine surface concentration patterns resulting from upwind plumes released in the upper layer and, more specifically, to test the adequacy of the assumption of a flat dividing-streamline surface. It was known from laboratory studies that, even well within the lower layer, streamline trajectories are not contained within horizontal planes. On the other hand, concentration measurements on hill surfaces (Snyder and Hunt, 1984) suggested that a flat-surface approximation may yield reasonable estimates. From a practical viewpoint, the mathematical models are vastly simplified if such an assumption yields reasonable estimates of surface concentration. Hence, Snyder and Lawson (1983) attempted to answer the question not from a detailed analysis of the shape of such a dividing-streamline surface, but from the more practical comparison of surface concentration patterns.

In the Snyder and Lawson experiments, linear density gradients were established in the tank and the hill was towed at a speed U such that the Froude number $F (=U/Nh)$ was 0.5 (establishing a dividing-streamline height of $H_D = 0.5$). Effluent was released at a height of 0.6 h and the resulting hill-surface concentration patterns were measured. A second tow was conducted wherein the entire model (hill, baseplate and stack, as a unit) was raised out of the water* to a point where the water surface was precisely

---

*The model is routinely mounted upside-down such that the baseplate is submerged a few millimeters below the water surface. In discussion of flow structure and plume behavior, however, we describe the results as if the model were right-side up.

at the dividing-streamline height, i.e., the water surface was at half the hill height. The model was towed at the same speed as in the full-immersion tows, so that the Froude number of this now half-height hill was unity, and all streamlines passed over the hill top. The flat water surface thus forced a flat dividing-streamline surface. The resulting surface concentration patterns were then compared with the full-immersion patterns to ascertain the effects of a flat dividing-streamline surface.

Figure 14 shows a comparison of the concentration distributions resulting from the full- and half-immersion tows. The plumes were spread broadly to cover essentially the entire surface above half the hill height (the dashed circle on the figure marks half the hill height). The contours on the windward side of the hill are roughly circular, but somewhat elongated in the streamwise direction. In the half-immersion case, no concentrations were observed below half the hill height, of course, because that portion of the model was out of the water. In the full-immersion case, the plume "hugged" the lee side of the hill and was swept to elevations considerably lower than half the hill height.

Figure 15 presents a scatter diagram comparing, on a point-by-point basis, the surface concentrations measured in the half- and fully-immersed cases. Within the region of large concentrations, the two cases compared quite favorably, the half-immersed case yielding concentrations approximately 10 to 20% larger than the fully-immersed case. In the region of low concentration, quite large differences occurred (worst case, a factor of 10). However, a close examination shows that in all cases where the concentrations differed by more than a factor of 2, the sampler locations were very close to half the hill height, i.e., either at 0.505 h or 0.51 h.

These results suggest that the assumption of a flat dividing-streamline surface in a mathematical model is a reasonable approximation to make, at least with regard to predicting the locations and values of maximum concentrations and areas of coverage.

3.4 Comparison with Field Studies

While much has been learned about flow and dispersion under stable conditions, the applicability of towing-tank results to the atmosphere is sometimes questioned. The height of the dividing-streamline is directly proportional to wind speed, and in a towing tank, the wind speed is constant. The unsteadiness of the wind in the field, however, can allow the plume to impac on the front surface at one moment, then sweep it over the top at the next moment. Similarly, a meandering wind will result in a plume being swept from one side of the hill to other, resulting in significant reductions of surface concentration. Carefully designed field studies are required to determine the effects of these aspects of atmospheric flows and to aid in the interpretation of towing-tank results so that full-scale predictions can be made.

In a major field study recently completed, Environmental Research and Technology, Inc., under contract to the EPA, collected six-weeks of data on flow and dispersion around a 100 m high, isolated hill (Cinder Cone Butte) in a broad, flat river basin in southwestern Idaho. Detailed measurements were made of wind, turbulence and temperature profiles in the approach flow and at other positions on the hill. SF_6 as a tracer and smoke for flow visualization were released from a mobile crane that allowed flexible positioning of the source height and location. One-hundred samples on the hill collected data on surface concentrations. LIDAR was used to obtain plume trajectories and dimensions. The study concentrated on stable plume impacti

One particular hour from that field-study data was selected for simulation in the stratified towing tank (Snyder and Lawson, 1981). That hour was 0500 to 0600, October 24, 1980 (Case 206), which may be characterized as very stable, i.e., light winds and strong stable temperature gradients. Measurements made during the towing-tank experiments included glc's under various stabilities and wind directions, vertical distributions of concentration at selected points, plume distributions in the absence of the hill, and visual observations of plume characteristics and trajectories.

This series of tows showed that the surface concentration distributions were extremely sensitive to changes in wind direction. For example, Figure 16 shows that the distribution shifted from the north side of the hill to the south side with a shift of only 5° in wind direction. Comparisons of individual distributions with field results showed very much larger maximum glc's and much narrower distributions in the model results. To account for this variability in the winds measured during the hour, a matrix of 18 tows (3 wind directions x 6 wind speeds) was conducted and the concentration patterns were superimposed. Figure 17, a scatter plot of superposed model concentrations versus field concentrations, showed a marked improvement over the previous single-tow comparisons. The largest model concentrations were within a factor of 2 of the highest field values.

3.5 Limitations of the Towing Tank

In our previous discussions of the dividing-streamline concept, we implicitly assumed that the portion of the flow with insufficient kinetic energy to surmount the hill top was able to pass around the sides of the hill. However, if the model is two-dimensional (spans the width of the tank), the fluid with insufficient kinetic energy is blocked or trapped upstream. Because of continuity, the blocked fluid must conserve its volume. Hence, as the distance between the model and the upstream endwall of the tank

323

decreases, the blocked fluid is "squashed", i.e., since the length of the blocked region must decrease with distance along the tow, the depth of the fluid that was initially blocked must increase. Eventually, all of the fluid upstream of the model will spill over the top. The fixed end on the tank, in effect, provides a uniform approach flow velocity profile at a distance upstream of the model that varies as the experiment progresses.

Snyder et al (1984) addressed this problem through a series of towing-tank studies. They used a series of "infinite" ridges of quite different cross-sectional shape to test the validity of the "steady-state" assumption of flow upwind of two-dimensional ridges under strongly stratified conditions. Three ridge shapes were used, a vertical fence, a triangular ridge with a slope of 63°, and a Witch of Agnesi $(1/(1+x^2))$ with a maximum slope of 39°.

Results will be shown here only for the triangular ridge. A density sampling rake was positioned 11 hill heights upwind of the ridge and towed with the ridge. Density samples were collected just prior to the start of the tow, during the first and last meter of the tow, and at the one-third and two-thirds points of the tow (the total length of the tow was approximately 20 m). The experiments were conducted at a Froude number of 0.5 (based on the tow speed and the undisturbed density profile), so that the calculated depth of blocked fluid (the dividing-streamline height) was 0.5h.

Figure 18 shows the density profiles collected during the tows. It is clear that the initial near-linear density profile was continuously modified during the tow at a position 11 h upstream and that upstream conditions did not reach a steady state. The profiles tended toward neutral at elevations below half the hill height.

The conclusion from this series of studies was that steady-state conditions are not established in strongly stratified flows (say, F < 1) over two-dimensional ridges in finite-length towing tanks. The "approach flow"

324

velocity and density profiles changed continuously during the tow. Thus, these experiments have no analog in the real atmosphere.

4. CONCLUSIONS

Laboratory measurements of the maximum glc's from elevated sources in the vicinity of two- and three-dimensional terrain features have been presented and compared. Stable conditions produced the highest glc's when the source was located near the dividing-streamline. In this case, the maximum glc was close to the concentration on the plume centerline from an identical release in the absence of the hill. Under neutral conditions, the maximum glc's occurred with the source located just downwind of a two-dimensional ridge. This maximum was found to be about 15 times the maximum glc observed for an identical release over flat terrain.

For practical purposes, a plume released in the upper layer (above the dividing-streamline) in strongly stratified flow over a hill can be treated as a release from a shorter stack upwind of a shorter hill, i.e., as if a ground-plane were inserted at the dividing-streamline height.

A comparison of field and laboratory observations of concentration patterns on a hill surface under strongly stratified conditions showed very good correspondence when wind speed and direction variability were accounted for in the model experiments. Concentration distributions on the hill surface were found to be extremely sensitive to slight changes in wind direction.

Finally, strongly stratified towing-tank experiments on flows over two-dimensional ridges were found to have no analog in the real atmosphere because of the unsteadiness created by the finite length of the tank.

325

REFERENCES

Arya, S.P.S. and Shipman, M.S., 1981: An Experimental Investigation of Flow and Diffusion in the Disturbed Boundary Layer over a Ridge; Part I: Mean Flow and Turbulence Structure, Atmos. Envir. v. 15, no. 7, p.1173-1184.

Arya, S.P.S., Shipman, M.S. and Courtney, L.Y., 1981: An Experimental Investigation of Flow and Diffusion in the Disturbed Boundary Layer over a Ridge; Part II: Diffusion from a Continuous Point Source, Atmos. Envir., v. 15, no. 7, p. 1185-1194.

Bass, A., Strimaitis, D.G. and Egan, B.A., 1981: Potential Flow Model for Gaussian Plume Interaction with Simple Terrain Features, Rpt. to Envir. Prot. Agcy. under Contract No. 68-02-2759, Res. Tri. Pk., NC, 201p.

Britter, R.E., Hunt, J.C.R., Marsh, G.L. and Snyder, W.H., 1983: The Effects of Stable Stratification on Turbulent Diffusion and the Decay of Grid Turbulence, J. Fluid Mech., v. 127, p. 27-44.

Castro, I.P. and Snyder, W.H., 1982: A Wind Tunnel Study of Dispersion from Sources Downwind of Three-Dimensional Hills, Atmos. Envir., v. 16, no. 8, p. 1869-87.

Castro, I.P., Snyder, W.H. and Marsh, G.L., 1983: Stratified Flow over Three-Dimensional Ridges, J. Fluid Mech., v. 135, p. 261-82.

Courtney, L.Y. and Arya, S.P.S., 1980: Boundary Layer Flow and Diffusion over a Two-dimensional Low Hill, Reprints Vol., Second Joint Conf. on Appl. of Air Poll. Meteorol., Mar. 24-28, New Orleans, LA, Amer. Meteorol. Soc., Boston, MA, p. 551-8.

Egan, B.A., 1975: Turbulent Diffusion in Complex Terrain, Workshop on Air Poll. Meteorol. and Envir. Assess., Amer. Meteorol. Soc., Sept. 30-Oct. 3, Boston, MA.

Hunt, J.C.R., Puttock, J.S. and Snyder, W.H., 1979: Turbulent Diffusion from a Point Source in Stratified and Neutral Flows around a Three-Dimensional Hill: Part I: Diffusion Equation Analysis, Atmos. Envir., v. 13, p. 1227-39.

Hunt, J.C.R., Richards, K.J. and Brighton, P.W.M., 1984: Stratified Shear Flow over Low Hills: II. Stratification Effects in the Outer Flow Region, To be submitted to Quart. J. Roy. Meteorol. Soc.

Hunt, J.C.R. and Snyder, W.H., 1980: Experiments on Stably and Neutrally Stratified Flow over a Model Three-Dimensional Hill, J. Fluid Mech., v. 96, pt. 4, p. 671-704.

Khurshudyan, L.H., Snyder, W.H. and Nekrasov, I.V., 1981: Flow and Dispersion of Pollutants over Two-Dimensional Hills: Summary Report on Joint Soviet-American Study, Envir. Prot. Agcy. Rpt. No. EPA-600/4-81-067, Res. Tri. Pk., NC., p. 143.

Pasquill, F., 1974: Atmospheric Diffusion, 2nd Ed., Chichester, Ellis Horwood Ltd., John Wiley and Sons, NY, NY, p. 429.

Snyder, W.H., 1979: The EPA Meteorological Wind Tunnel: Its Design, Construction, and Operating Characteristics, Envir. Prot. Agcy. Rpt. No. EPA-600/4-79-051, Res. Tri. Pk., NC, p. 78.

Snyder, W.H., 1981: Guideline for Fluid Modeling of Atmospheric Diffusion, Envir. Prot. Agcy. Rpt. No. EPA-600/8-81-009, Res. Tri. Pk., NC, p. 200.

Snyder, W.H. and Britter, R.E., 1979: Aspect Ratio Study, Unpublished Data Report, Fluid Modeling Facility, Envir. Prot. Agcy., Res. Tri. Pk., NC.

Snyder, W.H., Britter, R.E. and Hunt, J.C.R., 1980: A Fluid Modeling Study of the Flow Structure and Plume Impingement on a Three-Dimensional Hill in Stably Stratified Flow, Proc. Fifth Int. Conf. on Wind Engr. (J.E. Cermak, ed.), v. 1, p. 319-29, Pergamon Press, NY, NY.

Snyder, W.H. and Hunt, J.C.R., 1984: Turbulent Diffusion from a Point Source in Stratified and Neutral Flows around a Three-Dimensional Hill; Part II: Laboratory Measurements of Surface Concentrations, Atmos. Envir. (to appear).

Snyder, W.H. and Lawson, R.E. Jr., 1981: Laboratory Simulation of Stable Plume Dispersion over Cinder Cone Butte: Comparison with Field Data, in EPA Complex Terrain Modeling Program, First Milestone Report - 1982, Rpt. No. EPA-600/3/82-036, Envir. Prot. Agcy., Res. Tri. Pk., NC, p. 250-304.

Snyder, W.H. and Lawson, R.E., Jr., 1983: Stable Plume Dispersion over an Isolated Hill: Releases above the Dividing-Streamline Height, Unpublished report, Fluid Modeling Facility, Envir. Prot. Agcy., Res. Tri. Pk., NC.

Snyder, W.H., Thompson, R.S., Eskridge, R.E., Lawson, R.E., Jr., Castro, I.P., Lee, J.T., Hunt, J.C.R. and Ogawa, Y., 1983: The Structure of Strongly Stratified Flow over Hills: Dividing-Streamline Concept, Submitted to J. Fluid Mech.

Thompson, R.S. and Snyder, W.H., 1976: EPA Fluid Modeling Facility, Proc. Conf. on Modeling and Simulation, Rpt. No. EPA-600/9-76-016, Envir. Prot. Envir. Agcy., Wash. D.C., July.

Weil, J.C., Traugott, S.C. and Wong, D.K., 1981: Stack Plume Interaction and Flow Characteristics for a Notched Ridge, Rpt. No. PPRP-61, Maryland Power Plant Siting Program, Martin Marietta Corp., Baltimore, MD, p. 92.

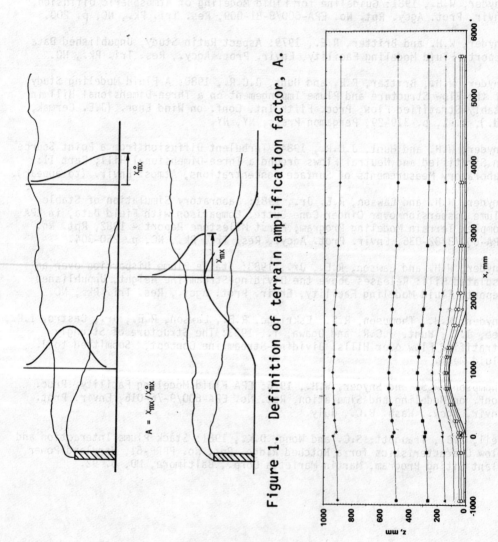

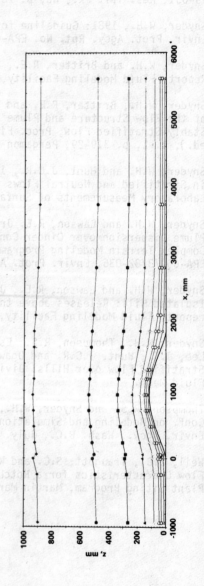

Figure 1. Definition of terrain amplification factor, A.

$$A = x_{mx}/x^{\circ}_{mx}$$

Figure 2. Streamline patterns over hill 2 from experimental data.

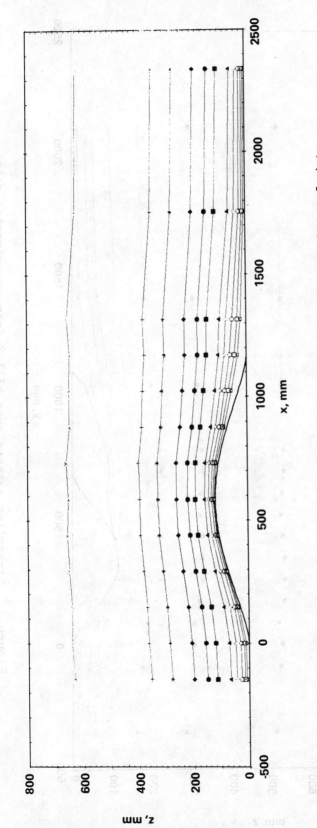

Figure 3. Streamline pattern over hill 5 from experimental data.

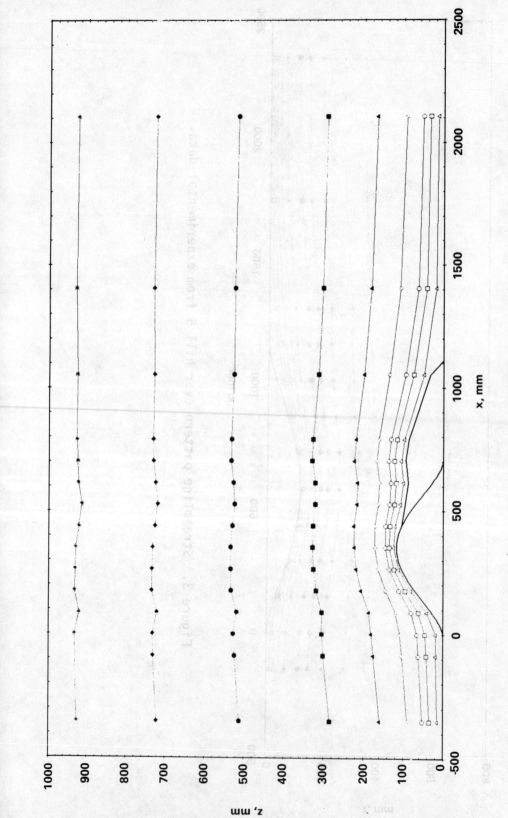

Figure 4. Streamline pattern over hill 3 from experimental data.

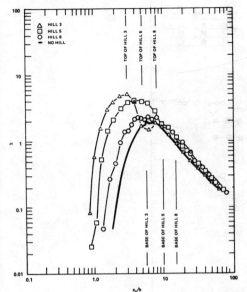

Figure 5. Surface concentration profiles over hills; Upwind base stack location; $H_S/h = 0.5$.

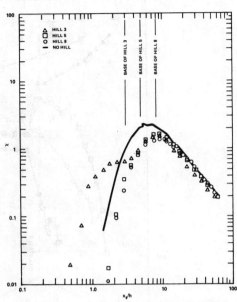

Figure 6. Surface concentration profiles over hills; Hill top stack location; $H_S/h = 0.5$.

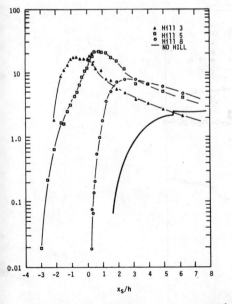

Figure 7. Surface concentration profiles over hills; Downwind base stack location; $H_S/h = 0.5$.

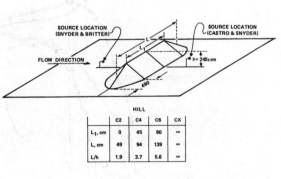

Figure 8. Geometry of triangular hills of different aspect ratio.

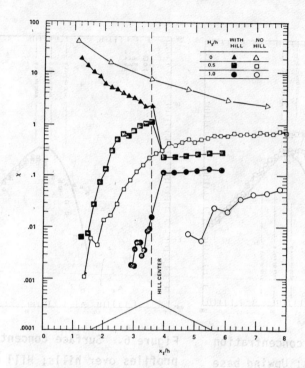

Figure 9. Surface concentration on hill C2.

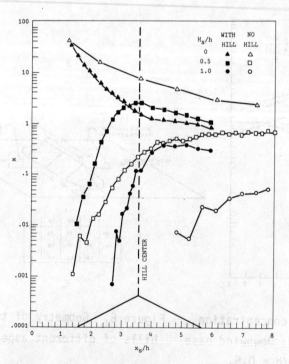

Figure 10. Surface concentrations on hill CX.

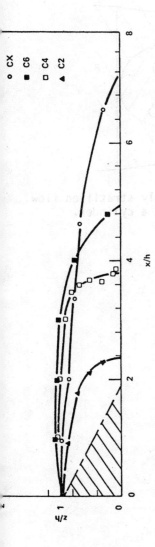

Figure 11. Loci of zero-longitudinal mean velocity on the centerline of hills of different crosswind aspect ratio.

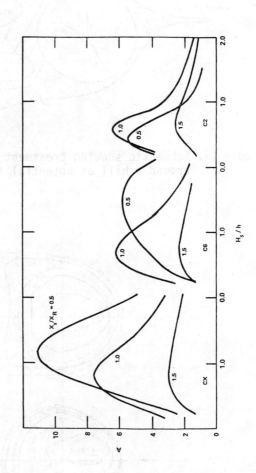

Figure 12. Variation of terrain amplification factor with stack height for hills of different crosswind aspect ratio and downwind sources.

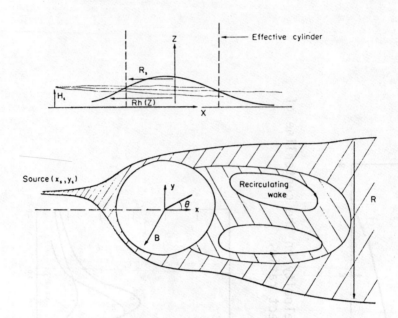

Figure 13. Schematic showing treatment of strongly stratified flow around a hill as potential flow about a cylinder.

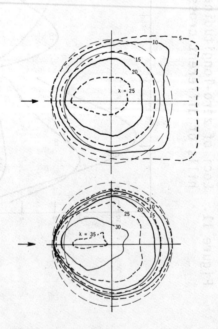

Figure 14. Concentration distributions measured on the hill surface. F = 0.5, H_s/h = 0.6. Top: fully submerged; Bottom: half submerged.

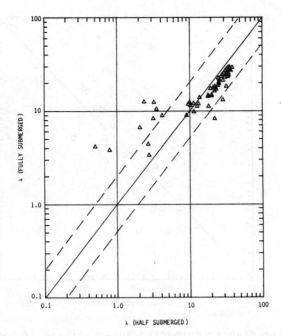

Figure 15. Comparison of concentrations. Half submerged versus fully submerged. F = 0.5, H_s/h = 0.6.

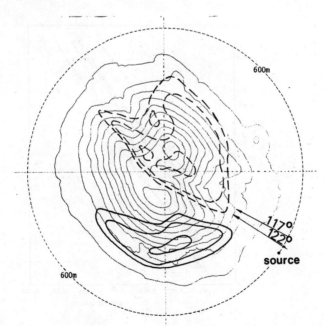

Figure 16. Concentration distributions measured during individual tows: H_s/h = 0.31, H_D/h = 0.38; Wind direction: —— 117°, ---122°.

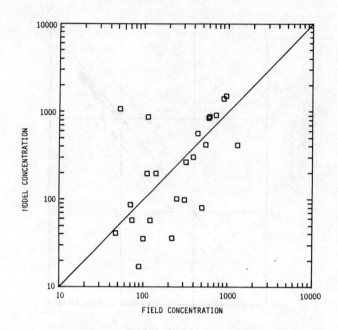

Figure 17. Scatter diagram comparing superposition of concentration distributions from 18 tows with field distributions.

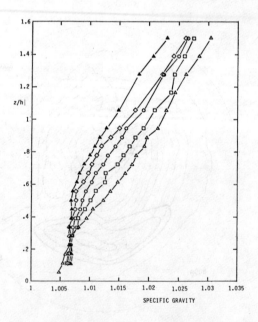

Figure 18. Density profiles measured 11 hill heights upstream of two-dimensional triangular ridge. $\triangle$, initial density profile. Sampling interval centered at x=: $\square$, 1.3m; $\bigcirc$, 6.7m; $\diamondsuit$, 11.8m; and $\blacktriangle$, 17.7m from start of tow.

Session 4: Observation and Analysis (3)

THE HIERARCHY OF MOTION SYSTEMS OVER LARGE PLATEAUS

Elmar R. Reiter
Department of Atmospheric Science
Colorado State University
Tang Maocang
Lanzhou Institute of Plateau Atmospheric Physics,
Chinese Academy of Sciences.
Shen Rujin
Institute of Atmospheric Physics,
Chinese Academy of Sciences.

1. INTRODUCTION

Large plateaus and mountain ranges not only exercise a barrier effect on atmospheric flow processes, but constitute elevated heat or cold sources which generate baroclinicity that results in a variety of circulation systems. Depending on the space and time scales over which these baroclinic processes are effective we can distinguish a hierarchy of evolving circulation systems. These systems interact with each other, sometimes in a way which makes it difficult to assess from diagnostic studies the separate impacts of any one scale.

The longest time scales which we will consider here are concerned with seasonal, or monsoonal variability. The thermal effects of the northern hemisphere plateaus, on this scale, interact with the seasonal variability of the global general circulation.

Even within the monsoonal time scale one can distinguish mesoscale quasi-permanent pressure systems over the Tibet an Plateau and over the Western Plateau of the United States. These pressure systems appear to be tied to prominent features in the topography (Reiter, 1982) and have a strong impact on regional climate characteristics, such as the frequent formation of convective precipitation systems, or the prevalence of desert conditions.

The interannual variability of the general circulation impacts on these monsoon systems, but so do the transient, synoptic disturbances, the sum total of which constitutes the seasonal climate. These disturbances, having a time scale of a few days, are strongly affected by topographic features and their associated surface heat budget distributions.

A major impact of these synoptic systems comes from the diurnal variability of the heat source and sink distributions over the large plateaus. A large, diurnal plateau circulation system can be identified theoretically as well as diagnostically, having a space scale of the order of 10^3 km, and a time scale of 24 hours. This circulation system undergoes monsoonal (seasonal) changes, but should not be confused with the monsoon system as such.

Finally, there are local wind systems, such as mountain and valley breezes, which operate on relatively small space scales and on a diurnal time scale. They are affected by the plateau circulation system the monsoonal system, and the synoptic systems.

The enumeration of these systems in a hierarchical order of space and time scales seemingly implies linear interactions, directed from larger to smaller scales. Such an implication might grow even stronger in the subsequent, more detailed description of some of these systems. We should emphasize, therefore, that all these interactions, a priori, have to be assumed as nonlinear. We will have to allow for the fact that local, diurnal scales can influence regional and large scales of longer duration. It is only our inadequate analytical and numerical modeling techniques that impose a preference of down-scale forcing.

2. THE MONSOONAL SCALE

It is impossible to separate in diagnostic studies the seasonal changes of the global, or hemispheric, general circulation from changes brought about solely by the thermal effects of the large plateaus. As these plateaus undergo their annual heating and cooling cycles the planetary wave configurations change accordingly and produce changing aspects of the wind fields over the plateaus, thus altering the dynamic plateau effects. Reiter and Westhoff (1981) have shown that the inter-annual variability of the ultralong planetary waves (waves No. 1 and 2) are subject to largest interannual variabilities in the latitude ranges and seasons in which the major, northern hemisphere plateaus change their character between heat and cold sources (see e.g. Ye,1982). Figure 1 shows the ratio between planetary wave amplitudes computed from the long-term mean 500 hPa height distributions, and the same amplitudes computed daily, and then averaged according to calendar date irrespective of the phase position of the waves. The results thus obtained become a measure of the interannual variability of these planetary waves. A value of 1.00 in Fig. 1 would indicate a perfect recurrence of wave position each year at the same time, whereas a value of 0.00 would stand for random variability of wave positions. Waves 1 and 2 appear to be strongly affected, at monsoonal time scales, by plateau and continental heat source characteristics, although oceanic temperature anomaly effects cannot be ruled out as a cause for interannual planetary wave variability.

Most numerical modeling studies have concerned themselves with the variability of sea-surface temperatures (SST) as a possible, long-acting perturbation mechanism (see e.g. Julian and Chervin, 1978; Hanna et al., 1984, and many other theoretical and diagnostic studies). Since little is known about the interannual variability of the intensity of the Tibetan and North American heat sources and sinks, we are not yet in a position to simulate the effects of such variations on monsoon circulations. Suggestions have been made by Hahn and Shukla (1976) and Chen and Yan (1978) that the interannual variability of show cover over the Tibetan Plateau might be the cause for variability in the Indian monsoon circulation. Reiter and Ding (1980/81) have pointed out that the snow cover over the Tibetan Plateau might be tied to circulation anomalies which are linked to anomalous SST distributions in Atlantic and/or Pacific. Obviously, a challenging problem awaits us in resolving this question. A solution will, undoubtedly, have a significant impact on the skill of long-range (seasonal) forecasts. Its achievement requires close coordination between the acquisition

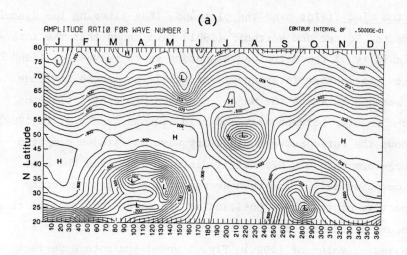

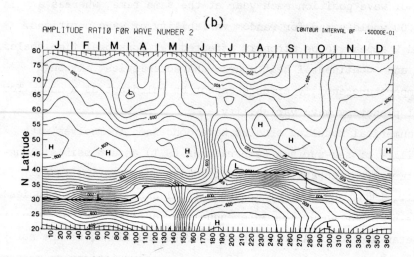

Fig. 1 Ratio between planetary wave amplitudes computed from cal-
endar-date averaged 500 hPa heights (1946—1979) to wave
amplitudes computed on a daily basis, then averaged by
calendar date irrespective of phase angle. High values
are indicative of interannual persistence, low values
of interannual variability. (a) Wave No. 1; (b) Wave
No. 2 (Reiter and Westhoff, 1981).

342

and analysis of a useful data base and the execution of numerical
models suitable for climate studies.

The monsoonal effects of the plateaus come to light from the
longterm mean geopotential height distributions characteristic of
the planetary boundary layer (PBL). Over the Western Plateau of North
America the 850 hPa surface provides a good indication of that layer,
whereas over the Tibetan Plateau the 600 hPa surface is more appropriate. Figures 2(a) and (b) show the January pressure patterns of the
PBL over the Western United States (Tang and Reiter, 1984) and over
the Tibetan Plateau (Gao et al., 1981; Ye et al., 1979). High pressure
systems dominate the plateau regions. The resultant wind fields agree
well with the details of the pressure distribution. In Figs. 3(a) and
(b) summer conditions (July) are displayed. Not only do we find the
dominance of low-pressure conditions over both plateaus, but there
are semipermanent mesoscale features whose impact is felt in the
development of convective precipitation systems. Notably, the anticyclonic shear line over Wyoming and Idaho marks the northern extent of
the monsoon-related occurrence of severe convection, as could be proven
from satellite composite pictures.

The layer characterized by a monsoonal wind reversal (>120 degrees
between winter and summer) is rather distinct over both plateaus
(Figs.4a and b), but thicker over the Tibetan Plateau than over the
United states. The higher elevation of the Tibetan Plateau obviously
generates a greater degree of baroclinicity with its surroundings,
causing a more vigorous circulation system to establish itself. It
should be pointed out that the low-level jet stream of Texas and
Oklahoma lies, at least in part, within the domain of the monsoonal
plateau influence. It is also dominated by the diurnal plateau
circulation, as will be demonstrated later.

The monsoonal plateau effects leave a strong imprint on the seasonal distribution of precipitation. During summer, relatively moist
conditions prevail to the south and the east of the plateaus (Figs.
5a and b).Especially along the eastern slopes of the Rocky Mountains
a summer precipitation peak is evident (Fig. 6). West of the Rocky
Mountains frontal disturbances cause a winter maximum and a more even,

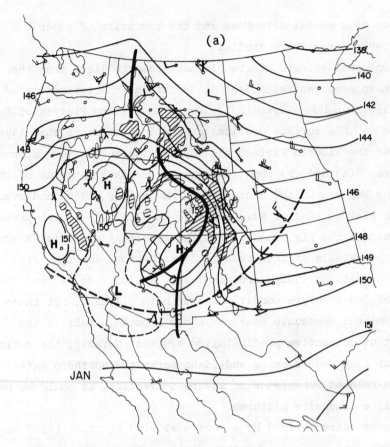

Fig. 2 (a) Mean 850 hPa heights at 1200 GMT in January (solid lines,
geopotential decameters) and resultant winds at the surface
(if only one arrow is drawn) and at the surface and 850 hPa
(if two arrows are drawn, the lighter one refers to 850 hPa).
Velocities indicated as follows: No bard <0.5 m/sec; short
barb 0.5-1.4 m/sec; long barb 1.5-2.4 m/sec, etc. Heavy,
full lines indicate axes of high pressure, dashed lines of
low pressure systems (Tang and Reiter, 1984).

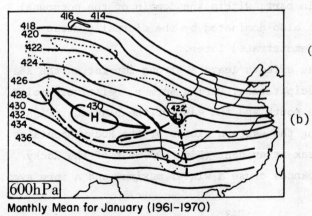

(b) Mean 600 hPa contours
in January over Asia
(dotted: the Tibetan
Plateau after Gao et
al., 1981 and Ye et
al., 1979).

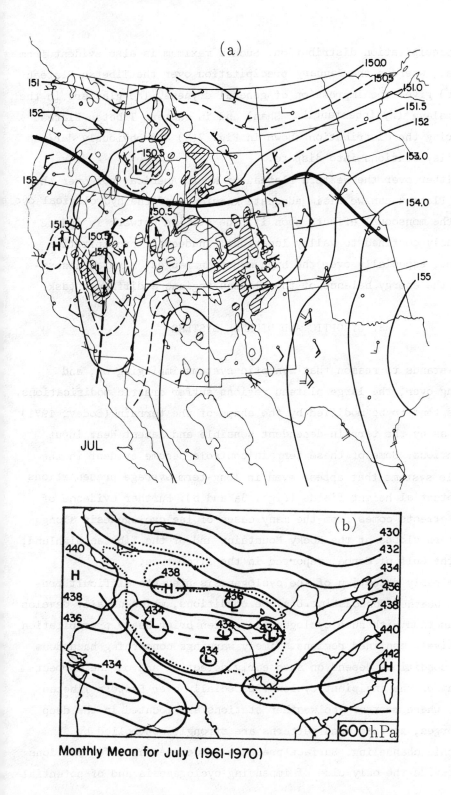

(a)

JUL

(b)

Monthly Mean for July (1961-1970)

600hPa

Fig. 3 Similar to Fig. 2, but for July.

annual precipitation distribution. Such a maximum is also evident from
Fig. 7(a), whereas the January precipitation over the Tibetan Plateau
(Fig. 7b) gives the impression of general dryness. The glaciers in the
Transhimalaya (Nyainqentanglha Shan) and in eastern Tibetan Plateau
might bring the distribution shown in Fig. 7(b) and obviously derived
from valley stations into dispute.

Neither over the Western Plateau of North America, nor over the
Tibetan Plateau can we claim adequate knowledge of the hydrological cycle
and of the monsoonal effects upon it. The station network is sparse
and usually confined to valley locations. Inadequate precipitation
estimates, especially over the high mountain ranges, make an assessment of the energy balance of the plateau regions a difficult task.

3. SYNOPTIC AND DIURNAL SYSTEMS

It stands to reason that synoptic systems impinging on, and
traveling over, the large plateau regions suffer drastic modifications.
Vortices tend to be modified by the shape of the terrain (Godev, 1971)
as well as by the terrain-dependent sensible and latent heat input
distributions. Some of these terrain controls become evident in the
mesoscale systems that appear even in long-term average presentations
of geopotential height fields (Figs. 3a and b). Further evidence of
terrain effects comes from the many cases of lee cyclogenesis along
the eastern slopes of the Rocky Mountains and of the Tibetan highland,
and in the Gulf of Genoa, reported in the literature.

The early detection of lee cyclogenesis presents difficult problems in weather forecasting.Upslope conditions, which usually develop
in conjunction with such cyclogenesis, often bring heavy precipitation
to relatively confined regions. Timely warnings concerning hazardous
weather conditions depend on such early detection and on the correct
placement of the incipient cyclone. Especially over Eastern Tibetan
Plateau, where most of the weather stations are located in the deep
river gorges, surface wind patterns are strongly controlled by
topographic channeling. Surface pressure tendencies at a few stations
often provide the only clue of impending cyclogenesis and of potential-

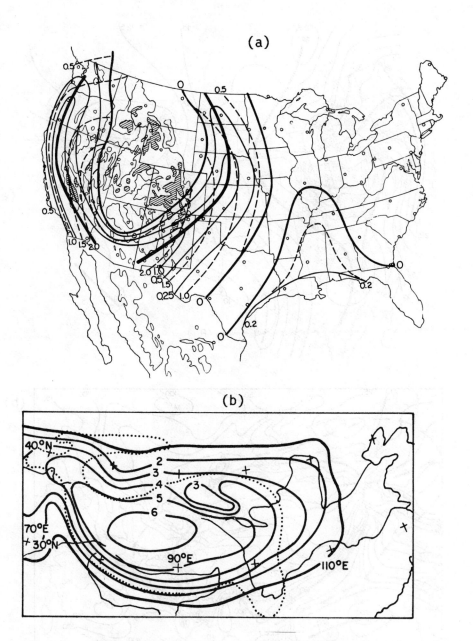

Fig. 4 (a) Dashed lines: height above sea level (km) of the top
of the layer with monsoonal wind reversal >120 degrees;
full lines: thickness of that layer (km) above terrain.
Data for 1200 GMT were used (Tang and Reiter, 1984).
(b) Height of top of monsoon layer over the Tibetan
Plateau. Heavy line delimits the extent of the monsoon
region (after Gao et al., 1981).

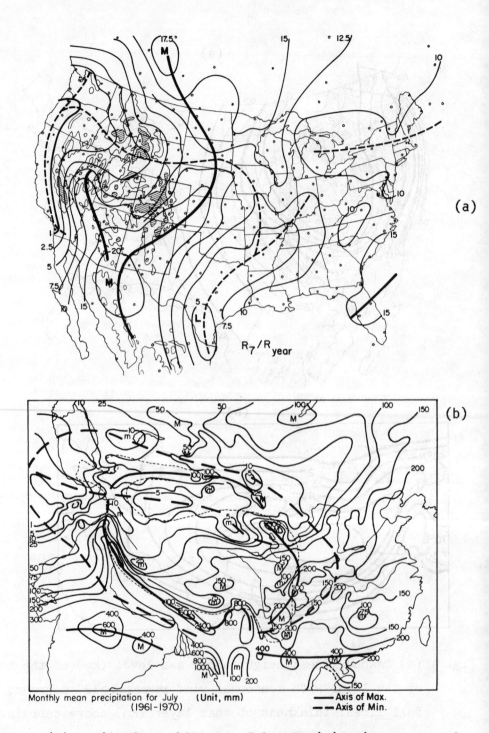

(a)

(b)

Monthly mean precipitation for July   (Unit, mm)
(1961-1970)

—— Axis of Max.
-- -- Axis of Min.

Fig. 5 (a) Ratio of monthly mean July precipitation to annual
precipitation over United States (Tang and Reiter,
1984). (b) Monthly mean July precipitation (mm) over
the Tibetan Plateau (after Ye et.al., 1979).

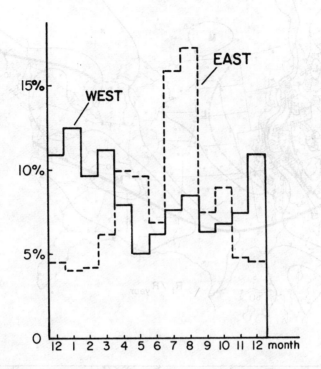

Fig. 6 Monthly precipitation, in percent of annual, at Crested
Butte ("west") and Buena Vista ("east"), Colorado (Tang
and Reiter, 1984).

ly heavy precipitation in the sparsely populated mountain regions.

Numerical modeling of synoptic systems in mountainous terrain
still suffers from a number of difficulties. As an example, we present
the case of Tibetan cyclogenesis of June 8, 1979. A prognostic model,
originally developed by Anthes and Warner (1978), was modified to
accommodate conditions as encountered over the high plateau. Table 1
contains an overview of the changes which have been made in the
original version of the model.

The relatively large domain of the model and the use of a σ-coordi-
nate system necessitated a drastic smoothing of the topographic features
used in the model. As Fig. 8 shows, the steep "cliffs" of the Himalayas
and Transhimalayas were all but eliminated in the model. Best results

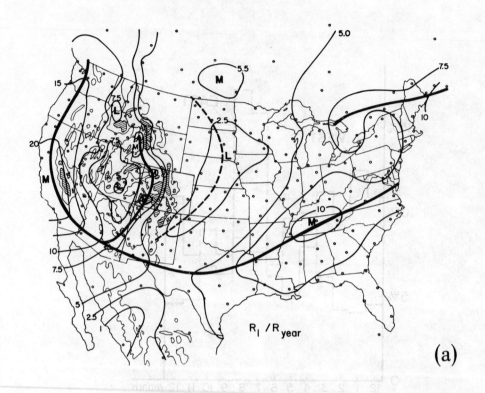

R_l /R_{year}

(a)

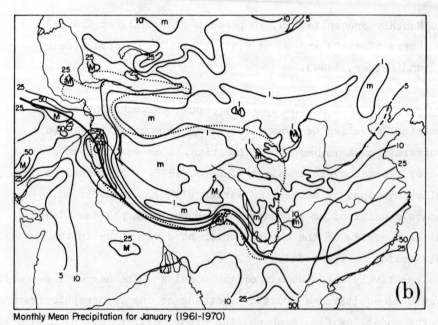

Monthly Mean Precipitation for January (1961-1970)

(b)

Fig. 7 Similar to Fig. 5, but for January.

Table 1 Comparison between Anthes and CSU mesoscale models.

	Anthes et al., 1982	C.S.U., Shen 1984
Input Data	LFM First Guess, Enhanced by Surface and Significant Level Data	Gridded Values From Subjectively Analyzed Fields
Initialization	Remove the Vertical Integral of Divergence	None
Delta s	90 km	96 km
Delta t	191 s	180 s
Horizontal Domain	41×41	31×41
Horizontal Grid Structure	Lambert Conformal, Staggered "B" Grid	Mercator, Nonstaggered Grid
Vertical Layers	10	6
Vertical Coordinate	Sigma-P	Sigma
Spatial Finite Difference	2nd Order	2nd Order
Temporal Finite Difference	Brown-Campana, Time Filter	Euler and Central
Lateral Boundary Condition	Time-Dependent, From Observations	Fixed
Terrain	Yès	Yes
Surface Heat Flux	Over Water Only	Over Land
Surface Evaporation	Over Water Only	Over Land
PBL (Including Surface Layer)	Bulk, $1.5 \times 10^{-3} < CD < 2.0 \times 10^{(-3)}$	Bulk, $(CD = (1 + 0.0001 \phi^*) \times 10^{(-3)}$
Shortwave Radiation	No	Incl. in Sfc. Heat Flux
Longwave Radiation	No	$T^* - Ta$
Convective Clouds & Precipitation	Kuo-Type (Anthes, 1977)	Kuo
Nonconvective Clouds & Precipitation	**Saturation** Criterion 100%	Upward Motion and 80% Saturation
Horizontal Internal Mixing	$K(Del)^4$	K prop. U.V; $K (Del)^4 T, Q$
Vertical (Above PBL) Mixing	None	Yes
Computer Time for 24 h Forecast	6.6 Min on Cray-1A	3 Min on Cray-1A

ϕ^* = geopotential = $(g \cdot z)$
z = height above sea level
T^* = ground surface temperature

351

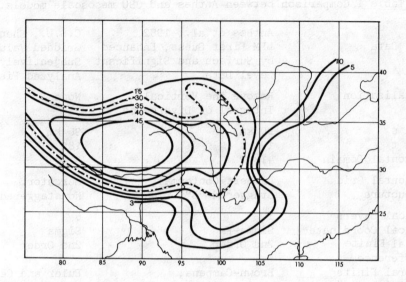

Fig. 8 Terrain height, in hundreds of meters, used in numerical model.

were achieved with a reduction of the surface elevations shown in Fig. 8 by a factor of 0.8. In spite of these shortcomings in the assumed terrain characteristics, the model performed reasonably well, leading us to believe that terrain details, in many cases, are not of overriding importance in the development of synoptic-scale systems.

It should be pointed out that our computations did not rely on surface heat fluxes parameterized from radiation balance considerations, but used the differences between air and soil surface temperatures. The latter are being measured directly at many station locations in China and seem to provide a more accurate input for heat flux estimates than the radiation balance parameterizations. Too many poorly accounted effects enter into the local radiation balances, such as vertical moisture and temperature profiles, cloud and soil conditions, etc., to provide a reliable estimate of soil surface temperatures. Furthermore, we relied on Chinese weather maps, subjectively analyzed, to define the initial conditions of our model run. In earlier work (Reiter and Gao, 1982) we found the objective analyses provided by the U.S. National

Meteorological Center (NMC) to be rather inadequate over the Tibetan Plateau. Also, of great importance in the specification of the initial conditions is the inclusion of the observed temperature fields instead of temperature fields obtained from the hydrostatic equation. The use of observed temperatures reduces errors introduced by the hydrostatic equation (Shen, 1983) and prevents the model from "blowing up".

Figures 9 and 10 show sequences of 700 hPa wind field and geopotential height prediction, to be compared with the observed wind fields depicted in Fig. 11. It should be noted that, with the exception of the river gorges of Eastern Tibetan Plateau, this isobaric surface is a figment of modeling. In reality it would lie beneath the ground surface. Nevertheless, the decay of an eastward moving vortex over the Tibetan Plateau, and the subsequent development of another vortex over the river gorge region were predicted with reasonable accuracy by the model. The heavy precipitation predicted by the model may have been an exaggeration (Fig. 12). Had it been used for flood warning, it would have served its purpose, however.

A number of difficulties still remain to be overcome. We already mentioned the inadequate representation of terrain features in the model. Estimates of convective precipitation need further refinement based on atmospheric physics rather than on the fine tuning of assumed vertical profiles of condensation heating. The role of evaporation from the soil may be significant and is not yet treated adequately.

Directly measured soil surface temperatures appear to constitute an improvement over parameterically derived values. Such temperatures are not measured in the United States and in Western Europe. Chinese measurements rely on a large-bulb thermometer, with half of the bulb buried in soil. One may criticize this method of measurement. It is, however, inexpensive and seems to be effective and might warrant imitation by weather services in other countries.

Diurnal surface heating variations were included in our model run by first estimating the daily average temperature differences between air and ground at each grid point from direct observations made at synoptic stations four times a day. Then a sinusoidal, diurnal varia-tion with twice the amount of the mean difference reached at noon, and

(a)

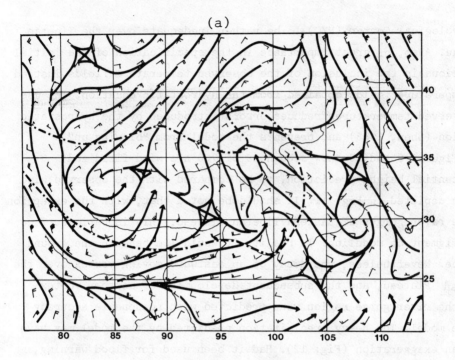

(a) 6-hour forecast, verifying 1800 GMT, June 8, 1979.

(b)

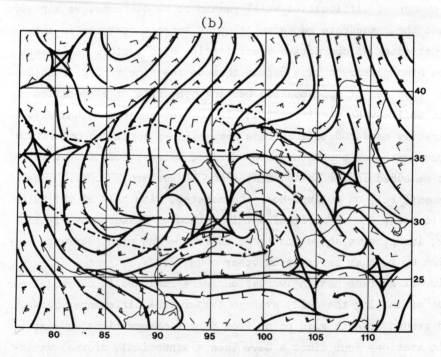

(b) 12-hour forecast, verifying 0000 GMT, June 9, 1979.

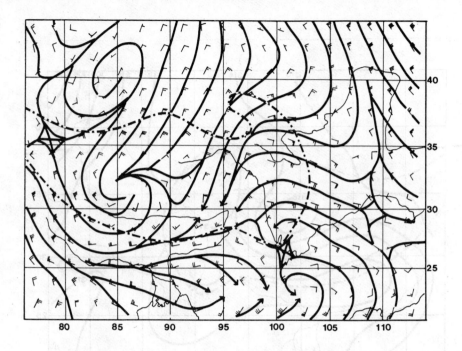

(c) 24-hour forecast, verifying 1200 GMT, June 9, 1979.

Fig. 9 Model predictions of 700 hPa wind field.

(a)

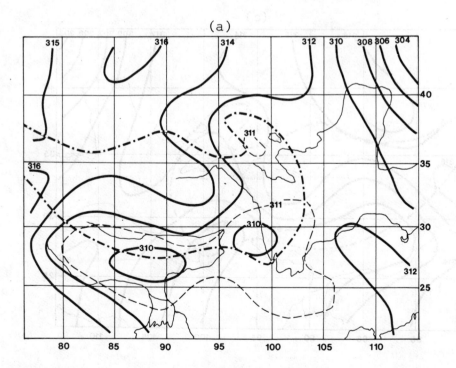

Fig. 10 Model predictions of 700 hPa geopotential heights for
same validation times as in Fig. 9.

(b)

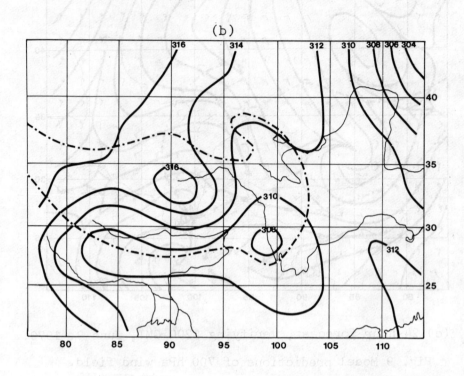

Fig. 10 Model predictions of 700 hPa wind field.

(c)

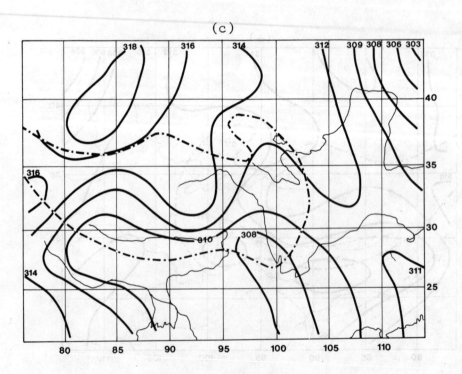

Fig. 10 (Continued)

356

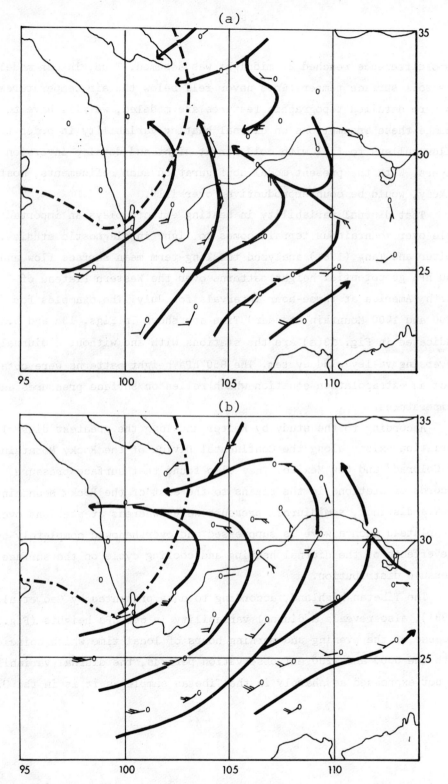

Fig. 11 Observed 700 hPa streamlines (a) 0000 GMT, June 9,
1979; (b) 1200 GMT, June 9, 1979.

zero difference reached at midnight was assumed. Thus, in our model the soil surface temperatures never fell below the air temperatures. As more detailed topographic features are modeled, we will have to change these assumptions on diurnal heating variability in order to allow valleys to fill with cold air at night and develop inversion layers. With the present model configuration such refinements, most likely, would be counterproductive "overkill".

That diurnal variability in heating effects plays an important role over mountainous terrain comes to light in diagnostic studies. Reiter and Tang (1984) analyzed the long-term mean surface flow and 850 hPa geopotential height patterns over the Western Plateau of North America at three-hour intervals for July. The examples for 0200 and 1400 Mountain Standard Time are shown in Figs. 13a and b. Also indicated in Fig. 13(a) are the stations with and without a diurnally reversing valley wind system. The 850 hPa height patterns were obtained from an extrapolation equation which relies on surface pressures and temperatures.

According to the study by Reiter and Tang the greatest diurnal variation exists along the Continental Divide of the Rocky Mountains of Colorado and New Mexico. They also found that surface pressure records of stations in the plains to the east of the Rocky Mountains show a distinct, semidiurnal pressure wave, whereas at stations over the plateau such a wave is suppressed and overshadowed completely by the effects of the diurnal heating and cooling cycle on the surface pressure distribution.

The Tibetan highland, according to data presented by Gao et al. (1981), also reveals a diurnal variability in 600 hPa heights (Fig.14). Because of the evening and morning hours of local time which coincide with the 0000 and 1200 GMT observation periods, the diurnal variability is not expressed as sharply in the Tibetan sample as it is in the U.S.

sample described above.

From Fig. 13 it appears as though the Texas low-level jet streams were embedded in southerly flow with only little diurnal variation. A different picture emerges, however, if one analyzes three-hour geopotential height variations and vector differences in resultant winds. As an example, we present the mean geopotential height changes in July from 1100 to 1400 MST and from 2300 to 0200 MST, together with the appropriate, three-hour vectorial changes in the resultant winds of these observation times (Fig. 15a, b). A clear, diurnal reversal in the plateau effects on the wind systems over the plains to the east of the Rocky Mountains can be seen. Thus, it appears that the Texas LLJ is not only part of the monsoon flow system described in the preceding chapter which causes a broad, southerly flow during summer,

(a)

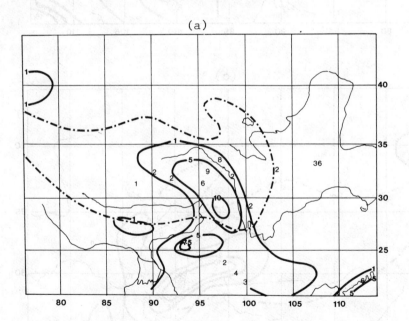

Fig. 12 Predicted (contour lines, mm) and observed (numerical values, mm) precipitation. (a) 6-hour period, ending at 1800 GMT, June 8, 1979. (b) 12-hour period, ending at 0000 GMT, June 9, 1979. (c) 6-hour period, ending at 1200 GMT, June 9, 1979.

(b)

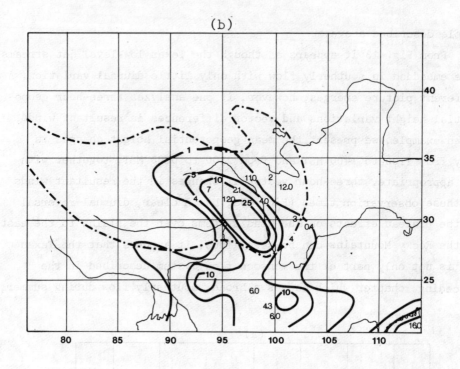

(c)

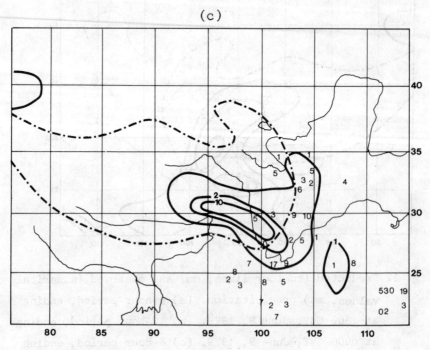

Fig. 12 (Continued)

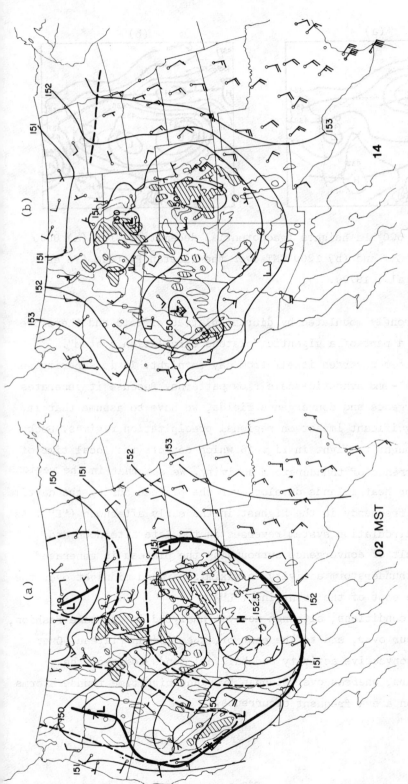

Fig. 13 Mean 850 hPa heights (geopotential decameters) in July at (a) 0200 MST (0900 GMT) and (b) 1400 MST (2100 GMT). Resultant winds are symbolized as in Fig.2. Trough is indicated by heavy solid line. Stations marked "x" show no diurnal wind reversal, stations with "v" do. Dashed line in Fig. 8 delimits the extent of the regions with diurnal valley-mountain breeze systems (Reiter and Tang, 1984). Terrain above 2750 m is hatched.

(a)   (b)

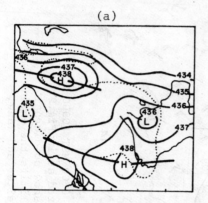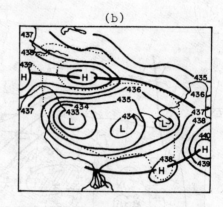

Fig. 14 600 hPa heights (geopotential decameters) for (a) 0000
        GMT and (b) 1200 GMT over the Tibetan Plateau (Ye et
        al., 1979).

but it is strongly modulated by diurnal plateau effects and therefore
becomes also a part of a gigantic "plateau circulation system".

This system reverses itself from day to night. It is superimposed
upon monsoonal- and synoptic-scale flow patterns. Because it generates
its own divergence and convergence fields, we have to assume that it
carries a significant impact on regional precipitation regimes. Such
impact is brought to light in Fig. 16 which depicts the local time of
maximum occurrence of thunderstorm activity. We see that in the regions
in which major heat islands develop over the plateau during the daytime
thunderstorm frequency is the highest in the early afternoon ("E"). As
the plateau circulation system reverses in the late afternoon and
evening, a belt of convergence surrounding the plateau is generated,
causing the thunderstorm activity to maximize around midnight over the
plains to the east of the mountains ("L").

Similar conditions, although not yet analyzed in the same fashion,
appear to occur over, and to the east of, the Tibetan Plateau. Over
the plateau convective activity and hailfall tends to favor the
afternoon hours, whereas over Sichuan Province nighttime thunderstorms
and convection are a frequent occurrence.

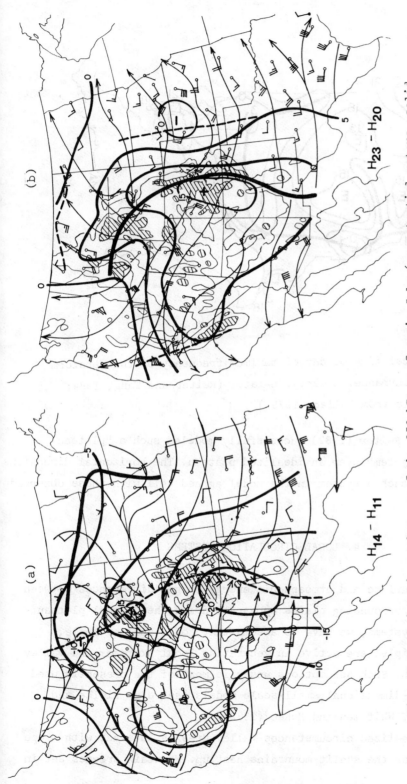

Fig. 15 Mean three-hour changes of the 850 hPa surface in July (geopotential meters, solid lines of medium thickness), three-hour changes of resultant winds (barb notation is exaggerated by a factor of five over explanation given in Fig. 2), and "streamlines" of these vector changes (thin, solid lines). Axes of cyclogenetic (convergent) and anticyclogenetic (divergent) centers are indicated by heavy dashed and solid lines, respectively. (a) 1400 MTS minus 1100 MST, (b) 2300 MST minus 2000 MST (Reiter and Tang, 1984)

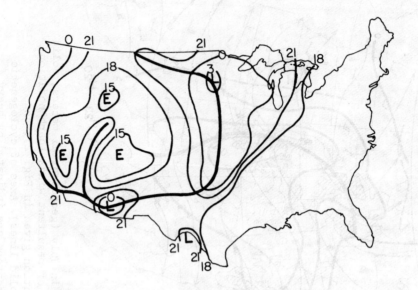

Fig. 16 Local time of day of maximum frequency of thunderstorm
occurrance. E=early. L=late. (Reiter and Tang, 1984;
data from Wallace 1975.)

Sang and Reiter (1982) successfully modeled such a "plateau
circulation system" over an idealized plateau. The horizontal dimensions
obtained for such a system in the model agreed well with those observed
in nature.

4. LOCAL CIRCULATION SYSTEMS

If the surface and atmospheric energy budget differences between
plateaus and surrounding plains appear to drive the diurnal plateau
circulation systems, we have to assume that, on a smaller scale,
similar principles are active in generating local mountain and valley
breeze systems. Such assumptions underlie most of the presently used
two-and three-dimensional,small-scale models [see e.g. Pielke and
Mahrer (1975), Whitemen and McKee (1982)].

Under idealized circumstances valley breezes interact with slope
wind systems as the sunlit mountains heat up. Mountain breezes set in

with cooling in the evening and during the night. In nature, however, conditions rarely are as simple as that. From the foregoing discussion we have to assume that horizontal pressure gradients, caused by differential heating of plateaus by monsoonal and diurnal effects, modulated by synoptic disturbances, will interact with the local effects of heat source distributions.

As an example of such an assumed interaction we show in Fig. 17 the diurnal variability of resultant winds for July, indicated by speed and direction, for Denver and Colorado Springs, Colorado. These two stations lie only about 100 km from each other. Both are to the east of the Continental Divide. Whereas Denver reveals a wind oscillation between southwest and northeast, conforming to the orientation of the South Platte River valley, the diurnal wind variation in Colorado Springs to the south of Denver appears to be strongly affected by a low ridge of hills to the north of the station (the Palmer Ridge). Both stations lie within the domain of the plateau circulation of the eastern slopes of the Rocky Mountains, described in the preceding chapter. The rather low Palmer Ridge, however, provides a significant, local perturbation which is superimposed on the larger-scale plateau

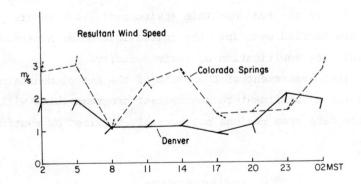

Fig. 17 Mean diurnal variation of resultant wind speed
(m/sec) and direction at Colorado Springs and
Denver, Colorado, for July (Reiter and Tang,
1984).

wind system.

From Fig. 17 one can see that convergence between the flow to the
north (Denver) and to the south (Colorado Springs) of the Palmer Ridge
dominates the daytime hours and maximizes during the early afternoon.
At night divergent flow conditions prevail. Thus, the Palmer Ridge,
insignificant as its elevation may be in comparison to the towering
Rocky Mountains rising only a short distance to the west, becomes a
significant, local focal area for thunderstorm development. The regional fauna attests to the fact: The "Black Forest" to the north of
Colorado Springs stretches a "finger" of forest land eastward into the
surrounding, treeless plains.

5. CONCLUSIONS

The foregoing discussion provided enough evidence for the profound
effects of plateaus on circulation systems of all scales. Most of
this evidence had to be compiled by patient work with insufficient
data. The harsh mountain environment usually is characterized by the
absence of population centers, hence by the absence of weather stations
with sufficiently long records to allow meaningful analyses. Most of
our analysis work, therefore, is heavily biased towards conditions
unrepresentative of the real mountain environment. Unfortunately,
such biases are carried over into the initialization of numerical
models and into the verification of their veracity.

The critical assessment of the state of the art of mountain meteorology calls for a commitment to measurement programs which will
provide better data from regions not normally serviced by routine
observations.

6. ACKNOWLEDGMENTS

The research reported in this paper is supported by NSF Grant ATM
83—13270, a new NASA Grant (no Grant Number received as of this date)
and Air Force Office of Scientific Research, Air Force Systems Command,

USAF, under Grant Number AFOSR 82-0162. The United States Government is authorized to reproduce and distribute reprints for Governmental purposes notwithstanding any copyright notation thereon.

REFERENCES

Anthes, R.A. and T.T. Warner, 1978: Development of hydrodynamic models suitable for air pollution and other mesometeorological studies. Mon. Wea. Rev., 106, 1045–1978.

Chen, L.T. and Z.X. Yan, 1978: A statistical analysis of the influence of anomalous snow cover over Qinghai-Tibetan Plateau during the winter-spring on the monsoon of early summer (in Chinese). Proceedings of the Conference on Medium and Long-Term Hydrometeorological Prediction in the Basin of the Yangzi River. May 1978, Vol. I, Hydro-Electric Press, Beijing.

Gao, Y.X., M.C. Tang, S.W. Luo, Z.B. Shen and C. Li, 1981: Some aspects of recent research on the Qinghai-Xizang Plateau meteorology. Bull. Amer. Meteor. Soc., 62 (1), 31–35.

Godev, N., 1971: Anticyclonic activity over Southern Europe and its relationship to orography. J. Appl. Meteor., 10, 1097–1102.

Hahn, D.G. and J. Shukla, 1976: An apparent relationship between Eurasian snow cover and Indian monsoon rainfall. J. Atmos. Sci., 33 (12), 2461–2462.

Hanna, A.F., D.E. Stevens and E.R. Reiter, 1984: Short-term climatic fluctuations forced by thermal anomalies. J. Atmos. Sci., January issue.

Julian, P.R. and Chervin, R.M., 1978: A study of southern oscillation and Walker circulation phenomenon. Mon. Wea. Rev., 106, 1433–1451.

Pielke, R.A. and Y. Mahrer, 1975: Representation of the heated planetary boundary layer in mesoscale models with coarse vertical resolution. J. Atmos. Sci, 32 (12), 2288–2308.

Reiter, E.R., 1982: Typical low-tropospheric pressure distributions over and around the Plateau of Tibet. Arch. Met. Geoph.Biokl., Ser. A, 31, 323–327.

———— and Y. -H Ding, 1980/81: The role of Qinghai-Xizang Plateau in

feedback mechanisms affecting the planetary circulation.
Scientia Atmospherica Sinica, 4 (4), 300—309, 5 (1), 9—22.

———— and D.R. Westhoff, 1981: A planetary wave climatology. J.
Atmos. Sci., 38 (4), 732—750.

———— and Deng-yi Gao, 1982: Heating of the Tibet Plateau and movements
of the South Asian high during spring. Mon. Wea. Rev., 110 (11),
1694—1711.

———— and Maocang Tang, 1984: Plateau effects on diurnal circulation
patterns. Accepted, Mon. Wea. Rev.

Sang, Jianguo and E.R. Reiter, 1982: Model-derived effects of large-
scale diurnal thermal forcing on meteorological fields. Arch.
Geoph. Biokl., Ser. A, 31 (3), 185—203.

Shen, Rujin, 1983: The rationality of the initialization of temperature
and the calculation of pressure gradient term . Scientia Atmospherica Sinica, 7 (2), 189—200.

Tang, Maocang and Elmar R. Reiter, 1984: Plateau monsoons of the
Northern Hemisphere: A comparison between North America and Tibet.
Accepted, Mon. Wea. Rev.

Wallace, J.M., 1975: Diurnal variations in precipitation and thunderstorm frequency over the conterminous United States. Mon. Wea.
Rev., 103, 406—419.

Whiteman, C.D. and T.B. McKee, 1982: Breakup of temperature inversions
in deep mountain valleys: Part II. Thermodynamic model. J. Appl.
Met., 21, 290—302.

Ye, D. Zh., 1982: Some aspects of the thermal influences of the
Qinghai-Tibetan Plateau on the atmospheric circulation. Arch.
Met. Geoph. Biokl., Ser. A, 31, 205—220.

————,Y.X. Gao, M.C. Tang, S.W. Lo, C.B. Shen, D.Y. Gao, Z.S. Song,
Y.F. Qian, F.M. Yuan, G.Q. Li, Y.H. Ding, Z.T. Chen, M.Y. Zhou,
K.J. Yang, and Q.Q. Wang, 1979: Meteorology of Qinghai-Xizang
(Tibet)Plateau. Science Press, Beijing (in Chinese).

AN INVESTIGATION OF THE SUMMER LOWS OVER THE QINGHAI-XIZANG PLATEAU

Lu Junning
Nanjing Institute
of Meteorology

Qian Zhengan
Lanzhou Institute of
Plateau Atmospheric
Physics, Chinese
Academy of Sciences

Shan Fumin
Gansu Provincial
Weather Center

Low pressure areas over the Qinghai-Xizang Plateau on the 500 hPa level in the summer half year are found to be one of the major systems producing rainfall there[1]. In our investigation, 11 similar cases of occurrence of these lows are sorted out from the observations made during the first QXPMEX. Their structure is studied, using the compositing method[2] and taking the mean 500 hPa low center (denoted in the Figures as ▲) as the reference low center. The climatological factors favorable for the development of these lows are also discussed.

I. THE COMPOSITE STRUCTURE OF THE QINGHAI-XIZANG LOW

1. Field of Flow (Wind Field)

During the initial stage of development of the 500 hPa low over the Plateau, an apparent cyclonic circulation or a convergence center is observed at and below the 500 hPa level in the vicinity of the reference center. Near this center above 400 hPa an anticyclonic circulation with the tilting axis nearly vertical is found Fig. 1(a). Fig. 1(b) shows that when the low moves to the eastern part of the Plateau, east of 95°E, in addition to cyclonic circulation around the 500 hPa reference center, a weak disturbance appears on the west side of the 400 hPa reference center, and the convergence zone to the east of the surface reference center is a reflection of the 500 hPa low at

369

the surface. In that case, the vertical axis of the low tilts towards northwest. It should also be noted that, in Fig. 1(b), to the west of the surface reference center, there exists a convergence zone, which, according to analysis made from the observed data, results from the convergence in the western part of a quasi-west-east-orientated shear line over the summer Plateau. Once sufficient heat energy is accumulated and dynamic condition favorable, a Plateau low develops.

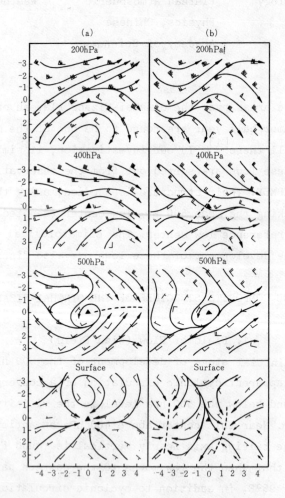

Fig. 1(a) Composite flow fields of different levels in the initial stage of the Plateau Low.

(b) Composite flow fields of different levels when the low moves to the eastern part of the Plateau.

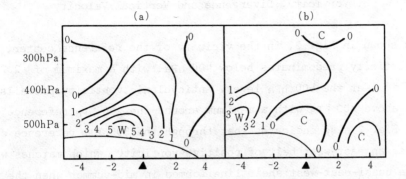

Fig. 2 Cross-sections of temperature departure from normal of
the low (°C). (a) The temperature structure at the beginning of the formation; (b) The temperature structure
when the low moves to the eastern part of the Plateau.

2. Thermal Structure

Most of the lows originate west of 95°E (over the western part of
the Plateau), where intense incident solar radiation heats up the ground
surface and, in turn, warms up the atmosphere, thus producing instability and disturbances[3,4]. The Plateau low, therefore, is of the warm-
core type during the initial stage (Fig. 2a). When the low moves to
the eastern part of the Plateau, precipitation increases and the air
at the lower levels is cooled by the heat loss due to evaporation of
raindrops and the incursion of cold air while the upper part gets
warmer on account of latent heat released through condensation. In
result, a thermal structure of a warm upper part and a cold lower part
occurs.

3. Vorticity, Divergence and Vertical Velocity

As shown in Fig. 3, in the vicinity of the reference center, cyclonic vorticity predominates below 400 hPa, with a maximum of 2.2 × $10^{-5} \cdot s^{-1}$ at the 500 hPa level. Anticyclonic vorticity prevails at heights above 400 hPa with a maximum around the 200 hPa reference center. Fig. 3 also indicates that through the 500 hPa reference center stretches an east-west belt of positive vorticity, which matches well with the quasi-east-west shear line formed in mid-summer when the easterlies on the southern side of the Iranian High comes into contact with the southwest monsoon over the southern part of the Plateau. This shows that the depressions formed over the western part of the Plateau often originate from disturbances over the western section of the shear line, and move eastward along the shear.

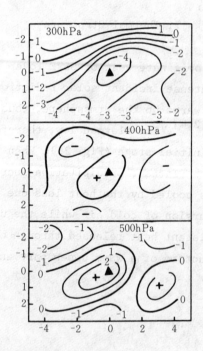

Fig. 3 Vorticity ($10^{-5} \cdot s^{-1}$) fields around the low at different
levels.

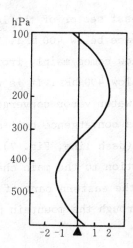

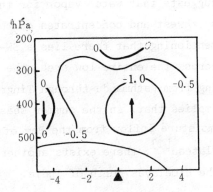

Fig. 4 Vertical divergence
($10^{-5} \cdot s^{-1}$) profile in
the vicinity of the
reference center.

Fig. 5 Zonal section of
vertical speeds around
the Plateau low
(10^{-3} hPa $\cdot s^{-1}$).

As to the divergence field in the neighborhood of the reference center, convergence prevails below 400 hPa with a maximum of $-1.6 \times 10^{-5} \cdot s^{-1}$ in the 500-400 hPa layer. A non-divergence layer exists between 400 and 300 hPa where as divergence prevails above 300 hPa.

Upward motions are found throughout the entire column from the surface up to about 100 hPa with maximum speeds occurring between the 400 and 500 hPa levels (Fig. 5). The ascending motion is stronger to the south of the low than to the north; this is consistent with the fact that more precipitation occurs in the south.

4. Moisture and Precipitation

Fig. 6 gives the moisture flux on the periphery around the low center at different levels between the 500 and the 300 hPa surface, indicating that the moisture flux entering the southwest sector is

greater than the outflow leaving the northeast sector of the low and
that a net water vapor convergence takes place below 400 hPa. This
suggests that water vapor for the Plateau low comes mainly from the
southwest and concentrates on the levels below 400 hPa. It is worth
mentioning that there lies a SW-NE belt of water vapor convergence
across the entire low area, the axis of this convergence belt stretching from Kathmandu through Tingri to Nagqu (dash line, Fig. 7). This
implies that, in the summer season, in addition to the main channel of
moisture inflow from the Bay of Bengal to the eastern part of the
Plateau[1], there exists another channel through the mountain passes of
the Himalayas north of India.

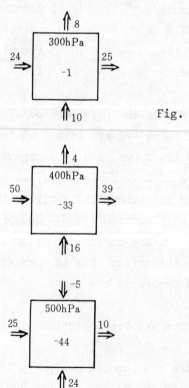

Fig. 6 Horizontal moisture budget
around the low area at
different levels (10^4t·
$(2.5° \cdot 100$ hPa $\cdot$ 12 h$)^{-1}$).

Precipitation produced by the low is centered mainly on an area
within 400 km around the reference center, the heavier rain area
(solid line, Fig. 7) coinciding with the strong moisture convergence

zone. The average daily rainfall amounts to 6-11 mm. The average daily rainfall in the east is twice as much as that in the west due to the lower elevation and hence thicker moist layer over the eastern part.

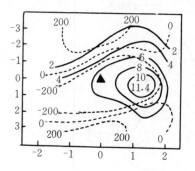

Fig. 7 Average rainfall (mm., solid line) around the low area and vapor flux divergence ($10^{-1} \cdot d^{-1}$) at the 500 hPa surface.

5. Energy Field

The zonal available potential energy A_z, eddy available potential energy A_E, zonal kinetic energy K_z, eddy kinetic energy K_E, and their transformations have been computed by the Krishnamurti energy formula[5] for conditions with and without low development.

When there is the development of the low, A_z is at its peak value at all levels between 500 and 300 hPa along the zonal belt through the reference center (32.5°N), with the maximum A_z value amounting to 134 $J.kg^{-1}$ at the 500 hPa level, which is twice the amount for conditions without cyclogenesis. It is obvious that the intensification of uneven heating of the underlying surface, thus the instability of stratification, is one of the causes leading to the development of the low.

K_E is closely related with the development of the low. Computation shows that K_E has a value of 70-90 $J.kg^{-1}$ to the northwest of the reference center with cyclogenesis while its value is significantly small without cyclogenesis.

A_E is usually an important component to be transformed to K_E. With the development of the low, at about 700 km to the west of the 500 hPa

375

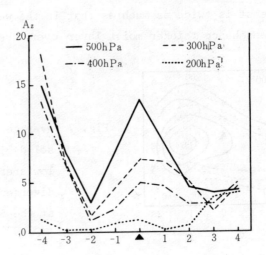

Fig. 8 Variation of A_z $(10 \text{ J} \cdot \text{kg}^{-1})$ in the zonal belt through
the reference center during the development of the low.

reference center is an area of high A_E values, the maximum being
300 J•kg^{-1}. This is also due to the intensification of the uneven heating of the western part of the Plateau and the resulting instability
of stratification. In the absence of cyclogenesis, the maximum A_E value
over the western Plateau at 500 hPa level only comes to 80 J•kg^{-1}.
Therefore, available potential energy also plays an important part in
the development of the low.

Transformation of A_z into A_E always takes place in the zonal belt
through the reference center at 500 hPa and a small amount of A_z is
converted into K_z and of K_z into K_E, no matter whether there is cyclogenesis or not. The significant characteristic is that when there
is cyclogenesis, conversion of A_E into K_E takes place in the belt at
all levels throughout (solid line, Fig. 9) with maximum transformation
at the 500 hPa level. Without cyclogenesis, conversion of K_E into A_E
occurs in the belt at all levels (dash line, Fig.9), a situation
unfavorable for the increase of K_E and hence unfavorable for development
of the low.

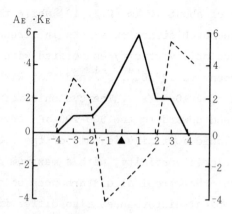

Fig. 9 Transformation of A_E and K_E (10^{-4} J·kg^{-1} s^{-1}) at
500 hPa in the zonal belt through the reference
center with cyclogenesis (solid line) and without
(dash line).

II. COMPARISON OF THE PLATEAU LOW WITH OTHER TROPICAL DISTURBANCES

The development of the low stems from thermal disturbance caused
by uneven surface heating of the Plateau. Is there any distinction between this and other tropical thermal disturbances? To answer this
question, the Indian Monsoon Depression, the African Disturbance and
the Western Pacific ITCZ Wave[2,5] are picked out for comparison.

It is well-known that these latter systems develop over uniform
ocean surface with abundant moisture supply whereas the Plateau low
forms over the arid Plateau with intense incident radiation and rugged
terrain over 4000 m in elevation above sea level. The Plateau Low ranges
from 400 km to 800 km across, with an eastward displacement of 2-3
degrees of longitude per day and a life span of 1-3 days. The Indian
Monsoon Depression and other tropical perturbations have horizontal
scales of a few thousand kilometers and a westward movement of 5.7
degrees of longitude a day, lasting for 3-7 days. All these disturbances

have vertical extent of about 10 km (Fig. 10) while the Plateau Low is only 2-3 km in thickness. Distinction exists in strength as well as in scale. the maximum cyclonic vorticity associated with the Indian Monsoon Depression may amount to $12 \times 10^{-4} \cdot s^{-1}$ and that with the weaker Western Pacific ITCZ Wave, $3 \times 10^{-5} \cdot s^{-1}$. The cyclonic vorticity of both disturbances may extend upward to the 300-200 hPa level. For the Plateau Low, the maximum cyclonic vorticity is only of the order of $2 \times 10^{-5} \cdot s^{-1}$ and the upper anticyclonic vorticity, with a maximum of $-4 \times 10^{-5} \cdot s^{-1}$, is much stronger than other tropical disturbances of the Northern Hemisphere. These tropical disturbances also differ from the Plateau Low in vertical motion. Disturbances, such as the Indian Monsoon Depression have their vertical motion concentrated ahead of them in the

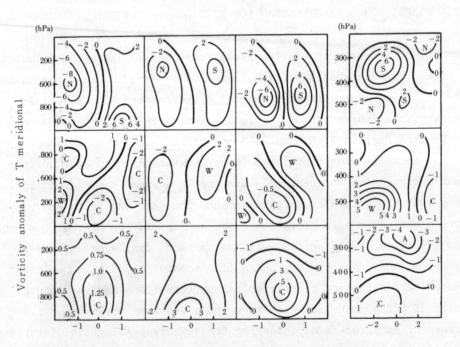

Fig. 10 Vertical cross-section of meridional wind
component, departure of temperature from
normal and vorticity.

direction of their horizontal movement whereas ascending air encircles the Plateau Low in all directions. As shown in Fig. 10, the Indian Monsoon Depression and the African Disturbance have a cold core in the lower troposphere and a warm core in the upper troposphere with roughly equal thickness; the Western Pacific ITCZ Wave is cold on the west side of the center and warm on the east side. However, the Plateau Low has a warm core throughout during the developing stage. Only when the low moves to the eastern part of the Plateau does the cold air take the place of the warm core in the lower troposphere with the warm core remaining in the upper troposphere (Fig. 2).

In short, the low over the Qinghai-Xizang Plateau is obviously a special weather system of small scale, shallow depth, weak intensity and short life, a system characterized by the thermal and dynamic features influenced by a particular underlying surface — the Qinghai-Xizang Plateau. Once the lows move out of the Plateau, they dissipate rapidly with the changed conditions of the underlying surface.

III. STUDY OF THE CLIMATOLOGICAL FACTORS AFFECTING THE DEVELOPMENT OF THE LOW

1. Factors Favoring Development

Six factors have been sieved out by employing Gray's screening method for typhoon development[6]:

1) Relative vorticity in the surface layers

Through the analysis of the data obtained in summer, 1979, it has been found that most of the lows generate at the west end of the shear line over the Plateau. Analysis of the vorticity budget around the low indicates that the horizontal convergence term contributes more to the positive vorticity than any other terms in the vorticity equation by one order of magnitude. This implies that a convergence positive vorticity field is essential to the generation of the low and, other things being equal, the greater the positive vorticity, the more likely the development of the system, that is, the frequency of occurrence of the

low is directly related to the relative vorticity in the surface layers
(the 500 hPa ζ_r value being used).

2) Effect of barotropic instability in the surface layers

Instability of the 500 hPa mean flow has been computed in the
case of cyclogenesis by using the criterion for barotropic instability
$\beta - \dfrac{\partial^2 u}{\partial y^2}$ ($\times 10^{-10} \cdot s^{-1}$). It can be seen from Table I that $\beta - \dfrac{\partial^2 u}{\partial y^2}$
$= 0$ somewhere between 32.5°N and 34.5°N, thus the flow is barotropically
unstable and cyclogenesis is facilitated in this zonal belt. The
barotropic instability of flow depends largely on the horizontal shear
while the criterion for baroclinic instability is closely related to
the vertical wind shear. According to observations made in summer,
1979, the vertical wind shear is very slight as the low develops. It
is quite obvious that the development of the low is independent of baroclinic instability. Hence barotropic instability of flow is one of the
factors affecting the development of lows over the Plateau. This may
be expressed by the reciprocal of the parameters at different latitudes
in Table 1.

Table 1 Composite values of criterion for barotropic instability
of the flow of the 500 hPa low at different geographical
locations.

	82°E	84°E	86°E	88°E	90°E	92°E
38.5°N	0.2325	0.7683	0.7177	0.8997	1.0968	0.8239
36.5°N	0.2576	-0.2833	-0.2782	-0.2833	-0.3035	-0.0912
34.5°N	-0.4315	-0.7504	-0.3961	-0.9016	-0.8156	-0.5124
32.5°N	0.0737	0.5185	-0.3158	0.3416	0.0889	-0.2295
30.5°N	0.5958	0.6817	0.2725	0.6969	0.6969	-0.4644
28.5°N	0.4003	0.2436	0.5671	0.2436	0.4003	0.5924
26.5°N	0.2386	0.5823	0.4051	0.4155	0.5772	0.4104

3) Effect of vertical shear of wind speeds

It is found that the vertical shear $|\frac{\partial v}{\partial p}|$ during the summer time over the Plateau is too small to meet with the baroclinic instability required for cyclogenesis but it helps accumulate heat and moisture in the air column over the Plateau, a favorable condition for the development. In fact, during the periods of prevalent cyclone activities in summer, 1979, the vertical shear of wind speeds between 500-200 hPa over stations located at the western part of the Plateau was found to be smaller than the mean value for the summer season as a whole, being smaller by 2-5 meters per second per 300 hPa with cyclogenesis than without[4]. Therefore $.|\frac{\partial v}{\partial p}|$ may be regarded as a factor affecting the development of the Plateau Low. The smaller the vertical wind shear is, the more likely the development of lows is, other things being equal. There seems to exist a negative correlation. Thus, the reciprocal $1/ |\frac{\partial v}{\partial p}|$ may be used to express this effect.

4) Effect of difference of surface- and air-temperature

Many researchers have found that increased temperature difference between the surface and the overlying air aids the development of the Plateau low. [1,3,4,7] In summer, 1979, the mean of the temperature difference between the surface and the air for the stations located in the western part was found to be 2°C higher with cyclogenesis than without. At places with large temperature difference between the surface and the air, more sensible heat is transferred from the surface upward, providing energy to develop instability, a situation favorable for the low development. Thus, it is more likely that the lows tend to develop over the areas with great difference of these two temperatures.

5) Instability of stratification

As a low develops, the uneven heating of the near-surface air over the western Plateau tends to increase the instability of stratification. This instability and the longitudinal and latitudinal temperature gradients affect directly the production of A_E and A_Z and their transformation into each other. The more unstable the stratification and the larger the temperature gradients, the more likely the development of lows. Thus, $1/(\gamma_d - \gamma)$ may be regarded as one of the factors closely related to the frequency of occurrence of the Plateau Low.

6) Relative humidity

According to observations made in the summer months of 1979, for the same areal dimensions, the cloud amount is 5% larger with the development of lows than without. The air over areas of larger cloud amounts tends to have higher relative humidity. The mean relative humidity $\overline{RH}$ for the 400-300 hPa layer may be used in our case. The higher the $\overline{RH}$, the larger the possibility of occurrence of the low. The parameter $(\overline{RH} - 30)/30$ may be regarded as one of the factors associated with the frequency of occurrence of the low.

2. Physical Parameters Determining the Frequency of Occurrence of the Plateau Low

Effects on the formation of the Plateau Low of six factors have been discussed. It may be assumed that the area where the product of these six parameters is the largest should, in the main, be the area where the frequency of occurrence of the low is the highest. Hence the product of these six parameters may be defined as the frequency parameter of the development of Plateau lows.

The formulations of these parameters and their units are: vorticity parameter $= (\zeta_r + 1.3) \times 10^{-5} \cdot s^{-1}$ barotropic instability parameter $= 2/(|\phi - 34| + 1)$ vertical wind shear parameter $= 1/|\frac{\partial v}{\partial p}|$ (m$\cdot$s^{-1}.300 hPa^{-1}) surface- and air-temperature difference parameter $= T_s - T_a$ (°C) instability parameter $= 1/(\gamma_d - \gamma)$ (100 m$\cdot$°C^{-1})

relative humidity parameter $= (\overline{RH} - 30)/30$ Thus, the frequency parameter of occurrence of the low should be

$$(\zeta_\gamma + 1.3) \times (\frac{2}{|\phi - 34| + 1}) \times \frac{1}{|\frac{\partial v}{\partial p}|} \times (T_s - T_a) \times (\frac{1}{\gamma_d - \gamma}) \times \frac{RH - 30}{30}$$

The computed frequency distribution (Fig. 11) for the period of June-August, 1979 matches quite well with the observed distribution (Fig. 12) except for the Yarlung Zangbo River Valley where the computed value is rather too high.

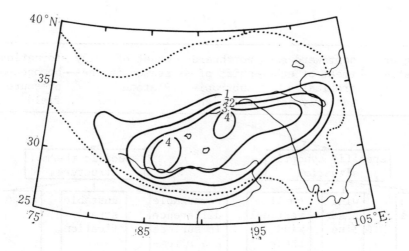

Fig. 11 The computed frequency of occurrence of the
        Plateau Low for June-August, 1979 (number of
        occasions).

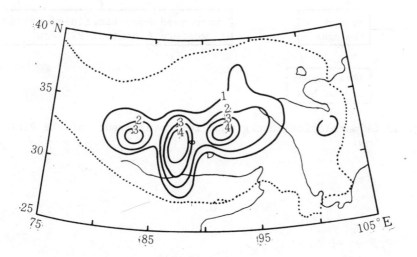

Fig. 12 The observed frequency of occurrence of the
        Plateau Low for June-August, 1979 (number of
        occasions).

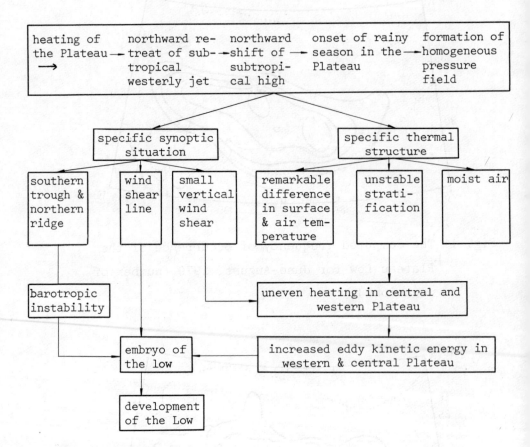

Fig. 13 Climatological background for development of the Plateau
Low.

3. A Schema of the Physical Process of the Development of the Plateau Low

The above analysis shows that the Plateau Low is a special weather system formed under particular temporal and spatial conditions.

After spring has set in, the thermal structure and subsequently the flow field over the Plateau and its periphery experience great changes. In the meantime, the subtropical westerly jet stream south of the Plateau weakens and the westerly jet to the north is appreciably strengthened with the subtropical high being steered northward, resulting in a homogeneous pressure field in this area. The summer circulation pattern comes to pass when the rainy season set in at the end of June. This is the climatological background from which the Plateau Low originates. On the basis of the analysis of the factors affecting the development of the low, a block diagram is presented herein as a preliminary physical schema.

REFERENCES

[1] Lhasa workshop on the Qinghai-Xizang Plateau meteorology, A study of vortices and shearlines on 500 mb over Qinghai-Xizang Plateau during summer time, Science Press, 1981 (in Chinese).

[2] Reed, R. J., The structure and properties of african wave disturbances as observed during phase III of the GATE, Mon. Wea. Rev., Vol. 105, No. 3, 317-323, 1977.

[3] Lu Junning and Zheng Changsheng, Research on the pre-rainy season vortex over the Qinghai-Xizang Plateau, Collected Papers of QXPMEX (1), Science Press, 218-228, 1984 (in Chinese).

[4] Qian Zhengan and Shan Fumin, Analysis of the newly established vortex over the western part of the Qinghai-Xizang Plateau during rainy season, Ibid, 292-242, 1984 (in Chinese).

[5] Krishnamurti, T. N., M. Kabamitsu, R. Godbole, C. Carr, J. H. Chow, Study of a monsoon depression (I), Journal of the Meteorological Society of Japan, Vol. 53, No.4, 227-240, 1975.

[6] Gray, W. M., Tropical cyclone genesis in the Western North Pacific,
    J. Met. Sec. Japan, 55, 465-482, 1977.

[7] Research group of Qinghai-Xizang Plateau low system, Preliminary
    study on the genesis and development of low vortex over the Qinghai-
    Xizang Plateau in mid-summer, Scientia Sinica, 341-350, 1978
    (in Chinese).

ON THE EASTWARD MOVEMENT OF QINGHAI WARM DEPRESSIONS

Ma Henian , Liu Zichen , Qin Ying and Liu Tianshi
Institute of Meteorological Science of Shaanxi Province

In the north of Qinghai-Xizang Plateau, warm depressions appear
frequently over the Qaidam Basin. Most of them do not move across
105°E, but some of them can move further east and change to dynamic
depressions, causing heavy rain in the northeast side of the Plateau.
What are the synoptic conditions to the Qinghai warm depressions moving
towards the east and changing to dynamic depressions? This is a
problem which forecasters are interested in. By using the data of
QXPMEX (1979), a case is studied in this paper of the Qinghai warm
depressions moving towards the east, and, by using the conventional
synoptic data from June to August 1963—1965, 1972—1980, the statistical
characters of this kind of depressions are also examined.

I. A TYPICAL CASE

From 21—23 June 1979, there was a Qinghai warm depression moving
from the Qaidam Basin towards the northeast area of the Qinghai-Xizang
Plateau. It is shown that a typical warm depression is located in the
Qinghai Basin at 0000 GMT, 21 June and a positive vorticity center
coincides with the warm center. The vorticity decreases with height
and approaches negative above 600 hPa. The area of depression coincides
with an area of convergence and ascent flow, but with weak intensity.
They decrease with height sharply and change to divergence and
subsidence flow above 600 hPa (Fig. 1(a)). In the area of the 700 hPa
depression the specific humidity is only 3—4g·kg^{-1}, so that the clear
or partly cloudy weather appears as a consequence.

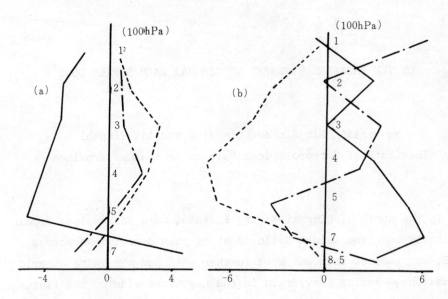

Fig. 1 The vertical profile of vorticity (solid line,
$\times 10^{-5} s^{-1}$), divergence (dotted and dashed line,
$\times 10^{-5} s^{-1}$) and vertical velocity (dashed line,
$\times 10^{-3} hPa \cdot s^{-1}$) at the centre of the 700 hPa
depression (a) 0000GMT , 21,June 1979.
(b) 1200 GMT. 22 June.

Thirty six hours later,it changes to a deep dynamic depression
with the centre located at 106°E, 36°N (Fig. 2). On the surface map,
it appears as a depression with a cold front invading in. On the
700 hPa map, the depression associates with a temperature trough to
its rear. Then there is an area of warm advection in front of the
depression centre and an area of cold advection behind it. Up to the
500 hPa map, there is a geopotential height trough with a temperature
ridge above the 700 hPa depression. The warm advection is very strong
in the front of the trough. Corresponding with this system, above the
depression, the positive vorticity with ascent flow remains upward
throughout the troposphere. The vorticity presents its maximum of
$6.3\times 10^{-5} s^{-1}$ at 700 hPa and the maximum vertical velocity is -7.4×10^{-3}
$hPa \cdot s^{-1}$ at 400 hPa (Fig. 1 b). In the area of the 700 hPa depression,

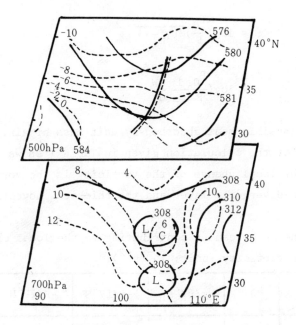

Fig. 2 1200GMT 22 June 700hPa, 500hPa map

the specific humidity is as high as 8–10 $g \cdot kg^{-1}$, then there is a larger area of precipitation in the west and south of the depression.

Is this dynamic depression generated locally or developed from the Qinghai warm depression mentioned earlier?

The following equation is used to calculate and analyse the cause of the local change of the vorticity where the dynamic depression appears:

$$\frac{\partial \zeta}{\partial t} = A_\Omega - g/f \nabla^2 A_T - R/f \nabla^2 S - R/f \nabla^2 H$$

Where A_Ω is the vorticity advection on the non-divergence layer (500 hPa) and A_T is the thickness advection from 700 hPa to the non-divergence layer:

$$S = \log(\frac{P_0}{P})\overline{\omega(\gamma_m - \gamma)}$$

$$H = \log(\frac{P_0}{P})\frac{1}{C_p}\overline{\frac{dW}{dt}}$$

where W is the sensible heat absorbed by unit mass of air.

The result of the calculation given in Table 1 shows that the main cause of the local change of the vorticity is the vorticity advection at non-divergence layer and the thickness advection.

Table 1 the major causes contributing to the local change of vorticity (unit: $10^{-9}s^{-2}$).

Term Time	A_Ω	$-g/f\nabla^2 A_\Omega$	$-R/f\nabla^2 S$	$-R/f\nabla^2 H$
1200GMT, 22	0.926	0.105	0.085	0.024

On the other hand, by using the surface synoptic data at intervals of 3 hours and the 700 hPa data at intervals of 12 hours to track back towards the west, it is found that this dynamic depression is the previous Qinghai warm depression as a result of its moving towards the east and denaturing itself gradually. Along the moving path of the depression, there are corresponding pressure trends at an appropriate time at every station (Fig. 3). The synoptic conditions of this warm depression moving to the east and becoming a dynamic depression are analysed. It is found that from 22–23 June there was a trough on the 500 hPa map moving from the Tarim Basin towards the Qaidam Basin. In front of it above the 700 hPa warm depression there was a warm ridge moving together with the trough. As the trough moving towards the east, there was a cold front moving into the Qaidam Basin and invading the depression from the west.

In front of the depression, however, there was a southely moist jet at the 700 hPa map on the east side of Qinghai-Xizang Plateau (Fig. 4). When the depression was moving towards the east, abundant

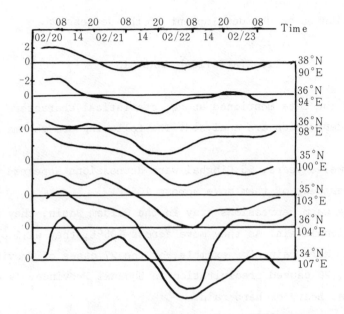

Fig. 3 About the evolution with time of surface pressure
departure along the stations on the moving path
of Qinghai warm depression.

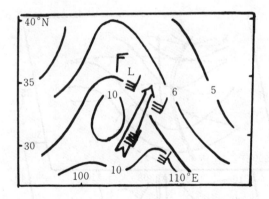

Fig. 4 The 700 hPa specific humidity field $(g \cdot kg^{-1})$ at
1200 GMT, 22 June. The arrow indicates the moist jet.

vapour was transferred into the depression by the jet. Then the dry
adiabatic process changes to a moist one. Diagnostic calculation for
1200 GMT, on 22 June shows that there is a beating centre of latent
heat coinciding with the centre of vertical motion. This shows that

it also contributes to the development of the depression.

II. STATISTICAL ANALYSIS

By using the data mentioned above, statistical characters of the Qinghai warm depression are analysed. The following are the main conclusions:

(1) In twelve years, 65 Qinghai warm depressions appeared in this area and 42 per cent of them move across 105°E.

(2) While warm depressions stay in the Qaidam Basin, they can't cause any regional rain. As they move across 105°E, they will often cause heavy or hard rain. For example, of the 27 cases of "moving across 105°E", 26 caused precipitation in Shaanxi Province. 14 cases caused regional heavy or hard rain.

(3) Moving across 105°E, the warm depressions change to dynamic depressions. By using more complete data of the 19 cases of "moving across 105°E", a composite analysis is made (Fig. 5). It is found that the composite structure is similar to the typical case analysed above,

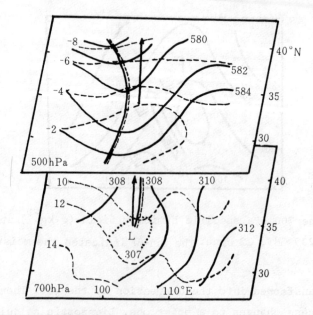

Fig. 5 The 700hPa and 500hPa composite map of 19
Qinghai warm depressions moving over 105°E.

392

Table 2 Statistical table of Qinghai
warm depressions.

months term	6	7	8	total
move to east	12	11	4	27
don't move to east	15	13	10	38
total	27	24	14	65
mean	2.3	2.0	1.2	5.4

that is, the 700 hPa depression associates with a temperature trough
and the 500 hPa trough accompanies a T ridge.Even if it is an average
structure, the temperature advection is still very distinct.

(4) By using the detail data of 19 stationary Qinghai warm depressions over the Qaidam Basin, another composite analysis is made. It is
found that on 500 hPa map, there is another trough to the west of 700 hpa
warm depression centre, but the distance between them is as far as
10° longitude. Only as the 500 hPa trough moving into the Qaidam Basin
is closer to the warm depression of 700 hPa, is it the case in which
the movement of the depression to the east starts. But if the 500 hPa
H trough is weaker and there isn't any cold front associated with it,
the Qinghai warm depression will become weaker as it moves towards
the east. Of the 27 cases of "moving across 105°E", 24 cases are
invaded by the cold front.

The composite analysis of the moisture and the wind field for
the 19 cases of "moving across 105°E" also shows that there is a
stronger southerly wind belt with a larger moisture near 105°E. It is
also in agreement with the case analysed above. But this stronger moist
wind belt doesn't exist in the composite field drawm from the Qinghai
warm depression cases which haven't moved across 105°E (Fig. is
omitted). Finally, for the purpose of comparison, a composite analysis
is also made for 16 cases of the 500 hPa troughs without corresponding
warm depression (Fig. 6). It is found that when this kind of troughs

393

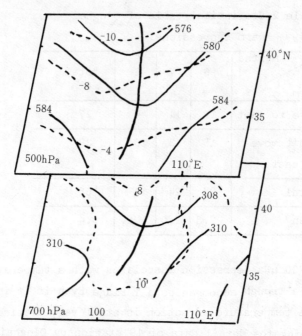

Fig. 6 The 700 hPa and 500 hPa composite maps of 16 cases
of the 500 hPa troughs moving over 105°E without corresponding warm depression.

moves across 105°E, the temperature advection is much weaker than
those shown in Fig. 5. It seems that the warm depressions may play
a part in contributing to the baroclinic development of the troughs.

To sum up, the main conditions of the Qinghai warm depression
moving towards the east and becoming dynamic depressions are as
follows:

1) To the west of the Qinghai warm depression, there is a 500 hPa
trough moving towards it. Corresponding with this, there is a cold
front invading in from the west.

2) To the east of the Qinghai warm depression, that is, on the
east of Qinghai-Xizang Plateau, there is a southerly moist low jet.

394

Characteristics of airflow over and around mountain
ranges according to "ALPEX"

Joachim P. Kuettner
National Center for Atmospheric Research

The influence of mountain ranges on the atmosphere covers practically all
scales from local effects to global circulation pattern. The main effort so
far has been on the two ends of these scales, namely on the gravity scale of
about 10 km and on the planetary scale of about 10,000 km. It is, however, in
the intermediate range (synoptic and subsynoptic) that orography produces the
most weather-effective processes and it is here that scale interactions are
least understood and most difficult to model.

I.   The Alpine Experiment (ALPEX)

The Global Atmospheric Research Program (GARP) focused on these problems
in the Alpine Experiment conducted during 1982 in the European and
Mediterranean area. The Alps produce practically all interesting mountain
phenomena. In fact, they have the highest frequency of cyclogenesis in the
northern hemisphere.

The scientific objectives of ALPEX were quite specific. They included:

- the mechanism of lee cyclones
- the deformation of fronts by mountains
- the total drag of the mountain range and its components including
  momentum transports
- special mountain wind systems, such as Foehn, Bora and Mistral
- radiative and latent heat effects
- orographic precipitation and floods.

There was also a large oceanographic component called MEDALPEX (for
Mediterranean ALPEX). It was focused on the effects of wind forcing on the
dynamics of the Mediterranean Sea, including storm surges and coastal pile-up.

Among the many special observing systems installed in the inner
experiment area, an upper-air network of unprecedented density (Fig. 1) and

395

the availability of long-range research aircraft with drop-wind sounding capability were most noteworthy.

Two months of intensive operations encountered extraordinary favorable weather events, making it possible to fulfill most, if not all scientific objectives. In addition, ALPEX has focused the interest of the international scientific community on the problem of airflow over mountains such that increased theoretical work has already produced results preceding those of the observational phase.

II. Some Preliminary Results

From the large body of preliminary scientific results obtained by various authors who participated in ALPEX, a few are singled out here which appear to be of broad interest.

The first refers to the airflow around mountins and related "flow splitting."

To an unexpected degree, low-level cold air (under 3 km) manages to flow around, rather than over, the mountains, whereby the area in which the "splitting" of the flow occurs remains rather stationary. This process affects practically all orographic phenomena. Aircraft tracks and upwind soundings show that the transition between the flow going over and that going around the mountain barrier occurs in a vertical layer of less than 1 km depth (Fig. 2, 3).

In the upper layer, atmospheric features such as frontal zones, jet streams, vorticity centers, move nearly unrestrained across the mountains but are deformed, in many cases, by vertical stretching on the lee side.

In the lower layer, the airstream in partly blocked, partly deflected and circumnavigates the mountain complex on both sides. The consequences are multiple: Fronts are deformed and regenerated, developing new baroclinic zones and dissolving existing ones. The lateral low-level airflow around the right side of the barrier (between the Alps and the Pyrenees) gains high speed ("Mistral"), see Fig. 4, after Blondin and Bret, and displays some characteristics of a hydraulic-catabatic current and the flow through a gap (Alps/Pyrenees).

Theoretical work following ALPEX has helped to somewhat clarify the mechanism of flow-blockage by a mountain barrier. For high Rossby numbers, the blocked region seems to be controlled by the Froude number and propogates progressively upstream. For low Rossby numbers, the width of the blocked region is limited by the Rossby radius of deformation (Pierre-Humbert).

Considerable theoretical and diagnostic work is being devoted to the problem of frontal passage across a mountain barrier and its proper numerical simulation. While post-frontal airmasses are retarded and deflected around the mountain in the manner just described, the deformation of the front itself, the secondary circulations set up by this deformation and their significance for the formation of cyclones are still obscure and are presently under study by many researchers. Observational work by Steinacker has clarified the movement of the airmasses and the decelerations and accelerations of the front near the surface and aloft (Fig. 5). He has also derived surprisingly large downward displacements of parcel trajectories which cross sharp Montgomery potential gradients developing on higher level isotropic surfaces (Fig. 6).

At the Alps, this mechanism leads to spectacular "North-Foehn" situations clearly visible on satellite pictures

Directly related to the blockage of cold airmasses on one side of the mountain barrier and the Foehn-heating on the other side is the drag exerted by the atmosphere on the mountain. Lines of micro-barographic stations

installed on three sections of the Alps and the Yugoslavian coastal mountains
have provided unique 10 minute data on the pressure distribution across the
mountain barriers enabling the computation of the total drag and its
components (Richner and Phillips; Hafner and Smith). Airflow traverses across
these pressure sections are expected to provide the contribution of the wave
drag from lee wave momentum flux measurements (Fiedler, Durran). Some
examples of very preliminary results will be shown. It is interesting that
rotation (Coriolis effect) makes a negative contribution to the non-rotational
drag (Pierre-Humbert) and so does low level moisture (Barzilon).

These flow phenomena, and especially the disruption of frontal zones
causing thermal unbalance of the high level jetstream, are intimately
connected with the development of lee cyclones.

Several distinctive types of lee cyclones have been observed, among them
the "advective" and "stationary" types. Definite concepts now exist for these
different types of lee cylcones (Bleck, Tibaldi, Buzzi, Smith and others).
The difficulties of existing global prediction models became quite evident
during the ALPEX field phase, especially with the advective type. New models
using higher resolution, "envelope orography" and the comprehensive ALPEX data
set have already produced remarkable improvement in the qualitative and
quantitative predictions of lee cyclones and their propagation (Figs. 7, 8,
after Dell'Osso and Tibaldi,). Many of these results may be applicable to
other mountain ranges.

A physical basis for the success of "envelope orography" has been given
by Pierre-Humbert. He shows that the usual smoothing of the mountain barrier
in low resolution models (by preserving its crossectional area) reduces the
blocking effect by decreasing both Rossby and Froude numbers. In the non-
rotating case the maximum barrier height must be preserved. In the rotating

398

case also the steepness of the slope h/l is important. If it becomes flatter than f/N ($\sim$10^{-2}) it may completely eliminate the blocked region (Fig. 9). This will have a profound effect on the propagation of fronts, lee-cyclogenesis and the formation of Rossby waves.

For a mountain range of the size of the Alps one would not expect direct planetary scale effects because the b changes are too small. Interestingly enough, however, the Alps appear to influence indirectly, during lee cyclogeneses, even the planetary waves (both amplitude and phase speed) and may therefore participate in the development of "blocking" situations. Such a case occurred during ALPEX in late March 1982.

Finally, we wish to mention the cold <u>Bora</u> downslope winds observed in many parts of the world and feared for their violence and gustiness. The classical (original) Bora occurs in an easterly current over the Yugoslavian coastline, which falls steeply into the Adriatic Sea. Repeated probing, especially by aircraft during ALPEX, has provided a comprehensive data set never available before.

Work on this subject is still in progress, but it is already clear that the Bora is a high reaching phenomenon causing large vertical displacements in the stratosphere. It has also been shown that much of the behavior of the lower flow with its severe turbulence (Fig. 10) can be interpreted by hydraulic theory (Petre). An interesting aspect of some of the cases is the existence of a "critical level" where the flow changes from east to west. It can be expected that the obtained data will support some interesting theoretical work presently going on on this subject.

III. Applications to Larger Mountain Complexes

The question arises whether the scientific results of ALPEX--obtained near a mountain range of 1,000 km length and 200 km width--can be generalized and to what extent they are of interest to atmospheric developments occurring near the Tibetan Plateu, the Himalayas, and the Rocky Mountains, whose characteristic dimensions are of the order of 3,000 by 1,500 km.

The primary difference lies, no doubt, in the role of the elevated heat sources and sinks presented by these huge plateaus. Here, the area factor of more than 20 practically excludes dynamic similarities (Fig. 11). However, with regard to the plateau edges and rims, many of the ALPEX results may apply, especially since the blocked flow upstream has a "mechanical" plateau effect. This is particularly true for the observed lee effects, such as the lee cyclones, downslope winds, channeled surface jets and lee waves.

In addition to the scientific results much has been learned in ALPEX about the needed observing systems. At some time in the future, a similar effort may possibly be considered for the Tibet-Himalayan area. Impressive advances can be achieved by filling in gaps in the upper air network, even on a temporary basis, by the use of long-range research aircraft releasing dropwindsondes (including a new light weight version presently under development) over data-sparse areas and by special techniques used in the programs of geostationary satellites.

Such an effort will require careful planning, several years in advance, and should be based on the definition of very specific scientific objectives. In view of the significance of the Tibetan Plateau and the Himalayas for the global atmospheric circulation, this should be a worthwhile undertaking.

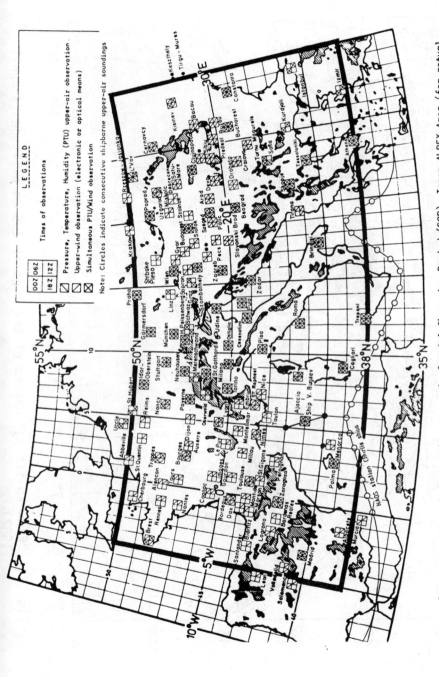

Fig. 1 Upper-air Network during the two months Special Observing Period (SOP), Inner ALPEX Area (for actual locations and times of soundings taken by the V. BUGAEV and CHARLIE Ships refer to Figure 4.23)

Fig. 1. ALPEX inner experiment area with special upper air observing network.

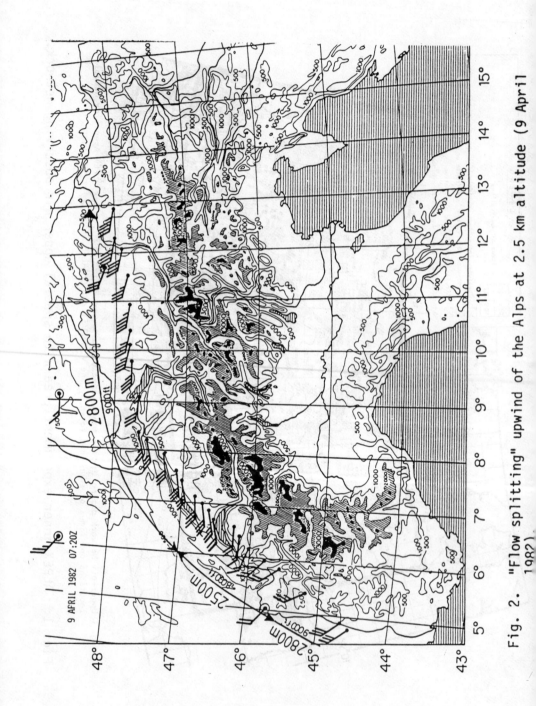

Fig. 2. "Flow splitting" upwind of the Alps at 2.5 km altitude (9 April 1982)

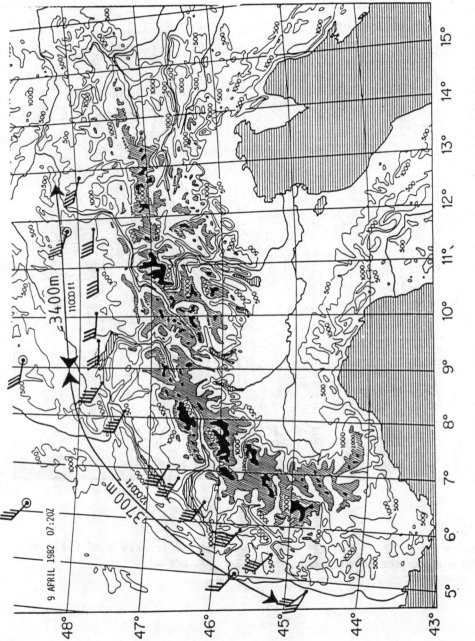

Fig. 3. As Fig. 2, but approximately 1 km higher: The flow crosses the mountain barrier without restraint.

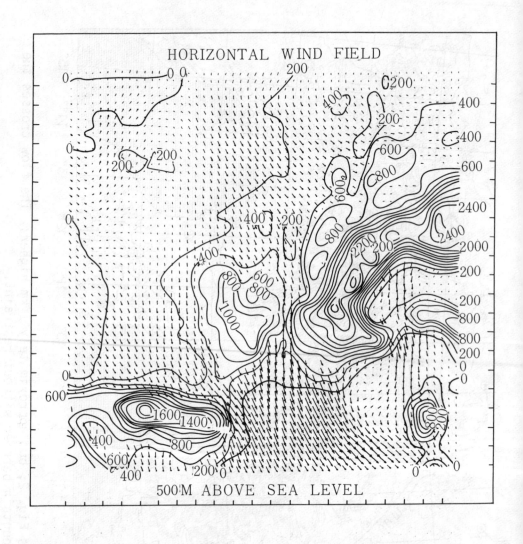

Fig. 4.

The "Mistral" on March 21, 1982: Computed horizontal wind field at 500 m above sea level. (After Blondin and Bret.)

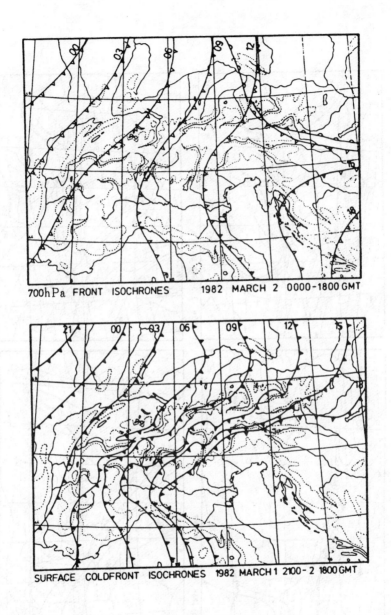

Fig. 5.
Cold front crossing the Alps at 700 hPa (top) and surface (bottom).
After Steinacker.

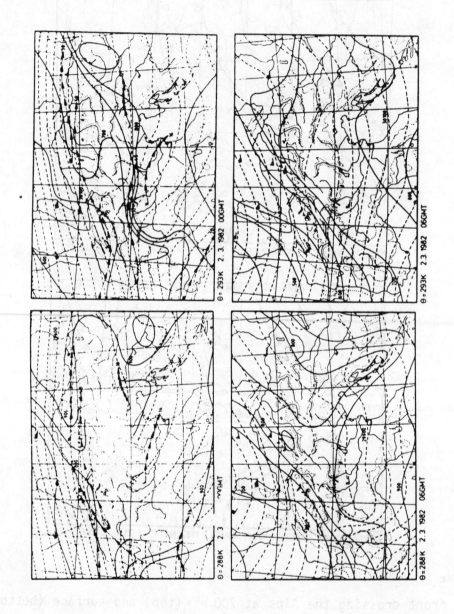

θ=293K   2.3.1982   00GMT

θ=293K   2.3.1982   06GMT

θ=288K   2.3.1982   00GMT

θ=288K   2.3.1982   06GMT

Fig.
Cold front crossing the Alps at 700 hPa (top) and surface (bottom).
After Svinback(?).

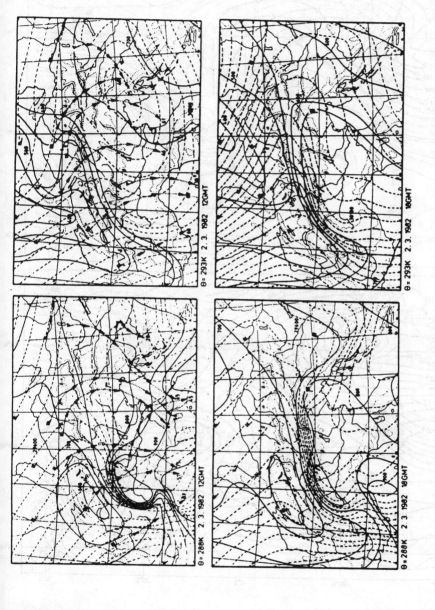

Fig. 6. Descending motion during "North-Foehn": Isentropic trajectory analysis by Steinacker indicating parcel descents of 3 to 4 km (600 to 950 hPa).

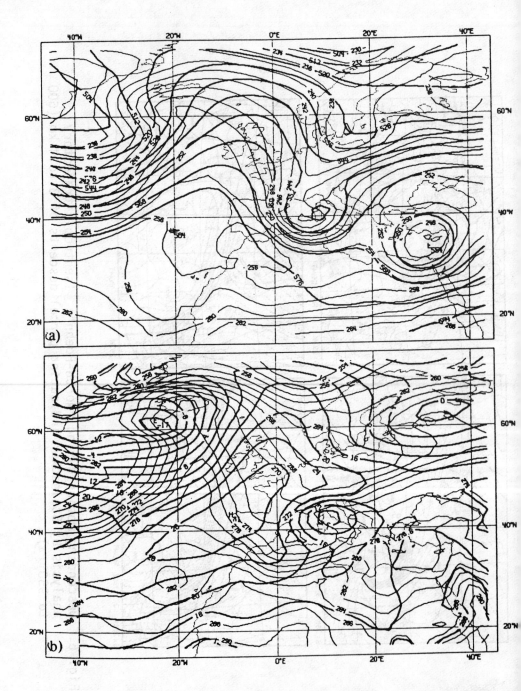

Fig. 7. ECMWF operational analysis of lee cyclone on 5 March 1982.

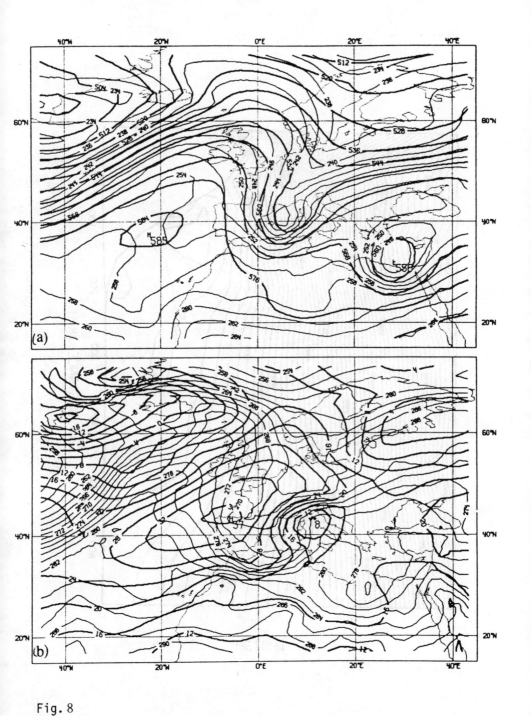

Fig. 8

48 hour forecast for 5 March 1982 using "half envelope" topography.
(After Dell'Osso and Tibaldi).

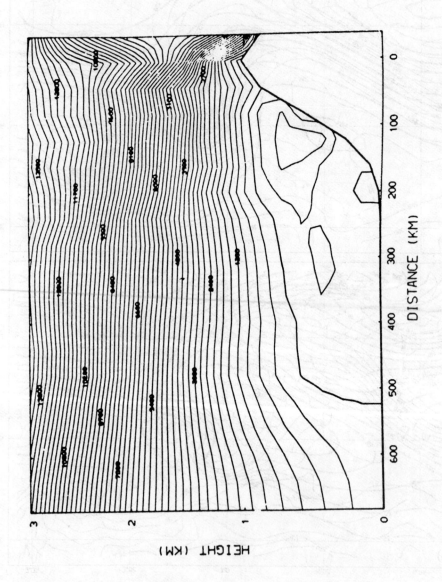

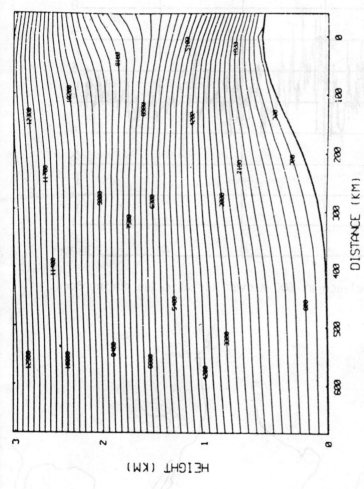

Fig. 9 . Numerical model of upstream flow blockage after Pierre-Humbert.

Top: Mountain of 1 km height and 100 km length scale produces blocked region corresponding to Rossby radius of deformation.

Bottom: Smoothed mountain profile (preserving crossectional area) eliminates blocked region reducing both Rossby and Froude numbers.

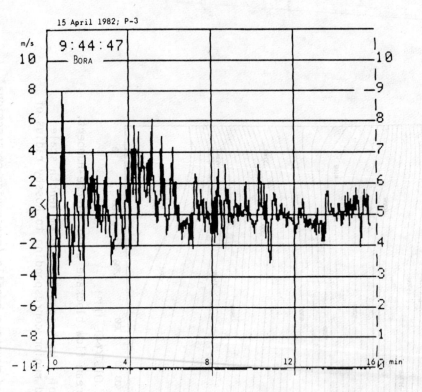

Fig. 10. Turbulence at 600 m over Adriatic See during Bora.

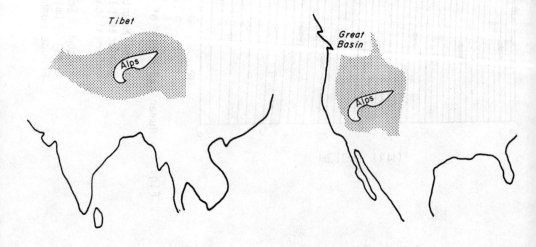

Fig. 11

Size comparison of Alps with Tibet (left) and the Great Basin of the Rocky Mountains (right).

References

Because of the preliminary nature of the results presented in this paper only the following references are quoted:

- "ALPEX Field Phase Report" (1932), GARP-ALPEX No. 6A. WMO, Geneva
- ALPEX Preliminary Scientific Results" (1982), GARP-ALPEX No. 7. WMO, Geneva

Most of the other sources are still in the process of publication or are due to personal communications.

413

600 hPa CIRCULATION SYSTEMS OVER THE Qinghai-Xizang
PLATEAU AND SOUTH ASIA IN THE SUMMER, 1979

Yin Daosheng

Qinghai Meteorological Service

I. INTRODUCTION

It is known that the Qinghai-Xizang Plateau is the greatest one in
the world. Although there are plenty of the highest mountain ranges on
it, this huge massif is rather flat on the whole. Owing to the necessity
of improving local weather forecasting as well as the research work in
Plateau meteorology, Chinese meteorologists have long been desirous of
revealing the circulation systems on the Plateau ground. But two obstacles should be surmounted at first, i.e., the dearth of data and the
lack of convenient means in analysis. At last it came a chance in
1979. At that time, two important Experiments, MONEX and QXPMEX were
carried out and a lot of useful data obtained. Further more, since 1976
a practical method in Plateau analysis had also been developed. This
work was based on the two combined.

Since the average height of the Qinghai-Xizang Plateau is about
4500 meters which is very near the 600 hPa level ($\sim$4200m), this isobaric chart may be used as surface chart in this region. But at the same
time it certainly is an upper-air chart outside the Plateau. Furthermore,
any upper-air systems can emerge at any part of the highland just as
the Plateau surface systems do. In order to make the chart reflect facts
objectively, we approached many cases at first and then this isobaric
chart was analysed in such a particular way as follows:

1) Within the scope of the Plateau, surface elements such as wind,
cloud, weather, temperature and pressure are all plotted and analysed

as we usually do on surface charts. These elements are chosen from
about 40 stations with heights 3500meters up above sea-level.

2) Fronts should be analysed where isotherm ribbons coordinate
well with flow fields. Otherwise, shearlines or trough lines were
analyzed only. Therefore, one can see on a single chart that the two
systems can co-exist or even be combined and at different times they
can transform themselves interchangeably.

3) Contours are analysed at a 20 geopotential meter interval and
isotherms at a 2°C interval. This is because that the Plateau systems
are much weaker than those in westerlies especially in summertime.

During QXPMEX, by analysing 600 hPa twice a day, it was found that
sometimes the circulation systems were quite different from those
known on 500 hPa previously. This fact seems to remind us that we
should not lose our sight of the importance of the Plateau boundary
layer no matter how shallow its depth may be.

II. WEATHER SYSTEMS

1. Squall Lines

Most students of Plateau meteorology are familiar with the frequency of squall lines during summertime, But no one has ever seen it on
a surface chart before, since isobars are not analysed over orographical
areas. This is true when we look into the case on 7, August 1979 (not
shown). We can hardly realize why there is a severe thunderstorm aris-
ing so suddenly over the eastern Plateau within a scope of 5°×10°.
Nothing was significant 12 hours earlier and 500 hPa (Fig. 1a) gives us
little information either. Neither deep troughs nor frontal zones are
there. Only a weak shearline is moving between the two powerful cells
(over 590) of a Qinghai-xizang high. But from Fig. 1(b), especially
Fig. 1(c), we can clearly tell that a squall is just developing on the
shearline. The wake depression is weak and an asymetric heat low has
developed well in front of the squall line. Temperature contrast looks
more strikingly the height. does How can it be interpreted? It is known
that the Qinghai-Xizang high is strengthening up greatly with height

415

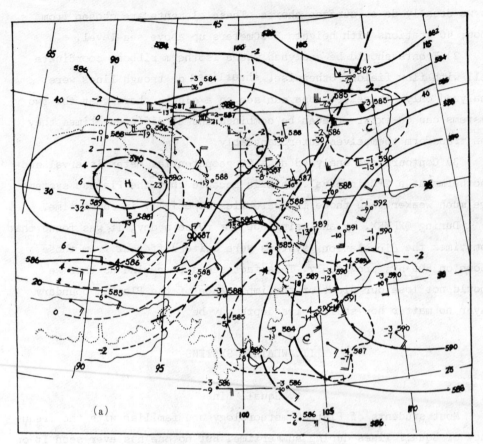

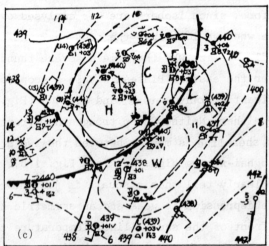

Fig. 1 A case of squall lines, 1200GMT, 7 Aug. 1979.

(a) 500 hPa; (c) 600 hPa in detail.

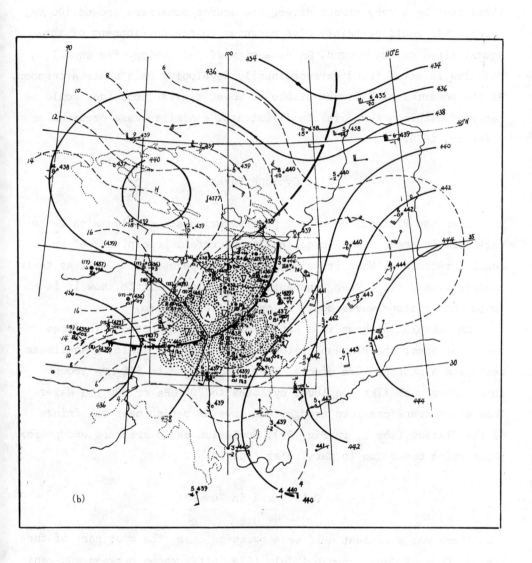

(b)

Fig. 1 (b) 600 hPa;

owing to its warm structure so from Figure 1(a) one can imagine that
there must be a very strong divergence source somewhere around 100 hPa
level. This would certainly give momentum to the development of the
squall lines on the ground. So as a sign of the coming of a squall, a
heat low is often found extraordinarily developing on a clear afternoon.
As the powerful high is too stable to move, similar processes could be
repeated within a few days. To illustrate, a similar case occurred on
8 August, 1979.

2. Cyclone Waves

Over the Plateau, sometimes precipitation accompanies with 24 hr
katallobaric areas. This is called "pressurefall precipitation" by
local forecasters. When it rains, surface temperatures as well as their
changes i. e., ΔT_{24}, keep constant or even rise a little. Now it is
found that cyclone waves could be one possible system.

An example is shown in a series of Fig. 2. Compared with those
sea level ones, the Plateau cyclones have a smaller scale, a shallower
depth and a shorter life circle. They are related to 500 hPa troughs
in a manner much like sea level cyclones to 700 hPa or 500 hPa waves.
They often transform into troughs and lows or perish near the fringe
of the Plateau (see L_2 in Fig. 2e). Occlusion cases are rare except for
those which occur due to particular topography.

3. Cold Waves in Summer

There was a violent cold wave sweeping over the most part of the
Plateau from 29 June through 1 July 1979. Differences between 600- and
500 hPa (not shown) are as prominent as those between the surface and
upper-air charts. But the cold air mass breaking out over the Plateau
later did not first appear in the west then gradually moved eastward
on 600 hPa level like what we have best known on the surface. It immediately came from above at the layer of 150-200 hPa over Tarim Basin
(the southern part of xinjiang) just to the north of the Plateau. As
ΔT_{24} is a good indicator for thermal advection and its effectiveness

418

Fig. 2 A course of the evolution of Qinghai-Xizang cyclone
waves on 600 hPa from (a) 1200GMT 13 July through (f)
0000GMT 16 July 1979, for every 12 hours.

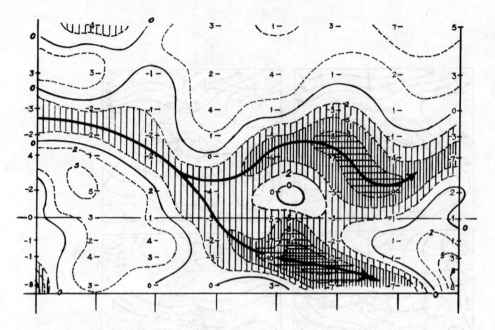

Fig. 3 Time section of ΔT_{24} ($^{\circ}C \cdot d^{-1}$) at Andihe Station (51848)

has widely been accepted in China, especially in Plateau meteorology[1] we give the time section of ΔT_{24} at Andihe Station (51848) in Fig.3. It may be seen that the cold air mass had first split into two branches before 29 June. While one branch still remained aloft and intensified continuously, the other one quickly descended to the ground and soon filled the basin up as a cold cushion. 12-24 hours later, the upper branch was beginning to pour down over the cushion and made its way into the Plateau. A summer cold wave was then breaking out. This processes can be modeled in Fig. 4.

Five rawinsonde stations over Tarim Basin reported that temperatures at 300 hPa in June 29 1200GMT varied from -33°C to -38°C. 24 hours later, at 1200GMT 30 June, a cold tongue (not shown) with the maximum temperature fall being -13°C. d^{-1}, was found in the central part of the Plateau (Table 1). Figure 5 might help better understand the process. In the first row there are cross sections for ΔT_{24}, second for the potential temperatures (θ), and third the corresponding 600 hpa. From this Figure the two branches of cold air mass are well displayed. One

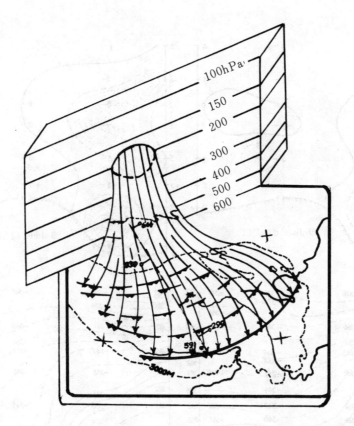

Fig. 4 A model for the Plateau cold outbreaks.

Table 1 Temperatures and their 24 hr changes (ΔT_{24} in parentheses) over the Plateau at 1200GMT 30 June

Tuotuohe (56004)	$4.8°C(-5°C.d^{-1})$
Gar (55248)	$11.3°C(-13°C.d^{-1})$
Shuanghu(Exp. Base)	$4.8°C(-15°C.d^{-1})$
Shiquanhe(55228)	$15.2°C(- 7°C.d^{-1})$
Xainza (55472)	$11.2°C(- 5°C.d^{-1})$
Nagqu (55299)	$9.4°C(- 4°C.d^{-1})$
Average	$9.5°C(- 8°C.d^{-1})$

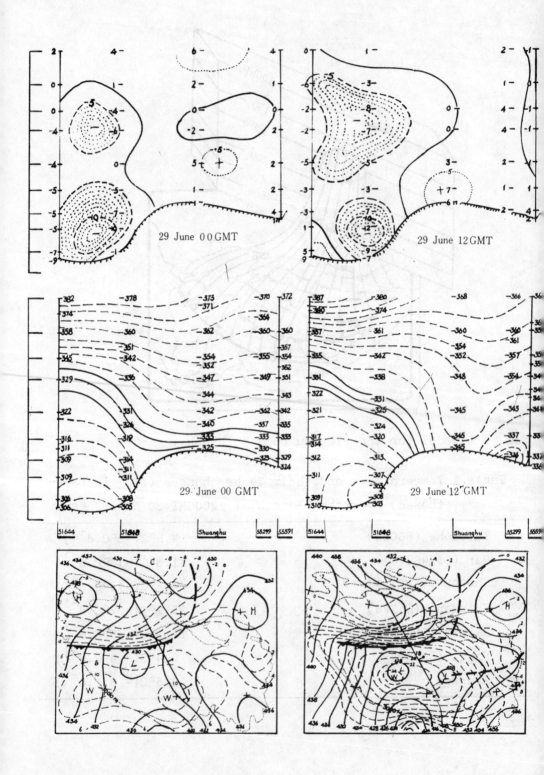

29 June 00 GMT

29 June 12 GMT

29 June 00 GMT

29 June 12 GMT

51644   51848   Shuanghu   55299   55591   51644   51848   Shuanghu   55299   55591

422

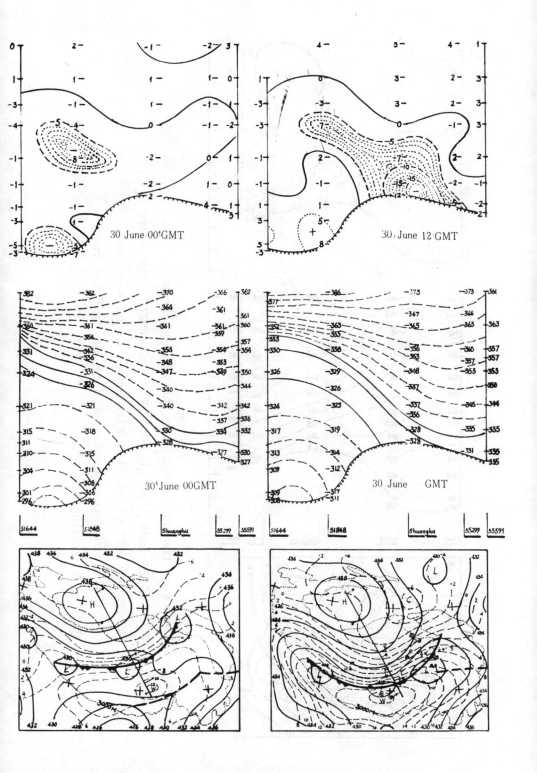

30 June 00 GMT

30 June 12 GMT

30 June 00GMT

30 June   GMT

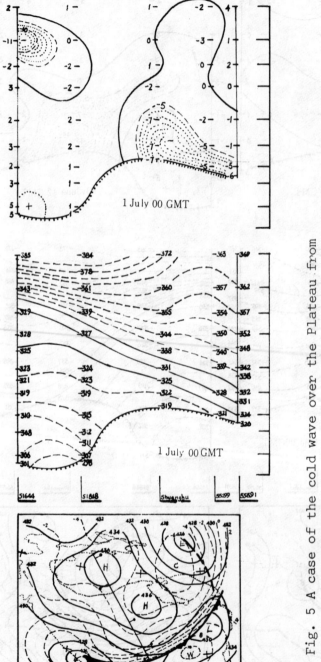

Fig. 5 A case of the cold wave over the Plateau from 0000GMT 29 June through 0000GMT 1 July 1979, for every 12 hours (left ot right), First row: cross sections of △T24; second row: cross sections of θ; Third row: the corresponding 600 hPa.

has served as a cushion before the other is pouring down along the frontal θ lines (streamlines also). Since the distance between Tarim Basin and the cold front at 1200GMT 30 June is about 1000km, assuming the cold air descending dry-adiabatically, the results in Table 1 could be obtained when there existed in the midtroposphere an average north-west flow at 10 $m.s^{-1}$ accompanying a downward motion at 5 $cm.s^{-1}$ in the rear part of a quasi-stationary trough lasted for 24 hours there. All these conditions are checked without much difficulty in summertime. Being a weather forecaster there for a long time, the author is often puzzled by the question why there are more cold processes observed over the Plateau region than in any other places at the same latitudes during the warm season (Fig. 6). From this example, a possible answer may be that the cold air mass is directly coming from above nearby not from west far away so there is less diabatic heating to take part in.

4. South Disturbances

Due to their huge sizes both the Qinghai-Xizang Plateau of China and India have their own circulation systems in summer. But how can they influence each other? Or of the author's interest, how can a south disturbance get across the Himalayas to enter the Plateau on 600 hPa? There are at least two possible ways to make it: one is passing through gaps either on the west or east of the mountains. Figure 7(a) shows a disturbance entering the Plateau from west. After 2.5 days when the disturbance drifted outside the Plateau and joined the wester-lies, it caused a heavy rainfall over a vast area in North China. The passing through east gaps which seems a little easy. Figure 7(b) shows the successive positions of a disturbance splitting from a monsoon low. The two kinds of disturbances are all coming from Indian lows well developed and quite steady. Disturbances can only be determined by wind shear and rainfall. No significant pressure or temperature changes occur.

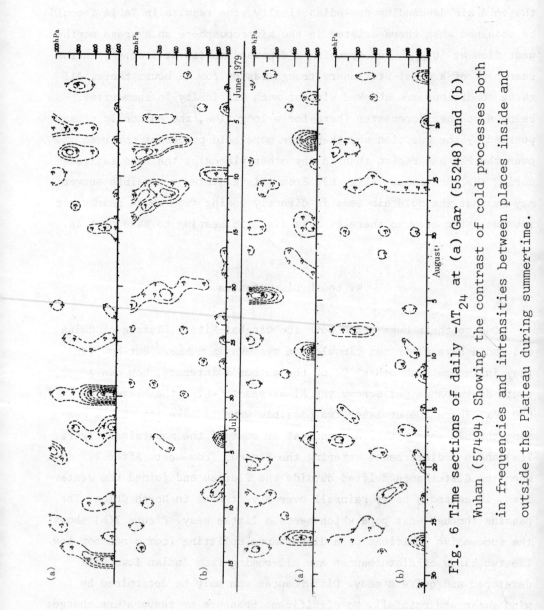

Fig. 6 Time sections of daily $-\Delta T_{24}$ at (a) Gar (55248) and (b) Wuhan (57494) Showing the contrast of cold processes both in frequencies and intensities between places inside and outside the Plateau during summertime.

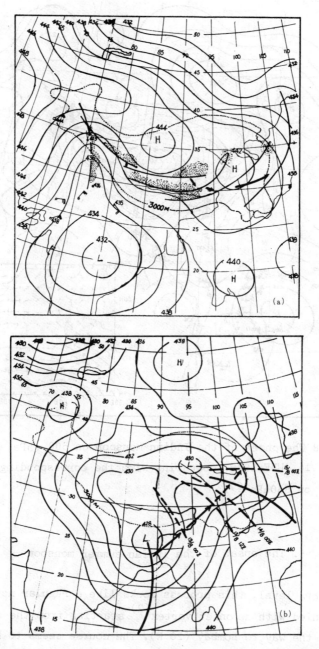

Fig. 7 South disturbances entering the Plateau on 600 hPa (a) from west, 0000GMT 9 July 1979; (b) from east, 1200GMT 14 June 1979.

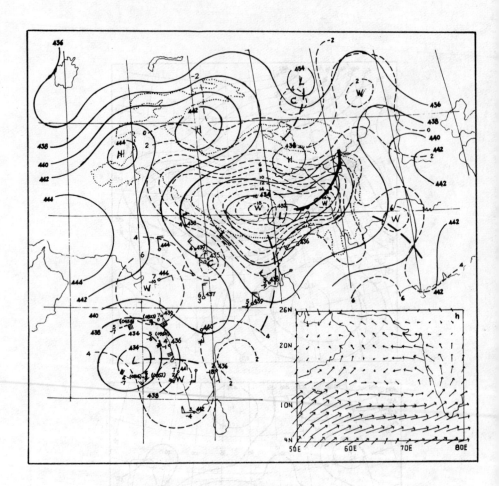

Fig. 8 The onset of the Indian monsoon on 600 hPa, 1200GMT
16 JUNE 1979. Comparing with the corresponding flow field
on 850hPa[2].

5. The onset of the indian summer monsoon

On 600 hPa level, this process looks like a regular movement of
ITCZ accompanied with an onset vortex[2] starting from the southern tip
of India all the way to Bombay (20 N). The course seems to have two
sub-periods:

1) From 13 through 15 of June. The subtropical high began to with-

draw from subcontinent and the onset vortex formed up over the Arabian Sea.

2) From 16 through 19 of June. ITCZ appeared at the southern tip of India. On moving northwardly, it led the strong zonal flow from the Arabian Sea gradually to prevail all over India. At the same time the high to the north of ITCZ quickly weakened and perished.

By using the MONEX data including the aircrafts dropwinsondes, the course was clearly analysed. One of these charts is given in Fig. 8. It may be seen that the onset vortex on 600 hPa level coincides well with flow field[2]. Its notable asymetric thermal structure is known as one of the typical characters of the midtropospheric cyclones whose maximum intensity is observed at 600 hPa[3].

III. CIRCULATIONS

1. Diurnal circulations

Figure 9 shows the 31-day mean differences of 600 hPa height between 00GMT and 12GMT defined by

$$\overline{\Delta H_{12}} = \overline{H}(1200GMT) - \overline{H}(0000GMT)$$

Three points can be derived from Fig. 9:

1) The sensible heating effect from the Plateau surface is far from uniform. Heating centers represented by the two maximum areas of $-\overline{\Delta H_{12}}$ are located in the western and eastern parts of the Plateau respectively, Heat lows coordinate with them well (see next section). On the other hand, it again confirms the conclusion drawn by Flohn[4] that the highs tend to stay and intensify between these two centers.

2) The larger positive values are found in the north rather than in the south and a pair of centers with different signs are situated over Tarim Basin and the Western Plateau. It means that the diurnal vertical circulations caused by sensible heating of the Plateau are related more with the westerlies than with subtropical circulations. The northern descending branch would probably be a constant trigger

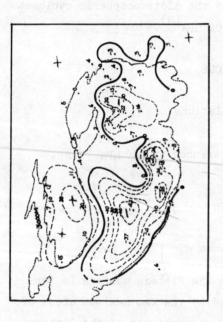

Fig. 9 The 31-day mean differences of 600 hPa height between 1200GMT and 0000GMT (in GPT Meter)

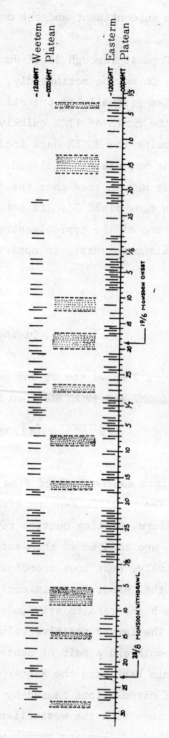

Fig. 10 The occurrences of heat LOWs on 600 hPa (bars) and HIGHs on 500 hPa (Colums) from 1May through 31 August 1979.

for the cold outbreak in summer.

3) Assuming the inward and outward flows are approximately equal in magnitudes, then from Figure 11 the mean penetrating flow along the Altay Shan could be estimated at an order of 1 m.s^{-1}. Similar results were obtained by other authors[4,5].

2. Heat lows

Depressions are often classified as Western and Eastern lows since they are greatly concentrated in these two regions. Their frequencies during the TIPMEX period are shown in Fig. 10 and three facts may be seen as follows:

1) Two thirds of Western lows forming at 1200GMT (18 Local Time) is a sign showing that the sensible heating is playing an important role there.

2) When the highs on 500 hPa are active the number of the lows on 600 hPa decreases.

3) The first appearance of Western low on 23 May is considered a regime of seasonal change over the Plateau (see next section). After the onset of the Indian monsoon, more western lows especially at 00GMT are observed. This might be another sign for the coming of rainy season to the western Plateau.

3. Time sequence from may through August 1979 (Fig. 11)

It can be roughly divided into five stages:

1) Before 17 May, the sinuous circulation prevailed. The whole Plateau was still under the control of the westerlies as in winter time.

2) From 18 May through 4 June, contours gradually became closed or half closed around the Plateau. The circulation was in the shape of a inverted "omega", reflecting the fact that the sensible heating was ʔ then taking place to a remarkable extent over the Plateau.

3) From 5 June through 24 June, the cell enlarged day-by day until it came to Bangladesh and her surroundings were included in the cell.

Fig. 11 Time sequence of the circulation on 600 hPa from 1 May
through 31 August 1979.(a) 1200GMT 7 May, standing
for the period before 17 May; (b) 1200GMT 23 May,
18 May to 4 June; (c) 1200GMT 14 June, June 5 to 24;
(d) 1200GMT 14 July, 25 June to 18 August for active
LOWs; (e) 0000GMT 20 July, same as (d) but for active
HIGHs; (f) 1200GMT 29 August, After 19 August.

4) From 15 June through 18 August, as a whole, the Plateau and subcontinent were all covered by a huge irregular cell. This is the main type on 600 hPa during the summertime in 1979. Three centers can usually be found in the cell. One was located in the central part of India and the other two were distributed in the western and eastern parts of the Plateau respectively. Cells were destroyed by the activity of the highs at times. Each break could last for 3-5 days. The expanding of the cell from the Plateau to subcontinent seemingly reflects the combining of the two heating sources in the midtroposphere in South Asia. At this stage, besides the Plateau sensible heating source there appeared a new one over India caused by monsoon rainfalls. As the beginning of rainy season in Bangladesh is about 10-15 days earlier than in India, the cell seemd first to have expanded there as mentioned above.

5) After 19 August, the westerlies intensified again and gradually drifted southward to about 40°N. The center of the Indian low disappeared as soon as summer monsoon withdrew. A number of transient highs and lows moved over the Plateau. This can be still considered a transition type as the Qinghai-Xizang rainy season did not end completely yet.

IV. A SUGGESTION

From every angle, practical and theoretical, it seems that the 600 hPa analysis is worth more testing to prove its usefulness. Once this is proved and acknowledged, we hope that WMO will help to make it routine operation.

REFERENCES

(1) Lhasa Workshop on the Qinghai-Xizang Plateau Meteorology, A study of vortices and shearlines on 500 mb over the Qinghai-Xizang Plateau during the summer time Science Press, 1981(in Chinese).
(2) Krishnamurti, T.N. et al., On the onset vortex of the summer monsoon, Mon. Wea. Rev., Vol. 109, No. 2, 345-363, 1981.
(3) JOC, The monsoon experiment, GARP Publications Series, No, 18, 1976.

(4) Flohn, H., Contributions to a meteorology of the Tibetan Highlands, Atm. Sci. Paper, No. 130, 1968.

(5) Luo, H. and Yanai, M., The large scale circulation and heat sources over the Tibetan Plateau of 1979. Part 1: precipitation and kinematic analyses, Mon. Wea. Rev., Vol. 111. No. 5, 922-944, 1983.

LEESIDE EXTRATROPICAL CYCLONE DEVELOPMENT AND THE RELATION

OF TOPOGRAPHY TO ACCELERATION OF CIRCULATION

Donald R. Johnson
Dept. of Meteorology
and Space Science and Engineering Center
University of Wisconsin-Madison

1.   Introduction

In an effort to gain insight into the evolution of extratropical cyclone from the incipient to the decaying stage, several diagnostic studies at the University of Wisconsin have utilized quasi-Lagrangian transport relations to study the mass circulation and angular momentum transport within cyclones (Johnson and Downey, 1975a and b). These studies have isolated two distinct types of development for extratropical cyclones over the continental United States, one of which is related intimately to the effects of mountainous terrain (Johnson, Wash and Petersen, 1976).

The main purposes of this paper are: 1) to review and summarize the diagnostic results showing the structure of the mass circulation and angular momentum transport within an extratropical cyclone developing in the lee of the Rockies, 2) to establish the two types of development, and 3) to discuss the forcing of the mass circulation and its relation to topography. Some of the diagnostic results summarized in this paper were previously presented by Johnson, Wash and Petersen (1976). Additional diagnostics results included herein are from a detailed comparison of isentropically and isobarically averaged transport statistics that are being analyzed at the University of Wisconsin.

During the 96-hour period of the case study for the Alberta cyclone of 30 March - 02 April (Johnson, Wash and Peterson, 1976), the first 36 hours displayed a mass circulation with an inward branch in upper isentropic layers and an outward branch in lower layers. The spin-up of the relative circulation occurs from the excess of the inward transport of storm absolute angular momentum in the upper layers over the outward transport in lower layers and a transfer of angular momentum from the upper to lower layers through pressure and inertial torques. In view of the light precipitation during this initial period relative to the latter period, the influence of topography, and importance of baroclinic processes, this stage is called the period of leeside dry baroclinic development.

After the period of initial development, the mass circulation undergoes transition to a structure with inward transport in lower layers and outward transport in upper layers. The excess of the inward transport of absolute angular momentum over the outward transport of the upper branch serves to spin up the relative circulation while the upward transfer of the angular momentum through the pressure torque and diabatic transport develops the vortex vertically and forces occlusion. In view of the marked increase in precipitation and importance of diabatic processes during this period, this stage is called the period of moist baroclinic development.

2.   Synoptic Discussion

The initial development at 1200 UT 30 March 1971 in the lee of the Rocky Mountains over Alberta occurred on a stationary polar front as a short wave 500 hPa trough approached the west coast of the United States (Fig. 1b). By 1200 UT 31 March (Fig. 1c) impressive surface cyclogenesis had occurred over the Dakotas as the system deepened to 990 hPa The warm sector was dominated by clear subsiding air while a band of clouds (Fig. 3a) and very light precipitation occurred north of the surface center. The 500 hPa trough was centered over Idaho with a jet wind maximum located to the south-southwest.

By 1200 UT 01 April (Fig. 2b) the amplitude of the 500 hPa trough had increased significantly through strong baroclinic development. By this time, the surface depression (Fig. 2a) was located in northern Wisconsin and although filling slightly, had developed a stronger pressure gradient and a stronger low-level circulation. Light precipitation was confined to an area northwest of the surface center. The satellite image in Fig. 3b reveals extensive cloudiness intruding into the surface center from the Gulf of Mexico.

At 1200 UT 02 April (Fig. 2c and 2d) the cyclone was located over northeastern Ontario and a 500 hPa cut-off developed over Lake Superior in conjunction with the occurrence of significant precipitation that developed within the cyclone's circulation. The clouds in the satellite image (Fig. 3c portray a typical pattern for an occluding cyclone. In the following time periods the surface and 500 hPa lows drifted slowly eastward and began to fill.

The overall life cycle of this storm is quite similar to Palmen and Newton's (1969) description of cyclones developing under the influence of orography. They state "when development is favored both by orographic influences and divergence aloft, a slackening occurs when the cyclone has moved eastward and the orographic influence ceases; and further development ensues when the upper level trough has moved forward relative to the sea-level cyclone, with intensified upper divergence over it."

3.   Mass and Angular Momentum Transport

The azimuthally averaged structure of the storm is analyzed within a quasi-spherical coordinate system that translates with the central pressure of the cyclone. The storm track for this cyclone and position of the budget volumes are illustrated in Fig. 4. The transport equations for mass and angular momentum of storms were presented by Johnson and Downey (1975a and b). See Appendix A.

a. An overview of Eliassen's theory for the forcing of the meridional mass
    circulation within a vortex

In 1951 Eliassen developed the theory of stable meridional motion forced

by a combination of heating and torques. In his theory these two factors

combine to force the transport of absolute angular momentum that is necessary

for the development and maintenance of the vortex. In a hydrodynamically

stable vortex, the role of the torques is to force the horizontal branches of

the mass circulation through the surfaces of angular momentum, and the role of

diabatic processes is to force the vertical branches through isentropic

surfaces. In this manner the mass circulation is maintained in the presence

of hydrodynamic stability and serves to supply the angular momentum and energy

needed to sustain the vortex. The basis for extension of these concepts to

the cyclone stems from Johnson's (1974) interpretation of Eliassen's theory

for the extratropical cyclone and from Gallimore and Johnson's (1981)

theoretical application to the isentropic zonally averaged circulation of the

circumpolar vortex.

In an interpretation of Eliassen's theory for the extratropical cyclone

(Johnson, 1974), basic components are the mass circulation and its forcing,

shown in the upper and lower schematics of Fig. 5. A meridional mass

circulation within the storm vortex is defined in the sense that the

azimuthally averaged circulation will display an inward branch towards the

axis of rotation (arrow 4), an outward branch (arrow 2) and a vertical branch

(arrow 1). It is also implicitly assumed that the storm environment contains

a vertical branch of opposite direction to that within the storm vortex (arrow

3). The meridional circulation and the sense of the circulation may be

opposite to that displayed in the schematic. As the diagnostic results suggest, the sense of the meridional circulation defines different types of development and this in turn controls the evolution of the extratropical cyclone.

Eliassen emphasized that the sense of a meridional circulation in the vicinity of a heat source ($\dot{\theta} > 0$) is to force motion along a surface of constant absolute angular momentum from higher to lower pressure (the upward branch in the schematics) and a heat sink ($\dot{\theta} < 0$) will force motion along a surface of angular momentum from lower to higher pressure (the downward branch in the schematics). In the vicinity of a point source of angular momentum ($\dot{g}_a > 0$), the sense of the meridional circulation is along an isentropic surface away from the axis of rotation (the outward branch in the schematics) while a point sink ($\dot{g}_a < 0$) forces motion towards the axis of rotation (the inward branch in the schematics). With respect to torques along a surface of constant angular momentum, the motion will be towards the axis of rotation where torque is a minimum and away from the axis where torque is a maximum. Eliassen also notes that the mass circulation will develop and be most intense where the hydrodynamic stability is minimized.

Although the schematics show motion through the isentropic surfaces, the meridional circulation may occur irrespective of the presence or absence of diabatic processes. If the motion were adiabatic, the inward and outward branches in the schematic would force destabilization within the lower isentropic layers and stabilization within the upper layers. An alternative would be adiabatic descent along the axis of rotation in conjunction with an inward branch forced by a negative torque in the upper layer and an outward branch forced by a positive torque in the lower layer. Within the vortex, destabilization would occur in the upper layer while stabilization would be found in the lower layer. With the presence of diabatic processes, the constraints on the tendency of stability are removed. An important aspect of absolute angular momentum of storms is that cyclonic development requires a net transport of this property from an environment towards the axis of rotation of an incipient vortex.

b. The observed mass circulation and angular momentum balance

Time sections of isentropic and isobaric azimuthally averaged mass transports for a storm volume with an equivalent radius of 7.5° latitude are presented in Figs. 6a and 7a. The first 36 hours of the time sections illustrate the leeside dry baroclinic development phase of this cyclone. During this period the cyclone deepened from a central pressure of 1002 hPa to 992 hPa. In the isentropic result a well-defined mass circulation with an inward branch above 300 K and an outward branch between 275 and 300 K is portrayed (Fig. 6a). The boundary near 300 K between the two branches tended to be displaced upward during the first 36 hours as a general warming of the entire troposphere of the cyclone vortex occurred. The second inward branch below 275 K is relatively weak. However, after 1200 UT 31 March this branch quickly became dominant in the lower layers while the outward branch was displaced to the upper isentropic layers. At the time of the mass circulation reversal, the upward diabatic mass transport intensified in association with latent heat release. Thus the storm was characterized by a reversal of the meridional mass circulation between the periods of dry and moist baroclinic development. In the isobaric azimuthally averaged mass transport, the mass circulation with an inward branch below the mid troposphere and an outward branch above simply intensified during the transition from the dry to the moist baroclinic stage of development.

Inspection of the surface and 500 hPa analyses in Fig. 2 reveals that the pressure gradient near the vortex center intensified and the cyclone occluded rapidly during the latter stage. During the initial 36-hour period (Fig. 1) development occurred with little tendency for occlusion. The 500 hPa wave amplified markedly during the transition stage. Note that the significant westward tilt of the pressure wave with height was maintained during the leeside dry baroclinic development. During the moist baroclinic stage an extensive closed 500 hPa low develops through occlusion.

The distribution of the azimuthally averaged angular momentum transport is shown in Fig. 8(a). The vertical structure of the circulation for three distinct times is portrayed through the areally averaged distribution of storm relative angular momentum in Fig. 8b. The profile for 0000 UT 30 March shows an extremely weak cyclonic circulation below 290 K. The profile for 1200 UT 31 March indicates that spin-up below and spin-down above 300 K has occurred by the end of the dry baroclinic stage. The profile for 0000 UT 02 April shows strong spin-up at all levels. These distributions show that the leeside dry baroclinic development is characterized by relatively shallow spin-up without the forcing of occlusion, while the moist baroclinic development is characterized by deep spin-up in conjunction with the forcing of occlusion.

The characteristic structure for the vertical distribution of angular momentum is also implicitly related to the time section of the isobarically area averaged potential temperature distribution shown in Fig. 9. During dry baroclinic development increasing temperatures are indicative of warm air advection into the cyclone vortex through which a warm core structure develops within an amplifying baroclinic wave. During the moist baroclinic development decreasing temperatures are indicative of cold air advection into the cyclone

vortex through which a cold core structure develops with occlusion. As
emphasized earlier, the 500 hPa wave simply amplified during dry baroclinic
development while a deep vortex circulation developed during the moist
baroclinic development. From thermal wind considerations, the decrease of
angular momentum with height at 1200 UT 31 March is consistent with the
circulation expected within a warm core vortex. The more intense and
vertically extended circulation at 0000 UT 02 April is consistent with that
expected within a cold core vortex (see Fig. 8b).

While one is tempted to simply explain the evolution of the temperature
structure and vertical extent of the vortex through quasi-geostrophic theory
and baroclinic temperature advection, analyses of angular momentum exchange i
conjunction with Eliassen's concepts provide additional insight.

In the first 36 hours of this cyclone, the inward transport of storm
absolute angular momentum occurs in the isentropic layers between 305 and
325 K (Fig. 8a) while transport outward in the layers below 305 K exceeds the
inward transport. In the absence of other processes, spin-down would occur i
lower isentropic layers in association with the outward transport of angular
momentum. Thus this distribution of transport dictates that the spin-up in
the lower layers during the leeside development of this storm is forced by th
transport of absolute angular momentum into the upper isentropic layers of th
storm volume. For the spin-up to occur in the isentropic layers below 305 K,
angular momentum must be transferred from upper to lower isentropic layers
through the combination of negative and positive torques. The total torque
distribution for the isentropic mass circulation, which is determined by the
sum of the pressure, inertial and frictional torques plus the convergence of

the eddy angular momentum transport, is presented in Fig. 6b. Note the remarkable similarity of the patterns for the mass circulation and the total torque. As such, in the dry baroclinic phase negative torques force the mass transport inward in the isentropic layer between 305 and 325 K, while positive torques force the mass transport outward below 305 K. Since the earth component of absolute angular momentum dominates the relative component, the sense and forcing of the angular momentum transport shown in Fig. 8a is the same as the mass transport. The downward transfer of angular momentum from the upper to the lower isentropic layers occurs in conjunction with the forcing of the mass circulation through the pressure, inertial and frictional torques. If one assumes the component of the frictional torque associated with lateral viscous stresses is small relative to vertical stresses, the vertical integral of the pressure, inertial and frictional torques reduces to a boundary integral at the earth's surface (Johnson and Downey, 1975b). Furthermore, with a symmetric surface pressure distribution and uniform terrain, the component of the boundary integral associated with pressure and inertial torques vanishes. In this case sinks of angular momentum by pressure and inertial torques in one layer must be balanced by corresponding sources of angular momentum in another layer (Johnson and Downey, 1975b). In the leeside dry baroclinic phase of the Alberta cyclone, the gain of angular momentum in the lower isentropic layers that spins up the storm's circulation must be by transfer of the angular momentum from the region of negative torque to the region of positive torque. Thus the combination of pressure and inertial torques embedded in the baroclinic structure simultaneously force lateral mass transport and vertically transfer angular momentum.

444

In the moist baroclinic phase, the negative total torque in the lower layers forces inward mass and angular momentum transport while positive total torque in the upper layers forces outward mass and angular momentum transport. In this phase, however, the occluding vortex tends to symmetry and the net transfer of angular momentum by pressure and inertial torques is reduced. This result is implicit in the partitioning of the mass transport into geostrophic and ageostrophic components shown in Figs. 6(c) and 6(d) (Vergin, 1979). Note during the dry baroclinic phase, the geostrophic component dominates the mass transport while during the moist baroclinic phase the ageostrophic mode dominates. Johnson and Downey (1975a and b) discuss the restrictions of lateral mass transport to the irrotational mode as a vortex becomes symmetric such as occurs during occlusion. They also discuss the unique relation between geostrophic mass transport and pressure torque. The existence of a geostrophic mode of transport and pressure torque requires asymmetric waves and vortices. Because of the symmetry that develops with occlusion the upward angular momentum transport occurs largely through diabatic mass flux associated with latent heat release (Johnson and Downey, 1976). The occlusion occurs through the convergence of angular momentum in upper layers and subsequent adjustment of the mass distribution to a gradient balanced vortex. These two types of development isolated within the absolute angular momentum approach are undoubtedly related to the synoptician's view that some cyclones remain shallow warm-core vortices with little vertical development while others occlude through extended vertical development throughout the troposphere.

c. The relation of topography and baroclinic structure to the acceleration of circulation

Since circulation is equal to the area integral of vorticity and the angular momentum of a circular ring is equal to the product of the circulation within the ring and the radius of a ring, the three properties -- angular momentum, circulation and vorticity -- are related by

$$G_a = RC_a = R \int_{z_s}^{Z} \int_0^{2\pi} \rho u_a R d\alpha dz = R \int_{z_s}^{Z} \int_0^{R} \int_0^{2\pi} \underline{k} \cdot (\nabla \times \rho u_a) R d\alpha dR dz \qquad (1)$$

where G_a is the absolute angular momentum of a cylindrical shell, C_a is the circulation, ρ is density, u_a is the azimuthal component of absolute velocity, R is the radius of the shell extending from the center of the storm, α is the azimuthal coordinate, z is height, and $\underline{k}$ is the vertical unit vector. For simplicity, cylindrical geometry is used in this discussion in lieu of the storm spherical geometry (Johnson and Downey, 1975a). In the schematic (Fig. 10a) the height of the earth's surface z_s ranges from sea level ($z_s=0$) to the top of the mountain ($z_s=Z$). The upper height surface of the cylindrical shell is equal to the height of the mountain Z. The time rate of change of angular momentum and circulation within this cylindrical shell is

$$\dot{G}_a = R\dot{C}_a = R \int_{z_s}^{Z} \int_0^{2\pi} \rho \dot{u}_a R d\alpha dz \quad . \qquad (2)$$

The effect of the pressure stress on the transfer of angular momentum through the earth's surface (Fig. 10b) is determined through scaling of the vertical height by

$$\eta = \frac{z - z_s}{Z - z_s} \quad ; \qquad \begin{array}{c} z_s < z < Z \\ 0 < \eta < 1 \end{array} \quad . \qquad (3)$$

Transformation of the vertical coordinate of the integral from z to η in (2), in combination with the simplification that the acceleration of the circulation is due solely to pressure forces, yields

$$\dot{G}_a = R\dot{C}_a = -R^2 \int_0^1 \int_0^{2\pi} \frac{\partial z}{\partial \eta} \frac{1}{R} \frac{\partial p}{\partial \alpha_z} \, d\alpha d\eta \qquad (4)$$

$$= R \int_0^1 \int_0^{2\pi} \frac{\partial}{\partial \eta} \left(p \frac{\partial z}{\partial \alpha_\eta} \right) d\alpha d\eta \qquad (5)$$

$$= -R \int_0^{2\pi} p_s \frac{\partial z_s}{\partial \alpha_\eta} \, d\alpha \quad . \qquad (6)$$

In the vertical integration of (5) expressed by (6), the condition that the pressure stress vanishes on the upper surface at the uniform height Z has been used. These results show that the direct effect of surface pressure stresses on the circulation is restricted to the portion of the atmosphere extending from sea level to the top of the mountain barrier. Insight into the net acceleration by the surface pressure stress is provided by expressing (6) as

$$\dot{G}_a = R\dot{C}_a = -R \left(\int_{\alpha_1}^{\alpha_2} p \frac{\partial z_s}{\partial \alpha_\eta} \, d\alpha + \int_{\alpha_2}^{\alpha_1} p \frac{\partial z_s}{\partial \alpha_\eta} \, d\alpha \right) \qquad (7)$$

where $\alpha_1 = \alpha(z_s=0)$ and $\alpha_2 = \alpha(z_s=Z)$. In both integrals the topography z_s is assumed to be a monotonic function of α. In the segment from α_1 to α_2, z_s is a monotonic non-decreasing function of α_1 (i.e., one is ascending the mountain as α increases) while in the segment from α_2 to α_1, z_s is a monotonic non-increasing function of α (i.e., one is descending the mountain as α increases). By the Mean Value Theorem, the integral (7) becomes

$$\dot{G}_a = R\dot{C}_a = -R(\bar{p}_{1 \to 2} - \bar{p}_{2 \to 1})Z \qquad (8)$$

where $\bar{p}_{1\rightarrow 2}$ and $\bar{p}_{2\rightarrow 1}$ are the appropriate mean values over their respective segments. These results show that the acceleration increases with the pressure difference between the two segments and with the height of the mountain relative to the height of the "leeside plain." Deceleration of the circulation occurs when the mean pressure of the upslope portion exceeds the mean pressure of the downslope portion, while acceleration occurs when the pressure of the downslope portion exceeds the pressure of the upslope portion.

In computations of the vertically integrated pressure torque for the Alberta cyclone from the surface pressure distribution, the results indicated that angular momentum was extracted from the cyclone and given to the earth. Note in Figs. 1a and 1b that the surface pressure north of the incipient low center is higher than the pressure south of the center. Thus the result is in agreement with the empirical evidence. Although one might conclude that this effect simply opposes development since the circulation is decelerated, the use of Eliassen's concept of the forcing of a mass circulation suggests otherwise. The effect of a negative pressure torque is to force mass transport towards the axis of rotation of the storm. As such the forced mass circulation transports angular momentum towards the storm's axis of rotation and induces convergence of angular momentum transport within the cyclone leading to spin-up of the relative circulation. The effect is analogous to the intensification of the circulation when a laboratory vortex couples frictionally with a lower boundary. In all likelihood the asymmetry of surface pressure with respect to an incipient leeside low pressure center developing over sloped topography is an important factor in understanding leeside cyclone development. The problem of accurate numerical simulation of leeside development is compounded by the difficulty of accurate computation of pressure gradient forces over sloped terrain and the need to insure that integral boundary constraints are satisfied (Arakawa, 1972; Arakawa and Lamb, 1977; Johnson and Uccellini, 1983). The distribution of the pressure asymmetries in leeside development involves frontal structure within the cylindrical shell, as well as the mass distribution of the larger scale wave structure of the overlying atmosphere. The evolution of these features involves other dynamical processes that are important and essential to an overall forcing of cyclone development. The transfer of angular momentum by the pressure field is only one element of the forcing of a vortex and its interaction with its environment.

The warming of the cyclone vortex during dry leeside development as reflected in the isobarically area averaged potential temperature distribution (Fig. 9) is also a consequence of the forcing of mass circulation by the torque distribution. Note that the average temperature increases during the leeside dry phase and decreases during the moist phase of development. For illustrative purposes in Fig. 10c, the schematic for the mountain pressure torque is modified by the addition of an isentropic surface ($\approx$295 K) that slopes upward to the northeast from a surface position in the vicinity of the front. This structure in conjunction with the observed distribution of isentropic pressure torque provides insight into the development of a warm core for the Alberta cyclone during its dry baroclinic phase. The dome of cold polar air to the northeast is associated with the surface high pressure center northeast of the low center, while the air to the south and southeast is relatively warm. Transfer of angular momentum by pressure torques from one

449

area averaged isentropic layer to another is associated with the transfer of angular momentum azimuthally in Cartesian or isobaric coordinates. In the dry phase the negative torque in the upper isentropic layer of the Alberta cyclone and a positive torque in the lower isentropic layer correspond with a negative torque in the warm sector of a cyclone wave and a positive torque in the cold sector, both of which are occurring within the same isobaric layer. This distribution of torque results in angular momentum being transferred azimuthally from the warm sector to the cold sector within a cylindrical ring circumscribing the cyclone's axis of rotation. The negative torque in the warm sector forces warmer air to the axis of rotation while the positive torque forces the colder air from the cyclone vortex, both of which are realized through general warm air advection within the domain of the cyclone. Through this process the cyclone becomes a warm core vortex.

This result is quantitatively determined by rescaling the cylindrical shell into two layers, one extending from the earth's surface to the 295 K isentropic surface in the schematic (Fig. 10C) and the other from this isentropic surface to the constant height surface Z. The transfer of angular momentum between the two domains would be given by the azimuthal integral of $(p\partial z/\partial\alpha_\theta)$ at 295 K. For angular momentum to be transferred from the warm to the cold air, the pressure on the upslope side of the isentropic dome of cold air should be greater in a mean value sense than the pressure on the downslope side. The dome of cold air delineated by the isentropic surface may be viewed as a "mountain" which permits transfer of angular momentum through its surface, but which in turn moves in response to the convergence of angular

450

momentum within its being. In the dry leeside baroclinic development the "mountain" of cold air gains angular momentum and is forced from the developing cyclone vortex, while in the moist baroclinic development the "mountain" of cold air loses angular momentum to the warm air and is forced towards the developing vortex. In quasi-Lagrangian analyses, motion, advection and/or transport processes are all determined relative to the cyclone's axis of rotation. Thus, in the case of the cyclone, both the movement of the cyclone east-southeastward and the northward warm air advection through the upper great plains are a result of the forcing of the mass circulation by the torques within the cyclone vortex. Palmén and Newton (1969) call attention to the maximum warming in low levels below 700 hPa during orographic cyclogenesis while little effect occurs at 700 hPa. They suggest that maximum convergence takes place in a relatively shallow layer. The application of angular momentum principles in conjunction with Eliassen's concepts provides a fundamental basis to explain both of these conditions.

4. Summary

Our understanding of the development of cyclones under the influence of topography is still incomplete. Analyses of the distribution of cyclogenesis show distinct maxima over warm seas and oceans and in the lee of mountainous terrain (Petterssen, 1956) that cannot be explained simply by baroclinic or barotropic instability. There can be little doubt that the boundary processes at the earth's surface are important in forcing cyclogenesis. The means by which boundary processes force cyclones over oceans and in the lee of mountains undoubtedly vary.

Isentropic analyses of the mass circulation and angular momentum balance show that the development processes within cyclones also vary. In the case study summarized herein, the Alberta cyclone revealed two distinct phases with an intermediate transition period. In the leeside dry baroclinic stage a sink of storm angular momentum within the lower troposphere was induced through the combined effects of surface pressure and topography. The net sink of angular momentum within the cyclone's circulation occurs downstream of the mountains within the portion of the troposphere beneath the level of the mountain's crest. This sink forces the convergence of mass and angular momentum that is essential to development of the circulation.

The mass circulation that occurred during the leeside dry baroclinic phase was quite similar to the adiabatic mass circulation analyzed within a numerically simulated linear baroclinic wave (Bates and Johnson, 1984). However, one important difference was observed. The vertically integrated circulation of the linear baroclinic wave with a uniform lower boundary developed through a convergence of the eddy mode of angular momentum transport. In the Alberta cyclone, the vertically integrated circulation during the leeside dry phase develops even though the eddy mode of angular momentum transport is divergent. In this case, a remaining degree of freedom for the development of the vertically integrated circulation is through vertically integrated mass convergence with a resulting convergence of the mean component of angular momentum; the usual explanation being spin up from conservation of potential vorticity and stretching of vortex tubes. Such mass convergence will occur with the movement of the cyclone towards lower elevation. The profile of angular momentum reveals, however, that a two layered structure of the mass circulation occurred and that the spin-up of the circulation primarily occurred in the lower isentropic layer. The failure of the cyclone to spin up within the upper isentropic layer of mass convergence negates the explanation of cyclogenesis based solely on conservation of potential vorticity and stretching of vortex tubes.

The evidence summarized in this paper suggests that the forcing of a two-layered isentropic mass circulation by torques is in part associated with topography as well as by pressure and inertial torques embedded in the baroclinic structure. Even for leeside cyclogenesis, a two layered circulation seems essential from mass and angular momentum principles. The development of a cyclonic circulation in the lee of a mountain in its early stages tends to be fixed geographically. In an attempt to explain cyclogenesis with net mass convergence of a one layered structure, the vertical stretching that occurs within the one layer must be accompanied by vertical shrinking in another layer. Otherwise the surface pressure tendency of the hydrostatic atmosphere would be positive. The negative surface pressure tendencies that are observed can only occur either through mass divergence within the single layer of spin up, a contradiction to the requirement for mass convergence needed to realize spin up from stretching of vortex tubes, or through the mass divergence that would occur within the layer of the vertical shrinking. These considerations suggest that a view of leeside cyclogenesis based on a two layered structure that embodies Dines compensation constitutes the simplest structure that is plausible. A two layered structure for the mass circulation also permits much larger values of absolute angular momentum to be transported from an environment to a storm vortex within the layer of inflow than can be realized by mass convergence within a model based on a one layered structure. In all probability a cyclone's circulation can only develop to the intensities that are observed by the transport of angular momentum from an extended environment.

This two layered mass circulation with inward transport of angular momentum in the upper isentropic layer and downward transfer of angular momentum also forces warm air advection and the development of a warm core vortex. Such development which occurs primarily through adiabatic exchange requires asymmetry of wave structure and of the low tropospheric vortex. In deep tropospheric development and occlusion of cyclones, the increasing symmetry of the vortex restricts these baroclinic modes associated with frontal structures. As such frictionally induced mass convergence and the transfer of angular momentum from the lower layers upward through diabatic mass flux within the mass circulation become primary components. Johnson and Downey (1976) showed that friction and the upward transport of absolute angular momentum by latent heat release must be important in the forcing of occlusion.

Professor Sverre Petterssen noted in one of his seminars at the University of Wisconsin that extratropical cyclones are born in a variety of ways but that their appearances in death are remarkably similar. The results from Johnson and Downey (1976) and the case study of the Alberta cyclone (Johnson, Wash and Petersen, 1976) have isolated two distinct types of development for development of continental cyclones, but only one type for the stages of the occlusion and decay. These results support Petterssen's observation.

ACKNOWLEDGMENTS

The diagnostic results summarized for this Alberta cyclone are extracted from a collaborative study being prepared for publication jointly with Prof. Carlyle Wash, Dr. Ralph Petersen, and Mr. James Vergin. Their contribution is gratefully acknowledged. I also express appreciation for typing the manuscript to Becky Riedel and Judy Mohr and to Jim Ferwerda for drafting. The research was funded by the National Science Foundation Grants ATM 77-22976 and ATM 81110678.

Appendix A

TABLE A1. Quasi-Lagrangian mass budget (Johnson and Downey, 1975a).

The Definition

$$M = \int_{V_n} \rho J_n r^2 \sin\beta \, dV_n$$

The Budget Equation

$$\frac{dM}{dt} = LT + VT$$

where

Lateral Transport

$$LT = - \int_{n_B}^{n_T} \int_0^{2\pi} \rho J_n (U - W)_\beta \, r \sin\beta \, d\alpha dn \Big|_{\beta_B}$$

Vertical Transport

$$VT = \int_0^{\beta_B} \int_0^{2\pi} \rho J_n \left(\frac{dn}{dt} - \frac{dn_B}{dt}\right) r^2 \sin\beta \, d\alpha d\beta \Big|_{n_B}$$

Lower Boundary Conditions

$$\frac{dn_B}{dt} = 0; \quad n_B > n_s(\alpha, \beta, t)$$

$$\frac{dn_B}{dt} = \frac{dn_s}{dt}(\alpha, \beta, t); \quad n_B < n_s$$

$$n_s = n(\alpha, \beta, r_s, t) = \text{earth surface}$$

Upper Boundary Condition

$$\frac{dn_T}{dt} = 0$$

TABLE A2. Quasi-Lagrangian angular momentum budget
(Johnson and Downey, 1975b).

The Definition

$$G_a = \int_{V_n} \rho J_n g_a r^2 \sin\beta \, dV_n$$

where

$$g_a = r \sin\beta \, (U_a - W_{o_a})_\alpha$$

The Budget Equation

$$\frac{d}{dt} G_a = LT(g_a) + VT(g_a) + S_\Pi(g_a) + S_I(g_a) + S_T(g_a)$$

where

Lateral Transport

$$LT(g_a) = - \int_{n_B}^{n_T} \int_0^{2\pi} \rho J_n g_a (U - W)_\beta \, r \sin\beta \, d\alpha dn \Big|_{\beta_B}$$

Vertical Transport

$$VT(g_a) = \int_0^{\beta_B} \int_0^{2\pi} \rho J_n g_a \left(\frac{dn}{dt} - \frac{dn_B}{dt}\right) r^2 \sin\beta \, d\alpha d\beta \Big|_{n_B}$$

Stress Torque (includes pressure and frictional torque)

$$S_\Pi(g_a) = \int_{V_n} r^3 \sin^2\beta \, J_n \underset{\sim}{\ell} \cdot (\nabla \cdot \underset{\sim}{\Pi}) \, dV_n$$

where $\nabla \cdot \underset{\sim}{\Pi} = -\nabla p + \rho \underset{\sim}{F}$

Inertial Torque

$$S_I(G_a) = -\int_{V_n} r^3 \sin^2\beta \, \rho J_n \underset{\sim}{\ell} \cdot (\dot{\underset{\sim}{W}}_{o_a}) \, dV_n$$

Translational Source

$$S_T(G_a) = \int_{V_n} \rho J_n \frac{d_a \underset{\sim}{k}_o}{dt} \cdot \underset{\sim}{g}_a r^2 \sin\beta \, dV_n$$

C	Circulation
G_a	Budget volume integral of storm angular momentum
J_η	Jacobian for transformation of vertical coordinate to generalized coordinate, $|\partial z/\partial \eta|$
LT	Lateral transport
M	Total mass in a volume
R	Radius of a shell extending from the center of the storm
S	Source/sink integral
V_η	Volume with limits of integration specified by (α, β, η)
VT	Vertical transport
Z	Height of the top of a mountain
g_a	Vertical component of specific absolute angular momentum
p	Pressure
r	Length of position vector, $\underline{r}$
t	Time
z	Height
α	Azimuthal coordinate
β	Angular radial coordinate
η	Generalized vertical coordinate, scaling of the vertical height
θ	Potential temperature
ρ	Density

Subscripts or superscripts

B	Bottom or boundary
I	Inertial torque
T	Top of the atmosphere; translational source
a	Absolute frame of reference
o	Surface center of storm coordinate system
s	Surface of the earth
z	Cartesian coordinate system
Π	Stress torque
α	Azimuthal component
β	Radial component
η	Generalized coordinate system; η surface
θ	Isentropic coordinate system; isentropic surface

Vectors and Tensors

$\underline{F}$	Frictional force
$\underline{U}$	Earth-relative wind velocity
$\underline{W}$	Earth-relative velocity of the lateral boundary
$\underline{W}_o$	Earth-relative velocity of position vector to surface center of the storm
$\underline{k}_o$	Unit vector along the reference axis
$\underline{m}, \underline{l}, \underline{k}$	Orthogonal unit vectors pointing in radial, azimuthal, and vertical directions
$\underline{r}$	Position vector
$\underline{\underline{\Pi}}$	Stress tensor

TABLE A3 (cont.)

Operators

$\partial/\partial\zeta$	Partial derivative with respect to ζ where ζ may be t, θ, or n.
d/dt	Total derivative with respect to time
$\cdot$	d/dt
∇	Three-dimensional del operator
∇_n	Horizontal del operator
$\int$	Riemann integral

REFERENCES

Arakawa, A., 1972: Design of the UCLA general circulation model. Tech. Rep. No. 7, Numerical Simulation of Weather and Climate, Dept. of Meteor., UCLA, [NTIS N73-21508], 116 pp.

_____, and R. V. Lamb, 1977: Computational design of the basic dynamical processes of the UCLA general circulation model. Methods in Computational Physics, 17, J. Chang (ed.), Academic Press, 337 pp.

Bates, G. T., and D. R. Johnson, 1984: Mass and angular momentum diagnostics of cyclones within numerically simulated linear baroclinic waves. Mon. Wea. Rev., 112, 246-258.

Gallimore, R. G., and D. R. Johnson, 1981: A numerical diagnostic model of the zonally averaged circulation in isentropic coordinates. J. Atmos. Sci., 38, 1870-1890.

Johnson, D. R., 1974: The absolute angular momentum of cyclones. Subsynoptic Extratropical Weather Systems: Observations, Analysis, Modeling, and Prediction, Volume II seminars and workshops, Colloquium notes of the advanced study program and small-scale analysis and prediction project, National Center for Atmospheric Research, Boulder, CO, [NTIS PB-247286], 821 pp.

_____, and W. K. Downey, 1975a: Azimuthally averaged transport and budget equations for storms: Quasi-Lagrangian diagnostics 1. Mon. Wea. Rev., 103, 967-979.

_____, and _____, 1975b: The absolute angular momentum of storms: Quasi-Lagrangian diagnostics 2. Mon. Wea. Rev., 103, 1063-1076.

_____, and _____, 1976: The absolute angular momentum budget of an extratropical cyclone: Quasi-Lagrangian diagnostics 3. Mon. Wea. Rev., 104, 3-14.

_____, and L. W. Uccellini, 1983: A comparison of methods for computing the sigma coordinate pressure gradient force for flow over sloped terrain in a hybrid theta-sigma model. Mon. Wea. Rev., 111, 870-886.

_____, C. H. Wash, and R. A. Petersen, 1976: The mass and absolute angular momentum budgets of the Alberta cyclone of 30 March - 2 April 1971. Preprints, 6th Conf. Weather Forecasting and Analysis, May 10-13, 1976, Albany, Amer. Meteor. Soc., 350-356.

Palmén, E., and C. W. Newton, 1969: Atmospheric Circulation Systems. Academic Press, New York, 603 pp.

Petterssen, S., 1956: Weather Analysis and Forecasting, 2nd ed., Vol. I. McGraw-Hill, New York, 428 pp.

Vergin, J. M., 1979: The relative importance of geostrophic and ageostrophic modes of transport during cyclone development and maturation. M.S. thesis, University of Wisconsin, Madison, 137 pp.

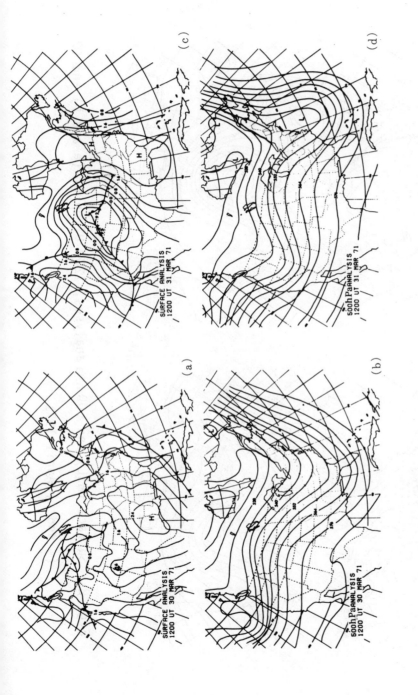

Fig. 1. Surface, (MSL) and 500 hPa analyses, 1200 UT on 30 (a and b) and 31 (c and d) March 1971.

463

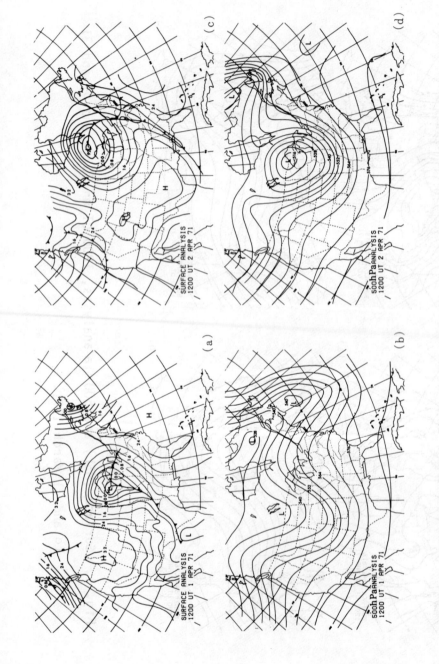

Fig. 2. Surface (MSL) and 500 hPa analyses, 1200 UT on 01 (a and b) and 02 (c and d) April 1971.

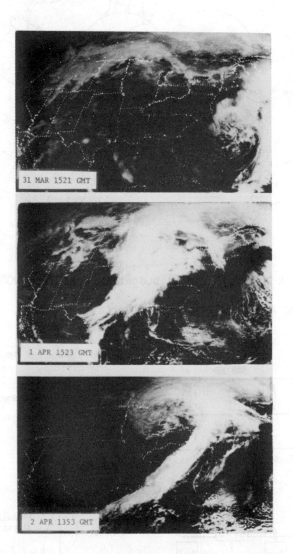

Fig. 3. Three successive ATS-III satellite images of the Alberta cyclone,
between 1500 and 1600 UT for 31 March (a), 01 April (b), and
02 April 1971 (c).

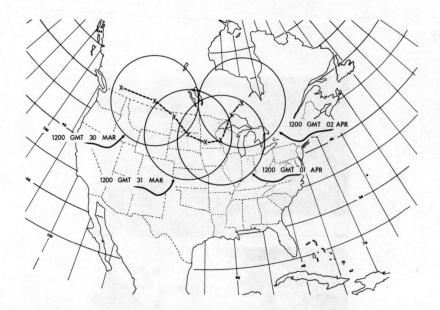

Fig. 4. Storm track and 7.5° radius budget volumes from 0000 UT 30 March t&
1200 UT 02 April 1971 for the Alberta cyclone.

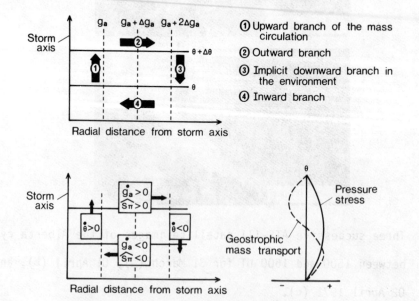

Fig. 5. Schematic of the radial mass circulation, its forcing and the
pressure stress profile within a vortex (from Johnson, 1974).

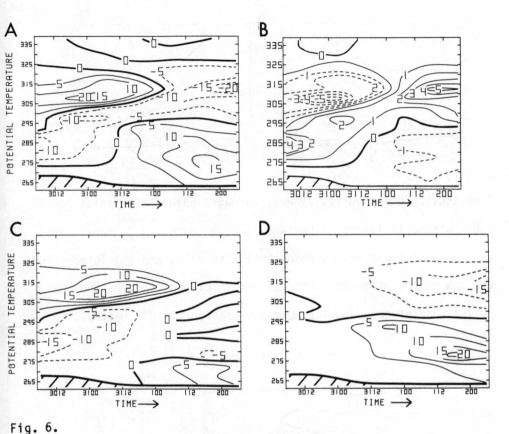

Fig. 6.

Time sections of isentropically area averaged mass budget statistics for 5 K layers, Alberta cyclone 30 March to 02 April 1971: (a) Inward lateral mass transport (10^9 Kg s^{-1}), (b) Total torque, the sum of the pressure, inertial and frictional torques and the eddy lateral angular momentum transport (10^{-1} Hadleys (1 Hadley = 10^{18} kg m^2 s^{-2}, Newton, 1971)), (c) Geostrophic component of mass transport (10^9 Kg s^{-1}), and (d) Ageostrophic component of mass transport (10^9 Kg s^{-1}).

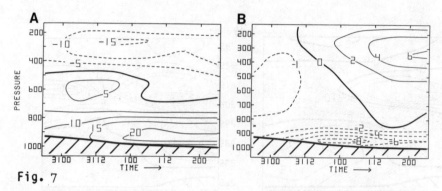

Fig. 7

Time section of isobarically area averaged budget statistics for
100 hPa isobaric layers, Alberta cyclone 30 March to 02 April 1971:
(a) Inward lateral mass transport (10^9 Kg s^{-1}), and (b) Total torque,
the sum of the pressure, inertial and frictional torques and the
convergence of eddy angular momentum transport (10^{-1} Hadley).

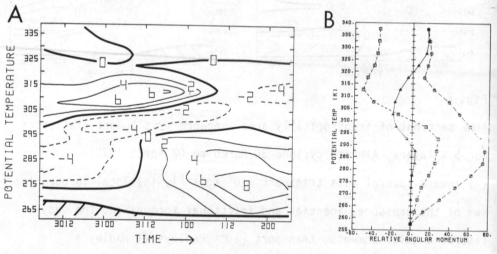

Fig. 8.

Time sections of storm absolute angular momentum budget statistics for
5 K isentropic layers, for Alberta cyclone, 30 March to 02 April 1971:
(a) Total lateral absolute angular momentum transport (10^{-1} Hadley),
and (b) Vertical profiles of area averaged specific relative angular
momentum (10^5 m^2 s^{-1}) for the incipient stage, the end of the leeside
dry baroclinic stage and the end of the moist baroclinic stage.

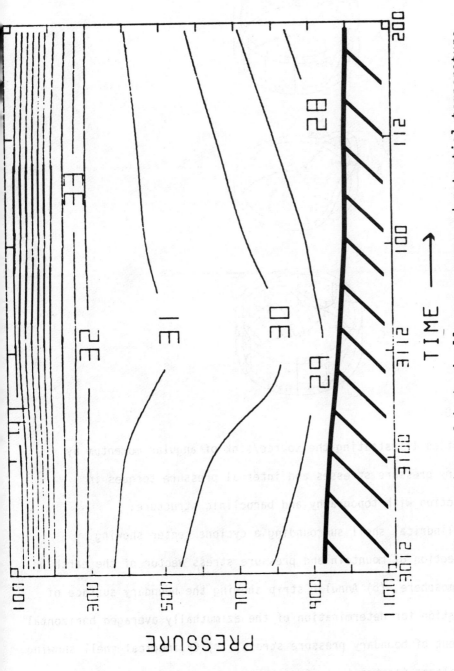

Fig. 9. Time sections of isobarically area averaged potential temperature (10 K) within the storm volume, 30 March to 02 April 1971.

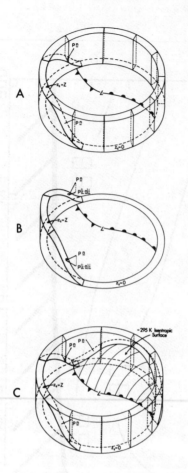

Fig. 10.

Schematics illustrating the source/sink of angular momentum by
boundary pressure stresses and internal pressure torques in
conjunction with topography and baroclinic structure:
(a) Cylindrical shell surrounding a cyclone center showing
intersection of mountain and pressure stress vector of the earth on
the atmosphere, (b) Annular strip showing the boundary surface of
integration for determination of the azimuthally averaged horizontal
component of boundary pressure stress, (c) Cylindrical shell showing
an idealized isentropic surface that separates the cold polar air to
the north-northeast of the Alberta cyclone low pressure center from
the warm subtropical air to the south.

AN ANALYSIS OF RADAR ECHOES FROM CONVECTIVE CLOUDS
OVER THE NAGQU REGION IN SUMMER, 1979

Qin Hongde

Academy of Meteorological Science,
State Meteorological Administration
Zhou Hesheng, Cui Youmin, Liu Jianxi and Yang Xiurong
Sichuan Provincial Institute of Meteorology
Qu Zhang
Lanzhou Institute of Plateau Atmospheric Physics,
Chinese Academy of Sciences

Radar echoes from convective clouds were analyzed statistically using the data obtained with a 3 cm wavelength radar sited at the Nagqu Meteorological Station during the Qinghai-Xizang Plateau Meteorological Experiment (May-Aug. 1979). It was found that the main characteristics of convective clouds were as follows.

1) The absolute heights of echo tops in Nagqu were higher than those in Chengdu. Monthly mean echo top heights above sea level in Nagqu were approximately 2—4 km higher than those in Chengdu, as shown in Fig. 1. However, the mean top height of shower and thunderstorm echoes above the ground (relative height) in Nagqu was about 3 km lower than in Chengdu.

2) The echoes in Nagqu were numerous. There might be as many as 13 echoes within a radius of 50 km around the radar. This is about three times the number in Chengdu, and six times the number of Cb cells estimated from satellite data over the Plateau by H. Flohn[1] (1968).

3) The size of echoes on a PPI display was much smaller in Nagqu than in Chengdu. The equivalent mean diameter of convective cells was almost 2.3 km. More than 90% of the echoes were smaller

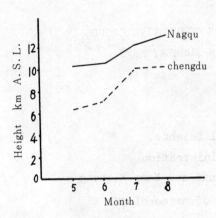

Fig. 1 Monthly variation of
mean echo top height.

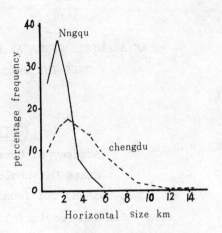

Fig. 2 Horizontal size distribution of echo cells.

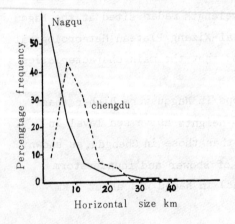

Fig. 3 Horizontal size distribution of echo
clusters.

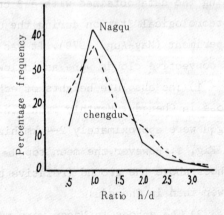

Fig. 4 Distribution of the ratio
of thickness to width of
echoes.

than 3 km in diameter; while in Chengdu the average diameter was 4.4
km, and the percentage of those echoes having a diameter over 3 km
might be up to 59%, as shown in Fig.2. In Nagqu, the larger the echo

472

cluster size, the lower the number of clusters found on the PPI display. Fifty-six percent of the echoes had a diameter smaller than 5 km. The mean diameter of clusters was equal to 9.2 km. Yet in Chengdu most of the clusters had a diameter of 5–10 km. Over 90% echoes were larger than 5 km in diameter, as shown in Fig. 3, and the equivalent mean diameter was 12.3 km.

4) In Nagqu, most echoes looked thin and tall. Echoes with a thickness to width (h/d) ratio smaller than unity were not so numerous as in Chengdu, as shown in Fig. 4.

5) The parameters mentioned above varied distinctly in different times of the day and between the dry and the rainy season. The mean echo top height had a minimum value in the early morning of only 4 km above the ground, both in the dry season and the rainy season; while in the afternoon it had a maximum of about 6 km above the ground in the dry season and 9 km in the rainy season. After sunset the echo tops descended gradually, about 6 km at night in the dry season and more in the rainy season. Mean size of the echoes appeared small in the morning but larger in the afternoon and evening. The echo size was usually larger in the rainy season than in the dry season. The echo number reached its maximum in the afternoon and its minimum in the early morning (Fig. 5–8).

Due to the influence of a Plateau vortex, Nagqu had its maximum rainfall in the year of 1979 during July 13–15. The precipitation was of a mixed type. The fallout was found in an arc-shaped band of echoes. With the development of the vortex, the raining process can be divided into four stages.

At the beginning, there appeared numerous echoes. In total,there were approximately 20–50 echoes within the range of a radius of 50 km around the radar, each with a diameter of about 2 km, as shown in Fig. 9. Then the small echoes gradually merged into a wide and dense band with a clear eddy structure, suggesting that the vortex had come into being (Fig. 10). By this time, it was difficult to find any small echo. Six hours later, the vortex took shape on the 500 hPa chart. In the southern warm sector of the vortex the echoes were mainly floccus in appearance, sometimes arranged in a few wide bands.

473

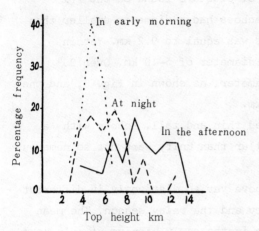

Fig. 5 Diurnal variation of
echo top height dis-
tribution.

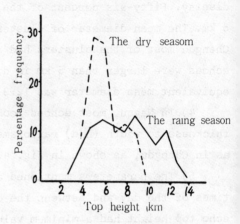

Fig. 6 Seasonal variation of
echo top height dis-
tribution.

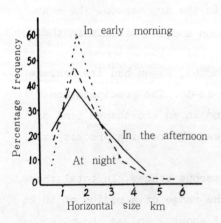

Fig. 7 Diurnal variation of
horizontal size dis-
tribution of echo
cells.

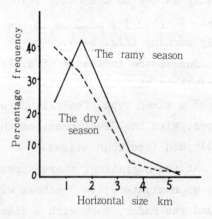

Fig. 8 Seasonal variation of
horizontal size dis-
tribution of echo cells.

In case of an outbreak of cold air, the echoes were chiefly in the
shape of lumps, extending to a high level and arranged in a long,
narrow band. To the south of a shear line, the echoes were low and
mostly floccus in form.

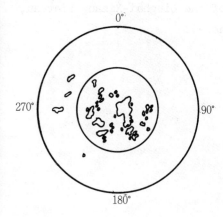 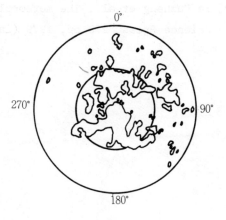

Fig. 9 Radar echoes on a PPI   Fig. 10 Radar echoes on a PPI
display at 1258 L.M.T.   display at 1350 L.M.T.
July 13, 1979. The angle   July 13, 1979. The
of elevation is 3°. Range   angle of elevation is
circles are 50 km apart.   3°. Range circles are
                                    50 km apart.

In short, the echoes of convective clouds in Nagqu had higher
tops, smaller sizes, and were greater in number and more variable in
shape than those in Chengdu, in agreement with the conclusion of Ye
Duzheng et al.[2] These features may be attributed to the high elevation, complicated topography and low humidity of Nagqu region. Owing
to the strong insolation over the Plateau, surface sensible heat
gives rise to instable stratification within the air layer below
400 hPa. Convective activities transport heat energy upwards, causing
the temperature of the air directly over the Plateau to be higher
than the environment at the same level, and so maintain the anticyclonic

circulation in the upper troposphere over the Plateau.

REFERENCES

[1] H. Flohn. Contribution to a meteorology of the Tibetan Highlands, Atmos. Sci. Paper, No. 130, Colorado State Univ. Fort Collins, 1968.
[2] Ye Duzheng et al., The meteorology of the Qinghai-Xizang Plateau, Science Press, 278 pp, 1979 (in Chinese).

LOW LEVEL AIRFLOW IN
SOUTHERN WYOMING DURING WINTERTIME
BY AIRCRAFT

John D. Marwitz and Paul J. Dawson[1]
Department of Atmospheric Science
University of Wyoming

1. Introduction and Topography

The winds persistently blow in southern Wyoming for the same reason
that the region was a popular route for early pioneers who migrated to
the western United States in the 1800's. Today it is still a popular
route with truck drivers and railroad operators. The continental divide
is very low in this section of the Rocky Mountain Barrier. The Great
Divide Basin in southern Wyoming at ~ 2000 m MSL is the lowest region
along the continental divide between Montana and New Mexico and provides
a natural passageway for the low level airflow through the Rocky Mountain
barrier. The rest of the Rockies in Colorado, Wyoming and Montana are
above 3000 m with numerous peaks above 4500 m. During the winter and
spring, pools of cold, stable air regularly collect in the extensive
Great Basin of Idaho, Nevada and Utah, west of the continental divide.
The cold air promotes the maintenance of a quasi-permanent anticyclone
during the winter months. The anticyclone, usually in conjunction with
the frequent cyclones in the lee of the central and northern Rockies,
provides a strong pressure gradient force which accelerates the low
level airflow over the continental divide, particularly through the
narrow gap or wind corridor in southern Wyoming. The topography of southern Wyoming (Fig. 1) channels the low level airflow through the Wyom-

---

[1]Present affiliation: Washington State University, Pullman, Washington.

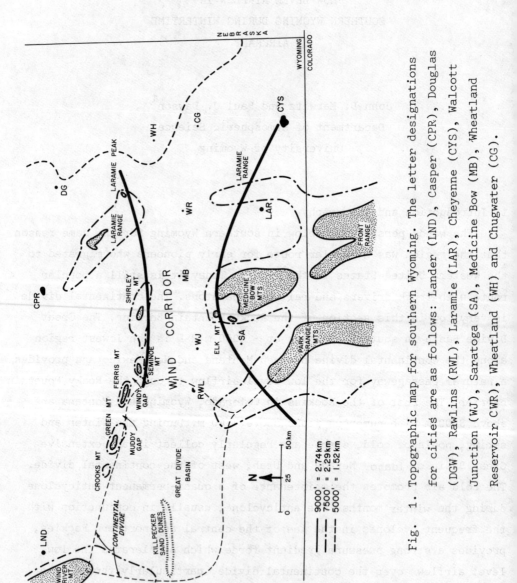

Fig. 1. Topographic map for southern Wyoming. The letter designations for cities are as follows: Lander (LND), Casper (CPR), Douglas (DGW), Rawlins (RWL), Laramie (LAR), Cheyenne (CYS), Walcott Junction (WJ), Saratoga (SA), Medicine Bow (MB), Wheatland Reservoir CWR), Wheatland (WH) and Chugwater (CG).

ing wind corridor. The Park Range Mountains south of Rawlins (RWL) and
the Medicine Bow Mountains south of Medicine Bow (MB) establish the
southern boundary, while the intermittent mountains between the Wind
River Mountains and Laramie Peak comprise the northern boundary of the
wind channel. The northern boundary contains a number of gaps through
which flows a significant amount of air from the Great Divide Basin.
The terrain through the wind channel was aptly dubbed the "wind corridor" by Kolm (1977) and is generally level at 2000 m MSL except for
a few ridges near RWL. Eastward from MB the terrain gradually ascends
$\sim$150 m to Wheatland Reservoir (WR). The WR area provides a natural gap
in the Laramie Range. From WR, the terrain descends abruptly into the
Great Plains region of eastern Wyoming.

The wind measurements in this study were made with fast response
air sensing probes and a litton LTN-51 inertial navigation system installed aboard the NCAR aircraft (304D). Basic aircraft capabilities
and instrumentation have been summarized by Burris et al. (1973) and
Lenschow et al. (1978). The air motion sensing system combined with the
inertial navigation system was capable of measuring the average horizontal air velocity to within ± 1 ms^{-1} and the average vertical air velocity to within ± 0.1 ms^{-1} (Kelley, 1973). The flights were typically
flown at 100 to 200 m AGL. Vertical profile soundings of the airflow
were also performed, consisting of climbs and descents plus level up-
wind/downwind flight legs at various altitudes within the planetary
boundary layer.

A listing of the 16 flights in 1976 with the primary eolian or
terrain feature investigated is given in Table 1. Interpretations of
the windflow characteristics from eolian features were described by
Marrs and Kolm (1982).

The wind characteristics in southern Wyoming based on a network
of anemometers and instrumented towers were reported by Martner and
Marwitz (1982). Seaman (1982) has simulated the airflow in this area
with a 3-D model and the results resemble the observations in a number
of significant ways. This article describes the airflow in southern
Wyoming based on the flights in the research aircraft. The case studies

TABLE 1. Wind study Flights of 1976. An X indicates the primary eolian
or terrain feature investigated and a double X indicates that
data from that flight are presented in this article.

DATE	WIND CORRIDOR	KILLPECKER DUNES	WINDY GAP	LARAMIE RANGE
5 Jan	X			
21 Jan	X		X	
22 Jan	XX		X	
23 Jan	XX	X		
26 Jan	X	X		
27 Jan				XX
10 Feb	XX		X	
11 Feb	XX	X		
12 Feb	X	X		
30 Nov	X			
1 Dec	X		XX	
2 Dec				X
8 Dec	X			
11 Dec	XX			
16 Dec				X
21 Dec	X	X		

selected for presentation are indicated in Table 1 and were chosen to
characterize the general winter airflow observed over various regions
of southern Wyoming. The observations from all 16 flights were incorporated into the general descriptions since similar tendencies and airflow
features were observed in most cases.

2. The Wind Corridor (11 December)

The flight on 11 December 1976 departed Laramie airport at 1206
LT and returned ~3.5 h later (1530 LT). The outer bounds of the wind
corridor were flown. Most of the flight (Fig. 2) was at ~100 m AGL
except for profile soundings near MB, ~35 km NW of RWL, ~50 km SW of
RWL, and near WJ. The flight track and observed wind velocities plus
analyzed streamlines and isentropes from this flight are presented in
Fig. 2. For those cases where the streamlines and isentropes are the
same, the streamlines are labeled in K. Where they are not the same
the isentropes are dashed lines and the streamlines are not labeled.
There are three small regions where the streamlines and isentropes are
not parallel. They are the region south of SA and west of the Medicine
Bow Mountains, the jet region downwind (northeast) of Elk Mountain,
and the region near Laramie. Since the isentropes and streamlines are
in general parallel, this indicates that there was little downward
transport of sensible heat in the corridor. Well-defined confluence
into the wind corridor and difluence downwind of the wind corridor is
apparent. Most of the low level baroclinity or thermal gradient upwind
of the wind corridor is present just west of the Medicine Bow Mountains.
In fact, the aircraft crossed the northern end of the Park Range at ~
1425 where θ was 296 K. Therefore, the strong baroclinic region south
of RWL was immediately west of the Park Range. The region over RWL is
quasi-barotropic. In the exit region of the wind corridor at the surface
(near MB) a strong thermal gradient existed uniformly across the corridor.

Another significant point to note in Fig. 2 is the low level jet
from the SSW immediately west of the Park Range. At 1415 the winds were
from the SW at 9 ms^{-1}. At 1423 the aircraft was due south of RWL and ~
20 km W of the continental divide. At this location the winds were from
the SSW at 15 ms^{-1}. This weak low level jet resembles the barrier jets

481

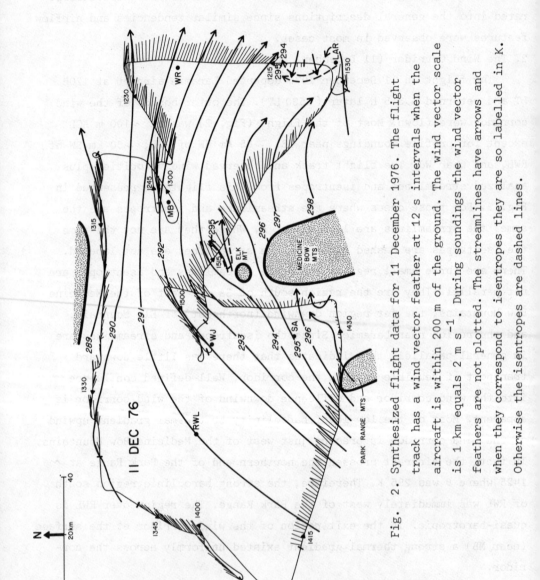

Fig. 2. Synthesized flight data for 11 December 1976. The flight track has a wind vector feather at 12 s intervals when the aircraft is within 200 m of the ground. The wind vector scale is 1 km equals 2 m s^{-1}. During soundings the wind vector feathers are not plotted. The streamlines have arrows and when they correspond to isentropes they are so labelled in K. Otherwise the isentropes are dashed lines.

observed upwind of the Sierra Nevada in California and numerically simulated by Parish (1982).

Vertical cross-sections were constructed normal to the streamlines across RWL (upwind of the wind corridor) and across MB (near the center of the wind corridor). The airflow across RWL (Fig. 3a) was moderately stable ($\frac{\partial\theta}{\partial z} \approx$ .005 K/m) with some moderate baroclinity ($\frac{\partial U_g}{\partial Z} \approx$ .03s^{-1}) in the lowest 0.6 km southeast of RWL. The mean wind speed was $\sim$ 15ms^{-1} The airflow across MB (Fig. 3b), on the other hand, was substantially different. There was strong static stability ($\frac{\partial\theta}{\partial Z} \approx$ .02K/m) at 3 km over MB. Below 3 km in the northern half of the wind corridor there was weak static stability ($\frac{\partial\theta}{\partial Z} \approx$ .002K/m). South of MB there was strong baroclinity ($\frac{\partial U_g}{\partial Z} \approx$ 0.1s^{-1}). The mean winds were $\sim$ 20 ms^{-1}. The development of the strong static stability at 3 km and the strong baroclinity resulted from an indirect circulation cell in the lowest 1 km through the wind corridor. The coldest air ($\theta \overset{\sim}{<}$ 292 K) flowed through the northern part of the wind corridor. In addition, there may have been some downward transport of sensible heat in the mountain wave in the lee of the Medicine Bow Mountains at the southern part of the wind corridor but it could not be directly resolved.

Assuming no N-S pressure gradient at 3 km, the hydrostatic pressure at the surface over MB exceeded that over LAR by $\sim$ 2.0 hPa. This results in a pressure gradient acceleration at the surface (a $\partial p/\partial y$) of $\sim$ .004 ms^{-2} acting toward the south. The centripetal acceleration (U^2/R;R $\approx$ 70 km) over Medicine Bow was $\sim$ .006 ms^{-2} acting to the north. Since the Coriolis acceleration (fU) was $\sim$ .002 ms^{-2} toward the south, the forces normal to the flow were in approximate gradient balance through the center of the wind corridor. Hess (1959) defines this type of flow as an anomalous anticylonic flow around a low.

The soundings from 50 km SW of RWL over the Great Divide Basin and over MB are presented in Fig. 4. As can be seen, the mean winds increased from $\sim$ 15 to $\sim$ 20 ms^{-1} as the air moved downstream. A weak low level jet of 22 ms^{-1} is present over MB at 600 m AGL. The RWL sounding is $\sim$ 1 K colder near the surface but this sounding was taken slightly north of the streamline through MB. The stable layer at MB is $\sim$ 200 m lower

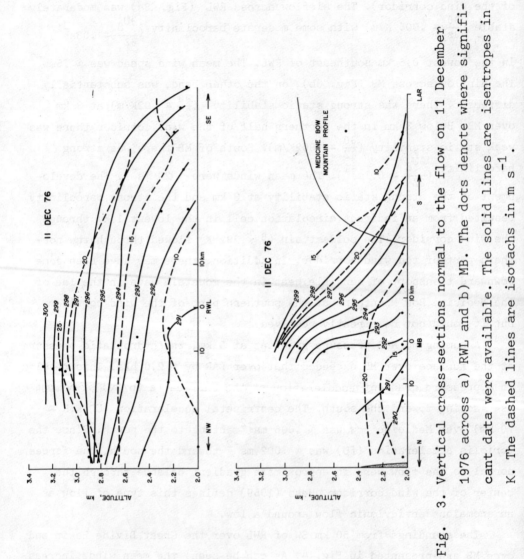

Fig. 3. Vertical cross-sections normal to the flow on 11 December 1976 across a) RWL and b) MB. The dots denote where significant data were available. The solid lines are isotachs in m s^{-1}. The dashed lines are isentropes in K.

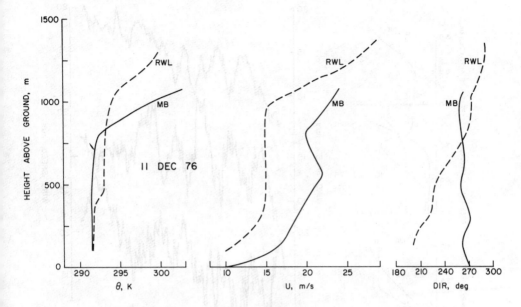

Fig. 4. Vertical profile of potential temperature (θ), horizontal
wind speed (U), and wind direction (DIR) over Medicine Bow
(MB) and Rawlins (RWL) on 11 December 1976.

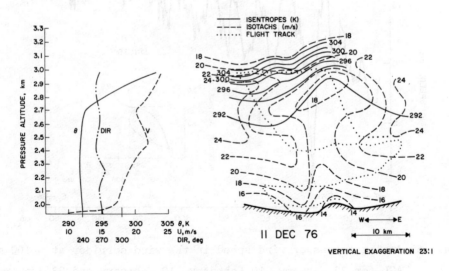

Fig. 5. Vertical profile of potential temperature (θ), horizontal wind
speed (U), and wind direction (DIR) over Medicine Bow on 11
December 1976 (left side). The corresponding vertical airflow
analysis from the profile sounding is shown on the right.

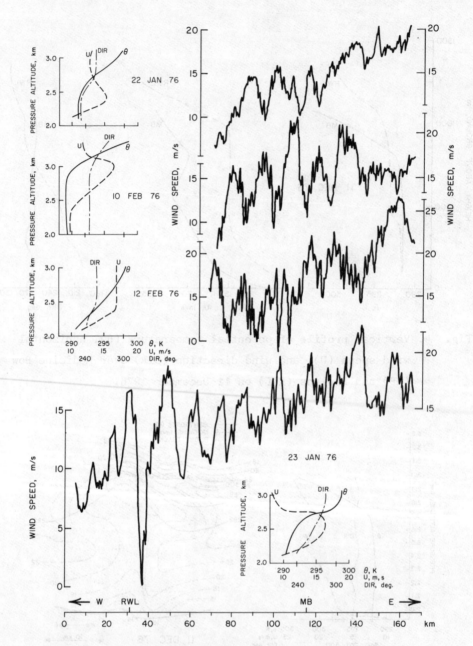

Fig. 6. Analog traces of wind speed in the wind corridor at ∿ 100 m AGL for 22 January, 10 February, 12 February and 23 January. Upwind Vertical profiles of potential temperature (θ), horizontal wind speed (U), and wind direction (DIR) are also shown.

than at RWL and is much more stable. The distance from RWL to MB is $\sim$ 100 km and with the observed mean winds the transport time was $\sim$ 1.5 h.

The data from the vertical profile sounding over MB are presented in Fig. 5. From the vertical structure of the flow a gravity wave is evident. The wind speed in the trough of the gravity wave varied from 16 to 24 m/s while the wind speed in the ridge of the gravity wave varied from 14 to 18 m/s. Mass continuity is indicated. The 24 m/s winds were at the base of the stable layer. Below the trough in the well-mixed layer, the Richardson number was low ($\sim$ .2). The gravity wave was tilted upwind implying a downward transport of momentum and sensible heat. Profile soundings on other days indicate that gravity waves with similar structure are common within the wind corridor. The gravity waves appear to be quasi-stationary in that a mass consistent tiled gravity wave can be synthesized from the θ and wind speed data.

During the latter part of the climb through the stable layer a breaking Kelvin-Helmholtz wave was encountered. In the upper left side of Fig. 5 the winds were observed to decrease with altitude from 24 to 16 ms^{-1} while θ increased from 290 to 304 K in 400 m. The Richardson number was $\sim$ 0.1 immediately downwind of this region moderate turbulence and rapidly fluctuating θ were observed indicating Kelvin-Helmholtz waves were present and breaking down to turbulence. The turbulence lasted for $\sim$ 2 km. The breaking K-H wave was in the trough of the much larger gravity wave.

Fig. 6 contains the analog traces of the air velocity near the center of the wind corridor at $\sim$ 100 m AGL for four flights. These data are typical of most flights through the wind corridor. The aircraft was flying upwind on each flight. Two characteristics are evident: the wind speed traces displayed a sinusoidal variation indicating the likely presence of gravity waves in the planetary boundary layer, and the mean speed increased downwind ($U \frac{\partial U}{\partial X} \cong .001$ ms^{-2}). Spectral analysis of these data indicate that the waves had wavelengths (λ) of 15 $\pm$ 2 km. The soundings upwind of the corridor are also presented for each case. Each of these soundings contain a low level jet near the base of the stable layer.

These waves appear to be trapped or resonant lee waves (Smith,

1980). The waves on 23 January 1976 were initiated by a pair of topographic ridges near RWL. The wind speed decreased from 17 to 0 m/s between the ridges just downwind of RWL. The ridges are separated by $\sim$ 15 km and acted to excite and reinforce the gravity waves if the natural wavelength was also $\sim$ 15 km. The trapped gravity waves on the other days may have also been initiated by these two ridges but the flight track veered off to Windy Gap before reaching RWL. From examining the data from flights one gets the impression that trapped gravity waves are typically present in the wind corridor during windy cases and that they are stationary with respect to the ground. To properly understand these gravity waves, additional flights are needed.

The natural wavelength of gravity waves can be estimated by multiplying the Brunt-Vaisala period by the wind speed.

$$\lambda = 2 \ \pi U (\frac{g}{\theta} \ \frac{\partial \theta}{\partial Z})^{-\frac{1}{2}} \tag{1}$$

If one uses the stability and mean winds of the stable layer, the natural wavelengths are $\sim$ 5 km. On the other hand by using the stability and mean winds in the well-mixed layer, the natural wavelenghts are $\sim$ 10 km. A good correspondence between the observed and natural wavelengths was obtained by using the stability in the well-mixed layer ($\frac{\partial \theta}{\partial Z} \cong$ .002 K/m) times the peak winds in the low level jet.

Scorer (1948) pointed out that a trapped or resonance wave occurs when the atmosphere contains a layer in which the Scorer parameter (ℓ^2) decreases rapidly with height. Smith (1980) has shown theoretically that the necessary condition for trapped waves is $\ell^2 < k^2$ aloft but $\ell^2 > k^2$ below, where $k = 2\pi/\lambda$, the wave number. The scorer parameter is

$$\ell^2 = \frac{g}{U^2\theta} \ \frac{\partial \theta}{\partial Z} - \frac{\partial^2 U/\partial Z^2}{U} \tag{2}$$

Examination of (2) indicates that through the wind corridor ℓ^2 becomes very large near the top of the planetary boundary layer because both the static stability increases and the low level jet develops near the top of the planetary boundary layer. Above the stable layer both the static stability and the vertical gradient of the wind shear typically

488

decrease significantly. Consequently, in the planetary boundary layer and extending through the stable layer $\ell^2 > k^2$ but above the stable layer $\ell^2 < k^2$.

Trapped lee waves should be vertically oriented (Smith, 1980) rather than tilted as above waves are observed (Fig. 5). Since there may be some leakage of kinetic energy through the stable layer and the surface friction is not accounted for in the trapped lee wave theory, these factors may account for the observed tilt.

The acceleration of the low level jet is caused by a combination of the confluence or mass continuity through the wind corridor and the thermal wind (N-S temperature gradient) effect. Although the residence time of the airflow in the wind corridor is much less than one half a pendulum day (~ 18 h), there is still enough time to cause some ageostrophic acceleration. Reexamining the vertical cross-sections across RWL and MB (Fig. 3), the confluence effect accounts for the 5 ms^{-1} increase in mean wind speed. The mass flux (Q) through the planetary boundary layer is $\rho \bar{U} WH$. Over RWL the mean winds ($\bar{U}$) were ~ 15 ms^{-1}, the width (W) was ~ 80 km and the height (H) was ~ 800 m. Over MB the corresponding values were $\bar{U} \approx 20$ ms^{-1}, $W \approx 60$ km and $H \approx 800$ m. Consequently the horizontal mass fluxes are about equal and the 5 ms^{-1} increase in mean wind speed resulted from the confluence effect. The low level jet, i.e., vertical wind shear, is assumed to result from surface friction and the thermal wind.

3. Windy Gap (1 December)

The 1 December flight was similar to that of 11 December except that several passes were made at various altitudes through a pair of gaps along the northern border of the wind corridor. Windy Gap was so named because of the striking appearance of the eolian features. Satellite and high altitude aircraft imagery of the area revealed pronounced sand and scour streaks. They are parallel to the dominant wind, converge into the gaps, and deflation areas are present downwind of the gaps (Kolm, 1977 and Marrs and Kolms, 1982). The Wyoming Highway Department routinely advises motorists traveling through Muddy Gap that "strong winds are present in Muddy Gap". Because of the pronounced eolian features and the possibilities of extreme turbulence, relatively light wind days were chosen to conduct aircraft investigations of the airflow through the gaps.

Fig. 7 is an enlarged diagram of the Windy Gap/Muddy Gap area. The figure contains the data for 1 December including velocity vectors along the 100 m AGL flight track through the gap regions and the streamline and isentrope analysis of the flight data in relation to the terrain. Similar data and results were observed on three other flights. The streamlines depict the splitting of the southwesterly flow south of Windy Gap. Some of the air curves to the right to remain in the main channel of the wind corridor.The airflow from the northern part of the Great Divide Basin accelerates through Windy Gap and Muddy Gap and then dramatically flares out downwind of the gaps. The cold air (θ=286 K) makes it most of the way through both gaps but is dramatically mixed with warmer air. Very warm air (θ=294 K) has subsided into the lee of Ferris Mt. Vortices are present in the lee of Green and Ferris Mt.

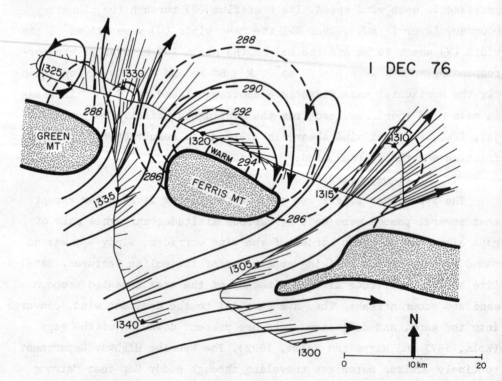

Fig. 7. Same as Fig. 2 except for Windy Gap (east of Ferris Mt.) and Muddy Gap (west of Ferris Mt.) for 1 December 1976.

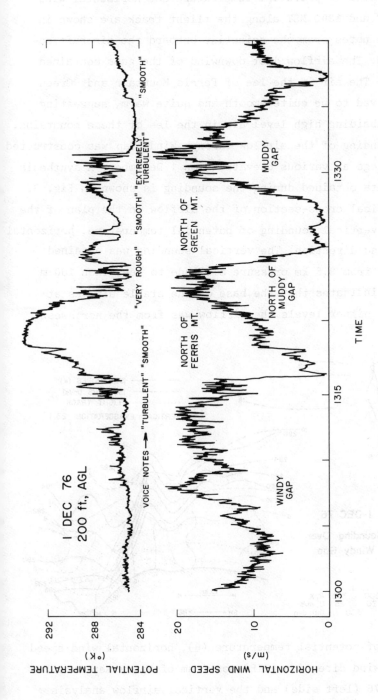

Fig. 8. Analog traces of horizontal wind speed and potential temperature for the 1 December 1976 flight segment through Windy
Gap and Muddy Gap between 1300 and 1340 MST.

491

The analog traces of potential temperature and horizontal wind speed between 1300 and 1340 MST along the flight track are shown in Fig. 8. Some voice notes from the scientist onboard the aircraft accompany the traces. The airflow just downwind of the gaps contained severe turbulence. The air in the lee of Ferris Mountain and Green Mountain was observed to be quite smooth and quite warm, suggesting the presence of subsiding high level air in the lee of these mountains.

A profile sounding of the airflow through Windy Gap was constructed from five flight legs at various elevations on 1 December. A synthesis of the aircraft data obtained during the sounding is shown in Fig. 9. It contains a vertical cross-section of the airflow in the plan of the flight legs and a vertical sounding of potential temperature, horizontal wind speed, and wind direction. The vertical sounding was obtained during the descent from 3.5 km pressure altitude to less than 100 m AGL. The sounding indicates that the base of the stable air was at 2.4 km and that at higher levels the airflow was from the northwest.

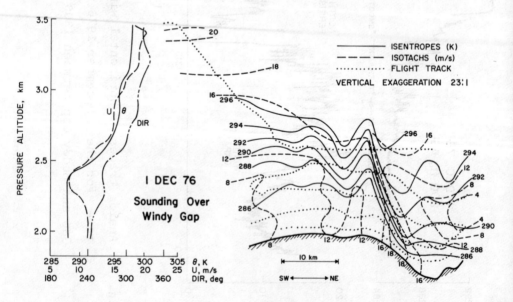

Fig. 9. Sounding of potential temperature (θ), horizontal wind speed (U), and wind direction (DIR) upstream of Windy Gap on 1 December 1976 (left side) and the vertical airflow analysis (right side) from the profile sounding over Windy Gap.

Thus the flow component through the gap vanished ~ 800 m above the upstream terrain sending the Scorer parameter to infinity. "Critical layers" may occur when the wave equation contains mathematical singularities such as the horizontal wind decreasing to zero or reversing in direction. Booker and Bretherton (1967) have shown that small amplitude waves are absorbed at such critical layers. Gossard and Hooke (1975) describe critical layers as energy and momentum sinks or sources for gravity waves and, as such, are sites of strong coupling between waves and background flows.

Windy Gap contains a small ridge at the mouth of the gap and a deflation area in the lee of the ridge (see Fig. 9). The isentropic analysis indicates that a dramatic gravity wave resembling a hydraulic jump occurred in the lee of the ridge. Some of the isentropes descended 600 m in a concentrated layer of high kinetic energy and then ascended into a standing wave in which much of the kinetic energy was dissipated as turbulent kinetic energy. This occurs over the deflation area in the lee of the ridge. At 100 m above the terrain the horizontal wind speed increased from 8 ms^{-1} upwind of the ridge to 19 ms^{-1} over the lee slope.

Contrary to the trapped lee waves which are small-amplitude two-dimensional mountain waves, the hydraulic jump is large-amplitude mountain wave (Smith, 1980). The hydraulic jump is, therefore, a non-linear phenomenon. Houghton and Kasahara (1968) obtained asymptotic solutions to the "shallow water" equations which describe flow over an isolated ridge. Their results were plotted in terms of M_c and Fo.

$$M_c = H_c/h_o \qquad (3)$$

where H_c is the height of the crest and h_o is the depth of the approaching fluid. Fo is the Frounde number

$$F_o = U_o \, (h_o g \, \Delta \, \theta/\theta)^{-\frac{1}{2}} \qquad (4)$$

where U_o is the velocity of the approaching fluid, g is gravity, $\Delta\theta$ is the strength of the stable layer above the approaching fluid and θ is the potential temperature in the approaching fluid. Values of M_c and

Fo for both the flow through Windy Gap and over the ridge at the upwind edge of the Killpecker Sand Dunes were evaluated and compared to the solutions derived by Houghton and Kasahara (1968). The values were found to indicate that hydraulic jumps were possible. The observed flow structures appeared to display hydraulic jumps in that the amplitude of the waves were greater than half the depth of the approaching flow and turbulence was present downstream of the large-amplitude wave.

Another interesting wave feature occurred just upwind of the critical flow column. This wave appeared to be a steady state phenomenon because it was observed between 2.3 and 2.6 km pressure altitude during the upper three flight legs. In this event as well as on three other hydraulic jump events documented by the aircraft in this area, both steady and unsteady waves were observed just upwind of the critical flow columnn.

Long (1972) has summarized several theoretical and experimental investigations of "upstream effects" occurring when lee waves begin to break internally and become strongly turbulent. The upstream influence may be in the form of a bore or surge which moves upstream of an obstacle and raises the depth of the upstream flow layer. Wong and Kao (1970) and other investigators theoretically obtained upstream influence effects in the form of multiple jets in the upstream velocity profile. Some of these features appeared to be present in the hydraulic jump events described in this ivestigation, although none were sufficiently well documented to be certain.

4. The Laramie Range (27 January)

The eastern portion of the wind ⟨corridor in southern Wyoming was investigated by the research aircraft on three occasions. The north-south oriented Laramie Range dominates southeastern Wyoming. The range serves as a topographic ramp over which the airflow spills onto the High Plains of eastern Wyoming (see Fig. 1). The Wheatland Reservoir (WR) area is the lowest region along the Laramie Range and is immediately downwind from the wind corridor. Pilots routinely report moderate or severe turbulence in the lee of the Laramie Range.

The first flight (27 January) was the most complete and most dramatic and will be the one presented. Fig. 10 contains the flight track,

494

wind velocity vectors and analyzed streamlines and isentrope fields at
∿ 100 m AGL. The flight consisted of four soundings and four N-S legs.
The soundings were made southwest of LAR, ∿ 30 km north of MB, over
Cheyenne (CYS), and a descent back onto LAR. The N-S legs were northbound in the lee of the Medicine Bow Mountains, southbound just upwind
of the Laramie Range, northbound in the immediate lee of the Laramie
Range and finally southbound ∿ 50 km downwind of the Laramie Range over
the High Plains north of CYS. Because of the extreme turbulence encountered in the vicinity of the rugged terrain in the immediate lee of the
Laramie Range, additional flight legs were not flow in the area. Drift
angles of 20 to 25 deg were required to fly in the 30 ms^{-1} crosswind
with a true air speed of 75 ms^{-1}.

Some of the more obvious characteristics of the airflow in southern
Wyoming are revealed by this case study. The airflow exiting the wind
corridor displayed a difluence factor of two to one as it expanded
from a width of ∿ 60 km in the wind corridor to ∿ 130 km over and north
of CYS. The mean velocity through the corridor was 20 m/s and increased
to 30 m/s in the lee of the Laramie Range just upwind of Chugwater
(CG). We will see later (Fig. 12) that in spite of the strong difluence
there was substantial downward transport of sensible heat and momentum
in the mountain wave from the Laramie Range to cause the increase in
winds over Chugwater. Over the High Plains north of CYS the wind speed
was 15 to 25 ms^{-1} with a mean value of ∿ 20 ms^{-1}. The streamlines and
isentropes were parallel upwind of the Laramie Range (except in the
flow east of LAR). This is in argreement with Fig. 2. The streamlines
and isentropes were also parallel along the Colo/Wyo border. In the
lee of the Laramie Range the streamlines and isentropes were nearly
orthogonal indicating substantial downward sensible heat flux in the
extreme turbulence region. Wake flow, i.e. easterly winds, was present
northwest of LAR in the lee of the Medicine Bow Mountains. In fact, a
distinct temperature perturbation, convergence and wind shift line was
encountered during the southbound leg east of LAR and even along the
northbound leg in the lee of the Laramie Range, a distance of ∿ 100 km
southeast of the Medicine Bow Mountains. When passing through the con-

vergence line ∿ 10 km east of LAR, the winds changed from 010° at 7 m/s to 250° at 11 m/s in a distance of 2.5 km (Conv = .005 s^{-1}). The temperature perturbation was ∿ + 1.5 K and there were no clouds, updrafts or turbulence in this region.

On the return flight to LAR from CYS evidence was obtained on the depth of the convergence line near LAR. About 20 km SE of LAR at 3.2 km pressure altitude (1 km AGL) the winds shifted from 310° to 275° and back to 300° over LAR. Simultaneously, the potential temperature changed from 304 K to 299 K and back to 303 K over LAR. Seaman (1982) simulated the airflow in southern Wyoming using the Anthes and Warner (1978) model and a stable sounding similar to that present on this day. His model result showed a distinct wake flow and convergence line in the same area.

Another characteristic illustrated by this case study is that the prevailing winter wind at LAR is from the WSW and the air comes from north central Colorado while the prevailing winter wind at CYS is WNW and the air comes from the southern or warmer half of the wind corridor. The airflow at LAR comes across the lower mountains between the Medicine Bow Mountains in Wyoming and the Front Range in Colorado (Fig. 1). The warmest air near the surface is typically near the Colo/Wyo state line south of LAR. Relatively cold (θ < 293 K) and turbulent air was present ∿ 30 km southwest of LAR but was mostly mixed out with warmer upper level air before reaching the Laramie Range. The air south of the convergence line and downwind of the cold air was slightly warmer than 294 K.

The vortices or interfaces on both the north and south side of the wake flow northwest of LAR were quite warm (θ < 299 K) while the center of the wake or return flow was cool (θ ≈ 294 K).

North of the wake flow the wind speed increased while the potential temperature decreased. A vertical cross-section (Fig. 11) was constructed through the wind corridor near MB. The base of the stable layer was at 2.6 km, well below the height of the Medicine Bow Mountains. The low level jet was at 2.5 km pressure altitude with a peak velocity of 25 ms^{-1} The very stable layer ($\frac{\partial\theta}{\partial Z}$ = 0.07 K/m) above 2.6 km was in fact a 7°C inversion. The cold air region was quasi-barotropic. Trapped lee waves were observed within the wind corridor. Very strong baroclinity was again present in the southern half of the wind corridor with values of $\frac{\partial U_g}{\partial Z}$ being .05 to .1 s^{-1}. The precise value could not be determined because no sounding was available in the critical region over MB.

The ascent sounding over the state line south of CYS (Fig. 12) had some interesting characteristics. This sounding utilized the surface wind and temperature at CYS at 1500 LT. A very stable layer was still present in the lowest 100 m with a 5°C temperature inversion present below the aircraft. A low level jet of 23 m/s was present $\sim$ 150 m AGL. The jet may have been near the ground because the flow was stable and descending on the lee side of the Laramie Range.

5. Conclusions and Summary

Several research flights were flow in southern Wyoming in the stablystratified low level airflow which is prevalent during the winter season. The results indicate that mesoscale airflow which is quasi-steady state can be studied effectively with an instrumented aircraft. Most of the flight time was spent in the valley areas at an altitude of $\sim$ 100 m AGL in order to study the characteristics of the wind for a wind energy/eolian landform study. Three or four soundings to $\sim$ 1 km AGL were made in order to document the vertical structure of the planetary boundary layer and the overlying stable layer.

The airflow through the 60 km wide gap in the continental divide, i.e., the Wyoming wind corridor, was of primary concern because this was the region of strong winds and the planned windfarm. Windy days in southern Wyoming typically display a surface high pressure area in western Colorado or Utah and a cyclone on the northern High Plains. The resulting pressure gradient forces the stably-stratified air in western Colorado and Wyoming to pile up on the western slope resulting in a barrier jet as modeled by Parish (1982). The airflow upwind of the wind corridor experiences significant confluence as it approaches the wind corridor over the Great Divide Basin of Wyoming. As the air passes through the wind corridor it experiences an indirect circulation cell in the lowest kilometer. The coldest air flows through the north side of the corridor and the warmest air flows through the south side. The

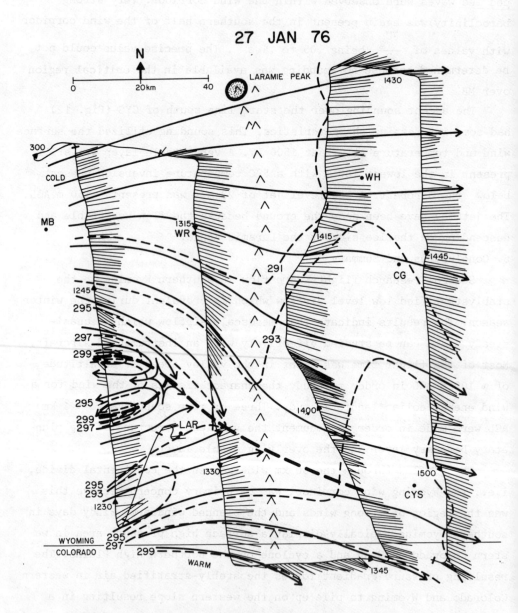

Fig. 10. Same as Fig. 2 except for Laramie Range flight on 27 January
1976. Note convergence line (dashed line) near LAR and extending to near CYS. Isentropes are thin dashed lines.

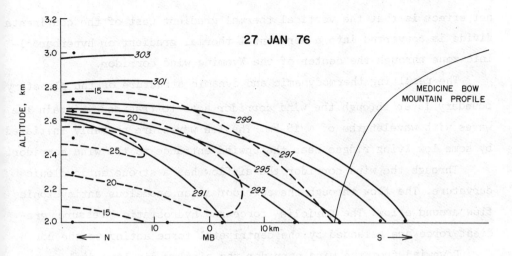

Fig. 11. Same as Fig. 3 except for 27 January 1976 across MB.

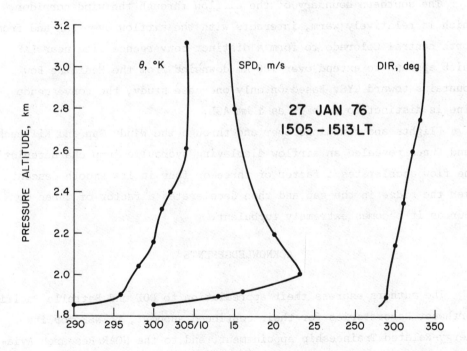

Fig. 12. Vertical profile of potential temperature (θ), horizontal
wind speed (U), and wind direction (DIR) over Cheyenne on 27
January 1976.

net effect is that the vertical thermal gradient west of the continenta\
divide is converted into a horizontal thermal gradient or hyperbarocl-\
inic zone through the center of the Wyoming wind corridor.

The resulting thermodynamic and dynamic structure of the planetary boundary layer through the wind corridor acts to trap and maintain lee waves with wavelengths of $\sim$ 15 km. The lee waves are probably initiated by some low lying ridges near the upwind entrance to the wind corridor.

Through the wind corridor the airflow has a strong anticyclonic curvature. The flow through the corridor is an anomalous anticyclonic flow around a low. The Coriolis force and hydrostatic pressure gradient force are balanced by the centripetal force acting to the north.

Downwind from the wind corridor the airflow displays difluence such that the width of the airflow through the wind corridor (60 km) doubles to $\sim$ 120 km.

The southern boundary of the airflow through the wind corridor, which is relatively warm, interacts with the airflow over LAR and from north central Colorado to form a distinct convergence line near LAR which appears to extend over 100 km downwind from the Medicine Bow Mountains toward CYS. Based on only one case study, the convergence line is distinct to as high as 1 km AGL.

Flights and soundings over and through the Windy Gap and Killpecker Sand Dunes revealed an airflow displaying hydraulic jump characteristics The flow accelerates a factor of three or four in its smooth descent over the ridge in the gap and then decelerates a factor of three or four as it becomes extremely turbulent.

ACKNOWLEDGEMENTS

The authors express their appreciation to DOE and Battelle Pacific Northwest Laboratories for their continued support, to NSF for its Energy-Related Traineeship appointment, and to the NCAR Research Aviation Facility for the use of its aircraft and data processing facilities.

The authors also wish to thank the pilots from NCAR for the safe rides, the staff members and graduate students from the Atmospheric

Science Department of the University of Wyoming who assisted in the flight operations, and the other staff members of the department for their photographic, drafting and typing services. The research was initially supported by Department of Energy under Contract No. EY-76-S-06-2343 and was completed through support by Department of Interior under Contract No. 9-07-70-S0104.

REFERENCES

Anthes, R. and T. Warner, 1978: The development of mesoscale models suitable for air pollution and other mesometeorological studies. Mon. Wea. Rev., 106, 1045-1078.

Booker, J. and F. Bretherton, 1967: The critical layer for internal gravity waves in a shear flow. J. Fluid Mech., 27, 513-539.

Burris, R. H., J. C. Corington and M. N. Zrubek, 1973: Beechcraft Queen Air aircraft. Atmospheric Technology, 1, NCAR, P.O. Box 3000, Boulder, CO 80307, pp. 25-30

Gossard, E. E. and W. H. Hooke, 1975: Waves in the Atmosphere: Atmospheric Infrasound and Gravity Waves-Their Generation and Propagation. Elsevier Scienctific Publishing Company, 456 pp.

Hess, S. L . 1959: Introduction to Theoretical Meterology. Holt, Rinehart and Winston, New York, 362 pp.

Houghton, H. and A. Kasahara, 1968: Nonlinear shallow fluid flow over an isolated ridge. Comm. Pure Appl. Math., 21, 1-23.

Kelley, N. D., 1973: Meteorological uses of inertial navigation. Atmospheric Technology, 1, NCAR, P.O. Box 3000, Boulder, CO 80307, pp. 37-39.

Kolm, K., 1977: Predicting surface wind characteristics of Wyoming from remote sensing of eolian geomorphology. Ph.D. Dissertation, Department of Geology, University of Wyoming, P.O. Box 3006, Laramie, Wyoming 82071, 115 pp.

Lenschow, D. H., C. A. Cullian, R. B. Friesen and E. N. Brown, 1978: The status of air motion measurements on NCAR aircraft. Preprints 4th Symp. on Meteorological Observations and Instrumentation. Denver, Co., American Meteorological Society, 433-438.

Long, R. R., 1972: Finite amplitude disturbances in the flow of inviscid rotating and stratified fluids over obstacles. Annual Rev. of Fluid Mech., 4, 69-92.

Marrs, R. and K. Kolm, 1982: Interpretation of windflow characteristics from eolian landforms. Special Paper #192. The Geological Society of America, P.O. Box 9140, 3300 Penrose Pl., Boulder, Co 80301. 109 pp.

Martner, B. E. and J. D. Marwitz, 1982: Wind characteristics in southern Wyoming, J. Appl. Meteor., 21, 1815-1827.

Parish, T., 1982: Barrier winds along the Sierra Nevada Mountains. J. Appl. Meteor., 21, 925-930.

Scorer, R., 1948: Theory of lee waves of mountains. QJRMS, 75, 41-56.

Seaman, N., 1982: A numerical simulation of three-dimensional mesoscale flows over mountainous terrain. Rep. AS 135, Department of Atmospheric Science, University of Wyoming, Laramie, 86 pp.

Smith, R., 1980: The influence of mountains on the atmosphere. Advances in Geophysics, Vol. 21, 87-223.

Wong, K. K. and T. W. Kao, 1970: Stratified flow over extended obstacles and its application to topographic effect on vertical wind shear. J. Atmos. Sci., 27, 884-889.

Session 5: Observation and Analysis (4)

SOME FACTS ABOUT THE SEASONAL TRANSITION
OF ATMOSPHERIC CIRCULATION IN EARLY SUMMER,1979

Chang Jijia and Zhu Fukang
State Meteorological Administration

Peng Yongqing and Wang Panxing
Nanjing Institute of Meteorology

The seasonal transition of atmospheric circulation is an important subjects of atmospheric circulation research. In this area of research, it is necessary to study not only the mean conditions of atmosphere during each of the seasons, but also the transition between the various mean conditions. As long as there exists seasonal transition the structure and internal properties of the global atmospheric circulation will would experience pronounced changes, and as a result a series of important weather phenomena may take place, for instance, the Indian SW monsoon, the Mei-yu in the lower and middle reaches of Changjiang River, etc. For these reasons the seasonal transition of atmospheric circulation has attracted many scientists' attention.

During 1979 in the FGGE, an unprecedented amount of observational data was obtained, including the data by both conventional and special observational systems. In May-August 1979, the Qinghai-Xizang Plateau Meteorological Science Experiment was carried out, providing conventional data and heat source observational data for the Qinghai-Xizang Plateau (QXP) and its neighboring regions. All these observations have created an improved data base for research into the seasonal transition of atmospheric circulation.

In 1979, the southern Indian Sw monsoon started on 12 June. The Bombay SW monsoon and the Mei-yu in the lower and middle reaches of Changjiang River bagan on 19 June. Before the onset of the monsoon, a

dramatic series of significant weather events occurred, and the seasonal transition of atmospheric circulation in the Northern Hemisphere
was accomplished. A brief review of the weather events is given as
follows.

In early May, the southern branch of westerly jet retreated northward and arrived on the QXP.

6-10 May. A cross-equatorial flow appears near Somalia.

11-15 May. The circulation at 30 hPa exhibits the transition from
winter to summer pattern, the tropical easterly jet appears over South
Asia, the South Asian high approaches the Indo-China peninsula for the
first time in this year.

16-20 May. The intensity of westerlies at 100 hPa decreases rapidly in the region of 30-50°N and 20-130°E.

26-31 May. The jet over China shifts northward, while the convection activity is mainly concentrated over the north of the equator.

1-5 June. The seasonal transition of the 30 hPa circulation pattern is accomplished and a polar anticyclone is established, the 50 hPa
circulation pattern starts the transition from winter to summer.

6-10 June. The intensity of westerlies at 100 hPa increases markedly in the region of 30-50°S and 20-130°E. The ground net radiation
at Shiquanhe increases sharply to its first peak value. The coefficient
of No. 2 eigenvector (monsoon component) of EOF of the flow pattern
over 30 N-30°S and 60-150°E increases rapidly.

11-15 June. The monsoon starts in southern India, the Somalia jet
sets up and strengthens on 12 June, strong cold air activities develop
along 60°E and 140°E in the Southern Hemisphere on 13 June, the monsoon
onset-vortex develops over the Arabian Sea, and a ridge forms from the
Mozambique Channel to the equator.

16-20 June. The South Asian high approaches the QXP on 16 June,
the rainy season begins in Lhasa on 17 June, the SW monsoon begins in
Bombay on 19 June and on the same day Mei-yu begins in Shanghai.

What is listed above are the major large scale weather events that
happened in 1979 during the transition process from winter to summer.
The process takes about 50 days from the northward retreat of the westerly jet in early May to the formation of the summer circulation

pattern in East Asia and the onset of Mei-yu in the lower and middle reaches of Changjiang River. The occurence of all these events should be considered as the product and concrete exhibition of the seasonal oscillation of atmospheric circulation in the Hemispheres. The events would not occur if the Hemisphere were independent of each other. There must be a connection between them. Looking upon the whole process as listed above, it would seem that the greatest number of events significant to the reasonal transition occur during the period from the second pentad to the fourth pentad of June. So it is probably reasonable to consider that period as the seasonal transition period of atmospheric circulation from winter to summer in 1979. Some major features can be seen during this period:

1) The transition is noted as a global characteristic. Fig. 1, in a sketchy way, shows the seasonal variation in the early summer of 1979, and stresses the following points:

i) The northward displacement and weakening of westerlies in the Northern Hemisphere set an important background for the transition of the planetary scale flow pattern in the Northern Hemisphere from spring to summer.

ii) While the strong cold air activity comes into existence on the east coast of Africa and Australia on 13 June, the Somalia jet and the crossequatorial flow near 105°E also begin.

iii) The Bombay SW monsoon, the Mei-yu in Shanghai and the rainy season in Lhasa start at almost the same time during the fourth pentad of June.

Hence it can be seen that, as the favourable background for the seasonal transition in the Northern Hemisphere is established, the cold air activity in the Southern Hemisphere and its related cross-equatorial flow come into existence and become intense. This process should play an important part in triggering the onset of the Indian monsoon and the Chinese rainy season.

2) The transition is initiated in the stratosphere, then extends into the troposphere. By 3 June, a polar anticyclone has been set up at the 30 hPa level, as seen in Fig. 2, and the circulation has changed into the summer pattern. But at the 50 hPa level at that time, the

change toward summer has just started and the European high is under
way already intruding into the polar high. On 10 June at the 50 hPa
level the high system extends to Greenland, and a polar-centering anticyclone is built up with its two strong ridges lying over North America
and Asia, indicating the establishment of the summer pattern (Fig. 3).
Hence, the transition to the summer pattern in the stratosphere is initiated from the upper level.

It is worth pointing out that there is a certain relationship betwe-
en the stratospheric circulation and that in the upper troposphere.As a
striking event in the lower stratosphere, the summer pattern sets up
previously at 30hPa then at the 50 hPa during the first dekad of June,
and the strong ridges from the polar-centering high stretch out over
Northern America and Asia. Under the control of the high ridge over
Asia, the South Asian high jumps onto the QXP (Fig. 4) The overlapping
of the high pressure systems at upper and lower levels forms a deep
system, which promotes the northward advection and maintenance of the
South Asian high, and strengthens the tropical easterly jet. This is
one of the important features of the year's summer circulation in Asia.

3) In the transitional process,the construction of No.2 eigenvector
also called a monsoon component) directly reflects the seasonal transition. The empirical orthogonal function (EOF) method is used to resolve
the series of the mean flow pattern at low latitudes from May through
July 1979. The separated No. 2 eigenvector can be referred to as an
important component for the seasonal transition of circulation. Fig. 5
shows the construction of No. 2 eigenvector at the 500 hPa level:
(1) a cyclone occupies the Arabian Sea and the associated monsoon trough
dominates the eastern Arabian Sea, the Indian subcontinent and the Bay
of Bengal (2) A cold cyclone controls eastern Australia (3) A ridge
from the anticyclone over the African continent stretches out to the
central Indian ocean. It is noteworthy that the values of the coefficient of No. 2 eigenvector at 100, 500 and 1000 hPa (Fig. 6) undergo a
variation from negative to positive during the second to the fourth
pentad of June with a marked linear increasing trend.

4) The sensible heating in the west part of the QXP may well play
a role in the seasonal transition of atmospheric circulation in early

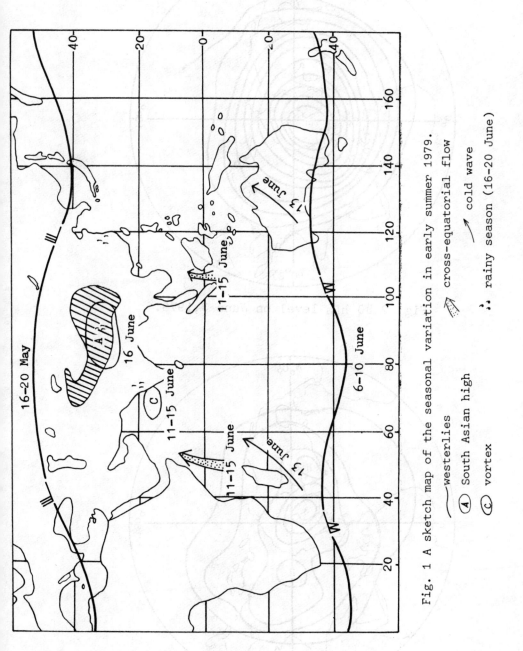

Fig. 1 A sketch map of the seasonal variation in early summer 1979.

— westerlies ⇨ cross-equatorial flow

Ⓐ South Asian high ⤳ cold wave

Ⓒ vortex ∴ rainy season (16-20 June)

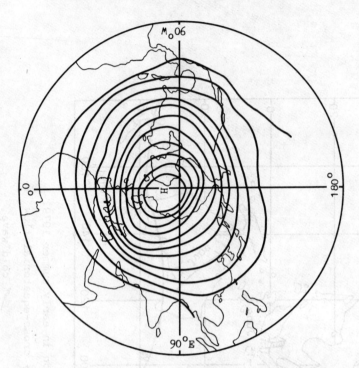

Fig. 2 30 hPa level on June 3,1979.

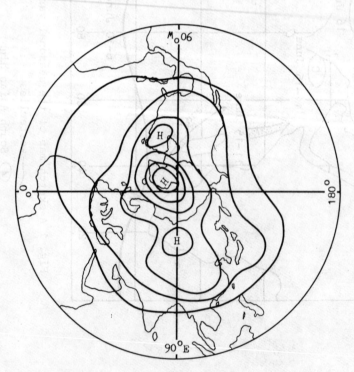

Fig. 3 50 hPa level on June 10, 1979.

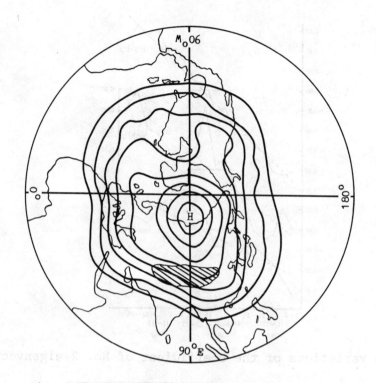

Fig. 4 The 50 hPa contour and the South Asian high
(shaded area) from June 26-30, 1979.

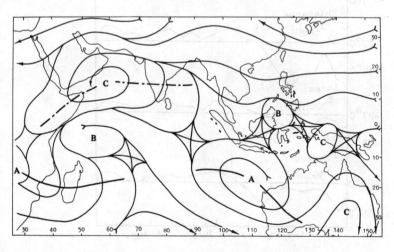

Fig. 5 The pentadly-mean flow pattern of No. 2 eigenvector of EOF at
500 hPa.

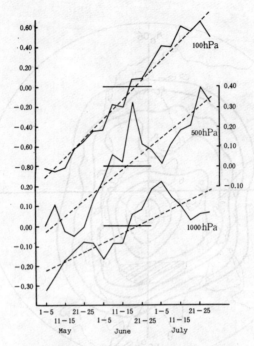

Fig. 6 The variations of the coefficient of No. 2 eigenvector.

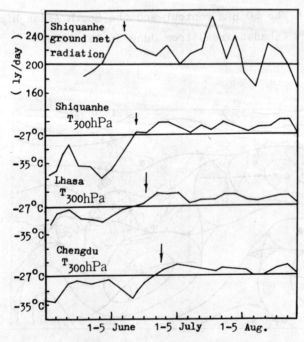

Fig. 7 The variations of the ground net radiation

at $T_{300\,hPa}$

summer Fig. 7 shows the variation of the ground net radiation at Shiquanhe and the 300 hPa temperature at Shiquanhe, Lhasa and Chengdu. All the values are averaged over each pentad. It is seen from the diagram that during 6-10 June the ground net radiation at Shiquanhe reached the first peak value which is a bit higher than that in the central plateau, and this occurred about two pentads before the rainy season comes in the central plateau on 17 June. The 300 hPa mean temperature has run up to -27°C and over since the third pentad of June with the greatest increase in the first dekad of June. Comparing this with the other two stations, for the 300 hPa mean temperature going to -27°C, it is a pentad later at Lhasa, and two pentads later at Chengdu. All these facts lead to a conclusion that in the early summer the local air in the western part of the QXP is warmed up after the net ground radiation there reaches the peak value. Then it takes about one pentad for the temperature to change into the summer pattern (300 hPa mean temperature $\geq$ -27°C), and finally the region with the greatest temperature increase gradually moves eastward and results in the temperature rising successively in Lhasa and Chengdu.

THE MEDIUM-RANGE OSCILLATIONS OF CIRCULATION
PARAMETERS OVER THE QINGHAI-XIZANG PLATEAU
IN THE SUMMER, 1979

Lu Longhau, Zhu Fukang and Chen Xianji

State Meteorological Administration

So far, there have been a lot of reports on the medium-range oscillation characteristics of the tropical and subtropical zone, and yet reports on those of the Qinghai-Xizang Plateau are very rare. The Qinghai-Xizang Plateau meteorological experiment (QXPMEX) was conducted in May-August 1979. During the period of the experiment 6 surface radiation stations and 4 radiosonde stations were set up on the plateau and a vast amount of data were obtained.

In this paper, we use multi-spectrum method to study the dominant scale, the vertical structure and horizontal propagation of the medium-range oscillations, and to discuss the interannual variation of the medium-range oscillations and its weather significance over the Qinghai-Xizang Plateau in the summer 1979.

The method used in this paper is the same as that is used in refer ence[1].The difference filter and 5 days running mean is applied for filtering the raw departure series are made for filtering small disturb ances and linear trends. In calculating spectrum estimates, Tukey windo is used, its maximum lag is 1/3 of the length of series, and according to the method of reference[2], the frequency points of calculating spectrum estimate are increased, with the spectrum estimates in low frequency having better resolution. Finally, we use Zangvil's method[3] to determine the dominant scale of all kinds, the product of power and frequency are expressed as spectrum characteristics.

I. DOMINANT SCALE OVER THE QINGHAI-XIZANG PLATEAU IN SUMMER 1979

In the summer 1979 a periodic oscillation of quasi-one week was shown in height, temperature and wind field of each isobaric level as well as in the ground thermal characteristics over the Qinghai-Xizang Plateau.

We can see from Fig. 1 that the major oscillations (the red noise confidence is 0.05) of height and temperature at various standard isobaric surfaces and tropopause over Shiquanhe, Gêrzê, Zhongba,

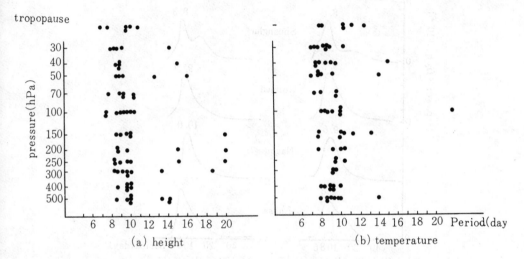

Fig. 1 The major oscillations of height and temperature at various isobaric surfaces over the eight stations on the Qinghai-Xizang Plateau.

Tingri, Shuanghu, Lhasa, Nagqu, Qamdo stations on the Qinghai-Xizang Plateau are about one week. We can see from Fig. 2 that the dominant scale of global radiation and net radiation over the six stations on the Plateau are also about one week. From Figs. 3 and 4 it could be seen that the ground-air temperature difference, total cloud cover, precipitation and 500 hPa u, v components over the plateau are also quasi-one week oscillation period. Therefore, quasi-one week oscilla-tion period is a dominant scale of the various meteorological elements and thermal parameters over the Qinghai-Xizang Plateau in the summer 1979[1].

The periodic oscillation of quasi-two week, which was shown in the monsoon region in 1967 found by T.N. Krishnamurti[4], became to be

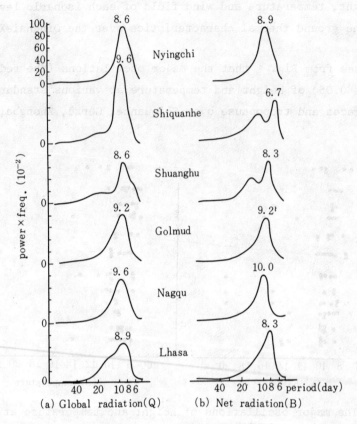

Fig. 2 The power spectrum of total radiation and net radiation in Lhasa etc. stations for June-August 1979.

secondary period in 1979. The periodic oscillation of 3—6 weeks is also shown in the Plateau region.

II. VERTICAL STRUCTURE AND HORIZONTAL PROPAGATION
OF THE QUASI-ONE WEEK OSCILLATION

The aerological station in the Qinghai-Xizang Plateau in summer 1979 are denser, which provide an favourable condition for discussing the vertical structure and horizontal propagation of quasi-one week oscillation.

1) Fig. 5 shows that there exists a quasi-one week oscillation in all layers from the middle troposphere to lower stratosphere over the

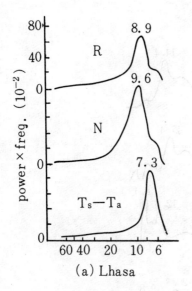 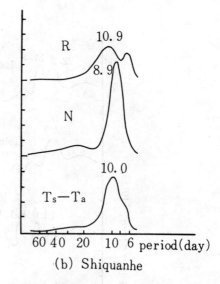

(a) Lhasa   (b) Shiquanhe

Fig. 3 The power spectrum of ground-air temperature difference,
(T_s-T_a), total cloud cover (N) and precipitation (R) in
Lhasa and Shiquanhe for June-August 1979.

Qinghai-Xizang Plateau and two maximum intensities occur in the middle
troposphere and lower stratosphere. Fig. 6 shows the mean phases of
quasi-one week oscillation of height and temperature at the various
isobaric surfaces of the eight stations given in Fig. 1. In making
Fig. 6, we only choose the points with mean coherence more than 50%
between 7.5—10.0 days oscillations at 100 hPa and those at other
layers. The phase difference with more than zero between 100 hPa and
the other level signifies that the oscillations at other levels lag
behind those at 100 hPa and vice versa. It can be seen from Fig. 6
that the axis of quasi-one week oscillation of height field is
fundamentally vertical, the phase lags between the various standard
isobaric surfaces and the 100 hPa level are less than 1 day; quasi-one
week oscillation of temperature at the various levels lead those at
100 hPa and the quasi-one week oscillations at 30 hPa and 500 hPa
lead those at 100 hPa by 2.3 and 6.0 days respectively. Since the
zonal propagation sense of quasi-one week oscillation of temperature
field at the upper and lower levels over the Qinghai-Xizang Plateau
is overturning, the sense in the middle-lower stratosphere is
eastward, that in the middle troposphere westward and the axis of

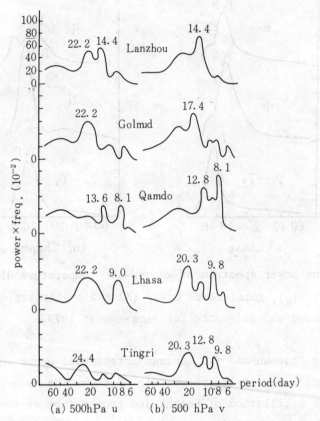

Fig. 4 The power spectrum of 500 hPa u, v components in
Lhasa etc. stations for May-October 1979.

quasi-one week oscillation system of temperature field tilts to the
west with height.

2) Table 1 gives the phase difference of quasi-one week oscillation (7.5—10.0 day) on the height and temperature field between
Lhasa station and the other 4 stations along 30°N.. In Table 1, the
"Φ" is the phase difference of each station relative to Lhasa station.
Its negative value means that the oscillation of Lhasa lags those of
other stations. "COH" is coherence (%). Table 2 illustrates the
relationship of oscillation phase between Lhasa and the other 4
stations along 90°E. Tables 1 and 2 show the characteristics of
longitudinal and latitudinal transportation.

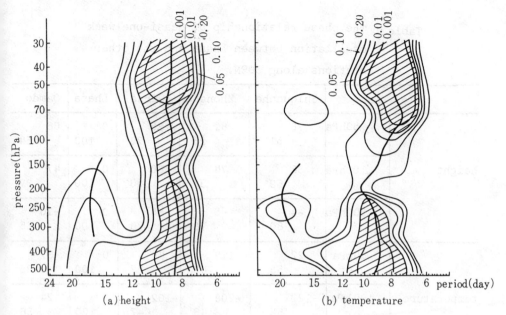

Fig. 5 The vertical structure of the quasi-one week oscillation in Zhongba.

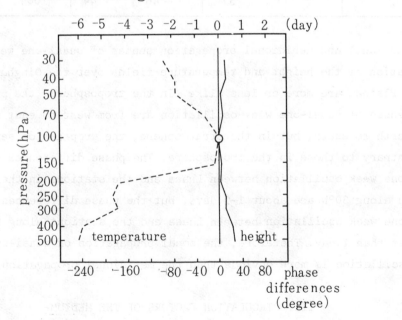

Fig. 6 The mean phases of the quasi-one week oscillation
of height and temperature at the various isobaric
surfaces over the Qinghai-Xizang Plateau.

Table 1 The phase relationship of quasi-one week oscillation between Lhasa and the other 4 stations along 30°N.

Φ COH		Shiquanhe	Zhongba	Gêrzê	Lhasa	Qamdo
height	50hPa	76 / 51	81 / 26	35 / 78	0 / 100	66 / 54
	100hPa	56 / 69	74 / 33	60 / 53	0 / 100	47 / 37
	500hPa	−4 / 80	−5 / 78	−5 / 78	0 / 100	−16 / 86
temperature	50hPa	22 / 55	119 / 43	46 / 67	0 / 100	−37 / 35
	100hPa	−122 / 55	−208 / 48	−102 / 47	0 / 100	24 / 56
	500hPa	−44 / 37	2 / 46	−17 / 44	0 / 100	−42 / 72

The zonal and meridional propagation senses of quasi-one week oscillation of the height and temperature fields over the Qinghai-Xizang Plateau are more or less alike. In the troposphere, the propagation senses of quasi-one week oscillation are from west to east and from north to south, but in the stratosphere, the propagation senses are contrary to those in the troposphere. The phase differences of quasi-one week oscillation between Lhasa and the stations in the western plateau along 30°N are about 1—2 days; but the phase differences of quasi-one week oscillation between Lhasa and the stations along 90°E are less than 1 day. Therefore, the zonal propagation of quasi-one week oscillation is more obvious than the meridional propagation.

III. CIRCULATION FACTORS OF THE MEDIUM-RANGE OSCILLATION

The South Asian high and tropical easterly jet are two important circulation systems in the upper troposphere in summer. Fig. 7 shows

Table 2 The phase relationship of quasi-one week
oscillation between Lhasa and the other
4 stations along 90°E.

Φ COH		Tingri	Lhasa	Nagqu	Shuanghu	Mangnai
height	50hPa	-32 / 46	0 / 100	-22 / 45	12 / 63	-16 / 63
	100hPa	-34 / 73	0 / 100	30 / 42	31 / 72	3 / 73
	500hPa	6 / 81	0 / 100	-1 / 97	3 / 96	-6 / 83
temperature	50hPa	0 / 73	0 / 100	60 / 60	21 / 58	1 / 75
	100hPa	-35 / 17	0 / 100	-33 / 75	-59 / 66	36 / 41
	500hPa	22 / 75	0 / 100	5 / 63	-51 / 40	6 / 37

that in the summer 1979, the two important circulation systems are also
obvious quasi-one week oscillation period. The phases of the quasi-one
week oscillation among the global radiation, net radiation, total cloud
cover, precipitation, ground-air temperature difference and the tropical
easterly jet, east-west and south-north shift of the South Asian high
are associated with each other: in June-August, the average location
of the South Asian high center is near 30°N, 85°E, when the center of
the South Asian high is to the west and the north, the tropical
easterly jet over Arabia is intensified, the total cloud cover and
precipitation over the Qinghai-Xizang Plateau are increased, but the
global radiation and net radiation are decreased and vice versa.

IV. INTERANNUAL VARIATION OF THE MEDIUM-
RANGE OSCILLATION CHARACTERISTICS

The results mentioned above are only the case in 1979. But in
other years, for example in 1967, the quasi-two week oscillation was

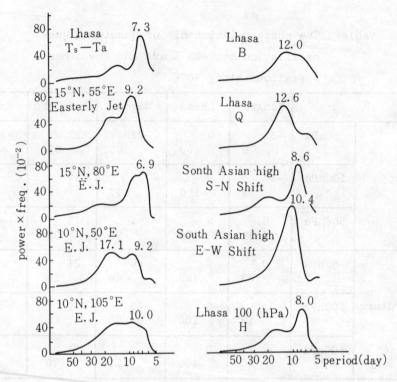

Fig. 7 The power spectrum of the South Asian high and
tropical easterly jet for May-July,1979.

dominant[4], in other words, the medium-range oscillation characterist
over the Qinghai-Xizang Plateau have obvious interannual variation.

The dominant scale length and the intensive variation of quasi-one
week and quasi-two week oscillations form quasitrienmial variation
period[5].

Fig. 1 shows the power spectrum of height at 100 hPa level in
Lhasa in June-August, 1966—1981. In Fig. 8., the solid line is the
various confidence limits of no dominant scale[3], the regions of the
oblique lines are those beyond confidence 0.05, the circle is the
major period of the medium-range oscillation, and the dot shows that
the major period is significant (the red noise confidence is 0.05).
It is showed in Fig. 8 that the interannual variation of oscillation
characteristics in Lhasa is very clear and that the periods of
dominant scale and the intensive variations of quasi-one week and
quasi-two week oscillations are about three years. The quasi-one week

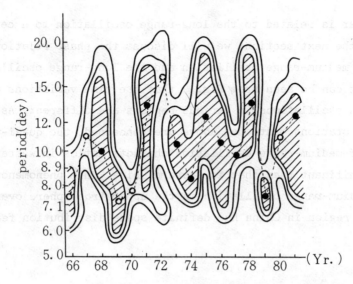

Fig. 8 The power spectrum of height at 100 hPa level in
Lhasa in June–August, 1966–1981.

and quasi-two week oscillation intensities in the upper troposphere
over the subtropical region in China also have this obvious interannual
variation.

In order to learn further the interannual variation of medium-
range oscillation over the subtropical region in China, we also
calculated the spectrum of the period of the dominant scale and the
intensities of quasi-one week and quasi-two week oscillations at
100 hPa level over Lhasa, Chengdu, Wuhan and Shanghai in 1966–1981.
In calculations, the intensity S_2 of quasi-two week oscillation is
denoted by the total variance percentage of 12–16 days, and the
intensity S_1 of quasi-one week oscillation by the total variance
percentage of 7.5–10 days. It can be seen from Fig. 9 that there
exists an obvious quasi-triennial oscillation period of the dominant
scales and the intensities of quasi-one week and quasi-two week
oscillations in Lhasa and Chengdu. The quasi-triennial oscillation
in Wuhan and Shanghai is not obvious, but the quasi-triennial oscillations of quasi-two week intensity in Wuhan and quasi-one week intensity
in Shanghai are still obvious. It indicates that the time variation
of the medium-range oscillations in the upper troposphere over the
subtropical region in China is regular, the medium-range oscillation

for each year is related to the long-range oscillation to a certain
extent. In the next section, we will discuss the phase relationship
between the medium-range oscillation and the long-range oscillation.
Moreover, it can be seen from Fig. 9 that the time variations of the
medium-range oscillations in various regions are different. As far
as the four stations mentioned above are concerned, the quasi-triennial
variation of medium-range oscillation characteristics in western China
is more significant than that in eastern China. This phenomenon shows
that the medium-range oscillation in the upper troposphere over the
subtropical region in China has definite space distribution feature.

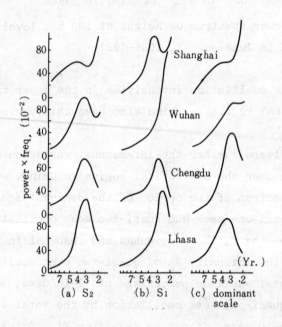

Fig. 9 The autospectrum characteristics of the medium-range
oscillations (cepstrum) in the upper troposphere
over the four subtropical stations in China.

V. WEATHER MEANING OF THE MEDIUM-
RANGE OSCILLATION CHARACTERISTICS

In discussing the oscillation of the summer monsoon system,

Krishnamurti suggested that the quasi-two week oscillation should be the inherent oscillation of the large scale monsoon system of the dynamical process and the quasi-one week oscillation may be the reflection due to the passage of local instability disturbances[4]. Murakami suggested that the quasi-5 day oscillations in the Indian monsoon region during 1962 should be a westward disturbance, and the 10—15 day oscillations seem to be related to the monsoon active and break cycle[6].

The quasi-one week and quasi-two week oscillations over the Qinghai-Xizang Plateau are closely related to the behaviors of the western Pacific subtropical high at 500hPa and the South Asian high at 100 hPa.When the phases of the medium-range oscillations or the oscillation characteristics are different, the behaviors of the 500 hPa western Pacific subtropical high and the 100 hPa South Asian high are also different. It can be seen from Fig. 10 and Table 3 that when the medium-range oscillation characteristics are different, the behaviors of the South Asian high are obviously different and the ridge line and main center are also different. When the quasi-two week oscillation dominant scale within 12—16 days in Lhasa is significant, the ridge line of the South Asian high over East Asia in July is to the north, and the mean ridge line between 110°—150°E is located in 35°N. At this time, the height of 100 hPa mean isobaric surface over Beijing region from June to August is higher (correlation coefficient $\gamma=0.52$; $\alpha=0.10$. $\gamma_\alpha=0.43$), and the major center of the South Asian high is to the east. When the quasi-one week oscillation (dominant scale within 7—9 days) in Lhasa is significant, the major center of the South Asian high is to the west, and the ridge line of the South Asian high over East Asia is to the south of 30°N, and the height of 100 hPa mean isobaric surface over the subtropical region in China from June to August is generally lower (in Table 3, the correlation coefficients of the intensity of the quasi-one week oscillation with the mean height at 100 hPa isobaric surface over Beijing and other four stations during June-August are all significant except for Shanghai).

In different phases of the quasi-two week and the quasi-one week oscillations, the behaviors of the South Asian high are also different.

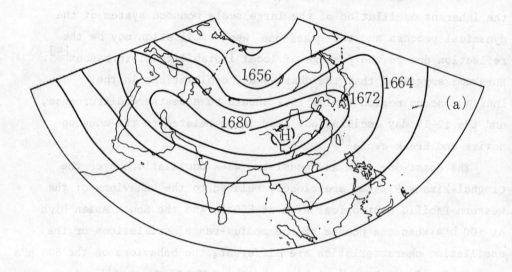

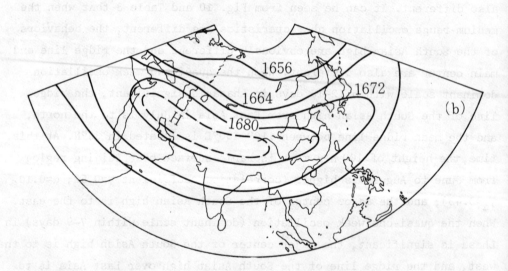

Fig. 10 Composite map of the different medium-range oscillation characteristics at 100 hPa in Lhasa in July.
(a) Obvious quasi-two week oscillation;
(b) Obvious quasi-one week oscillation.

Figs. 11 and 12 show 100 hPa mean height fields during the major
dominant scale for quasi-two week (July, 1975) and quasi-one week

526

(July, 1979) oscillations over Lhasa, respectively. The high phase
and low phase are expressed by three days including one day before
and after the peak or the valley, respectively.

It can be seen from Fig. 11 that in the high phase of the quasi-
two week oscillation, the ridge line of the South Asian high obviously
extends north-eastward as far as 35°—40°N over the western Pacific,
and the major center is to the east; in the low phase of the quasi-two
week oscillation, the ridge line of the South Asian high is at about
30°N over the western Pacific, and the major center is to the west and
south. It can be seen from Fig. 12 that in high phase of the quasi-one
week oscillation, the major center of the South Asian high is to the
west. The ridge line of the South Asian high over East Asia is located
to the south of 30°N; in the low phase of the quasi-one week oscillation, the major center of the South Asian high is to the east; the
ridge line of the South Asian high over East Asia is located at about
30°N.

It can be also seen from Figs. 11 and 12 that 100 hPa mean situations at different phases of quasi-two week and quasi-one week oscillations are almost opposite. In high phase of quasi-two week oscillation, the South Asian high over East Asia is stronger, while in high
phase of the quasi-one week oscillation, the South Asian high over
East Asia is weaker. In the quasi-one week oscillation, the mean ridge
lines of the South Asian high over East Asia, either high phase or low
phase, are to the south of that in quasi-two weeks oscillation. During
the years of the dominant quasi-one week oscillation shown in Fig. 10,
the mean ridge line of the South Asian high over East Asia is to the
south (5°) of that during the years of the dominant quasi-two week
oscillation. It shows that the influence of westerlies on the East
Asia region in quasi-one oscillation is more significant than that in
quasi-two week oscillation. Therefore, in discussing the physical
cause of quasi-two week and quasi-one week oscillations, this difference should be taken into consideration.

Fig. 8 shows that in some years the quasi-two week oscillation
is dominant whereas quasi-one week oscillation is dominant in some
other years. It can be seen from Table 3 that the correlation

Table 3 The relationships of the medium-range oscillation characteristics in the upper troposphere in Lhasa with the circulation features.

circulation features		quasi-two week oscillation	quasi-one week oscillation
100hPa South Asian high (July)	mean ridge line (110—150°E)	35.6°N	29.6°N
	major center (longitude)	101°E	54°E
500hPa western Pacific subtropical high (July)	mean ridge line (110—150°E)	28.2°N	24.6°N
	area index	13.4	25.0
correlation coefficients with the mean height at 100hPa level for June-August	Beijing	0.52	-0.43
	Shanghai	0.24	-0.35
	Wuhan	0.07	-0.44
	Chengdu	0.08	-0.58
	Lhasa	-0.40	-0.58
year		1971,1972,1975, 1978,1981	1966,1969,1970, 1974,1979

coefficients between the quasi-two week oscillation and the quasi-one week oscillation in the Lhasa station and other three stations are small, which shows on the whole that the two oscillations are independent of one an other.

Owing to the different oscillation characteristics in the Lhasa region, the behavior of the South Asian high is obviously different,

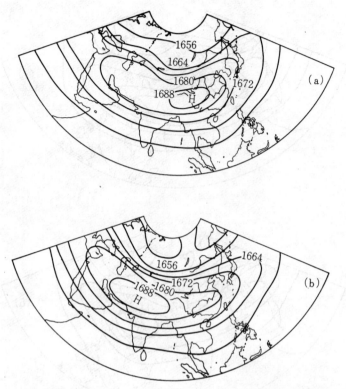

Fig. 11 100 hPa mean map in the different phases of the quasi-
        two week oscillation (July 1975).
        (a) high phase; (b) low phase.

Table 4 The correlation coefficients between the quasi-
        two week oscillation and the quasi-one week
        oscillation.

Lhasa	Chengdu	Wuhan	Shanghai
0.042	-0.060	0.006	-0.184

the influence of westerlies on East Asia is also different, and the two
oscillations are independent. Therefore, we consider the physical
causes leading to the two oscillations may be different. The quasi-two
week oscillation in the upper troposphere over the subtropical region
in China is probably a reflection of the inherent oscillation of the

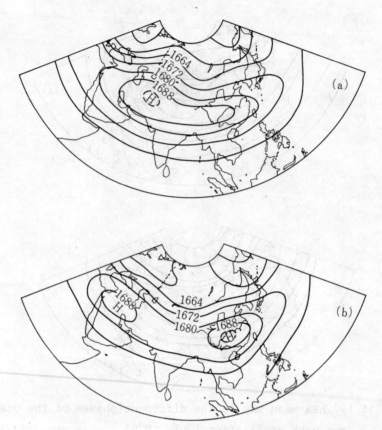

Fig. 12 100 hPa mean map in the different phases of the quasi-
one week oscillation (July 1979).
(a) high phase; (b) low phase.

South Asian high itself, while the quasi-one week oscillation is
probably a reflection of the forced oscillation from westerlies
disturbance. Thus, when the South Asian high is in east pattern, its
place is to the north and has stronger intensity over East Asia, and
the quasi-two week oscillation is dominant; when the South Asian
high is in west pattern, its place is to the south, the intensity
is weaker, the influence of the forced oscillation from westerlies
disturbance is larger, and the quasi-one week oscillation is dominant.
Whether this tentative idea is right or not still remains to be proved.

   In the eastern part of China in summer, because the South Asian

high in the upper troposphere tends to change synchronously with 500 hPa
western Pacific subtropical high, the western Pacific subtropical high
behavior is different when the oscillation characteristic in Lhasa is
different. It can be seen in Table 3 that when the height at 100 hPa
level over Lhasa is mainly the quasi-two week oscillation, the ridge
line of 500 hPa western Pacific subtropical high is to the north and
its area is smaller; when the height at 100 hPa level over Lhasa exhibits
mainly the quasi-one week oscillation, the ridge line of 500 hPa western
Pacific subtropical high is to the south and its area is larger.
Fig. 13 shows 500 hPa mean field in July for the different medium-range
oscillations in the upper troposphere over Lhasa. It can be seen from
Fig. 13 that in the different oscillation characteristics over Lhasa,
the behavior of the Indian depressure is also obviously difference.
The extent and intensity of the Indian depressure in the quasi-two week
oscillation are larger than those in the quasi-one week oscillation.

The medium-range oscillation characteristics in the upper troposphere are related to the summer precipitation in China. For example,
the variations of medium-range oscillation characteristics in Lhasa
during June-August 1966—1981 are closely related to the precipitation
of the seven regions from fifteen regions in China (refer to Table 5),
and the correlation coefficients are larger than the critical value.
When the quasi-one week oscillation is strong and the quasi-two week
oscillation is weak in Lhasa, the precipitations in Xinjiang-Gansu,
south of the lower reaches of the Changjiang River, South China
Sichuan and Guizhou, the southern Xinjiang, Qinghai-Xizang Plateau
in June-August are more than normal, while the precipitation in the
northern Xinjiang tends to be less than normal. The medium-range
oscillation characteristics in Chengdu, Wuhan and Shanghai are
related to the summer precipitation in some regions, but the significant
correlation regions are less than that in Lhasa. To study the medium-
range oscillation characteristics in the upper troposphere is useful
for the medium-long range forecast of summer precipitation in China.

In recenl years, more and more attention has been paid to the
quasi-triennial oscillation which acts as a major oscillation period
in the subtropical region. In reference[7], we have made a preliminary

531

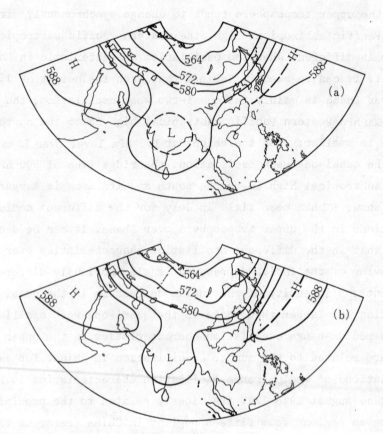

Fig. 13 The 500 hPa composite map of the 100 hPa different
oscillation characteristics in Lhasa in July.
(a) quasi-two week oscillation;
(b) quasi-one week oscillation.

summary in the quasi-triennial oscillation of the South Asian high. Not
only the height anomaly at low latitudes in the Northern Hemisphere
and the area and ridge of the South Asian high have the quasi-triennial
oscillations, but also the area and ridge of 500hPa western Pacific
subtropical high, the sea surface temperature in the equatorial eastern
Pacific and the frequency of the northwestern Pacific typhoon have
ones. In addition, the major oscillations of precipitation in the
western provinces of China[8] and temperature in the equatorial
regions[9] are quasi-triennial, too.

Table 5 The correlation coefficients between the
medium-range oscillations and the precipitation of 15 regions in China during June-August*.

	Intensity of quasi-two week oscillation	Intensity of quasi-one week oscillation	Dominant scale
1 Xinjiang-Gansu	Lhasa -0.59	Wuhan -0.46	Wuhan 0.50
2 Songliao Plain			
3 Nei Mongol		Chengdu 0.46	
4 North China			
5 Huaihe Valley			Shanghai 0.48
6 Changjiang Valley			
7 South of the lower reaches of Changjiang River	Chengdu -0.48	Lhasa 0.60	
8 South China		Lhasa 0.67	Wuhan -0.55
9 Yunnan			Chengdu 0.51
10 Sichuan-Guizhou	Shanghai -0.44		Lhasa -0.66
11 Hetao			
12 Hexi Corridor		Chengdu 0.61	
13 North Xinjiang		Lhasa -0.48	Wuhan 0.43
14 South Xinjiang		Lhasa 0.63	
15 Qinghai-Xizang Plateau		Lhasa 0.45	Lhasa -0.58

* The precipitation indexes of 15 regions are obtained from Central
Forecasting Office, State Meteorological Administration of China.

From the discussion mentioned above,the quasi-triennial variation (of the medium-range oscillation characteristic is correspondingly related to the quasi-triennial oscillation of the South Asian high. When the quasi-two week oscillation in Lhasa is strong, the area and ridge of the South Asian high and 500 hPa subtropical high are small and to the north; the sea surface temperature in the equatorial eastern Pacific is lower than normal.

That is to say, the strong quasi-two week oscillation tends to be the high phase of the triennical oscillation of the ridge line of the South Asian high and the 500 hPa western Pacific subtropical high, and tends to be the low phase of the triennical oscillation of the area indexes of the South Asian high and the subtropical high and the equatorial SST. When the quasi-one week oscillation is strong, things are contrary to that mentioned above. This shows that the quasi-triennial oscillation is an important phenomenon in the tropical and subtropical regions, not only the temperature, pressure and weather system have quasi-triennial oscillations, but the medium-range oscillation characteristic has one. The mechanics of the quasi-triennial oscillation seems to be related to the interaction between the ocean and the atmosphere[7].

REFERENCES

[1] Lu Longhua, Zhu Fukang, Chen Xianji and Zhu Yunlai, The medium-range oscillation over the Qinghai-Xizang Plateau in summer 1979, Collected Papers of the Qinghai-Xizang Plateau Meteorological Science Experiment (2), Science Press, 140—151, 1984 (in Chinese).
[2] Jenkins, G.M. and Wetts, D.C., Spectral analysis and its applications, San Francisco, Holden-Day, 525 pp 1968.
[3] Zangvil, A., On the presentation and interpretation of spectra of large-scale disturbances, Mon. Wea. Rev., 105, 1469—1472, 1977.
[4] Krishnamurti, T.N., and Bhalme, H.N., Oscillations of a monsoon system, Part 1, observational aspects, J. Atmos. Sci., 33, 1937—1954, 1976.
[5] Lu Longhua, Chen Xianji and Zhu Fukang, The interannual variation

of medium-range oscillation characteristics in the upper troposphere over the subtropical region in China in summer, Kexue
Tongbao, 28, 798-800, 1983 (in Chinese).

[6] Murakami, M., Analysis of summer monsoon fluctuations over India,
     J. Met. Soc. Japan, 54, 15-31, 1976.
[7] Zhu Fukang, Lu Longhua, Chen Xianji and Zhao Wei, South Asia
     high, Science Press, 95 pp., 1980 (in Chinese).
[8] Xu Guochang, Dong Anxiang, The quasi-three year period of precipita-
     tion in the West of China, Plateau Meteorology, 1, 2:11-17,1982.
[9] Zhang Mingli, Fu congbin, et al., A study of global surface
     temperature field in 70'S (1)—the characteristics of global
     temperature and summer cold disaster in the Northeast China in
     70'S, Scientia Atmospherica Sinica, 6, 229-236,1982.

A STUDY OF THE FIRST ARRIVAL OF THE SOUTH ASIA
HIGH ON THE QINGHAI-XIZANG PLATEAU

Sun Guowu

Institute of Meteorological Science of Gansu Province

As seen from the 100 hPa mean monthly charts, in summer the middle
and lower latitudes of the Northern Hemisphere are occupied by an
ultra-long wave system of wave number 2 with ridges located over two
continents and troughs over two oceans. One of the ridges situated in
South Asia develops a upper troposphere anticyclone, which is called
the South Asia high, over the Qinghai-Xizang Plateau.

The South Asia high undergoes distinct seasonal change[1]. From
winter to summer, the center of the South Asia high moves northwestward
from the region east of the Philippines in April via the Indo-China
Peninsula in May onto the Qinghai-Xizang Plateau in June. Consequently,
the South Asia high arriving at the Qinghai-Xizang Plateau may be considered as an important feature of the seasonal evolution of atmospheric
circulation in early summer, and it plays an important part in the
change of China's weather in this season.

I. THE FIRST ARRIVAL OF THE SOUTH ASIA HIGH ONTO
THE QINGHAI-XIZANG PLATEAU

We take the 100 hPa upper air wind at the Lhasa station as a
signal of the first arrival of the South Asia high on the Qinghai-
Xizang Plateau. The high arrives the moment the wind direction reaches
10°–90° (to the front of the high) and 90°–170° (to the back of the
high), Table 1.

From Table 1, we can see that each year the South Asia high takes
a different path and arrives at a different time.

Table 1 The dates and paths of the South Asia high arriving on the Qinghai-Xizang Plateau.

Year	Date of Arrival	Direction of Movement	Path
1965	June 13	SE--NW	E
1966	June 15	S--N	W
1967	June 5	SE--NW	E
1968	June 2	W--N	W
1969	May 26	W--E	W
1970	June 10	W--E	W
1971	June 4	W--E	W
1972	June 10	SE--NW	E
1973	June 8	W--E	W
1974	May 31	SE--NW	E
1975	June 10	SE--NW	E
1976	May 25	W--E	W
1977	June 21	W--E	W
1978	June 3	SE--NW	E
1979	June 17	S--N	W

Table 1 shows that the earliest date of arrival was May 25 and the latest was June 21; the mean date was June 8. Table 1 also suggests that there are three paths: the first path passes through the Indo-China Peninsula and then turns northwestward to ward the Qinghai-Xizang Plateau (40%); the second path passes through the Peninsula and then runs westward via North India onto the Qinghai-Xizang Plateau (13%); the third path passes through the Peninsula and runs westward to India and then northwestward to the Iranian Highland and finally moves eastward onto the Qinghai-Xizang Plateau(47%). For convenience we call the first path the East path (E) and the other two the West paths (W).

Due to the different paths the South Asia high takes, the 100 hPa circulation patterns vary when the South Asia high arrives on the Qinghai-Xizang Plateau. Fig. 1 shows that from 1965 to 1979 when the South Asia high arrived on the Qinghai-Xizang Plateau, its centre remained in the central part of the Qinghai-Xizang Plateau for three years (Fig. 1a); stayed in the southwest part of the Plateau for seven years (Fig. 1b), and then moved to the southeast part for five years (Fig. 1c).

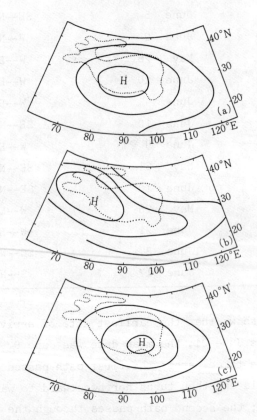

Fig. 1 The 100 hPa isobaric chart.

II. THE CAUSE OF THE FIRST ARRIVAL OF THE SOUTH ASIA HIGH ON THE QINGHAI-XIZANG PLATEAU

During the first pentad of June, 1979, the major center of the South Asia high moving slowly westward from the Yunnan-Guizhou Plateau to the south of the Qinghai-Xizang Plateau was accompanied by a sub-center arriving on the same side of the Qinghai-Xizang Plateau. The combined center moved over the Qinghai-Xizang Plateau in the fourth

pentad of the month (Fig. 2). From the data for 1965—1979 it seems
that such combining processes of ten took place in the neighborhood
before the South Asia high arrived at the Qinghai-Xizang Plateau. This
leads us to believe that the heating by the Qinghai-Xizang Plateau has
an effect on the South Asia high's arrival on the Plateau. Fig. 3
traces the surface net radiation at the Shiquanhe station in the west

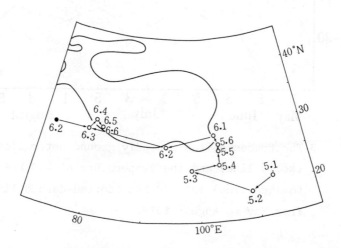

Fig. 2 The pentadly displacement of the South Asia high's
center for May and June 1979. The circle is the major
center and the dot is the sub-center of the high's.

of the Qinghai-Xizang Plateau in June 1979. The amount of net radiation
was more than 100 $W \cdot m^{-2}$ during the first pentad and in the next pentad
it increased to its peak value, 120 $W \cdot m^{-2}$. The increase occurred two
pentads before the South Asia high's arrival. At the same time, the
temperature at 300 hPa and 400 hPa increase appreciable from the first
pentad of June.

The status of the air column being heated from the ground to the
top of the atmosphere over East Asia during the first, third and fifth
pentads of June, 1979 is shown in Fig. 4. In the first pentad a heating
center is evident over the Qinghai-Xizang Plateau; the center still
exiated until the third pentad, at which point another heating center
was formed in the north of the Indian Peninsula; in the fifth pentad

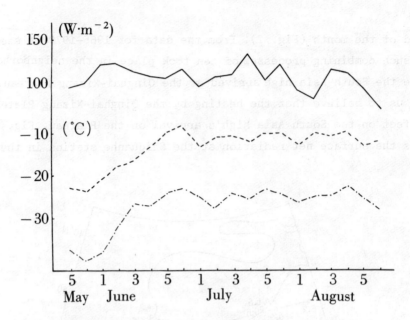

Fig. 3 The traces of the pentadly ground net radiation
(solid line) and the temperature at 400 hPa
(dashed line) and 300 hPa (dotted-dashed line)
from May to August 1979.

the two centers were intensified. As we know, in height the Qinghai-
Xizang Plateau reaches the middle troposphere, and its heating effect
can lift the air to a high-level and make the isobaric surface increas
its height. This may explain how South Asia high is formed and maintained

  To sum up, during the first, second and third pentads before the
arrival of the South Asia high over the Qinghai-Xizang Plateau, the
ground net radiation reached its peak value; the temperature at 300
hPa and 400 hPa increased considerably; and the heating centers over
the Qinghai-Xizang Plateau and the north of India were formed and
strengthened. This is the period when the South Asia high developed on
the southern side of the Qinghai-Xizang Plateau. Its major center and
sub-center came closer and overlapped up, and then the high moved
north to the Qinghai-Xizang Plateau. It can be seen that the heating
effect of the Qinghai-Xizang Plateau made important contributions to
the arrival of the South Asia high.

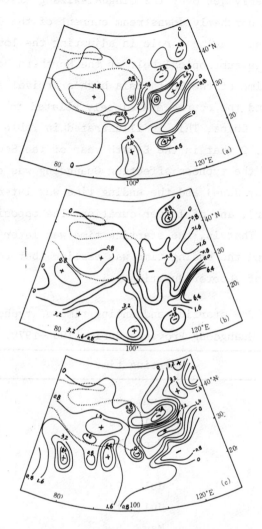

Fig. 4 The heating status throughout the air column over
East Asia ($^{\circ}C \cdot d^{-1}$). Charts a, b and c are for the
first, third and fifth pentads of June, respectively[2].

III. THE EFFECT OF THE SOUTH ASIA HIGH'S ARRIVAL OVER THE
QINGHAI-XIZANG PLATEAU ON THE WEATHER OF CHINA

During the summer the South Asia high is one of the dominating
systems in the upper troposphere over the Northern Hemisphere. The
South Asia high arrival time and course influences the position of the

subtropical westerly jet over the Qinghai-Xizang Plateau and also the
intensity of the northerly downstream current of the Qinghai-Xizang
Plateau as well as playing a role in adjusting the long wave ridges
and troughs in the subtropical belt of the Northern Hemisphere. Consequently, the time of the South Asia high's arrival at the Qinghai-
Xizang Plateau and the course are closely related to the onset of
summer weather in China, This is illustrated in Table 2. It can be seen
from Tables 1 and 2 that in the E-path year of the South Asia high the
starting time of the typhoon affecting Guangdong was earlier (before
the third dekad of June) and the ending time was later (after the third
decade of October), and of longer duration. The opposite is true for
the W-path year. That is, the starting time was later (after the first
dekad of July) and the ending time was earlier (before the second dekad
of October) and of shorter duration.

Table 2 The starting and ending time of typhoons in
Guangdong Province during 1965—1979.

Year	Starting time	Ending time
1965	Jun. 4	Nov. 13
1966	Jul. 13	Aug. 1
1967	Jun. 26	Nov. 8
1968	Aug. 5	Oct. 1
1969	Jul. 23	Sep. 28
1970	Jul. 16	Nov. 8
1971	May. 3	Oct. 8
1972	Jun. 27	Nov. 8
1973	Jul. 3	Oct. 18
1974	Jun. 8	Dec. 2
1975	Jun. 19	Oct. 23
1976	Jul. 22	Sep. 26
1977	Jul. 6	Sep. 25
1978	Jun. 26	Oct. 17
1979	Jul. 6	Sep. 23

The reason for such distinctions is that there exists s considerabl difference between flow fields at higher and lower levels over low-latitudes for the E-path and W-path year. The first case is marked by the occurrence of more typhoons or a stronger ITCZ circulation pattern, while the opposite holds true for the second case. And the main difference in these circulation patterns is manifested in the varied intensity of the cross-equatorial flow at higher and lower levels and of monsoon circulation: the cross-equatorial flow and the monsoon circulation are stronger in the E-path year and weaker in the W-path year.

To illustrate this point, we give here a chart of the anomaly of the atmospheric heating field in June in the subtropical belt (20°–35°N, 50°–180°E, Fig. 5) for the E-path year minus the W-path year. we find that in the east regions of the Asian Mainland and in the west Pacific, the atmospheric heating field is stronger in the E-path year than in the W-path year. As the atmospheric heating field is strong in the E-path year, the ascending movement and the monsoon circulation becomes correspondingly strong.

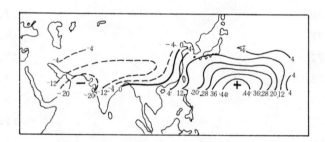

Fig. 5 A chart of the anomlay of the atmospheric heating
field in June (unit: 10^{-7} W·g^{-1}) for the E-path
year minus W-path year.

We take the discrepancy in the V-component between 850 hPa and 100 hPa in June at the Singapore station located near 105°E, over which the cross-equatorial flow must pass, to show the difference in the intensity of the cross-equatorial flow at higher and lower levels and the difference in the intensity of the monsoon circulation. The greater the discrepancy, the stronger the cross-equatorial south wind at lower levels and the stronger the cross-equatorial north wind at higher levels. The monsoon circulation is also strong and vice versa.

Table 3 shows the discrepancy in the V-component between 850 hPa and 100 hPa in June, 1965–1979 at the Singapore station. It's clear that the V-component in the E-path year is ≥ 9.3 m.s^{-1} and < 9.3 m.s^{-1} in the W-path year. The conclusion is the same as for the study of the atmospheric heating field.

Table 3 $V_{850-100}$ for the cross-equatorial flow in June 1965–1979.

Year	$V_{850-100}$ (m·s^{-1})
1965	9.7
1966	8.8
1967	11.7
1968	5.3
1969	8.7
1970	7.9
1971	4.9
1972	13.1
1973	8.7
1974	9.3
1975	11.1
1976	8.8
1977	8.4
1978	8.8
1979	8.2

REFERENCES

[1] Zhu Fukang, et al., South Asia high, Science Press, 1—95, 1980 (in Chinese).

[2] Yao Lanchang, et al., Studies of the mean atmospheric heat sources over the Qinghai-Xizang Plateau and its surrounding regions in summer, Collected Papers of the Qinghai-Xizang Plateau Meteorological Science Experiment (1), Science Press, 291—302, 1984 (in Chinese).

[3] Sun Guowu and Zhou Yi., The relation ship between the seasonal formation of the South Asia high on Qinghai-Xizang Plateau and the typhoon movement, collected papers of meteorological science and technology (6), Meteorological Press,(To be Published in Chinese).

[4] Long-Range Numerical Forecast Cooperation Group, The figures and data of monthly average of the heating field in the Northern Hemisphere, Meteorological Press, 1982 (in Chinese).

A CORRELATION ANALYSIS OF SURFACE NET RADIATION IN THE WEST QINGHAI-XIZANG PLATEAU WITH THE CLIMATIC ANNUAL VARIATION

Ji Guoliang[1]

Lanzhou Institute of Plateau Atmospheric Physics,

Chinese Academy of Sciences

Studies made by both Chinese and foreign meteorologists have shown that the Qinghai-Xizang Plateau surface heating field has an important influence upon the atmospheric circulation, weather and climate of China and East Asia and upon drought or water-logging over a large area[1-3]. In order to study the heating effect of the Qinghai-Xizang Plateau, in this paper we attempt to derive the surface net radiation values for the Shiquanhe and Nagqu stations from May to August of 1961—1978 by using the radiation observational data from the Qinghai-Xizang Plateau Meteorological Experiment (QXPMEX) in the same period of 1979 in the western Plateau. Through a study of the annual variation of monthly anomalies for surface net radiation at the two stations, the surface heating effect of the western Plateau upon the climatic change of China and East Asia is discussed.

I. DATA PROCESSING

The surface heat balance equation is as follows,

$$B - H = p + LE \qquad (1)$$

where (B-H) is the intensity of the surface heating field, B is the

---

1 Mr. Chen Youyu and Miss Li Bin assisted the calculational work.

surface net radiation, H is the heat flux entered into the soil, p is the heat flux in turbulent exchange, and LE is the latent heat of evaporation. For the Qinghai-Xizang Plateau, the heat flux entered into the soil in summer is less than 10% of the surface net radiation value, which is small compared to the heat flux in turbulent exchange*. Therefore, the surface net radiation B may be considered approximately as the surface heating field.

Most meteorological observatories and stations in China do not observe surface net radiation. However using the observed result of surface net radiation from the Shiquanhe and Nagqu stations from May to August of 1979, an empirical equation for the relationship between the surface net radiation and the absorbed short wave radiation was obtained in the form of eq.(2).

$$B = a + b \cdot Q \cdot (1-A) \qquad (2)$$

where B is the required daily total of surface net radiation in $w \cdot m^{-2}$, Q is the observed daily total of global radiation, A is the observed daily mean of surface albedo, and a and b are the empirical coefficients. From equation(2), the monthly total of surface net radiation is computed. Its theoretical basis has been discussed in reference[4].

Equation(2) shows that the surface net radiation is not only determined by the global radiation but is also correlated with the total cloud cover and the ground conditions. In the rainy season, because the total cloud cover increases and the ground is damp, the absolute values of coefficients a and b decrease appreciably (see Tab.1).

Because the instruments used in China for the observation of radiation have not been calibrated for a number of years, and for better use of the original data of above-mentioned two stations, we have made a comparison between the radiative data from QXPMEX and the data of the two stations during the same period, and then obtained a revised formula:

---

* Ji Guoliang Shen Zhibao, et al., The calculation method and its climatic extent for heat flux of soil in the Qinghai-Xizang Plateau, 1981.

$$Q = c + d \cdot Q' \tag{3}$$

where Q is the daily total of global radiation after revision, Q' is the original daily total of global radiation of the two stations, and c and d are the empirical coefficients, representing the differences due to different sites, time and instruments for observation.

For the years without solar radiational data, we use the coventional meteorological data and obtain the empirical equations for calculating the global radiation and surface albedo respectively:

$$Q = S_0 \cdot (e + f \cdot S_1) \tag{4}$$

$$A = m + n \cdot T_0 + p \cdot R \tag{5}$$

Eq.(4) is the commonly adopted equation for calculating the global radiation, where S_0 is the monthly mean of the daily astronomical radiation, S_1 is the monthly sunshine (in percentage), A is the monthly mean of the surface albedo, T_0 is the monthly mean of ground temperatures at 0 cm, R is the monthly precipitation and e, f, m, n and p are some empirical coefficients.

From the significance test it can be seen that all degrees of confidence of equations (1), (2), (3), (4) are 0.001 and that of equation (5) may reach 0.10. The computed surface net radiation values for the Shiquanhe and Nagqu stations from May to August of 1961—1978 are therefore trustworthy. The coefficients and computed relative errors are given in Tab. 1. From Tab. 1 it can be seen that the computed precision of equations meet our requirements.

II. THE RELATIONSHIP BETWEEN SURFACE NET RADIATION AND ATMOSPHERIC CIRCULATION

Fig. 1 shows the variation curves for the surface net radiation from May to August over the years at Shiquanhe. Curve (a) shows the surface net radiation in May and (b) shows the accumulated monthly anomalies of surface net radiation from May to August. From Fig. 1 it can be seen that the interannual variation of accumulated monthly anomalies of surface net radiation from May to August in the West

Tab. 1 The coefficients and computed errors of the equations.

month	formula	Shiquanhe correlation coefficient	Shiquanhe coefficient (1)	(2)	monthly relative error	Nagqu formula	Nagqu correlation coefficient	Nagqu coefficient (1)	(2)	monthly relative error
5	$B=a+b\cdot Q(1-A)$	0.94	-65.88	0.55		$B=a+b\cdot Q(1-A)$	0.84	-134.95	0.74	
6		0.82	-21.06	0.46	0.1%		0.76	-0.96	0.56	1.0%
7		0.68	-3.86	0.45	0.4%		0.94	-23.64	0.69	0.4%
8		0.75	-73.74	0.55	2.5%		0.97	-51.27	0.72	0.2%
5	$Q=c+d\cdot Q'$	0.81	315.03	0.55	0.1%	$Q=c+d\cdot Q'$	0.80	240.05	0.75	
6		0.85	375.64	0.49	0.1%		0.79	223.57	0.73	1.6%
7		0.79	272.53	0.66	0.1%		0.80	214.75	0.77	0.2%
8		0.76	354.62	0.51	4.3%		0.84	36.80	0.98	0.2%
5	$Q=S_0(e+fS_1)$	0.69	0.32	0.48						
6		0.66	0.47	0.37	0.8%					
7		0.86	0.23	0.65	0.4%					
8		0.53	0.46	0.36	2.1%					
$A=m+nT_0+pR$		0.56	m=0.25315 n=-0.00025 p=-0.00093		4.5%					

Qinghai-Xizang Plateau has a quasi-three-year period. It can be seen from the harmonic analysis that the confidence degree of the quasi-three-year period is 0.05 (Fig. 2). Further analysis shows that the interannual variation of surface net radiation is closely related to the interannual variation of sunshine (see Fig. 1b), with the correlation coefficient and confidence degree being 0.74 and 0.001 respectively It is, therfore, clear that for the western Plateau, the interannual variation of surface net radiation is mainly controlled by the sunshine percentage, i.e. the interannual variation period of surface net radiation is mainly controlled by the interannual variation period of sunshine percentage. Of course, the effect of ground conditions should not be ignored.

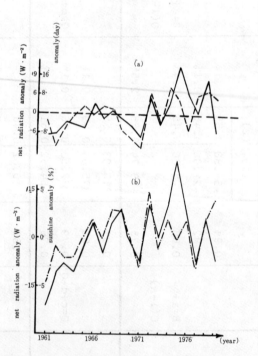

Fig. 1 Variation curves for the monthly anomalies of surface
net radiation in May(a) and for the accumulated monthly
anomalies of surface net radiation from May to August
(b) over the years at Shiquanhe.
————net radiation
- - - - - -anomaly of the onset data of the South-West
Monsoon, Bombay
- - · - · - · -monthly anomaly of sunshine(%).

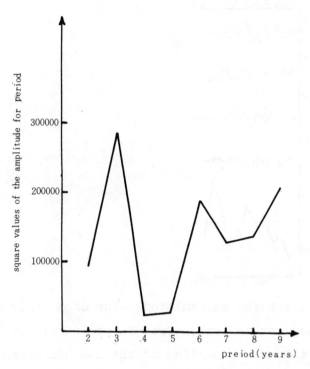

Fig. 2 The chart of the period for the accumulated monthly
anomalies of surface net radiation from May to
August at Shiquánhe.

By comparing the variation curves for the surface net radiation
in the western plateau and for the area anomaly of the 100 hPa
Qinghai-Xizang high (see Fig. 3), we find that the quasi-three year
period for the interannual variation of the accumulation of monthly
anomalies of surface net radiation from May to August in the western
Plateau coincides with the quasi-three-year period of oscillation for
the Qinghai-Xizang high of 100 hPa. In years when the accumulation of
monthly anomalies of the summer surface net radiation in the western
Plateau reaches a peak, the area anomaly of the 100 hPa Qinghai-Xizang
high is positive i.e. it is large in area and great in intensity,
and the mean position of the four ridge lines at 90°E, 100°E, 110°E
and 120°E is to the south. The reason for this is that the surface
heating field in the western Plateau is great in intensity, the ground

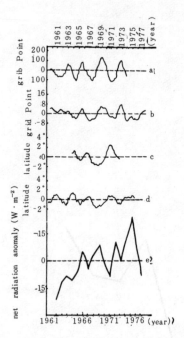

Variation curves for the accumulated value of monthly anomalies of surface net radiation from May to August at Shiquanhe and for the area anomalies of the 100 hPa Qinghai-Xizang high and 500 hPa West Pacific subtropical high.

a. Area anomaly for the 100 hPa Qinghai-Xizang high in the Northern Hemisphere (areas expressed in number of 5°×5° grid points within the range of 16560 geopotential meters).

b. Area anomaly for the 500 hPa West Pacific subtropical high.

c. Position anomaly for the ridge line of the 100 hPa Qinghai-Xizang high.

d. Position anomaly for the ridge line of the 500 hPa West Pacific subtropical high.

e. Accumulation of monthly anomalies of surface net radiation from May to August at Shiquanhe (Fig. 3a-d is from[2]).

is heated strongly, the warm air at the lower part of troposphere is heated and raised and the surrounding air converges towards the Plateau, thus the Qinghai-Xizang high is large in area and the mean position of the four ridge lines is to the south. From the analysis it can be seen that the monthly anomalies of surface net radiation at Shiquanhe show a negative correlation with the mean position of the ridge line of the 100 hPa Qinghai-Xizang high at 120°E in June[5], the correlation coefficient being -0.65 and the degree of confidence being 0.02. That is to say, in years when the monthly anomaly of surface net radiation in the western Plateau in May is positive, the mean position of the 120°E ridge line of the 100 hPa Qinghai-Xizang high in June is to the south. In years when the monthly anomaly of surface net radiation in May is negative, the mean position of the 120°E ridge line of the 100hPa Qinghai-Xizang high in June is to the north. In the same way, when the surface net radiation in the western Plateau in summer reaches a peak, the 500 hPa West Pacific subtropical anticyclone has a large area and the ridge line is to the south, too.

Through analysis, we also find that the interannual variation of the surface net radiation in May in the Plateau main (represented by the sum of net radiation at Shiquanhe and Nagqu stations) correlates Well with the date of the northward jumping to 30°N for the ridge lines of the 100 hPa Qinghai-Xizang high at 90°E and 120°E (see Fig. 4). In years when the anomaly is positive, the ridge line of the 100 hPa Qinghai-Xizang high jumps later than the mean date (probability of the same symbol is 79%) and vice versa. Obviously, this is closely related to the effect of heating, too.

III. CLIMATE AND THE INTERANNUAL VARIATION OF SURFACE NET RADIATION IN THE PLATEAU

The above analysis shows that the interannual variation of surface net radiation in the western Plateau has a great influence on the intensity and the position of the ridge line of the 100 hPa Qinghai-Xizang high. Therefore, it is bound to influence the weather and climate of China, East Asia and south Asia. In the following, we shall briefly

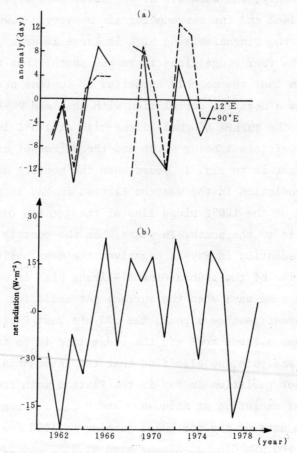

Fig. 4 Variation curves for the anomaly of surface net radiation in May in the Plateau and the northward jumping
date for the ridge line of the 100 hPa Qinghai-Xizang
high.

(a) The anomaly curves for the northward jumping date
of the 90°E and 120°E ridge line of the 100 hPa
Qinghai-Xizang high;

(b) The anomaly curves of the sum of surface net
radiation at the Shiquanhe and Nagqu stations in
May.

discuss on the intensity of the surface heating field and the climatic
interannual variation.

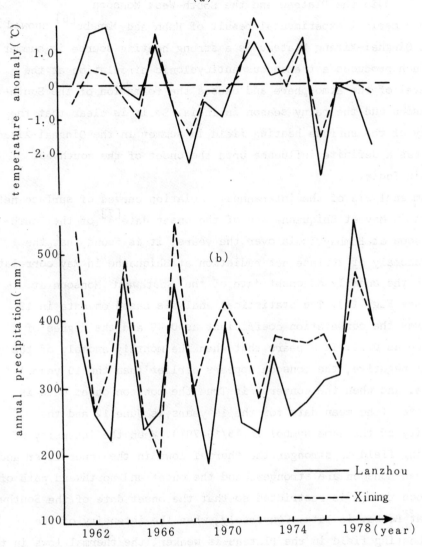

(a)

(b)

——— Lanzhou

- - - - - Xining

Fig.5 Variation curves for the temperature in August(a)
and the annual precipitation(b) from 1961–1979 at
Lanzhou and Xining. (The mean temperature of 19
years is 21.°C in Lanzhou and 16.6°C in Xining.
The mean annual precipitation for 20 years is
320.0mm in Lanzhou and 366.7mm in Xining.)

1. The Relationship Between the Surface Net radiation in the Plateau and the South-West Monsoon

The numerical experimental result of Hahn and Manabe[6] showed that the Qinghai-Xizang Plateau is a strong heating source in summer and as such produces a tremendous anticyclonic circulation at the upper level of the atmosphere and causes the formation of the South-West Monsoon and the rainy season in India. So it is clear that the intensity of the surface heating field in summer in the Qinghai-Xizang Plateau has a definite influence upon the onset of the southwest Monsoon in India.

From analysis of the interannual variation curves of surface net radiation in May at Shiquanhe and of the onset date[7] of the South-west Monsoon at Bombay,India over the years, it is found that the monthly anomaly of surface net radiation at Shiquanhe in May correlates well with the anomaly of onset date of the Southwest Monsoon at Bombay (see Fig. 1a). The statistical analysis based on data in 19 years gives the correlation coefficient as 0.49 and the degree of confidence as 0.05. It appears that when the monthly anomaly of the former is negative, the monsoon appears earlier than the 19 years' mean date, and when the converse is True the monsoon onset date is rather late (the mean date for the 19 years is June 12 and the probability of the same symbol is 15/19=79%). When the intensity of the heating field is stronger the thermal lows in the lower layer and the Plateau Monsoon are stronger, and the onset and northward path of the monsoon trough is restricted so that the onset date of the Southwes Monsoon at Bombay is late. Conversely, when the intensity of the surface heating field in the Plateau is weaker, the thermal lows in the lower layer and on the Plateau Monsoon are also weaker, therefore the Southwest Monsoon appears early. Wada[8] came to a similar conclusion from his analysis on the relationship between the Qinghai-Xizang high and the precipitation in India.

2. The Relationship Between Surface Net radiation on the Plateau and Precipitation in Gansu and Qinghai

In order to understand the influence of the surface heating field of the Plateau on the climatic change of Gansu and Qinghai to the

north-east of the Plateau, this paper presents an analysis of surface
net radiation in the Plateau, of the temperature of Gansu and Qinghai
in summer over the years, and of their precipitation in May and June
as well as the annual precipitation. Fig. 5 shows the variation curves
of the temperature in August and the annual precipitation during
1961—1979 at Lanzhou and Xining. It can be seen that: 1) the interannual
variations of both annual precipitation and the sum of the monthly
precipitation in May and June have a quasi-three-year period. (The
interannual variation curve of precipitation in May and June is omitted.)
The result of periodogram analysis shows that the degrees of confidence for annual precipitation and the sum of the precipitation in
May and June are 0.05 each month for Lanzhou and 0.10 and 0.05 for
Xining. This conforms with a related paper[9]. It has to be pointed
out that for the quasi-three year period of precipitation at the above
two stations as compared with the quasi-three-year period of the
accumulated monthly anomaly of surface net radiation from May to August
in the western Qinghai-Xizang Plateau, the phase at the two stations
lags behind by one year. 2) The temperature in August for Lanzhou and
Xining has a quasi-three-year period, too. (the degree of confidence
in each case is 0.05), and their phases are the same as the phase of
accumulated monthly anomaly of surface net radiation from May to August
at Shiquanhe.

The above analysis shows that in years when the accumulation of
monthly anomalies for surface net radiation from May to August in the
western Plateau presents a peak, the mean temperature in August for
Lanzhou and Xining stations is comparatively high (probability of the
same symbol is 5/6=83%), and for the following year, the sum of the
precipitation in May and June and the annual precipitation will present
a peak, too (probability of the same symbol is 5/6=83%). From the
analysis combined with the related data* it can be seen that when the
accumulation of monthly anomalies of surface net radiation from May to
August in the western Plateau shows a peak value, there will be no

---

* Xu Guochang, The climatic characteristics of drought in the northeast of the Qinghai-Xizang Plateau, 1981.

drought from the end of spring to the beginning of summer the followin
year(probability of the same symbol is 5/6=83%).In addition, there will
also be no drought or only a moderate drought in hot summer days (probability of the same symbol is 5/6=83%). Evidently, this is closely
related to the seasonal northward jump and the position of the ridge
lines of the 100 hPa Qinghai-Xizang high. Therefore, the intensity
of the surface heating field in the western Plateau has an indicative
significance to the prediction of drought or water-logging and
precipitation on the north-eastern side of the Plateau.

IV. CONCLUSION

From the above analysis, we can come to the following preliminary
conclusions.

1) The interannual variation of accumulated monthly anomalies of
surface net radiation from May to August at Shiquanhe has a quasi-
three-year period. This period coincides with the quasi-three-year
period of oscillation for the Qinghai-Xizang high of 100 hPa. In the
years when the summer surface net radiation accumulation of monthly
anomalies in the western Plateau presents a peak, the 100 hPa Qinghai-
Xizang high is large in area and great in intensity, and the mean
position of ridge lines is to the soutn. In the meantime, the 500 hPa
West Pacific subtropical anticyclone covers a larger area and has its
ridge line is to the south.

2) The surface net radiation at Shiquanhe is May has accords well
with the onset date of the South-West Monsoon at Bombay, India, with
the correlation coefficient being 0.49 and the degree of confidence
being 0.05. When the monthly anomaly of the former is positive, the
monsoon appears late, and vice versa.

The sum of the surface net radiations at the Shiquanhe and Nagqu
stations in May has also correlates well with the northward jumping
date of the 90°E and 120°E ridge lines. In the years when the anomaly
is positive, the ridge line of 100 hPa Qinghai-Xizang high jumps late,
and vice versa.

3) The interannual variation of the accumulation of monthly

anomalies of surface net radiation at Shiquanhe from May to August
has an indicative significance to the prediction of drought or water-
logging and precipitation for Gansu and Qinghai provinces on the northeastern side of the Plateau. In the years when the Shiquanhe's accumulation of monthly anomalies of surface net radiations from May to August
shows a peak value, the mean temperature in August for the Lanzhou and
Xining stations in that year is comparatively high, and for the following year the sum of the precipitation in May and June and the annual
precipitation will be peak values. Consequently for the following year,
too, there will be no drought from the end of spring to the beginning
of summer, in addition, there will be no drought or only a moderate
drought on hot summer days.

Owing to the limitation of data, this paper presents only a discussion of the relationship between the surface net radiation at the
Shiquanhe station in the western Plateau from May to August and the
interannual variation of climate. We shall make a further study on
the annual conditions for the whole Plateau in future.

REFERENCES

[1] Ye Duzheng, Gao Youxi, et al., Meteorology of Qinghai-Xizang
    Plateau, Science Press, 1979.
[2] Zhu Fukang, Lu Longhua, et al., South Asia high, Science Press,
    1980.
[3] Flohn. H., Contributions to a meteorology of the Tibetan Highlands,
    Atmos. Sci. Paper, No. 130, 1968, Colorado State University Ft.
    Collins.
[4] Weng Duming, Chen Wanlong, Shen Jiao-cheng, et al., The calculation
    method for arbitrarily period sum of global radiation in the
    Qinghai-Xizang Plateau initial approach, The Cellected papers of
    QXPMEX (3), 1986.
[5] Luo Siwei, Qian Zhengan, Wang Qianqian, The climatic and synoptical
    study about the relation between Qinghai-Xizang high pressure on
    the 100 mb surface and the flood and drought in East China in
    summer, Plateau Meteorology, Vol. 1, No. 2, 1982.

[6] Hahn, D.G. and Manabe, S., The role of mountains in the South Asian monsoon circulation, J. Atmos. Sci., 32, 1515—1541, 1975.

[7] Rao. Y.P., South west monsoon, India, Met. Dep., 1—367, 1976.

[8] Wada, H., Characteristic features of general circulation in the atmosphere and their relation to the anomalies of summer precipitation in monsoon Asia, Water Balance of Monsoon Asia, University of Hawaii Press, 111—130, 1971.

[9] Xu Guochang, Dong Anxiang, The quasi-three-year period of precipitation in the West of China, Plateau Meteorology, Vol. 1, No. 2, 11—17, 1982.

FUNDAMENTAL LARGE-SCALE STRUCTURE OF MEAN PRECIPITATION IN THE MONTANE UNITED STATES: TECHNIQUE

Reid A. Bryson and Val Mitchell
Center for Climatic Research
Institute for Environmental Studies
University of Wisconsin-Madison

INTRODUCTION

Mean precipitation charts for mountainous terrain or terrain of considerable relief are usually very similar in appearance to topographic charts. This is, of course, because of the well-known variation of precipitation with elevation (e.g. Figure 1). Unfortunately, this strong topographic signal obscures the large-scale precipitation patterns of dynamic origin and the fluctuations of those patterns. It is the purpose of this paper to describe a method for removing the topographic signal from the precipitation data in order to reveal the large-scale dynamic patterns.

It is not sufficient to simply assume that there is a single simple "lapse-rate of precipitation" and to use a single elevation correction factor over an entire large mountainous region. Consider for a moment the plot of mean annual precipitation versus elevation for the state of Arizona in the southwestern United States (Figure 2). It is clear from the scatter of the points that the correlation of precipitation with elevation is low and that a simple regression of one variable on the other would give very poor estimates. However, political divisions do not provide a logical basis for delineating a related set of climatic parameters. In the course of plotting the data in Figure 2 it became evident to the authors that the data points were grouped in very distinctive sets characterized by the physiographic region in which the stations were located. This is quite clear when one examines the data of Figure 2 in groups identified by distinctive symbols. Data for the low Sonoran Desert with scattered mountain ranges fall along a quite different regression line than data for the high, deeply dissected, Colorado Plateau or the mountainous border of the plateau, which is called t Mogollon Rim. The linear regressions of precipitation on elevation, for the data pertaining to each physiographic region taken separately, are given in Table 1, along with the precipitation at an elevation of 1500 m estimated fr the equations. If one were to use the "lapse rate of precipitation" specifi to each physiographic region for each station, region by region, one could obtain a map of "1500 m equivalent precipitation." It will be demonstrated below that such a map effectively removes the elevation effect and reveals t dynamically induced pattern.

TABLE 1 ELEVATION DEPENDENCE OF PRECIPITATION BY PHYSIOGRAPHIC REGION,
ARIZONA, U.S.A.

Sonoran Desert (basin and range topography)

P=82.870+ 0.3108 Z

P at 1500 m = 549 mm

Mogollon Rim (mountainous plateau edge)

P=102.794+ 0.2425 Z

P at 1500 m = 466 mm

Colorado Plateau (high, dissected plateau)

P=101.270+ 0.1592 Z

P at 1500 m = 340 mm

Where P is in millimeters per year and Z is the elevation in meters above sea level.

DEFINING PHYSIOGRAPHIC REGIONS

One cannot simply assume, without evidence, that the physiographic regions defined by geomorphologists are the best divisions for climatic purposes. Thus an experiment was performed which did not fully answer the question implicit in the above observation, but which did define the regions objectively. The regions thus defined appeared to be sufficient for the purposes of this paper.

One of the factors that would appear to be significant in the orographic effect on precipitation is the relief. Another is the steepness of the terrain. Still another would be the scale of the terrain features, for as Lettau (1969) has demonstrated, some large scale features are so configured as to be effectively streamlined and thus do not constitute roughness elements. He has also shown that vegetation height and spacing may be as important as topography in terms of aerodynamic roughness. One might also reason that the lapse rate of precipitation would be different in generally low regions with isolated mountains than in high plateaus with deep valleys.

Studies of the variance spectra of terrain profiles have shown that the spectra of most terrain types may be divided into three parts (Bryson and Dutton, 1968). On a log-log plot of spectral density against the inverse of distance "wavelength" there is a middle section that has a minus three slope ("cubic sub-range"). At shorter space scales the spectrum usually has an approximate minus one slope, and at larger space scales than the cubic sub-range the slope is nearly zero. Dimensional analysis shows that the cubic section indicates a range of space scales in which the slope of the land is essentially constant (Lettau, 1967). The variance at the ends of the cubic sub-range gives a measure of regional or local relief, while the horizontal scales are a measure of texture and length of slope. The rather uniform structure of the variance spectra for many types of terrain suggests that the measures of terrain characteristics described above are quite universal and could be applied over large areas.

Fortunately, maps of various terrain characteristics were available for the western United States (Hammond, 1964). It was thus possible to read off the values of the appropriate parameters at one degree latitude and longitude intersections for the entire western 40% of the country. An arbitrary estimate of the aerodynamic roughness of the vegetation could be estimated from the map of Küchler (1964). Principal components or "eigenvectors" of the normalized parameter arrays could then be calculated. The loadings on these eigenvectors at each grid intersection then gave a characterization of the terrain type.

Figure 3 gives the relative weighting of the various parameters for each eigenvector. For 23% of the 363 grid intersections a single eigenvector explained 80% or more of the variance. At 50% of the points only two eigenvectors were needed, and at 26% three were needed. Only at 1% were four needed. Outlining the areas characterized by a particular combination of a very few eigenvectors then gave an objective division of the country into quantitatively described physiographic regions (Mitchell, 1969) (Figure 4 and Table 2). The authors were pleased to note that the regions so defined were essentially indistinguishable from those arrived at by geomorphologists in the field.

TABLE 2

LEGEND FOR FIGURE 4 AND PHYSICAL INTERPRETATION OF EIGENVECTORS

Combination (dominant eigenvectors)	Representation
1 X	– Plain with grassland
1 2	– Low altitude, flat, with grass
1-2	– Moderate altitude, somewhat dissected, gentle slope on uplands, grass
1 3	– Average altitude flatland
1-4	– Moderate altitude and slope, low relief grassland
1 5	– Average altitude and slope, very low relief, average profile, mixed grass and shrub
-1 X	– Mountains with high elevation
-1 2	– Moderate altitude mountains with gentle slope in lowlands but high vegetation
-1-2	– High altitude mountains with gentle slope on uplands
-1 3	– High altitude, average slope, fairly high relief, gentle slopes in lowland, fairly high vegetation
-1-3	– Low altitude mountains with steep slope and high relief and high vegetation
2 X	– Low altitude, moderate slope, average relief, shrubs
2 3	– Average altitude, fairly low slope, average relief, gentle slope in lowland

(Further combinations may be interpreted by referring to the weightings of
individual parameters in the eigenvectors depicted graphically in Figure 3.)

REMOVING THE TOPOGRAPHIC EFFECT

Once the western montane region had been divided into physiographic regions, then one could begin to remove the detailed topographic effect to reveal the broad dynamic patterns There are, of course, a number of factors other than elevation alone that are pertinent, but in general they are less important. In specific areas they may be more important. For example, in the extreme northwestern portion of the study region we studied the effect of coastal proximity relative to elevation, using the forward selection method of Draper and Smith (1968). For that particular region which lies directly in the path of the zonal westerlies, coastal proximity explained more variance than did elevation. Such was not true in the continental interior, however, and probably would be unimportant in the interior of Asia.

Another factor of importance in some mountainous areas, especially if coastal, is trans-flow stress-differential-induced divergence (Bryson and Kuhn, 1961). This dynamic operates on a scale small enough to impact the data from individual stations but that it will obscure the dynamically induced pattern on a regional scale has not been demonstrated. To see this larger scale pattern clearly still requires the removal of the topograhic effect.

Part of the topographic effect on precipitation in mountainous regions is the result of convective cells forming selectively over elevated heat sources and part is the result of evaporation of the precipitation as it falls. The latter is particularly obvious in arid regions in the hot season. We shall not discuss the relative importance of these mechanisms at this point, but shall rely on the end results of the empirical study to indicate whether separation of these effects is necessary for the present purpose.

Once the region was divided into physiographically homogeneous regions the measured precipitation at the available stations was regressed against the elevation of the station. In regions of sparse data, some application of experience and judgment is required, as well as considerable experimentation, in order to obtain a relatively accurate estimating equation. Once that is done, however, one may then reduce the precipitation to the "equivalent precipitation" at some arbitrarily chosen convenient height. For our study we chose a height of 5000 feet (near 1500 meters) and used the following equation:

$$Pe=Po+(Zn-Zs)A$$

where Pe is the estimated "equivalent precipitation" at the nominal height Zn, Po is the precipitation observed at a particular station within the region at height Zs, and A is the regression coefficient for the region, or, in other words, the slope of the regression line for the region. In our case we found that we sometimes had to use data from similar regions, or in case the region had little relief and was close to 5000 feet in elevation we used the observed precipitation without modification. This situation is fairly common in the western United States, which in many respects is like a 5000 foot high plateau with a few deep valleys and a number of superimposed mountain ranges (Blodget, 1857).

THE LARGE SCALE PRECIPITATION PATTERN

The "equivalent 1500 meter precipitation" maps for the western United States are shown in Figures 5-7. It is clear that the strong dependence upon local topography has been removed, but that there is still a broad scale, fundamental pattern. This pattern can be directly interpreted in terms of the general circulation of the atmosphere as it changes from season to season.

The map for January, Figure 5, shows a very strong gradient of
precipitation inland from the west coast, from the Canadian border to the
Mexican border. The one exception is where the mountains are quite high close
to the coast near 35 degrees North. This is in accord with the fact that the
westerlies impinge directly on the entire coast during that month. The
gradient in the coastal region is so strong that the isohyets are not shown
for equivalent amounts greater than 6 inches (about 152 mm). East of the
Sierra Nevada-Cascade Range line the gradients of precipitation are weak, the
amount decreasing away from the Pacific Ocean source. The two exceptions are
the region of somewhat heavier precipitation in central Arizona, and the other
being the Columbia Plateau of Washington and the western face of the Northern
Rockies. The former can be understood as the result of moist air sweeping
around the southern end of the Sierra Nevada and impinging on the southwest
face of the Colorado Plateau, the Mogollon Rim. The latter is a consequence
of much moist air flowing through the gap in the Coast Ranges and Cascade
Mountains that is the valley of the Columbia River and then moving north and
east in the quasi-antitriptic flow that characterizes the northern
intermontane region in winter (Bryson and Hare, 1974). This air must
eventually impinge on the higher terrain in that direction, and result in
higher "equivalent precipitation" there.

The July pattern of "equivalent precipitation" (Figure 6) shows clearly
the northward shift of the westerlies that characterizes high summer. This
to be seen in the high coastal rainfall and strong gradient away from the
oceanic source region being only in the extreme northwest. Another change
from winter is to be seen in the evidence of the Gulf of California being a

source of moisture, with a gradient away from the source towards the north and east. Unlike winter, when the Mogollon Rim appears to be an obstacle to the high kinetic energy flow and to force uplift, the summer pattern suggests that it is simply a barrier to the northeastward movement of moisture. Over the remainder of the intermontane region there is little variation in "equivalent precipitation." However, east of the mountains, where the mean flow introduces moisture from the Gulf of Mexico to the south, the rainfall is greater. The annual "equivalent precipitation" pattern (Figure 7) is a composite, of course, and obscures the most interesting dynamic features that can be seen on the monthly maps.

SUMMARY AND DISCUSSION

It is clear, when one compares Figures 1 and 7, that the method described in the preceding paragraphs has effectively removed the overwhelming imprint of the mountainous terrain and revealed the fundamental spatial structure of the precipitation pattern. No claim can be made that this is the optimum method for extracting the dynamically produced pattern from the topographically obscured pattern. No experimentation was carried out to see whether a different selection of terrain parameters might work as well or better than the group used. This would appear to be a worthwhile experiment to do for a region of much data, like the western United States, especially if one were interested in applying the general method to a data sparse region where the fundamental terrain information was to be obtained by remote sensing. In such a situation one might also wish to experiment with regions defined by terrain parameters, but not necessarily aggregated by physiographic region. It might be possible, for example, to regress the precipitation at a suite of stations against elevation and several pertinent terrain parameters. Careful exploration along these lines might make possible the reconstruction of the dynamically produced pattern even in areas with one or fewer stations per physiographic province.

This study has clearly shown that in the western United States the fundamental precipitation pattern is closely related to well-defined moisture sources, the relative importance of which changes with the seasonal changes in the general circulation. Judging by the seasonal circulation changes, one would suspect that the same might be true in the mountainous core of Asia.

REFERENCES

Blodget, Lorin, 1857: CLIMATOLOGY OF THE UNITED STATES, Philadelphia, J.B. Lippincott and Co., 536pp.

Bryson, R.A. and J.A. Dutton, 1968: "The Variance Spectra of Certain Natural Series", from SYMPOSIUM ON QUANTITATIVE METHODS IN GEOGRAPHY, CHICAGO, MAY 1960 in QUANTITATIVE GEOGRAPHY, Vol. II: PHYSICAL AND CARTOGRAPHIC TOPICS, Evanston, Ill., Northwestern University Press, 1968.

Bryson, R.A. and F.K. Hare, 1974: "The Climates of North America" in CLIMATES OF NORTH AMERICA, Bryson and Hare, eds., Vol. II of WORLD SURVEY OF CLIMATOLOGY, Amsterdam-London-New York, Elsevier Sci. Publ. Co., 420pp.

Bryson, R.A. and P.M. Kuhn, 1961: "Stress-Differential Induced Divergence with Appication to Littoral Precipitation", ERDKUNDE 15: 287-294.

Draper, N.R. and H. Smith, 1968: APPLIED REGRESSION ANALYSIS, New York, John-Wiley and Sons, 407pp.

Hammond, E.H., 1964: "Classes of Land-Surface Form in the Forty-Eight States, U.S.A.", Map Supplement No. 4, ANNALS ASSOC. AMER. GEOGRAPHERS, Vol. 54.

Lettau, H., 1957: "Small to Large-Scale Features of the Boundary Layer Structure over Mountain Slopes", PROC. SYMPOSIUM ON MOUNTAIN METEOROLOGY, Fort Collins, Colorado, Colorado State University, Dept. of Atmos. Sci., Paper No. 122 pp. 1-74.

Lettau, H., 1969: "Note on Aerodynamic Roughness-Parameter Estimation on the Basis of Roughness-Element Description", JOUR. OF APPLIED METEOR. 8(5): 828-832.

Mitchell, V. L., 1969: THE REGIONALIZATION OF CLIMATE IN MONTANE AREAS, Ph.D. Thesis, Department of Meteorology, University of Wisconsin-Madison, 147 pp.

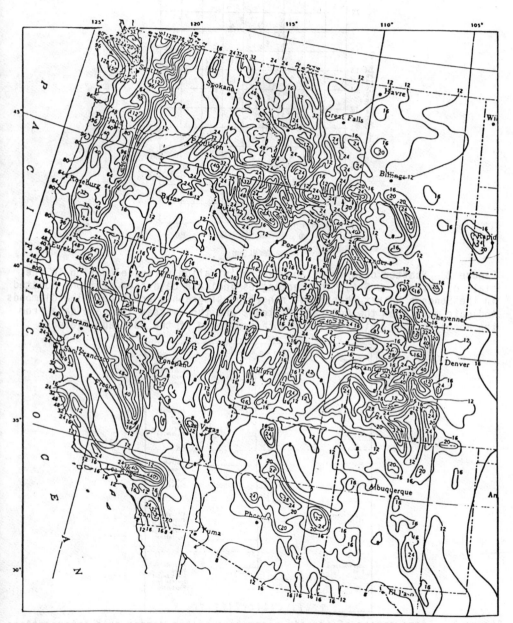

Figure 1 - Annual precipitation pattern in the western United States. The isohyets very much resemble the topographic contours, and a clear large scale pattern is not evident.

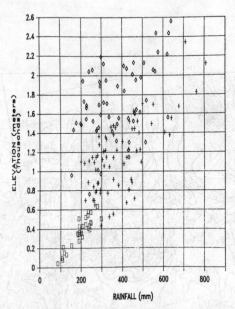

Figure 2 - Scatter diagram of annual precipitation versus elevation in the state of Arizona, U.S.A. The rectangles indicate data from stations located in the Sonoran Desert "basin and range" physiographic province. The diamonds indicate data from the Colorado Plateau, and the crosses indicate data from the Mogollon Rim, which is the mountainous edge of the Colorado Plateau.

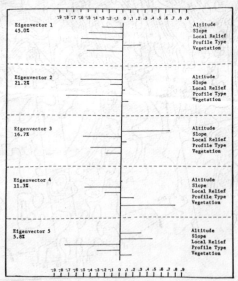

Figure 3 - Eigenvectors of terrain parameters at one degree grid intersections showing the weightings of each factor and the percent of the total variance reduced by each eigenvector.

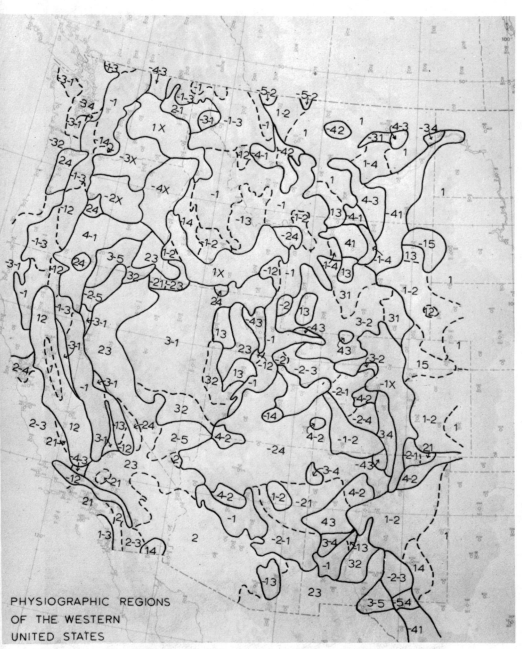

Figure 4 - Physiographic regions of the western United States defined
quantitatively and objectively in terms of the terrain parameter
eigenvectors. For the meaning of the numbers identifying each region see
Table 2.

PHYSIOGRAPHIC REGIONS
OF THE WESTERN
UNITED STATES

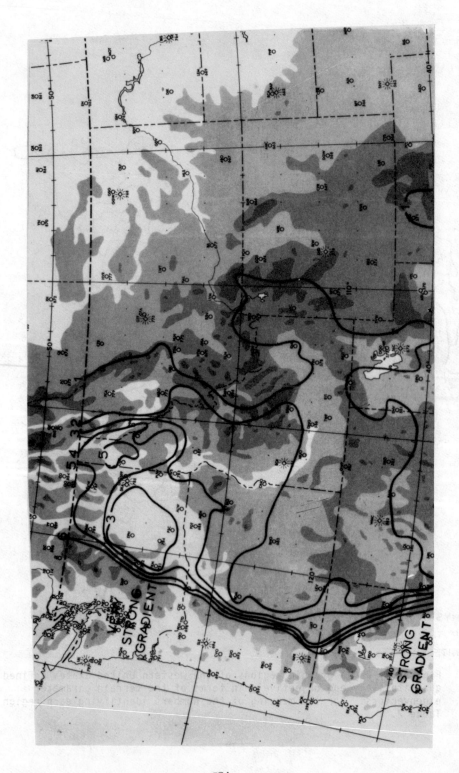

Figure 5 – January "equivalent precipitation" at 1500 m (5000 ft) elevation after removal of the topographic effect region by region, western United States, in inches.

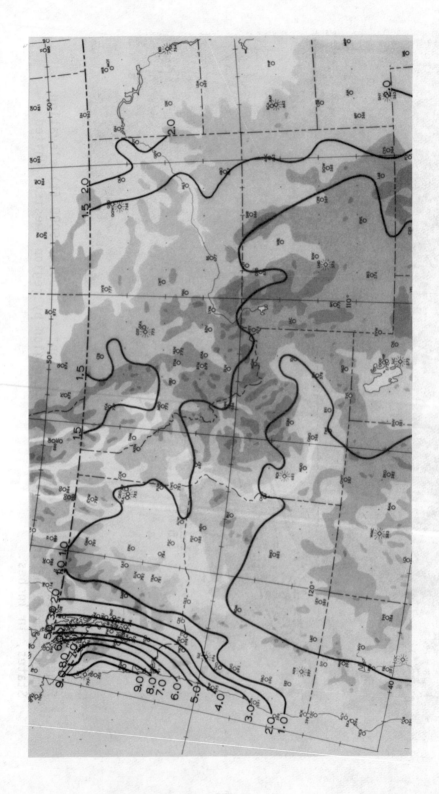

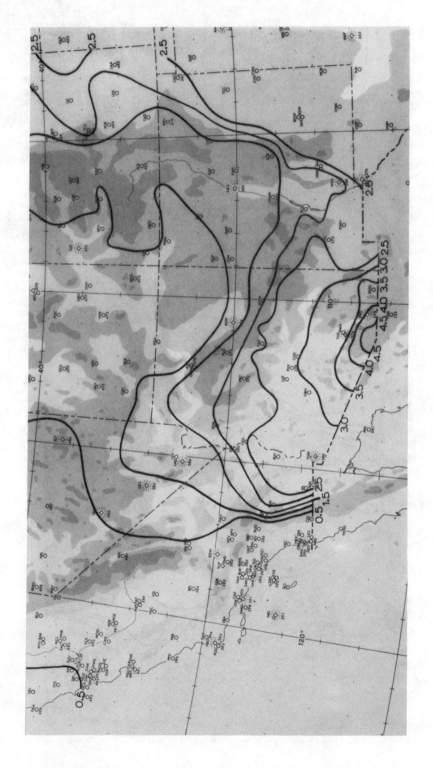

Figure 6 – July "equivalent precipitation" at 1500 m, in inches.

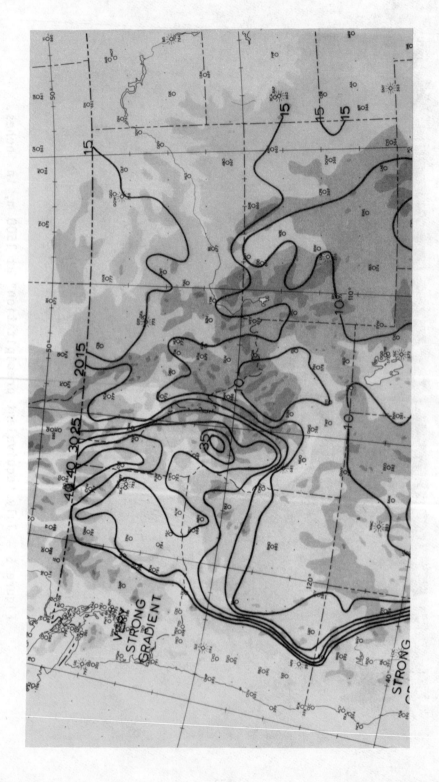

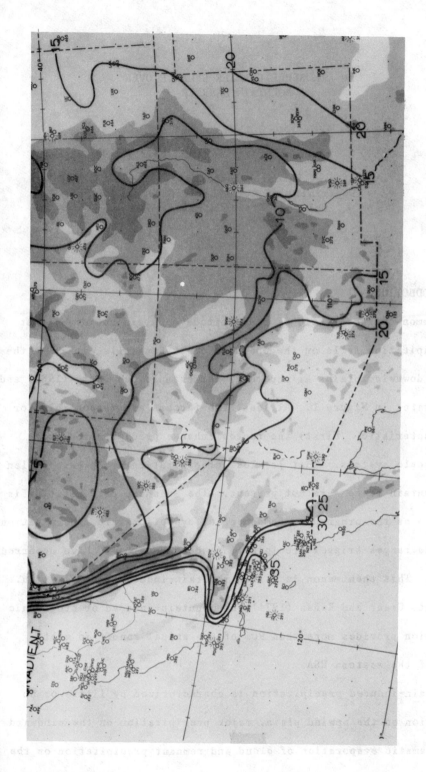

Figure 7 - Annual "equivalent precipitation" at 1500 m, in inches.

AN ATMOSPHERIC WATER BALANCE OVER
A MOUNTAIN BARRIER

Edward E. Hindman

Research Scientist
Department of Atmospheric Science
Colorado State University

1.0   INTRODUCTION

A common characteristic of mid-latitude mountain ranges is that
more precipitation falls on the upwind slopes than falls either on the
upwind or downwind plains as recorded, for example, by Barry (1981) and
as illustrated in Figure 1. At least two factors are responsible for
this characteristic. First, the ranges enhance precipitation from
synoptic-scale systems. That is, less precipitation would have fallen
if the mountain range were not present. The enhanced precipitation is
referred to as the orographic component of synoptic-scale precipitation.
Second, the ranges trigger precipitation where none would have occurred
otherwise. This phenomenon is called mountain-induced precipitation.
According to Grant and Kahan (1974) the mountain-induced or orographic
precipitation provides more than 90% of the annual runoff in most
sections of the western USA.

Mountain-induced precipitation is characterized by little or no
precipitation on the upwind plain, major precipitation on the windward
slopes, dramatic evaporation of cloud and remnant precipitation on the

leeward slopes with the attendant warming of the downwind air (due to latent heat released during cloud and precipitation formation). Thus, in the formation of mountain-induced precipitation, moisture flows through the cloud system with a fraction condensing and precipitating.

Rhea (1978) has developed a numerical simulation model which describes the general characteristics of orographic precipitation in the Colorado Rockies. He recommends that the model be combined with an atmospheric water balance box model; the box model would be used to specify the quantity of precipitation minus evaporation and the precipitation model would be used to arealy distribute the precipitation. The work reported in this paper is a first step toward developing the required box model.

The moisture flow through the box model is illustrated in Figure 2. The box is defined by a thin, vertical slab of air bounded on the upwind edge by a rawinsonde ascent from the surface to the tropopause, and on the downwind edge by the mountain crest. Then, the water balance of the precipitating orographic cloud consists of three components: inflow vapor (M_1), precipitation (M_2), and outflow moisture (M_3). The box model permits defining the following relationship between the moisture components:

$$M_1 = M_2 + M_3;$$

(1)

inflow moisture equals precipitation (no evaporation is assumed) on the upwind slopes of the mountain barrier plus the amount of moisture flowing over the barrier. The purpose of this paper is to estimate the components of the atmospheric water balance and discuss the significance of the relative magnitudes of the components.

2.0 APPROACH

The approach to this investigation was to use atmospheric measurements to estimate the three moisture components. Extensive surface and airborne measurements of winter Rocky Mountain orographic clouds are utilized. The Department of Atmospheric Science at Colorado State University has collected these data upwind, over and downwind of the Park Range in northwest Colorado at irregular intervals over the past eight years. The long, north-south oriented range is ideal for water balance studies.

The moisture components were estimated from the following measurements: M_1 - upwind rawinsonde temperature, moisture and wind, M_2 - surface precipitation, M_3 [condensate] - mountaintop incloud condensate, M_3 [vapor] - estimated, using (1). The values of the components were obtained by averaging measurements from a number of storms. Thus, a detailed analysis of storms or storm-stages was not undertaken. Rather, average water balance estimates with upper and lower bounds were derived. These values were used to give limits to the moisture components and, hence, provide a basis for more detailed studies using the 2-D numerical simulation model of Cotton et al. (1982, 1984).

3.0 MOISTURE COMPONENTS

3.1 Inflow Vapor [M_1]

Lee (1984) has reported vertical soundings of temperature, pressure, moisture and winds from Craig, Colorado which is 80 km upwind of the Park Range. These soundings were taken during periods when snow was falling on the Park Range. The soundings, taken at 3 hr intervals, were analyzed to define the inflow moisture. The inflow moisture was defined by the following expression:

$$M_1(g\ hr^{-1}) = Zy\bar{v}\ \bar{q}\ \bar{\rho}_a\ 3600 \qquad (2)$$

where Z (cm) is the distance from the surface to the tropopause, y (cm) is the unit width of the box, $\bar{v}$ (cm s^{-1}) is the moisture-weighted airspeed for the column where,

$$\bar{v} = \sum_{i=1}^{n} v_i(q_i/q_{total}) \qquad (3)$$

and i is a height interval on the sounding, q is the mixing ratio and, $q_{total} = \sum_{i=1}^{n} q_i$, $\bar{q}(g\ kg^{-1})$ is the average mixing ratio of the column, $\bar{\rho}_a$ (kg cm^{-3}) is the average air density of the column and 3600 has the units of s h $^{-1}$. Figure 3 illustrates the procedure for determining the inflow moisture value.

Soundings from two snowfall periods (11 Dec 79, 00Z-18Z and 23 Feb 79, 00Z-21Z) were analyzed (total of 12 soundings). The M$_1$ values, using (2) from the two periods were, respectively, 4.2 ($\pm$ 1.2) x 10^6 g h $^{-1}$ and 1.8 ($\pm$ 0.21) x 10^6 g h $^{-1}$. These values are significantly different due to the fact the uncertainties of the means do not overlap. Nevertheless, the values are well within the same order of magnitude, so combining all the soundings from the two periods will produce a reasonable value for our "order-of-magnitude" study. The resulting value of M$_1$ is 3.2 ($\pm$ 0.740) x 10^6 g h $^{-1}$.

3.2 Precipitation [M$_2$]

Precipitation measurements (24 hr new snowfall and water equivalent) are routinely taken at the Steamboat Ski Area located on the western slopes of the Park Range. The water-equivalent measurements have been used by Hindman et al (1983) to define an average snowfall rate of 0.38 ($\pm$ 0.12) mm hr^{-1}. Further, mountain-induced snowfall, on average, ceases about 25 km upwind of the ski area. Using the snowfall rate and the distance over which snow falls, the value of M_2 is

$$M_2 \text{ (g hr}^{-1}) = xyP\rho$$ (4

where x (cm) is the distance precipitation falls upwind of the barrier, y (cm) is the unit width, P (mm hr^{-1}) is the snowfall rate and ρ is the density of water (g cm^{-3}). Substituting the appropriate values into (4 results in a M_2 value of 9.5 ($\pm$ 3.0) x 10^4 g hr^{-1}. An implicit assumption in deriving this value is that negligible evaporation takes place in the subcloud layer. This assumption may be reasonable because cloud bases during storm periods are only ~ 700 m above the valley floor. Figure 4 illustrates the procedure for determining the precipitation value.

3.3 Outflow Moisture [M_3]

The outflow moisture consists of three components: M_3 [liquid], M [ice], M_3 [vapor]. The liquid component is defined from measurements b Hindman et al. (1983) of the average liquid flux (R) during storm periods at the crest of the Park Range, R = 5.1 ($\pm$ 1.1) mm hr^{-1}. Cloud top measurements from Lee (1983) provide an average value of 4 km AGL o 2730 m above the mountain crest (Z'). The values of M_3 are defined by,

$$M_3[\text{liquid}] \text{ (g hr}^{-1}) = Z'yR\rho$$ (5

Substituting the appropriate values into (5) results in a M_3 [liquid] value of 1.4 ($\pm$ 0.30) x 10^5 g h^{-1}. Figure 5a illustrates the procedur for determining the liquid outflow value.

The M_3 [ice] value is approximated by assuming that the precipitation "blow-over" rate at the crest is equivalent to the upwind pre-
that R is replaced by P. Substituting the appropriate values into (5)
R is replaced by P. Substituting the appropriate values into (5)
results in a M_3 [ice] value of 1.03 ($\pm$ 0.32) x 10^4 g hr^{-1}. Figure 5b
illustrates the procedure for determining the ice outflow [M_3 (ice)].

The value of M_3 [vapor] is approximated from (1) where M_3 = M_3
[liquid] + M_3 [ice] + M_3 [vapor]. Substituting the appropriate values
into this expression results in an M_3 [vapor] value of 2.96 ($\pm$ 0.677) $\times$
10^6 g hr^{-1}.

A summary of the moisture components and their values is given in
the Table.

TABLE   MOISTURE COMPONENT VALUES AND
THEIR UNCERTAINTIES

COMPONENT	VALUE (g hr^{-1})
M_1 - inflow vapor	3.20 ($\pm$ 0.740) x 10^6
M_2 - precipitation	0.095 ($\pm$ 0.030) x 10^6
M_3 [liquid]	0.140 ($\pm$ 0.030) x 10^6
M_3 [ice] outflow moisture	0.0103 ($\pm$ 0.0032) x 10^6
*M_3 [vapor]	2.96 ($\pm$ 0.677) x 10^6

* Inferred from M_1 = M_2 + (M_3 [liquid] + M_3 [ice] + M_3 [vapor])

4.0 DISCUSSION

Using the mean values and their ranges from the table, the
following values were calculated: the fraction of inflow moisture that

precipitates, $M_2/M_1 = 0.030$ with a range between 0.016 and 0.051 and th

fraction of inflow moisture that flows over the barrier, $M_3/M_1 = 0.97$

with a range between 0.95 and 0.98. It can be seen from the results

that between 2 and 5% of the inflow moisture actually precipitates (see

Figure 6). Conversely, between 95 and 98% of the inflow moisture flows

over the barrier.

These results indicate that, on average, a small amount of

atmospheric moisture precipitates on a mountain barrier; most of the

moisture flows over the barrier. This result is for an entire

atmospheric column. It may well be that more than 5% of the moisture i

the lower levels of the column precipitates and less than 5% in the

upper levels precipitates, averaging 5% for the entire column. If this

is so, then precipitation formation on downwind barriers may be

affected. Careful numerical simulation studies using the Cotton et al.

model are required to test this point.

Snowfall augmentation by cloud seeding upwind of mountain barriers

has been reported to increase wintertime precipitation by about 10 to

15% by Grant and Kahan (1974). Thus, cloud seeding activities, at best

will increase the amount of inflow moisture which precipitates by 0.3%.

Conversely, the outflow moisture will be reduced by 0.3%. Consequently

cloud seeding activities on upwind barrier should not "rob" moisture

from downwind barriers as illustrated in Figure 7.

5.0 CONCLUSIONS

A first-order water balance of an orographic cloud was produced

using a box model. The model accounted for inflow moisture,

precipitation, and outflow moisture. The values for these moisture

components were determined from atmospheric measurements made upwind an

586

over the Park Range in northwestern Colorado during a number of storm periods. Hence, the measurements represented values for a "typical" orographic cloud.

It was found that, on average, a small amount of atmospheric moisture precipitates on a mountain barrier (2 to 5%). Cloud seeding activities are estimated to increase these values 0.3%. Thus, cloud seeding activities on upwind barriers should not rob moisture from downwind barriers.

The empirical water balance developed herein is suitable for comparing with theoretical water balances generated from the 2-D numerical simulation model of Cotton et al. (1982, 1984). Such comparison will assist in determining the realism of the model for a variety of application studies (e.g., aircraft icing, snow crystal riming ....).

6.0 ACKNOWLEDGEMENTS

This paper is a portion of the research sponsored by NSF Grant ATM-8109590 and Air Force Contract F19628-84-C-0005.

7.0 REFERENCES

Barry, R.G., 1981: <u>Mountain Weather and Climate</u>, Methuen, London, 305

Cotton, W.R., G.J. Tripoli, and R.R. Blumenstein, 1982: The simulation of orographic clouds with a nonlinear, time-dependent model. <u>Preprints, Conf. Cloud Physics</u>, Amer. Meteor. Soc., Boston, 322–324.

Cotton, W.R., G.J. Tripoli, and R.M. Rauber, 1984: A numerical simulation of the effects of small scale tropographical variation on the generation of aggregate snowflakes. <u>Proc. 9th International Cloud Physics Conference</u>, 21–28 August 1984, Tallinn, U.S.S.R.

Grant, L.O. and A.M. Kahan, 1974: Weather modification for augmenting orographic precipitation. In <u>Weather and Climate Modification</u>, W.N. Hess, Editor, John Wiley and Sons, 282–317.

Hindman, E.E., R.D. Borys and P.J. DeMott, 1983: Hydrometeorlogical significance of rime ice deposits in the Colorado Rockies. <u>Water Res. Bul.</u>, <u>19</u>, 619–624.

Lee. R.R., 1984: Two case studies of wintertime cloud systems over the Colorado Rockies. <u>J. Atmos. Sci.</u>, <u>41</u>, Submitted.

Rhea, J.O., 1978: Orographic precipitation model for hydrometeorological use. Atmos. Sci. Paper 287, Colorado State Univ., Ft. Collins, CO 80523, 198 pp.

Mean Annual Precipitation, Inches

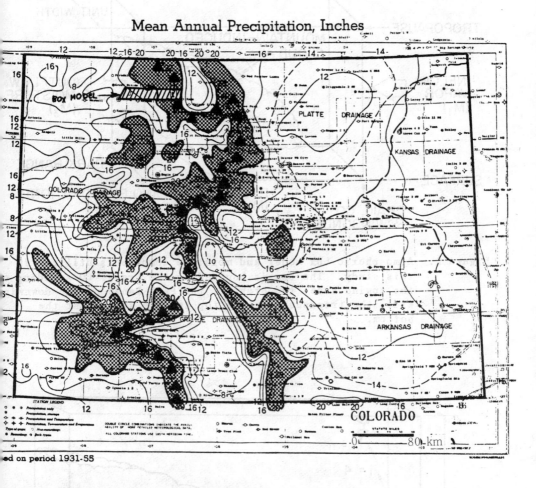

ed on period 1931-55

gure 1. Mean annual precipitation (inches) in Colorado, U.S.A. The
shaded regions represent ≥ 20 inches (61 mm) of
precipitation. The location of the box model used in this
paper is illustrated. The location of the continental divide
is illustrated by the line of triangles.

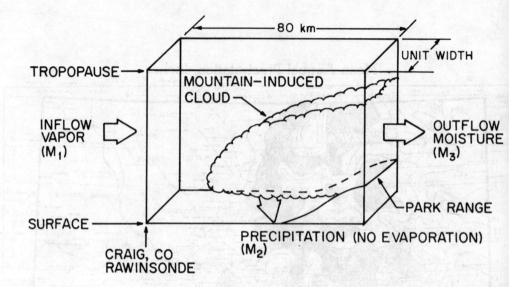

Figure 2.   Atmospheric box model and moisture components utilized in
this study. The box is bounded by the Craig, CO sounding o
the upwind side, the crest of the Park Range on the downwin
side, the tropopause at the top and the surface at the
bottom.

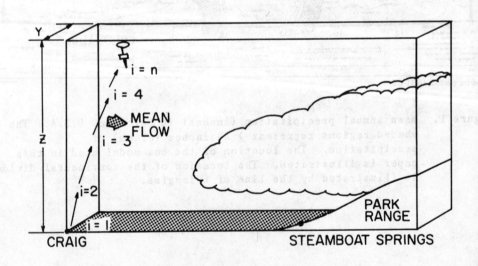

Figure 3.   Schematic of the procedure for estimating the inflow moistur
(M_1). Z represents the distance from the surface to the
tropopause, y is the unit width and i represents the
measurements from the Craig rawinsonde.

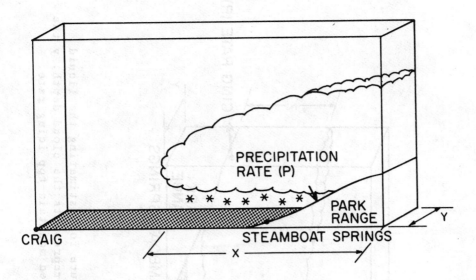

Figure 4. Schematic of the procedure for estimating the precipitation
(M_2). x represents the distance snow typically falls upwind
of the barrier from a mountain-induced cloud, y is the unit
width and P is the precipitation rate measured on the slopes
of the Park Range.

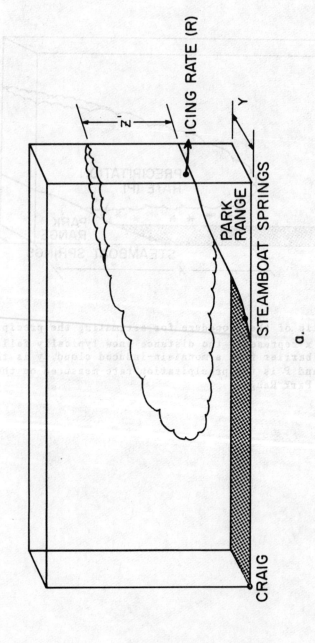

Figure 5. (a) Schematic of the procedure for estimating the liquid outflow [M_3 (liquid)]. Z' represents the cloud depth, y the unit width and R the measured mountain-top icing rate.

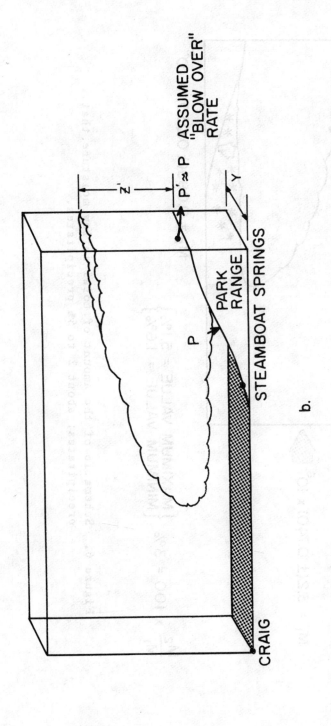

(b) Schematic of the procedure for estimating the ice outflow [M_3 (ice)]. The ice "blow over" rate is assumed to be equivalent to the upwind precipitation rate at any point.

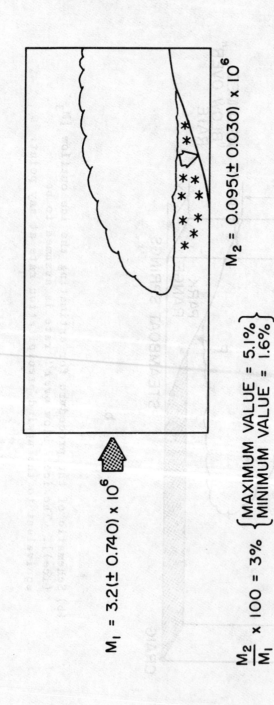

$M_1 = 3.2(\pm 0.740) \times 10^6$

$M_2 = 0.095(\pm 0.030) \times 10^6$

$\dfrac{M_2}{M_1} \times 100 = 3\% \begin{cases} \text{MAXIMUM VALUE} = 5.1\% \\ \text{MINIMUM VALUE} = 1.6\% \end{cases}$

Figure 6.   Schematic of the amount of total inflow moisture that
precipitates; about 2 to 5% precipitates.

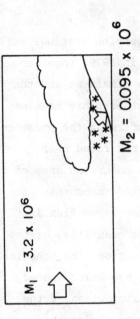

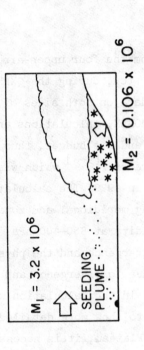

UNSEEDED CASE

$M_1 = 3.2 \times 10^6$

$M_2 = 0.095 \times 10^6$

$\dfrac{M_2}{M_1} \times 100 = 3.0\%$

SEEDED CASE

$M_1 = 3.2 \times 10^6$

SEEDING PLUME

$M_2 = 0.106 \times 10^6$

$\dfrac{M_2}{M_1} \times 100 = 3.3\%$

• SEEDING INCREASES 0.3% INFLOW MOISTURE WHICH PRECIPITATES

• SEEDING SHOULD NOT AFFECT DOWNWIND PRECIPITATION

Figure 7. Estimation of the effect of cloud seeding on the moisture balance.

THE SOURCE OF WATER VAPOR AND ITS DISTRIBUTION
OVER THE Qinghai-Xizang PLATEAU DURING
THE PERIOD OF SUMMER MONSOON

Huang Fujun
Institute of Meteorological
Science of Sichuan Province

Shen Rujin
Institute of Atmospheric Physics,
Chinese Academy of Sciences

By using the data from the four upper-air observing stations set up at Zhongba, Shiquanhe, etc., during the QXPMEX in 1979, together with the routine weather data on both sides of the Himalayas and the weather reports from the MONEX, calculations and studies are made of the source of water vapor and its budget, thus confirming the presence of a concentrated vapor transfer[1-3] which was first found near Zhongba during the QXPMEX in 1979. The calculation covers an area of 5—45°N, 40—110°E, including meridional and zonal wind components at 850—200 hPa, specific humidity at 850—400 hPa, water vapor flux in the x,y direction, flux divergence and the physical quantities of the variation in water vapor due to divergence and advection. The spherical coordinate formula is used in the calculation with 459 grids spaced 2.5°×2.5° apart. In order to show more detailed distribution of the water vapor budget on the Plateau, it is necessary to divide the Plateau and its neighborhood into five regions marked by A-E, as shown in Fig.1.

At the time when the Indian SW monsoon reached the vicinity of Bombay, the Meiyu season along the middle and lower reaches of the Changjiang River and the rainy season on the Plateau began simultaneously in China. According to the circulation regime and its precipitation distribution, the period of June 6—10 is defined as the pre-stage, June 16—20 as the initial stage, July 21—31 as the peak stage and August 21—31 as the ending stage of the rainy season on the Plateau.

The following is a detailed analysis of the source of water vapor and

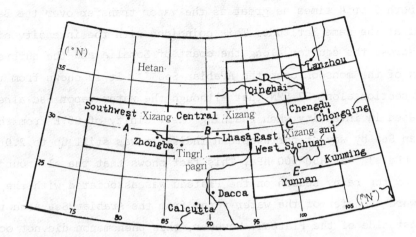

Fig.1 Diagram showing the division of regions for water vapor
budget.

its budget over the Plateau and its neighboring areas during the four
stages of summer monsoon.

1) The SW flow over the Bay of Bengal had transferred water vapor
to the eastern Plateau and Yunnan province before the rainy season
began. The flow originated from north of the equatorial buffer zone
across the sea area southeast of India into the Bay of Bengal. When
it encountered the northerly flow east of the Plateau, a transversal
shearline occurred at about 30°N, east of 95°E, causing the onset of
the rainy season to the east of the Plateau and over Yunnan province.
During the outbreak of the Indian monsoon, the SW flow over the Bay of
Bengal was displaced northeastward from the Arabian Sea across the
central Indian Peninsula, its intensity remaining the same as in the
pre-stage of the rainy season. In the meantime the shearline along the
Yarlung Zanbo River valley on the Plateau remained stable and extended
westward, followed by the rain band east of the Plateau[4]. It is
noteworthy that there was water vapor over the Arabian Sea area moving
across the western part of India to the western Plateau, thus facilitat-
ing the all-round onset of the rainy season on the Plateau. Fig.2(a)
shows the meridional transfer and flux divergence of the water vapor,

597

coming from the Arabian Sea and the Bay of Bengal, on the cross section of vapor flux along 15°N during the initial stage (June 16—20). They were both 3 or 4 times as great as the vapor transfer over the Bay of Bengal at the same latitude. This coincides with the intensity of the flow across the equator along the coast of Somalia and the outbreaking vortex of the monsoon over the Arabian Sea[5]. It is shown from the cross section along 27.5°N that although the water vapor had already decreased in intensity when reaching the vicinity of 85°E from the Arabian Sea by way of northwestern India, it was still up to 2.0 $g \cdot (cm \cdot hPa \cdot s)^{-1}$ at 850-600 hPa. This fact shows that the all-round onset of the rainy season on the Plateau was associated with the northward transfer of the water vapor from the Arabian Sea area to the west side of the Plateau. However, this phenomenon did not occur during the other stages of summer monsoon.

   2) During the peak stage of the rainy season of 1979, a typical break monsoon dominated a great part of India and consequently the high over the Plateau disappeared and was replaced by the frequent activity of low vortex, that is, the Indian monsoon trough was found lying at the foot of the Himalayas. Meantime a concentrated vapor transfer was observed over the southern Plateau as shown by the meridional transfer of water vapor in Fig.3, being as great as 4.8 $g \cdot (cm \cdot hPa \cdot s)^{-1}$ at 600 hPa along 85°E. Concurrently the vapor transfer was 2.4 $g \cdot (cm \cdot hPa \cdot s)^{-1}$ at 600 hPa at about 27.5°N and 87.5°N (southwest of Zhongba) and was 2.4-4.3 $g \cdot (cm \cdot hPa \cdot s)^{-1}$ at 600-500 hPa east of the Plateau, especially a vapor flux was easily seen over the Plateau at 30°N, 85-100°E. This indicates that in the peak stage of the rainy season there did exist a concentrated vapor transfer along the area from Zhongba to Tingri. Most probably this is one of the most important reasons for the intensification of precipitation due to the activity of low vortex. It is also found from the three diagrams in Fig. 4 that the zero line of vapor flux was displaced from 85° in the pre-stage to 80° in the initial stage and then to west of 80°E in the peak stage of the rainy season, with its center displaced from Pagri to Zhongba and then to Tingri and its intensity increasing from 2.2 to more than 4.0 $g \cdot (cm \cdot hPa \cdot s)^{-1}$. All this confirms

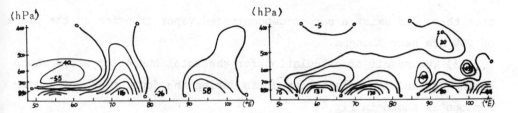

Fig.2(a) Water vapor flux along   Water vapor flux divergence

15.0°N, 50-105°E   along 15.0°N, 50-105°E

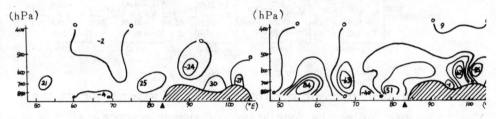

Fig.2(b) Water vapor flux along   Water vapor flux divergence

27.5°N, 50-105°E   along 27.5°N, 50-105°E

Fig.2 Meridional transfer of water vapor (June 16-20, 1979)

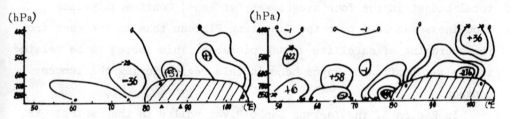

Cross section of meridional transfer   Cross section of meridiona

along 27.5°N, 50-105°E.   vapor flux divergence alon

                                          27.5°N, 50-105°E.

Fig.3 Meridional transfer of water vapor (July 21-31, 1979).

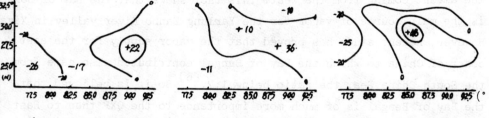

(June 6-10, 1979)   (June 16-20, 1979)   (July 21-31, 1979)

Fig.4 Meridional vapor transfer over the Plateau at 600 hPa

during the three stages of the summer monsoon.

that there did exist a rather concentrated vapor transfer on the plane of 600 hPa near Zhongba.

3) The result of calculation for the total budget of water vapor on the QXP and its neighboring areas during the four stages of summer monsoon is shown in Fig.4.

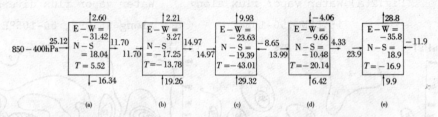

(a)   (b)   (c)   (d)   (e)

Fig.5 Distribution of the total budget of water vapor on the QXP during the four stages of summer monsoon.

In Region A, the southwestern QXP, water vapor came mainly from the west and there was outflow in both the south and the east. The total budget in the four stages was net loss. Considerably less cloudiness was seen over the Qiangtang Plateau than in the east from the averaging of satellite cloud pictures. This seemed to be related to the net outflow of vapor here in this region during the summer monsoon.

In Region B, the Yarlung Zanbo River valley in the Central QXP, water vapor transfer was also from west to east, the ratio of inflow to outflow being 2:1. The vapor entering into this region from the south, i.e., from the Bay of Bengal, accounted for more than half of the total inflow, amounting to about $19.26 g \cdot (cm \cdot hPa \cdot s)^{-1}$, almost twice the amount coming from the west. This fact shows that the Bay of Bongal is the main source of vapor for the Yarlung Zanbo River valley in Xizan However, recent study has proved that the vapor supply for the east coast of China to which the Bay of Bengal contributed also came from the South China Sea, the ratio being 1:2.[6] So it is believed that the Bay of Bengal is of much more importance to the QXP than to East China.

In Region C, the eastern QXP and West Sichuan, the total budget of water vapor was net gain during the four stages of the summer monsoon, so great in amount as to be the greatest in the five regions. The vapor transfer was chiefly from the southwest, with the Bay of Bengal being the greatest source for vapor supply, amounting to $29.32g \cdot (cm \cdot hPa \cdot s)^{-1}$, about twice the amount coming from the west. Besides, calculations show that vapor also came from the east, the net inflow being $8.65g \cdot (cm \cdot hPa \cdot s)^{-1}$. As a result the net gain was as great as $43.01g \cdot (cm \cdot hPa \cdot s)^{-1}$ in this region, a "basin" in which a continuous supply of water vapor sinks. That is why abundant rainfall occurs in Sichuan during the summer monsoon. This happens, of course, due to the unusual topography of West Sichuan situated east of the QXP.

In Region D, Qinghai province, the vapor transfer was both from north to south and from west to east in May and June, the budget remaining in the main balanced. In late July and August the vapor transfer from the south, as a result of the intensification of the SW flow on the QXP, changed from net outflow to net inflow with net gain amounting to $20.14g \cdot (cm \cdot hPa \cdot s)^{-1}$. Thus the vapor entering into this regions was sufficient for the rainfall in the peak stage of the rainy season.

In Region E, Yunnan province, there was net vapor gain in all stages of summer monsoon, being $6.42g \cdot (cm \cdot hPa \cdot s)^{-1}$ from the south and $13.49g \cdot (cm \cdot hPa \cdot s)^{-1}$ from the west. It may be concluded that the Bay of Bengal was the chief source for the vapor supply and the vapor coming from the south, the Indochina Peninsula, played a secondary role. However, it should be pointed out that although the rainy season began simultaneously in Yunnan and the eastern QXP and had the north-moving monsoon from the Bay of Bengal as the same vapor source, owing to the great difference of about 2000 m in elevation above sea level between the two areas the water vapor climbing over the south slope into the eastern QXP was only half as much as that into Yunnan, and consequently the two areas differed greatly in rainfall.

4) Analysis is also made of the values of water vapor flux convergence (divergence) at the seven points crossing at from 82.5°E, 85.0°E to 105°E along 30°N on the heartland of the Plateau. It is found that the chief physical process affecting the vapor budget on the heartland is vapor convergence (divergence) due to air mass convergence (divergence) and the role of vapor advection is always

the secondary.

In brief, The Bay of Bengal is the chief source responsible for the vapor supply in Xizang, Qinghai, Sichuan and Yunnan in summer, But this is not the case with the east coast of China where the SE monsoon originating in South China Sea is mainly responsible for the vapor supp Analysis indicates that it would take about 10 days for the rainy season to begin extensively on the QXP. Water vapor from the Arabian Sea can travel and reach the western QXP, contributing to the west-east extension of the rain band on the QXP. During the peak of the rainy season there exists a vapor transfer from Zhongba to Tingri, exerting a far-reaching influence upon the precipitation on the western QXP. It is found from the calculation of the vapor budget in the five regions that water vapor is not only transferred from the south by the monsoon originating in the Bay of Bengal but also is supplied from the western QXP, the ratio being 2:1. However, the vapor coming from the southwest is carried continuously by the westerly and SW flow to the central and eastern QXP and Qinghai province and when reaching the east side of the QXP, it is easily concentrated in Sichuan under the influence of the unusual topography and the constant low vortex in Southwest Sichuan resulting in plenty of rainfall. The summer monsoon is indeed of vital importance to Sichuan, "the land of abundance".

REFERENCES

[1] Qian Zhengan et al, Analysis on the initial vortex over the western Qinghai-Xizang Plateau during the rainy season, Collected Papers of QXPMEX. (1), Science Press, 1984 (in Chinese).

[2] Lu Junning et al, Analysis on the low vortex over the Qinghai-Xizang Plateau prior to the rainy season, Collected Papers of QXPMEX. (1), Science Press, 1984 (in Chinese).

[3] Huang Fujun et al, Relationship of the break and reactivation of the rainy season on the Qinghai-Xizang Plateau to the Indian Monsoon, Collected Papers of QXPMEX. (1), Science Press, 1984 (in Chinese)

[4] Huang Fujun, Characteristic features of the mean circulation and

precipitation distribution over the Qinghai-Xizang Plateau, To be published.

[5] Krishnamurti, T. N., Onset of summer monsoon-vortex, FGGE-The Global Weather Experiment First GARP Global Experiment Operations Report Series VOL 9, Part B.

[6] Chen Longxun, Some scientific questions about the monsoon studies in East Asia Collected Papers of Sino-American Monsoon Symposium.

AIRSTREAM CLIMATOLOGY OF ASIA

Reid A. Bryson
Center for Climatic Research
Institute for Environmental Studies
University of Wisconsin

INTRODUCTION

In 1966 an extension of the airmass concept was proposed for purposes of climatic analysis (Bryson, 1966). In essence the principle was that the air diverging from source regions•(usually anticyclonic regions) might go through many stages of modification in its surface path from source to eventual convergent zone where it might leave the surface or become unidentifiable. For example, air subsiding in the eastern part of the North Atlantic anticyclone might be called stable maritime tropical air (mTw). Farther downstream in the tradewinds it might be modified to an unstable form (mTk), then after rounding the western end of the anticyclone and moving east across the colder North Atlantic waters enter Europe as a stable maritime airmass (mPw). After entering central Asia the air might be cooled and dried enough to be called a variety of continental Polar air, even though its history might be traced back to the tropical Atlantic.

While concerning oneself with such distant antecedents has little value in synoptic meteorology, it has been found useful in interpretive climatology. (Wendland and Bryson,1981; Bryson and Hare,1974) In these works on airstream climatology, monthly resultant surface streamlines were necessary, and were constructed for the entire northern hemisphere (see Wendland and Bryson, op. cit.) Close analysis of the streamline patterns for Asia seems not to have been done. It is the purpose of this short paper to introduce the application of airstream climatology to the Asian region, despite many data problems and the world's most complex topography.

The reader must be warned at the outset that the data for which the streamlines were drawn represent a rather short time sample of a few years, mostly in the early 1960s period. Each reader may know of better data for some part of the region, of which the author would be grateful to be informed. There are also, of course, alternative analyses of the data, even with the data plotted at approximately two degree grid intersections.

THE STREAMLINE CHARTS

Figures 1-12 present monthly streamline patterns for the portion of the Northern Hemisphere centered on Asia. The spacing of the streamlines has no significance, but the streamlines are drawn exactly parallel to the wind vectors that were available. There is significance to the dashed confluence lines drawn on the charts, for they represent very nearly the mean monthly frontal positions. It is for this reason that the charts are arranged in the non-standard order starting with November. One notes that there is a northeast-southwest trending "mean front" in the Pacific in November. This confluence is between airstreams originating in Asia and in the Pacific anticyclone. The "mean front" is not present in the same general location in October as in November, but it remains fairly fixed from November through March. The same is true of the North American boundary between the airstream of Arctic origin and airstreams of other origins in the same time period. This is related to the natural, as opposed to nominal, seasons. Other boundaries, or confluences, between airstreams may be observed that are quite constant in location for several months. This fixed location of a "mean frontal zone" may be seen as a characteristic of a particular natural season.

As might be expected from the vast literature on the subject, the advancing and retreating edges of the summer monsoon airstream from the southern hemisphere is quasi-stationary only along the African-Arabian coastal area and along certain topographic features. In terms of the airstreams the parallelism of monthly movement in India and China is notable.

A feature of the streamlines that is often striking to one not familiar with the Asian general circulation is the alternation of flow to and from the Arctic. The flow from the continent in winter and the flow from the Arctic in summer constitutes a kind of vast northern monsoon. There is no month, however, when airstreams from the Arctic, Pacific or Atlantic penetrate the region between Gansu on the east to the Pamirs on the west, or between Mongolia and India.

AIRSTREAM DURATIONS

Figures 13 through 18 depict the duration, in months, of the presence of a particular airstream. The persistent exclusion of all maritime airstreams from the high western reaches of the People's

Republic of China is clear in figure 13, in which is shown the 12 months per year duration of an airstream we call Central Asian. The name denotes a difluent source region largely in the Xizang, Xinjiang, and Qinghai Regions. For three winter months that airstream reaches to the Arctic Ocean.

The flow southward from the Central Asian source region is shown as a different airstream (Fig.14) because of the enormous elevation difference between Xizang of China and the plains of India. We denote this flow southward from the core region the Trans-Himalayan airstream. It might be called the "winter monsoon" in India but for clarity in a more global context we prefer a more specific name to distinguish it from a quite different airstream from the vicinity of Burma. Figure 14 also shows that the cool Arctic airstream, which should be shallow, penetrates southward to the Himalayas only west of the Pamirs (enough to dominate a whole month).

Figure 15 shows that the eastward extent of the South Indian airstream is limited mostly to the west of a line from Hanoi to Singapore. East of that line the monsoon air is of the Australian or South Pacific airstreams.(Fig.16) In July South China is occupied by the former type and North China by the latter.(Fig.9) From year to year the relative dominance of these two sub-types seems to vary.

The maximum incursion of the North Atlantic airstream into Asia is in autumn(Fig.11). The duration of that flow is short(Fig.17). In contrast with that autumn flow, the North Pacific influx precedes the summer monsoon. Still earlier, in March through May, there is an airstream source located over Korea that has a quite large region of influence (Fig.18).

AIRSTREAM DOMINANCE AND CLIMATIC REGIONS

In those regions where there has been intensive study of airstream calendars and seasonally dominant frontal positions, it has been possible to define genetic climatic regions on the basis of the atmospheric circulation alone. These genetically defined regions in turn have been found to be closely related to biotic regions, even though no biological data was used in the definition of the regions.(Bryson, 1966 op.cit.; Bryson, Baerreis and Wendland,1970) If one examines the regions of half-year or longer duration of airstreams (Fig.19), along with the other charts in this paper, one can construct genetic climatic regions for Asia without the use of other climatic data. Whether these regions are closely related to biotic regions or are useful in Asia remains for future study.

It is interesting to note that all of the regions not shown as being occupied by a single airstream or group of related airstreams are regions of seasonally progressive monsoon phenomena, including the "Arctic monsoon" described above.

REFERENCES

Bryson, R.A.,1966:"Airmasses, Streamlines, and the Boreal Forest", GEOGRAPHICAL BULLETIN (Canada), 8(3):228-269.

Bryson, R.A. and F.K.Hare,1974:"The Climates of North America" in CLIMATES OF NORTH AMERICA, Bryson and Hare eds., Vol.11 of WORLD SURVEY OF CLIMATOLOGY, Amsterdam-London-New York, Elsevier Sci. Publ. Co., 420pp.

Bryson,R.A., D.A.Baerreis and W.M.Wendland,1970:"The Character of

Late- and Post-Glacial Climatic Changes", in PLEISTOCENE AND RECENT ENVIRONMENTS OF THE CENTRAL PLAINS (Symposium at Lawrence, Kansas, Oct.25-26, 1968) University Press of Kansas, pp.53-74.

Wendland, W.M. and R.A.Bryson,1981:"Northern Hemisphere Airstream Regions",MONTHLY WEATHER REVIEW 109(2):255-270

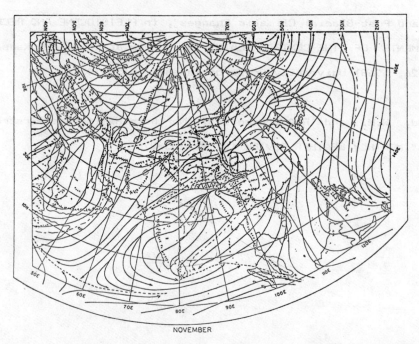

NOVEMBER

Fig.1 November resultant surface wind streamlines.

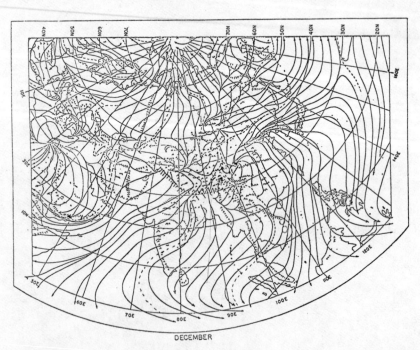

DECEMBER

Fig.2 December resultant surface wind streamlines.

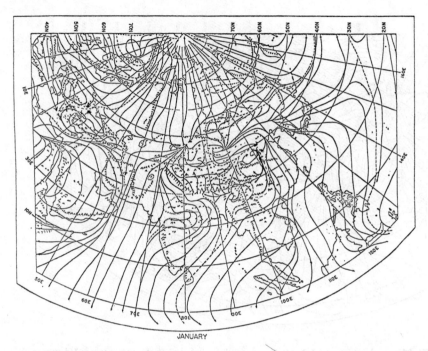

JANUARY

Fig.3 January resultant surface wind streamlines.

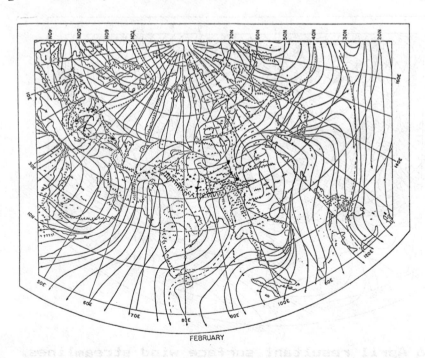

FEBRUARY

Fig.4 February resultant surface wind streamlines.

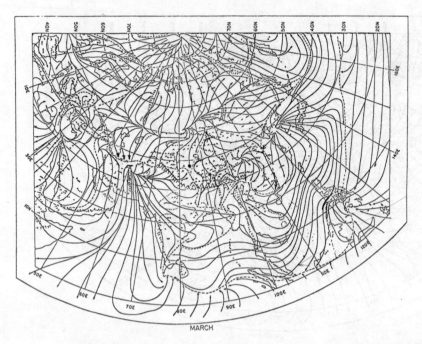

Fig.5 March resultant surface wind streamlines.

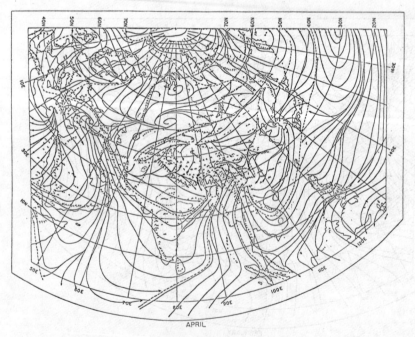

Fig.6 April resultant surface wind streamlines.

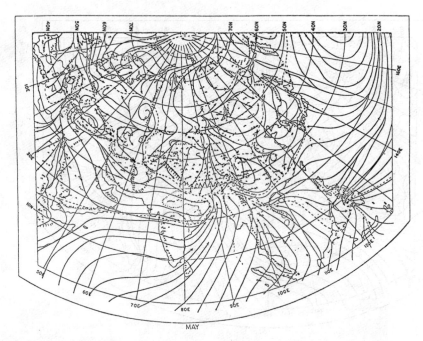

Fig.7 May resultant surface wind streamlines.

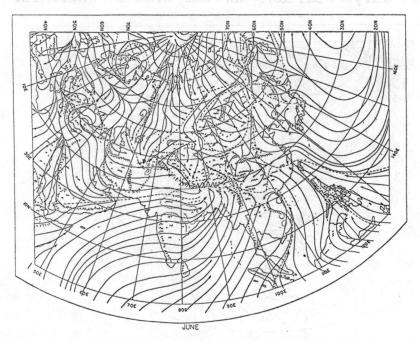

Fig.8 June resultant surface wind streamlines.

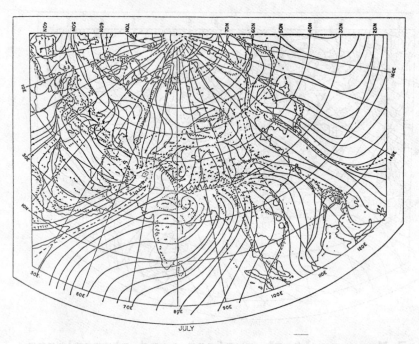

Fig.9 July resultant surface wind streamlines.

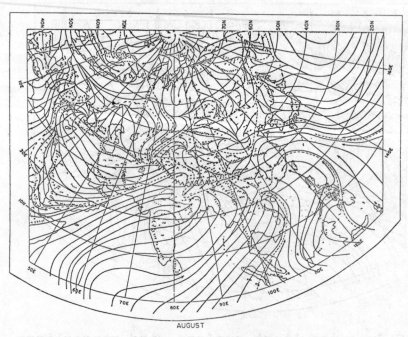

Fig.10 August resultant surface wind streamlines.

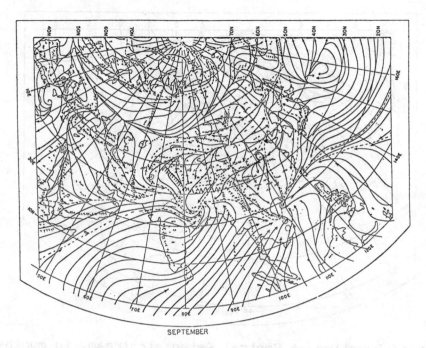

SEPTEMBER

Fig.11 September resultant surface wind streamlines.

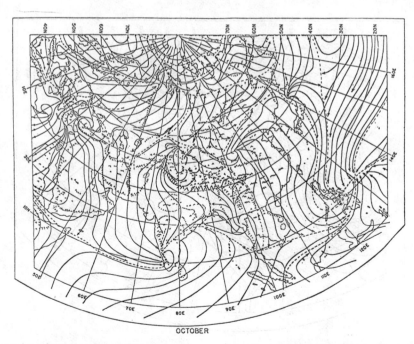

OCTOBER

Fig.12 October resultant surface wind streamlines.

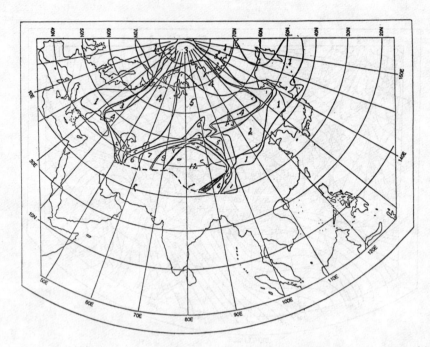

Fig.13 Duration of Central Asian airstream, in months.

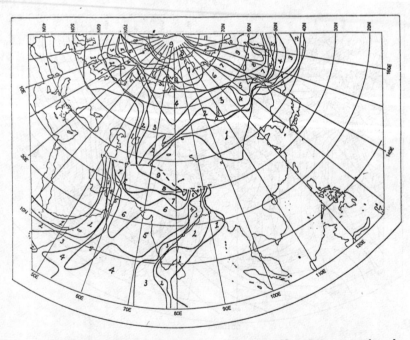

Fig.14 Duration of Arctic and Trans-Himalayan airstreams,

in months.

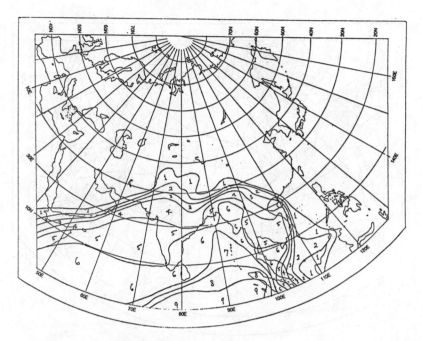

Fig.15 Duration of South Indian airstream, in months.

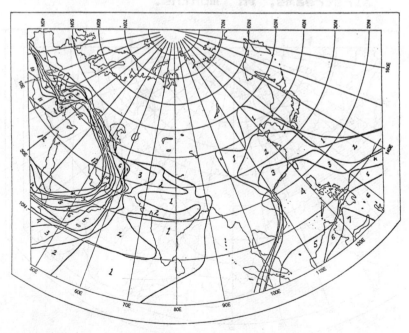

Fig.16 Duration of Australian or South Pacific, and Turkic or Saharan airstreams, in months.

617

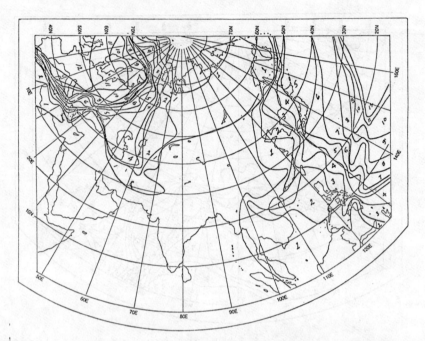

Fig.17   Duration   of North Pacific and North Atlantic

airstreams, in   months.

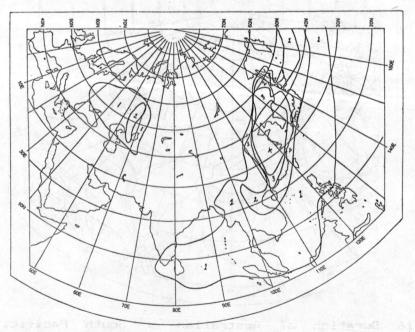

Fig.18   Duration   of Korean and Western Kazakhstan

airstreams, in   months.

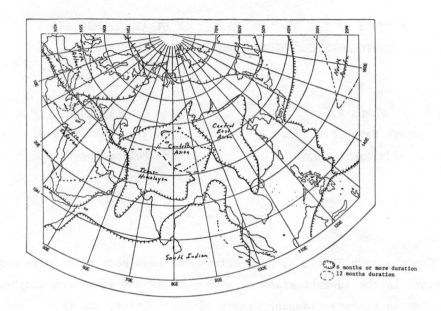

Fig.19 Regions of six months or more dominance by single
airstreams or closely related groups of airstreams.

A STUDY OF THE COUPLING EFFECT OF STRATOSPHERIC AND TROPOSPHERIC CIRCULATIONS IN EARLY SUMMER,1979

Zhang Jijia
State Meteorological
Administration

Peng Yongqing
Nanjing Institute of
Meteorology

I. INTRODUCTION

For the past 30 years, much effort has been made by Chinese meteorologists in the investigation of such medium- and long-term weather
processes as the pre-flooding season of South China, the Meiyu over the
middle and lower reach of the Changjiang River, and the rainy season
of North China. It has been recognized that the early or late onset of
the rainy season and its duration are usually related to the general
circulation pattern over the whole of Eurasia. As early as the 1950's,
the significance of the blocking action in the westerlies at 500hPa
and the activities of the western North Pacific subtropical high were
recognized[1]. In recent years, further investigation has drawn our
attention to the transition of the 100 hPa flow pattern, affirming that
the behavior of the South Asia high has great significance in relation
to the above-mentioned major weather systems over China in the summer
months[2,3]. Through our investigation of the seasonal variation of the
general circulation of the Northern Hemisphere for the period of the
QXPMEX in 1979, we have come to realize that the variation of the general circulation pattern over the Northern Hemisphere from the troposphere to the stralosphere and the coupling effect between the lower
stratosphere and the upper troposphere must be dealt with before we
can turn our attention to the major atmospheric events occurring over
Chian in summer.

II. VARIATION OF CIRCULATION PATTERNS AT
VARIOUS LEVELS OVER THE NORTHERN
HEMISPHERE DURING THE QXPMEX

The progression of variation of the circulation patterns at various levels during this period is first analysed with an aim to find the general features of the variations in the middle and upper atmosphere and the lower stratosphere.

1. At the 500 hPa Level

Changes in the circulation at 500 hPa involve phenomena of various scales. To single out the large-scale effect, we apply the global harmonics analysis to the 500 hPa contour field, using 5-day mean charts composited with 0—3 waves. The main features of changes in the 500 hPa circulation during QXPMEX are: 1) the number of waves in the zonal flow around the polar whirl changing from a normal of three in May and June to two in July; 2) three high-pressure systems occurring at the subtropical latitudes, i.e., the North American high, the North Africa high and the western North Pacific high. Another important feature at this level is a blocking action over the northern part of the North Atlantic, which developed and persisted from June 26 to July 6, as shown in Fig.1. For the particular year of 1979, therefore, the formation, maintenance

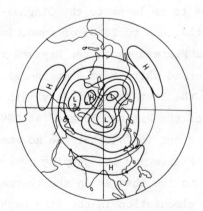

Fig.1 Blocking action over the North Atlantic at
500hPa (June 26 to July 6, 1979).

and the eventual dissipation of the blocking situation manifested
itself as a phase in the changes of the summer circulation in mid-tro-
posphere, for it is an essential circulation structure for the north-
south transport of thermal energy by large-scale eddies during the
adjustment of the super-long-wave and long-wave systems. It may present
itself as a decisive turning-point in the change of flow pattern at
500 hPa.

2. At the 100 hPa Level

In summer, a polar whirl predominates over the middle and high latitudes in the upper troposphere at 100 hPa. Certain regularities may be
found about the behaviour of the super-long-wave systems superimposed
on the polar whirl. The observed data of the summer of 1979 indicates
that the prevalant 3-wave flow around the polar whirl in May changes
into a 2-wave pattern in early June, and the 3-wave circulation is
restored only after the assymmetrical shift of the conter of the polar
whirl in late June.

Over the subtropical latitudes, three high systems persist throughout, corresponding to the high pressure centers at the 500 hPa level.
Of course, the behavior of the high over South Asia commands most of
our attention(Fig.2). The sequence of events may be described as follows. In the middle of May, the South Asia high lies over the area from
the South China Sea to the Phillipines; it stretches westward, reaching the western rim of the Indo-China Peninsula in the first dekad
of June and moves close to or leaps to the Qinghai-Xizang Plateau
in the fourth pentad (the 16th to 20th) of June. In the last dekad of
June, the high over the Plateau is of the Western type and then changes
into Eastern type. A second shift from over the Western Plateau to the
eastern part occurs in the last dekad of July with the high center
moving northward out of the Plateau located at 37°N in the last pentad.
The authors have pointed out in [4] that the movement of the South
Asia high close to the Plateau or its jumping over to the region is a
long-periodic response to the changes in the thermal structure and
temperature field, the circulation in the stratosphere being also affected by this heat source.

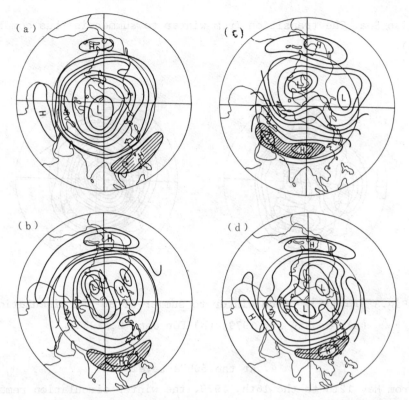

Fig.2 Progressive development of the South Asia high(shaded
area) at 100 hPa. (a) Over the South China Sea and the
Phillipines; (b) over the Indo-China Peninsula;(c)over
the Qinghai-Xizang Plateau (western type), and(d)with
center located east of the Plateau (eastern type).

3. At the 30 hPa Level

In the middle of May (from the 12th to 16th) the 30 hPa charts exhibit the transition of the winter circulation to the summer pattern.
This transition is of a typical character. The summer regime is initiated by the deep incursion of the North Pacific high into the polar
area through the Bering Sea and by progressive displacement of the
strong low pressure system over the Atlantic to Europe and the
Mediterranean. In the first pentad of June, the polar anticyclone is
established with its center deviated to the European side over the

623

Norwegian Sea. The transition from winter to summer is thus completed (Fig.3).

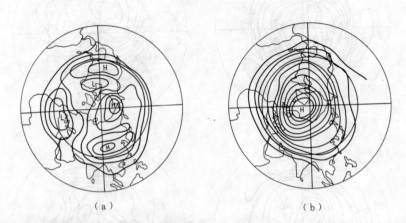

Fig.3 Transition from winter to summer of 30 hPa circulation.
(a) For May 15, 1979; (b) For June 3, 1979.

4. At the 50hPa Level

From May 12th to the 16th, 1979, the winter circulation remained prevalent at 50 hPa, with flow over the middle and high latitudes still under the control of the polar low and a high pressure belt over the subtropical latitudes with two spearate centers, one lying over West Asi stretching eastward across South China to the North Pacific; the other over the United States and the adjacent Atlantic. During the first pentad, the transition to the summer pattern at 50 hPa began with the invasion of the anticyclone into the polar region through Europe. At that time the center of the 50 hPa European high coincides with the center of the 30 hPa anticyclone center and this spurs the development of the summer circulation at the 50 hPa level. Subsequently, on the 10th of June, the 50 hPa European high advanced into Greenalnd forming a polar anticyclone centered near the North pole, thus establishing the summer circulation regime (Fig.4).

It can be seen from the above that, when the 30hPa flow starts its transition toward the summer circulation regime in the middle of May, the 50hPa circulation remains in its winter pattern. Only when the

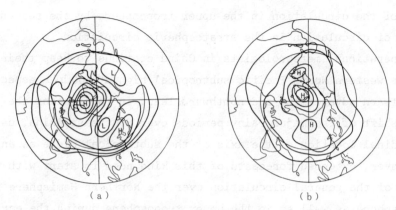

(a)   (b)

Fig. 4 Establishment of the summer polar anticyclone at 50 hPa.
    (a) For June 6, 1979; (b) For June 10, 1979.

summer circulation has been established in early June at the 30 hPa
level, does the 50 hPa flow start its transition to summer conditions.
That is to say, the establishment of the summer regime in the stratosphere proceeds from the upper levels downwards.

    It should be noted that there exists an interrelation between the
circulation in the stratosphere and that in the upper troposphere. The
most striking feature is that after the summer regime is set up successively in the first and second pentads of June at the 30 hPa and
50 hPa levels in the lower stratosphere,ridges from the polar anticyclone stretch toward North America and Asia, the latter, which reaches
to the Qinghai Xizang Plateau, being stronger. Ten days later, the
South Asia high at the 100 hPa level jumps or moves over the Plateau,
and stays under the control of the upper ridge. Afterwards the high
attains its peak intensity. This coupling of the lower and upper circulations not only facilitates the northward advancement of the South

625

Asia high and its maintenance there but assists the propagation of the stratospheric easterlies downward to help set up the tropical westerly jet stream [5]. This is of significance in predicting the seasonal variation of the circulation in the upper troposphere by the seasonal transition of circulation in the stratospheric circulation.

The operational meteorologists in China continue to keep their eyes on the western North Pacific subtropical high at 500 hPa, especiall on its westward advancement and northward leap, in forecasting the onset and shift of the major rainy periods over Southeast China, using the latitudinal location of the axis of the subtropical ridge as an index. However, extended forecasts of this kind need to start with the transition of the general circulation over the Northern Hemisphere in the stratosphere as well as in the upper troposphere during the early summer. The character of the stratospheric flows and the circulation in the upper troposphere is closely linked to the rainy season over East China, this phenomenon being of prognostic value.

III. Summary

The transition from the winter to the summer regime of the general circulation over the Northern Hemisphere has the following character-istic features:

1) The seasonal transition of the circulations in the lower strat-osphere and the upper troposphere is clearly made up of a series of successive steps, the first step being the onset of transformation to summer conditions at the 30 hPa level (middle of May). The second stage commences when the polar anticyclone is set up at the 30 hPa level and 50 hPa circulation starts its transformation (first pentad in June). The predominance of the polar anticyclone at 50 hPa and the establish-ment of the summer regime throughout the whole stratosphere may be considered as the third stage (end of the second pentad in June). Finally, when the South Asia high in the upper troposphere moves near (or jumps over) the Qinghai-Xizang Plateau and couples with the ridge stretching over Asia from the stratospheric polar anticyclone,, the fourth stage comes to pass (fourth pentad of Junw). The commencing

and ending dates may vary from year to year. However, the division into four stages itself is representative and is correspons with the northward shift of the rain belt in East China in summer.

2) The transition to summer regime of the circulation in the lower stratosphere is initiated from the higher levels. There is an evident time lag of the transition at the 50 hPa level in comparison with that at 30 hPa, a lag of one to two weeks. If the approach (or jumping over) of the Qinghai-Xizang Plateau by the 100 hPa South Asia high is considered as the establishment of summer circulation at that level, then the transformation at 100 hPa lags behind that of 50 hPa by about 10 days.

3) The ridge stretching over Asia from the polar anticyclone in the lower stratosphere in summer is coupled with the South Asia high in the upper troposphere, forming a high pressure system of large vertical extent over the Plateau, which upholds the persistence and northward advancement of the high and strengthens the tropical easterly jet stream over South Asia. This is a feature of great significance in the summer circulation over Asia.

REFERENCES

[1] Tao Shiyan, etc., The relationship between may-yu in the Far East and the behaviour of the circulation over Asia, Acta Meteorologica Sinica, Vol. 29, 119— 134, 1958 (In Chinese).

[2] Tao Shiyan, Zhu Fukang, The 100hPa flow patterns in the South Asian summer and its relation with the advance and retreat of the West Pacific subtropical Anticyclone over the Far East, Acta Meteorologica Sinica, Vol. 34, 385 —396, 1964 (In Chinese).

[3] Gao Ronglan, South Asia high, Mon. Met., Vol. 10, 1979 (In Chinese

[4] Chang Jijia, Pong Yongqing, Wang Dingliang, The structures of the South Asia high and its Characteristics of the time-frequency fields, Acta Meteorologica Sinica, Vol. 41,348— 352,1983(In Chinese

[5] Chang Jijia, et al., Some facts of the seasonal transition of the atmospheric circulation in early summer 1979, compiled in this volume.

Session 6: Modeling and Theory (2)

SOURCES OF FLOW VARIABILITY IN BAROTROPIC
PRIMITIVE EQUATION MODELS

J.N. Paegle[1], Zhen Zhao[3], Yan Hong[2], J.Paegle[1]

1. INTRODUCTION

Numerous observational studies of the Northern Hemisphere
circulation have discussed the regional characteristics of transient
eddies (paegle and Paegle, 1975, 1976; Blackmon et al., 1976; Blackmon
et al., 1977, 1979; Lau, 1979; Lau and Wallace, 1979; Wallace and
Gutzler, 1981; Blackmon and White, 1982; Blackmon et al., 1983a and b)
Atmospheric fluctuations are separated in these studies in short,
intermediate and long scales. The studies of Blackmon, Lau, Wallace
and collaborators have identified perturbations in the height field in
the eastern oceans that have a highly barotropic structure with strong
vertical coherence, while those in the interior of continents are
characterized by a more baroclinic structure. They also found that
transients with periods between 2.5 and 6 days have some of the
characteristics associated with baroclinic eddies while fluctuations
on longer time scales appear to be more barotropic. The latter have
been interpreted by some of the above cited authors in terms of two-
dimensional Rossby wave propagation. These and other works (from
Rossby, 1939, on) have shown the relevance of barotropic processes
to interpret some of the observed features of large scale circulations
    It is the purpose of the current paper to review the sources
of flow variability in barotropic flows, and present a description of

---

[1]Department of Meteorology, University of Utah.
[2]Lanzhou Institute of Plateau Atmospheric Physics, Chinese
 Academy of Sciences
[3]State Meteorological Administration Beijing, P.R. China.

observed transient activity during January and February 1979 that
can be interpreted by integrations of forced global barotropic primitive equation models.

2. POSSIBLE CAUSES OF TIME VARIABILITY

The linearized primitive equations on a basic state at rest
can be separated into horizontal and vertical structure equations.
Free oscillations obtained from these equations are referred to as
normal modes. The horizontal structure of these modes can be obtained
from the linearized shallow water equation with an equivalent depth
h_e which is obtained as the eigenvalue of the vertical structure
equation (Dikii, 1965; Longuet-Higgins, 1968; Kasahara, 1976). The
present paper reports on results obtained for h_e=10 km. Two types
of oscillations are found for the non-zonal modes with distinct frequencies: eastward and westward propagating gravity modes and westward
propagating Rossby-Haurwitz rotational modes. Examples of propagation
of rotational modes in the atmosphere are given by Madden, 1979.
Kasahara, 1980, showed that a latitudinally varying zonal flow could
substantially change the frequency of the rotational modes, but gravity-
inertia mode frequency as well as the horizontal mode structure were
little affected.

The structure of the free oscillations of the system depends
on the basic flow chosen. When the basic flow is zonally invariant,
atmospheric perturbations are found to be barotropically stable
(Phillips, 1954). In a recent article, Simmons et al.,1983, considered
a longitudinally asymmetric basic state for a barotropic model. They
found that the most rapidly growing mode associated with instability
of the basic state resembled some of the observed features of low-
frequency variability of the Northern Hemisphere. In this approach, the
forcing required to keep the basic flow steady is computed and perturbations from the longitudinally dependent basic state are considered.

A different approach considers a zonal flow which may change in
time. Stationary waves forced by fixed topography or heating sources
may acquire resonant growth if the basic flow changes into a state

632

with a free stationary mode (i.e., Tung and Lindzen, 1979a, b). The non-linear problem of a time dependent zonal flow could evolve into different flow configurations as a result of mulitple solutions.

Charney and DeVore, 1979, Paegle, 1979, among many others, studied the stability of the different flow equilibria for highly truncated models. Smaller scale processes not explicitly included in these formulations may be considered responsible in driving the atmosphere from one type of stable equilibrium to another. This process may be used to interpret changes in atmospheric flow from quasi-stationary high zonal index to strongly blocked situations. Unstable equilibria solutions are found when the basic state is close to resonance. In this case, topographically forced waves interact with the zonal flow through a form drag mechanism. Through this interaction the zonal flow evolves closer to the resonance state of Rossby waves forced by orography.

This type of instability for an inviscid case is described in Fig. 1 for an idealized channel flow truncated to one wave mode and one zonal flow mode. Then:

$$\psi = \sqrt{2}\, \psi_A \cos (my) + 2\psi_K \sin (my) \cos(nx) + 2\psi_L \sin (my) \sin (nx)$$

Fig. 1a shows schematically the structure of the mode forced by the orography and Fig. 1b illustrates the case for the

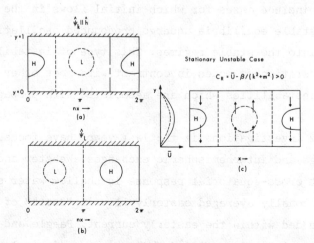

Fig.1. Illustration of orographic instability for a truncated channel flow model. See text for details.

perturbation extracting energy from the zonal flow which leads to
further amplification of the wave. For the inviscid case this
process leads to oscillations of the zonal flow and disturbances
with periods shown in Table 1.

Table 1

Period of oscillation induced by interaction of mean flow and
wave in the presence of mountains; shown for

$$\psi_{LO}=.025 \quad \psi_{KO}=0.0$$

$$W=5 \times 10^{6}m$$

$$\psi_{AO}=.05 \Rightarrow 10m/s$$

Wavelength $10^{6}m$	Period (Days)			
	$h/H=.1$	.2	.3	
~ 5	9.1	5.4	2.6	n=2
~ 10	7.1	5.6	3.1	n=1
~ 20	8.0	7.13	4.7	$n=\frac{1}{2}$

Extensions of this model to include frictional dissipation
and a zonally averaged forcing (as in Charney and De Vore, 1979)
can be shown to include cases for which initial flows in the neighborhood of unstable equilibria undergo pronounced oscillations
as they settle into the stable regimes. This type of instability is
associated with stationary modes in contrast with most other types
of atmospheric instabilities which are associated with propagating
disturbances.

A number of investigations in the last years have focussed on
tropical forcings and interhemispheric exchanges. Webster and Holton
(1982) find that cross-equatorial response in shallow water equation
take place in a zonally averaged easterly flow if "ducts" of westerlies are embedded within the easterly current. Paegle and Paegle
(1983), Paegle et al. (1984) studied the case of vanishingly small
absolute vorticity gradients and also found substantial interhemi-

spheric response for a zonally averaged easterly flow. This special
type of configuration is of interest in regions of strong convective
activity where the upper levels display strongly divergent motions
and the absolute vorticity tends to vanish.

Atmospheric linear response to transient forcing has been
studied by Paegle (1978), Lim and Chang (1981, 1983), Lau and Lim
(1982), and Paegle and Paegle (1983). Linearized analysis of the
shallow water equations were done by, Silva-Dias et al. (1983) for
the β-plane and Kasahara (1983) for a global primitive equation model.
For a vertical parabolic heating profile it is found that internal
modes are favourably generated and that excitation of rotational
or gravitational modes depend on the time scale of the heating. It
appears that even if the forcing favourably triggers internal modes,
external modes may be apparent in multi-layer baroclinic models due
to vertical variations of the zonal wind (Simmons, 1982) for the
case of stationary forcing.

The above cited works link atmospheric response to changes in
the boundary conditions. Another possibility is to consider that
atmospheric variability is due to internal dynamical processes. Egger
(1983) has shown that the low-frequency variance observed in the
atmosphere can be explained by the interaction of large scale systems
with synoptic waves in a global linearized equivalent barotropic model.
In this analysis the synoptic scale forcing is independent of the
planetary wave structure. On the other hand Frederiksen, 1979, has
shown that baroclinic disturbances are organized by the planetary
scales. The feedback between these two processes is explored by Reinhold
and Pierrehumbert, 1983, as a possible mechanism to explain the quasi-
stationary behaviour of the atmosphere. Some general circulation model
integrations (Lau, 1981) have also reproduced some of the observed
features of the planetary scales without transient boundary conditions,
lending support to the conjecture that these patterns may result from
the internal dynamics of the system.

We see from the above cited work that a multitude of processes
have been considered in the past as possible sources of variability in
primitive equation barotropic models. Linearized analyses have been

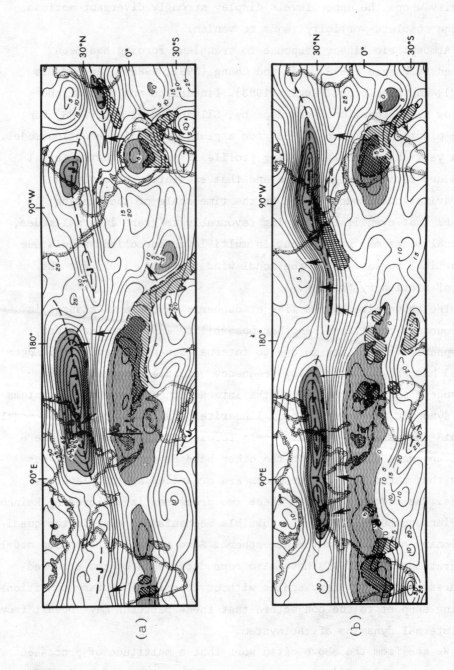

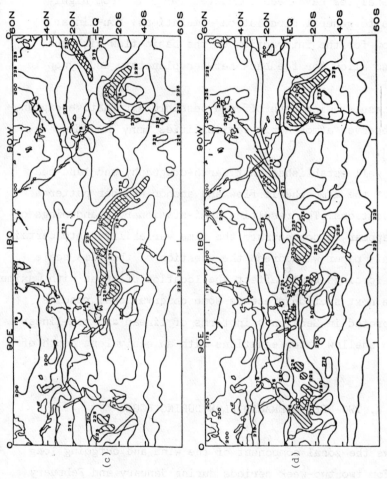

Fig.2. Isotachs of the zonal components of the wind analyzed every 5 m s^{-1} for two week averages denoted as Period 1(a) and 2(b) in the text. Values stronger than 45 m s^{-1} are shaded. Easterly flows are stippled. Outgoing long wave radiation is given in (c) and (d) in Watt m^{-2}. Values less than 225 Watt m^{-2} are shaded.

performed on stratified and homogenous atmospheres, with basic states
at rest; varying latitudinally, vertically or horizontally. Forcing
of external modes has been considered either as a result of barotropic
forcing, or baroclinic forcing which may project on the external mode
or appear as such due to the vertical shear of the zonal flow.

Nonlinear analyses have been conducted ranging from highly
truncated models to general circulation simulations. Non-linear
processes   allow for the interaction of the basic flow with the
perturbations and allow for feedback between planetary and synoptic
scales.

Although these studies have offered possible alternative explanations for the observed atmospheric variabilities many questions are
still unresolved.

It appears well established that earth-orography and land-sea
contrasts deflect the atmospheric flow and anchor weather patterns
to particular locations. The resulting semi-stationary meanderings of
the jet-stream appear to concentrate the time variability into certain
locations. In this paper we address the question of the time scale
in which remote forcings could affect the location of areas with maximum
variability. The next section presents some observational results
which are interpreted in Section 4 in terms of linear and non-linear
solutions of the shallow water equations with an equivalent depth of
10 km.

3. TWO-WEEK AVERAGED FLOW DURING FGGE

Fig. 2 shows the zonal component of the wind and outgoing long
wave radiation for two two-week periods during January and February
1979. These two periods (Paegle, Paegle and Lewis, 1984) have been
studied using the objectively analyzed level III-b data sets obtained
from ECMWF. These periods are characterized by a longitudinal shift
of the Pacific tropical convection. There is an increase of convection
over Africa and the Indian Ocean for Period 2 and an eastward displacement of the Asian jet. Fig. 3 and 4 suggest that the subtropical
jet is at least partly responding to the shift in tropical convection.

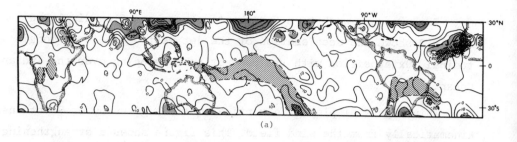

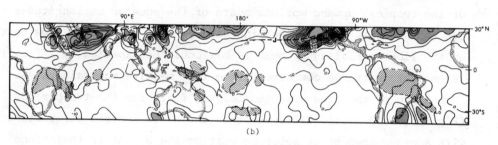

Fig.3. Isolines of $\dfrac{1}{100}\displaystyle\int_{0}^{100\ \text{hPa}} \nabla\ \nabla\bar{\phi}\ \dfrac{dp}{q}$ where a bar represents a two-week time average for period 1(a) and 2(b). Areas with small values of outgoing long wave radiation displayed in Fig.2 are shown with

$\boxed{///}$

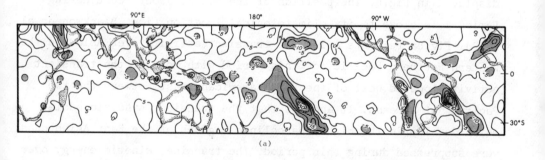

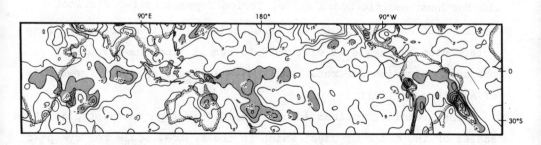

Fig.4. Averaged ω field at 400 hPa analyzed every 5×10^{-4} hPa s^{-1} for period 1(a) and 2(b).

The pressure work term for the time averaged flow displayed in Fig.3 shows the acceleration of the jet west of India in longitudinal juxtaposition with the convection developing over the Indian Ocean.

This figure displays the vertical motions given by $\omega = \frac{dp}{dt}$ obtained kinematically from the wind field. This figure shows a strengthening of the tropical upward motions south of the equator at longitudes of jet enhancement. The observed increase of the mean zonal flow east of the Tibetan Plateau of over 10 m/s could be expected to affect the large scale patterns downstream from this massive mountain range. Inspection of the time averaged height field for the two periods (Fig.5) show dramatic differences in the North Pacific circulations, with a pronounced block apparent west of the U.S.A. in the second period. For the present discussion the time averaged flow displayed in Fig. 2-5 can be interpreted as a basic state on which transient circulations are superimposed.

The transient kinetic energy for these two time periods is displayed in Fig.6. The position of the jet at about 130°E during Period 1 is close to its climatological location (i.e. Blackmon et al., 1977). Fig.6a reflects a lack of transient activity over Asia and the western Northern Pacific for this period with most of the transient activity located west of the North American coast between the two Pacific jet streams. It appears that the transient synoptic eddies which develop in the highly baroclinic regions of the East Asian coast were suppressed during this period. The transient kinetic energy over the Northern Pacific Ocean during Period 2 presents two distinct maxima east and west of the blocked region. This transiency has been interpreted by Paegle et al., 1983 to result from the favourable propagation medium for rotational waves in areas of strong gradients of absolute vorticity.

The transient motions identified in this figure have typical time scales of the order of days, which is the typical scale for synoptic disturbances commonly associated with baroclinic instability. The numerical integrations discussed in the next section suggest that at

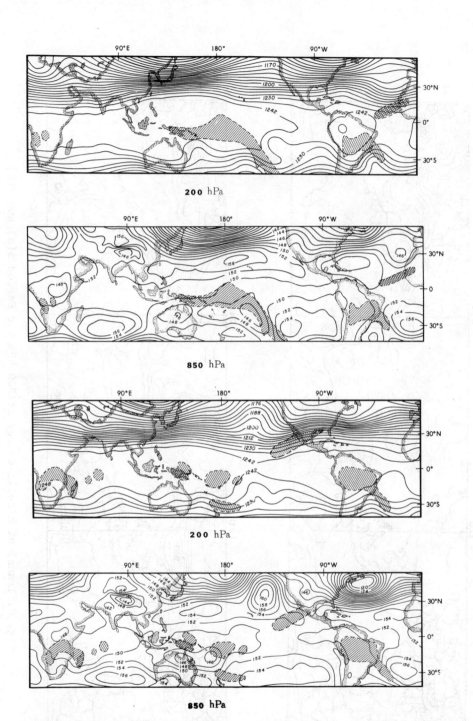

200 hPa

850 hPa

200 hPa

850 hPa

Fig.5. Averaged height field (in decameters) for Period, (upper two diagrams) and Period 2 (lower two diagrams) Areas [⧄] as in Fig.3.

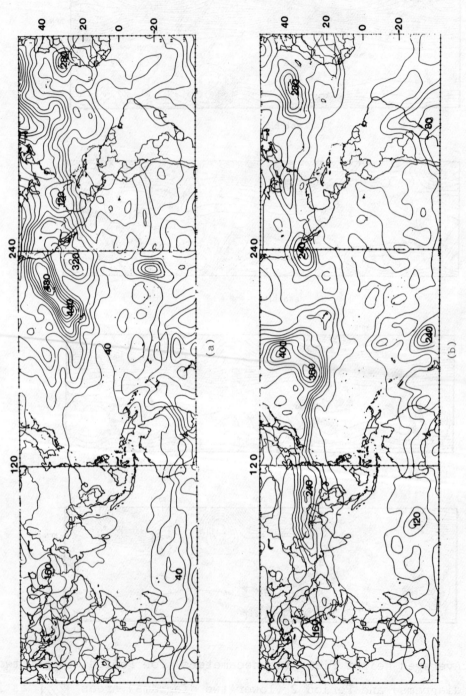

Fig. 6. Transient kinetic energy at 200 hPa for Period 1(a) and 2(b) in units of $m^2 s^{-2}$.

least partly, these maxima in the transient eddy activity can be explained in terms of barotropic processes. Numerical integrations of a shallow water equation model suggest that the local acceleration of the zonal wind due to Hadley type circulations which originate in the tropics affect the location of transient eddy activity in the Northern Hemisphere. Differences in model response depend on the longitudinal specification of the divergent tropical circulations in relation to the earth orography.

4. MODEL RESULTS

The global shallow water equation model is described in Paegle et al., 1983. It uses the Arakawa potential enstrophy, energy conserving scheme, an equivalent depth of 10 km; a 5° staggered latitude-longitude scheme; smooth topography used by the National Meteorological Center, a drag coefficient of 1/15 day^{-1} and it imposes a relaxation to the zonally averaged 300 hPa winter flow. Some runs include a mass source in the continuity equation to simulate heating. These sources are presently described as the divergence that is required to provide local balance in this equation.

The model is run to 40 days to study the adjustment of the flow to realistic orography starting from a zonally averaged flow.

The mountain induced perturbations are depicted in Fig. 7 in

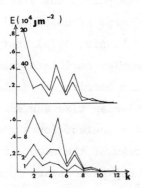

Fig. 7. Total energy for the twelve lowest wavenumbers. Each curve is labeled with a number indicating integration time in days.

terms of the energy spectra for the twelve longest atmospheric waves.
Initially, the energy perturbation appears in wavenumber 2. By day 8
another relative maximum appears in wavenumber 5 which remains
throughout the subsequent integration. From day 20 on, most of the
changes are observed in wavenumbers 1 and 2.

Fig. 8. Initial zonally averaged
          u (dashed-dotted line)
          and that after 24 days of
          integration (dashed line).

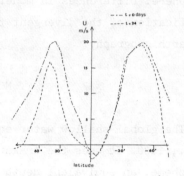

The zonally averaged flow after 24 days is considerably weakened
in the Northern Hemisphere (Fig. 8) by extraction of momentum by the
perturbations induced by orography in the presence of frictional
dissipation.

Most of the adjustment has taken place by day 24. Fig. 9 shows
the maximum kinetic energy changes between days 24 and 36 downstream of
major mountain ranges in the Northern Hemisphere. The stationary
response from the linear model counterpart of the numerical model is
shown in Fig. 10a. The tilt of the trough on the east Asian coast as
well as trough and ridge configurations in the linear and non-linear
runs are quite different. The variability of the points downstream of
the main mountain ranges can be interpreted in terms of conservation of
potential vorticity. We have argued (Paegle and Paegle, 1976) that if
the if the zonal flow upstream of a mountain barrier oscillates slowly,
the induced wavelength downstream of the mountain barrier would change
accordingly. These changes would be noted in eulerian time series
downstream of the mountain ranges in smaller time scales. For a zonal
flow oscillating with a period of 32 days upstream of the mountain barriers considerable contributions in the 2 to 8 day periods are found do-

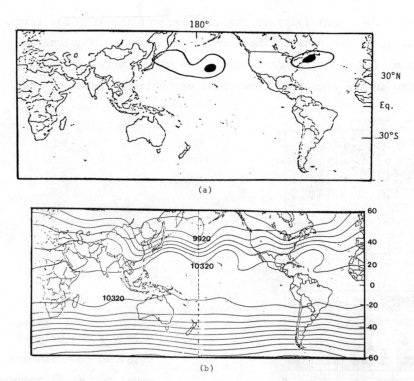

(a)

(b)

Fig. 9. Maximum kinetic energy change between integration day 24 and
36, analyzed every 10 $m^2 s^{-2}$ (a). Areas with values larger
than 20 $m^2 s^{-2}$ and shaded. Free surface height in meters at
day 24 analyzed every 80 m.

wnstream. This reasoning predicts a shift to higher frequencies east
of mid-latitude mountain ranges as shown in observations (Paegle and
Paegle, 1975, 1976; Blackmon et al., 1977).

Fig. 10 also shows the stationary response to two prespecified
mass sources designed to model tropical divergence fields with
magnitude similar to those observed in the tropical upper atmosphere.
The two forcing functions have no zonally averaged component and
differ in the longitudinal juxtaposition of the divergence and convergence fields with the earth orography. The maximum divergence value
are located at 150°E and 150°W for the two forcing functions. These
longitudes are used to identify the two cases. The stationary solutions

display large sensitivity to the placement of the tropical forcing in the Southern Hemisphere. The Northern Hemisphere solutions are identical for the three cases since the mass sources are entirely confined in the region of equatorial easterlies. Therefore the

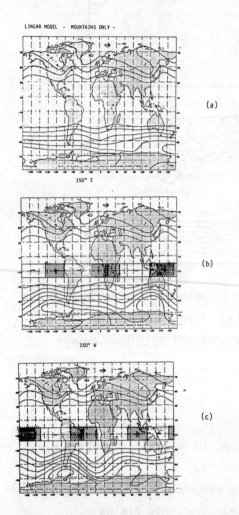

(a)

(b)

(c)

Fig. 10. Steady state heights obtained from linearized model forced by orography (a); and orography and tropical forcing (b), (c). Heights analyzed every 50 m. Tropical forcing is shown in (b) and (c) between the equator and 20°S with an analysis interval of 2. $10^{-6}s^{-1}$, positive values are shaded.

linear long term response of this model will show no impact to the tropical forcing in the Northern Hemisphere. The current emphasis is in the flow variability that results from inclusion of the tropical forcing and the linear stationary solutions are displayed as a guidance to the linearized model "climate".

Two non-linear simulations are integrated with the tropical forcing discussed above. These are started from the conditions given

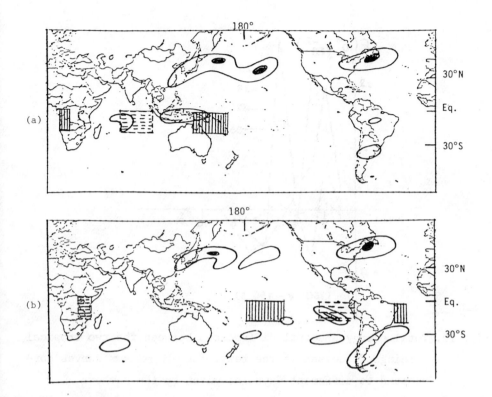

Fig. 11. Maximum kinetic energy change between day 24 and 36, analyzed
every 10 $m^2 s^{-2}$ (areas with value larger than 20 $m^2 s^{-2}$ are
shaded) for the tropical divergence maximizing at 150°E (a)
and 150°W (b). Areas shaded $\boxed{\equiv\equiv}$ indicate divergence
$<-2\times10^{-6} s^{-1}$ and $\boxed{||||||}$ $>2\times10^{-6} s^{-1}$.

for the day 24 of the run forced only by orography. The integrations
are continued for another 12 days. The maximum kinetic energy changes
in these 12 days are shown in Fig. 11 for comparison with that obtained
when only orography is present (Fig. 9a). The runs with tropical
forcing display a variability maximum east of Asia not present in the
run with orography only. The case with the maximum forcing at 150°W
also shows a center of maximum variability in the Eastern Pacific,
west of the Peruvian coast.

The tropical forcing projects only in wavenumbers 1 and 3. The
linear response of the model would be confined to these modes.
Excitation of other modes results from non-linear interactions.

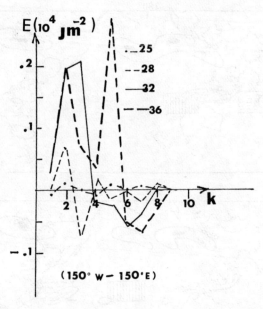

Fig. 12. Total energy spectral differences between the two tropical
forcings discussed in the text. The different curves correspond to different days, as given by the numbers.

The effect of the tropical sources in the global energy patterns
can be inferred from the difference fields presented in Fig.12. It
is of interest that the largest differences after 2 days appear in
wavenumber 2 and 5; which have been previously linked with topographic
effects.

Output from the numerical simulations are projected in the
normal modes of a barotropic primitive equation model with a basic
state at rest and an equivalent height of 10 km (Kasahara, 1976).
Fig. 13 compares the amplitude and phase response for four days after
the tropical heating is turned on for the highest amplitude Rossby
modes and for the Kelvin mode for the two numerical simulations. We
interpret modes with different amplitude and/or phase for the two
different cases resulting from the introduction of the tropical forcing
The phase for k=2 and 5 is the same for both cases while rotational
modes and the Kelvin mode for k=3 are clearly excited by the tropical
forcing. Although the initial response may be interpreted in terms of
propagation of waves on a horizontally varying medium produced by the

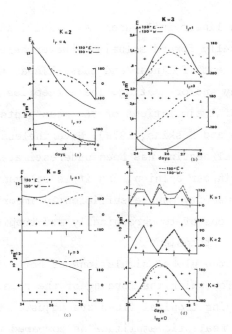

Fig. 13. Phase(+-) and amplitade (dashed and solid curves) of Selected
Ross by modes ((a), (b) and(c),and Kelvin mode (d) from the
time the tropical forcing is switched on at day 24.

earth orography the present results suggest that switching on tropical
forcing produces modifications of the basic state through non-linear
interactions which will affect the preferred regions of concentration
of atmospheric variability.

5. SUMMARY AND CONCLUSIONS

In completing this review it became apparent that the variety of
proposed mechanisms for large-scale atmospheric variability is almost
as varied as is the real atmosphere. Within this complexity it is
helpful to search for simple idealizations at the exclusion of many
complexities.

Rossby's original barotropic model is a successful prototype
of such an effort and continues to provide inspiration for many
studies. Numerous investigations have demonstrated the first order
influence of topography within this model and recent research has

suggested that non-linear influences may significantly modify the
linear topographically forced solutions. Nevertheless, there is a
continuing development of linear and quasi-linear theories of flow
shaped by topograhy and/or heating. These theories are supported by
remarkably successful simulations of the mean winter state by linear
models (e.g. Grose and Hoskins, 1979). Nevertheless, the extra-tropical
response to tropical heating has been underestimated in most linear
barotropic models in our opinion, and it is relevant to inquire
whether the non-linearity may be sufficiently important to be necessary in superpositions of topographically and tropically forced waves,
as well as for local transients.

We have presented global simulations of both linear and nonlinear
barotropic models for which the forcing of the zonal component and
longitudinally varying tropical flows are adjusted to produce atmospheric flows with realistic amplitude as compared to two 2-weekly
averages of FGGE data. Our analyses suggest that non-linear feedback
is essential to both the transient and quasi-adjusted solutions.

The introduction of tropical forcing excites a variety of modes
with a full frequency spectrum. Some of these modes decay as the flow
adjusts to the specified forcing. Linear stationary solutions for the
tropical forcing cases display no impact of the forcing in the Northern Hemisphere due to the existance of equatorial easterlies. The
nonlinear integrations present differences in the northern hemisphere
which appear to be related to the specification of the tropical forcing.
The excitation of global modes is studied for four days after the
heating is turned on in the orographically forced model. While the
forcing due to orography is geographically fixed that due to tropical
heating is not. It is natural to inquire whether the atmosheric response
would appear locked at particular geographical locations or whether
the phasing and amplitude of the response depend on the longitudinal
prescription of the tropical forcing. Study of the global modes helps
to clarify this point by showing that tropical heating excites modes
linked with orography which are clearly independent of the location
of the forcing. Such modes can only result in this model through non-
linear interactions. On the other hand, those modes which are due to

650

excitation by tropical heating alone are strongly dependent on the longitudinal specification of the tropical forcing and reflect to a certain extent the initial linear response of the model.

Interpretation of the resulting configurations in terms of globally excited modes must be done with care. Modes that propagate slowly relative to the dissipation time scale may be damped before the fluid inertia allows for a full range of interactions to take place. Although past investigations have emphasized the linear propagation of modes on a basic flow with variable refractive index, the current work suggests the importance of non-linear interactions to explain some of the observed flow variabilities in terms of barotropic processes.

Other studies have shown that such changes may be interpreted as local linear instabilities (Simmons et al., 1983). Since the tropically initiated variability concentrated in different regions (some of which appear to be locally stable for the time dependent problem forced by orography) it is necessary to assume that the non-linear model develops different propagation characteristics than the linear model or that the non-linearity is important to local transitions. It remains to be determined whether local variabilities in the present Rossby-gravity wave model have reasonable linear analogues when the linearization is performed about locally adjusted states.

ACKNOWLEDGMENTS

This material is based upon work supported jointly by the National Science Foundation, the National Oceanic and Atmospheric Administration, and the National Aeronautics and Space Administration under Grant Numbers ATM 8219198 ATM 8018158 and NAG 5-127.

REFERENCES

Blackmon, M.L., 1976: A climatological spectral study of the 500 mb geopotential height of the Northern Hemisphere. J.Atmos. Sci., 33 1607-1623.

Blackmon, J.M. Wallace, N.-C. Lau and S.L. Mullen, 1977: An obser-
    Vational study of the Northern Hemisphere wintertime circulation.
    J.Atmos. Sci., 34, 1040-1053.

Blackmon, R.A. Madden, J.M. Wallace and D.S. Gutzler, 1979: Geographical
    variations in the vertical structure of geopotential height
    fluctuations. J.Atmos. Sci., 36, 2450-2466.

Blackmon, M.L. and G.H. White, 1982: Zonal wavenumber characteristics
    of Northern Hemisphere transient eddies. J.Atmos. Sci., 39,
    1985-1998.

Blackmon, M.L., Y.-H. Lee and J.M. Wallace, 1983a: Horizontal structure
    of 500 mb height fluctuations with long, medium and short periods.
    Submitted to J. Atmos. Sci.

Blackmon, M.L., Y.-H. Lee and H.-H. Hsu, 1983b: Time evolution of 500
    mb height fluctuations with long, medium and short periods.
    Submitted to J. Atmos. Sci.

Charney, J. and J.G. De Vore, 1979: Multiple flow equilibria in the
    atmosphere and blocking. J. Atmos Sci., 36, 1205-1216.

Dikii, L.A., 1965: The terrestrial atmosphere as an oscillating system.
    Izv. Atmos. Ocean Phys., 1, 275-286.

Egger, J. and H.-D. Schilling, 1983: On the theory of the long term
    variability of the atmosphere. J. Atmos. Sci., 40, 1073-1085.

Frederiksen, J.S., 1979: The effects of long planetary waves on the
    regions of cyclogenesis: Linear theory. J. Atmos. Sci.,36,
    195-204.

Grose, W.L. and B.J. Hoskins, 1979: On the influence of orography on
    large scale atmospheric flow. J.Atmos. Sci., 36, 223-234.

Kasahara, A., 1976: Normal modes of ultralong waves in the atmosphere.
    Mon. Wea. Rev., 104, 669-690.

Kasahara, A. 1980: Effect of zonal flows on the free oscillations of a
    barotropic atmosphere. J. Atmos. Sci., 37, 917-929.

Kasahara, A., 1984: The linear response of a stratified global atmoshere
    to tropical forcing. Submitted to J. Atmos. Sci.

Lau, N.-C., 1979: The observed structure of tropospheric stationary
    waves and local balances of vorticity and heat. J. Atmos. Sci.,
    36, 996-1016.

Lau, N.-C., 1981: A diagnostic study of recurrent meteorological anomalies appearing in a 15-year simulation with a GFDL general circulation model. Mon. Wea. Rev., 109, 2287–2311.

Lau, and J.M. Wallace, 1979: On the distribution of horizontal transports by transient eddies in the Northern Hemisphere wintertime circulation. J. Atmos. Sci., 36, 1844–1861.

Lau, K.-M., and H. Lim, 1982: Thermally driven motions in an equatorial-plane: Hadley and Walker circulations during the winter monsoon. Mon. Wea. Rev., 110, 336–353

Lim, H., and C.P. Chang, 1981: A theory for midlatitude forcing of tropical motions during winter monsoons. J. Atmos. Sci., 38, 2377–2392.

Lim, H., and C.P. Chang, 1983: Dynamics of teleconnections and Walker circulations forced by equatorial heating. J. Atmos., Sci., 40, 1897–1915.

Longuet-Higgins, M.S., 1968: The eigenfunctions of Laplace's tidal equations over a sphere. Phil. Trans. Roy. Soc. London, A262, 511–607.

Madden, R.A., 1979: Observations of large-scale traveling Rossby waves. J. Geophys. Res., 17, 1935–1949.

Paegle, J.N. and J. Paegle, 1975: On the observed characteristics of quasi-geostrophic turbulence. Mon. Wea. Rev., 103, 1055–1062.

Paegle, J.N. and J. Paegle, 1976: ON the frequency spectra of atmospheric motions in the vicinity of a mountain barrier. J. Atmos. Sci., 33, 499–506.

Paegle, J., 1978: The transient mass-flow adjustment of heated atmospheric circulations. J. Atmos. Sci., 35, 1678–1688.

Paegle, J.N., 1979: The effect of topography on a Rossby wave. J. Atmos. Sci., 36, 2267–2271.

Paegle, J., J.N. Paegle and H. Yan, 1983: The role of barotropic oscillations within atmospheres of highly variable refractive index. J. Atmos. Sci., 40, 2251–2265.

Paegle, J. and W.E. Baker, 1982: Planetary-scale characteristics of the atmospheric circulation during January and February 1979. J. Atmos. Sci., 39, 2521–2538.

Paegle, J., J.N. Paegle and F.P. Lewis, 1984: Large scale motions of the tropics in observation and theory. Submitted to Pure and Applied Geophysics.

Phillips, N., 1954: Energy transformations and meridional circulations associated with simple baroclinic waves in a two-level quasi-geostrophic model. Tellus, 6, 273–286.

Reinhold, B.B. and R.T. Pierrehumbert, 1982: Dynamics of weather regimes: quasi-stationary waves and blocking. Mon. Wea. Rev., 110, 1105–1145.

Rossby, C.-G., et al., 1939: Relations between variations in the intensity of the zonal circulation of the atmosphere and the displacements of the semi-permanent centers of action. J. Mar. Res., 2, 38–55.

Silva Dias, P.L., W.H. Schubert, and M. DeMaria, 1983: Largescale response of the tropical atmosphere to transient convection. J. Atmos. Sci., 40, 2689–2707.

Simmons, A.J., 1982: The forcing of stationary wave motion by tropical diabatic heating. Quart. J. Roy. Meteor. Soc., 108, 503–534.

Simmons, A.J., J.M. Wallace and G.W. Branstator, 1983: Barotropic wave propagation and instability, and atmospheric teleconnection patterns. J. Atmos. Sci., 40, 1363–1392.

Tung, K.K. and R.S. Lindzen, 1979a: A theory of stationary long waves. Part I: A simple theory of blocking. Mon. Wea. Rev., 107, 714–734. K.K. and R.S.Lindzen

—— 1979b: A theory of stationary long waves. Part II: Resonant Rossby waves in the presence of realistic vertical shears. Mon. Wea. Rev., 107, 735–750.

Wallace, J.M. and D.S. Gutzler, 1981: Teleconnections in the geopotential height field during the Northern Hemisphere winter. Mon. Wea. Rev., 109, 785–812.

Webster, P. and J. Holton, 1982: Cross-equatorial response to middle latitude forcing in a zonally varying basic state. J. Atmos. Sci., 39, 722–733.

A NUMERICAL EXPERIMENT OF THE DYNAMIC EFFECT OF LARGE SCALE TOPOGRAPHY ON THE FORMATION OF SUBTROPICAL HIGH IN THE NORTHERN SUMMER

Luo Meixia and Zhu Baozhen

Institute of Atmospheric Physics, Chinese Academy of Sciences

I. INTRODUCTION

It is found that the zonal subtropical high belt in the Northern Hemisphere summer always breaks into some distinct anticyclonic circulation systems. These are the Qinghai-Xizang high, the Rocky high, the Iranian high, the North Africa high, the Pacific high and the Atlantic high. Three dimensional structure of these anticyclones was obtained in terms of the observed mean circulation charts by Van de Boogaard (1977). The features of vertical structures of these highs are different from each other.

Why does the zonal symmetric subtropical high belt break into a number of anticyclones in some particular regions? Why do these anticyclones possess distinct vertical structures? In this paper, we try to study the above problems by considering the effect of the topographic forcing on the basic zonal wind belt in a three-level primitive equation model with σ-coordinates.

II. NUMERICAL MODEL

The model has been developed from a tropical limited area model by Ji et al. (1982) to a spherical belt model. It is a three-level primitive equation model in the σ-coordinates. The governing equations can be written as follows:

$$\frac{\partial}{\partial t}\frac{P_* u}{m} = -m[\frac{\partial}{\partial x}\frac{P_* u}{m}u + \frac{\partial}{\partial y}\frac{P_* v}{m}u + \frac{\partial}{\partial \sigma}\frac{P_* \dot{\sigma}}{m}u]$$

$$+ \frac{P_*}{m}f(v-v_g) + \frac{P_*}{m}F_u \tag{1}$$

$$\frac{\partial}{\partial t}\frac{P_* v}{m} = -m[\frac{\partial}{\partial x}\frac{P_* u}{m}v + \frac{\partial}{\partial y}\frac{P_* v}{m}v + \frac{\partial}{\partial \sigma}\frac{P_* \dot{\sigma}}{m}v]$$

$$- \frac{P_*}{m}f(u-u_g) + \frac{P_*}{m}F_v \tag{2}$$

$$\frac{\partial}{\partial \sigma}\frac{P_* \dot{\sigma}}{m} = -\frac{1}{m}[m^2(\frac{\partial}{\partial x}\frac{P_* u}{m} + \frac{\partial}{\partial y}\frac{P_* v}{m}) + \frac{\partial P_*}{\partial t}] \tag{3}$$

$$\frac{\partial P_*}{\partial t} = -\int_0^1 m^2(\frac{\partial P_* u}{\partial x} + \frac{\partial P_* v}{\partial y})d\sigma \tag{4}$$

$$\frac{\partial}{\partial t}\frac{c_p P_* T}{m} = -m[\frac{\partial}{\partial x}\frac{P_* u}{m}(c_p T+\phi) + \frac{\partial}{\partial y}\frac{P_* v}{m}(c_p T+\phi)$$

$$+ \frac{\partial}{\partial \sigma}\frac{P_* \dot{\sigma}}{m}(c_p T+\phi] - \frac{1}{m}\frac{\partial \sigma \phi}{\partial \sigma}\frac{\partial P_*}{\partial t}$$

$$+ \frac{P_*}{m}f(uv_g - vu_g) + \frac{P_* H}{m} + \frac{P_*}{m}D_T \tag{5}$$

$$\frac{\partial \phi}{\partial \ln \sigma} = -RT \tag{6}$$

where F and D are the vertical and horizontal diffusion respectively, P_* is the surface pressure, m is map factor of Mercator projection, while

$$U_g = -\frac{m}{f}(\frac{\partial \phi}{\partial y} + RT\frac{\partial \ln P_*}{\partial y}) \tag{7}$$

$$V_g = \frac{m}{f}(\frac{\partial \phi}{\partial x} + RT\frac{\partial \ln P_*}{\partial x}) \tag{8}$$

The model used here includes only the mechanic forcing, surface friction caused by topography and the diffusion effect. The energy conservation difference scheme is employed by Zhu et al.(1982). The time difference scheme is Euler-backward. The integration domain is a spherical belt from 37.1°S to 62.2°N. The boundary condition on the southern and northern walls is fixed.

The computation of initial value and pressure gradient force is
same as Ji (1982). The initial mean zonal flow is indicated as Fig.1.
The geopotential height field is obtained by means of the balance equation on the south side of 22.5°N, and the geostrophic equation on the
north side of 22.5°N.

The model includes a real topography with smoothed mountains as
shown at Fig.2. The maximum elevation of the Qinghai-Xizang Plateau
and the Rocky mountains reaches 3800m and 1600m respectively.

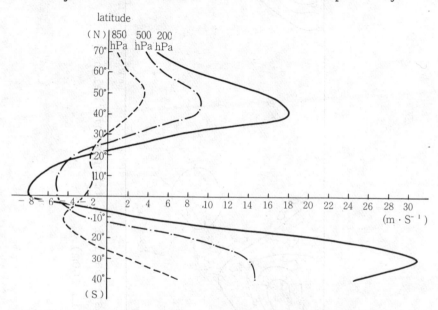

Fig.1 Latitudinal mean profiles of wind speed in summer.

III. NUMERICAL EXPERIMENT RESULTS

1. The Process of Formation of the Large-scale Anticyclonic
   Circulation

The numerical integration starts from the mean zonal flow of
summer and is performed up to 6 days. On the 1st day, the perturbation
caused by purely dynamical forcing of orography on the zonal mean
current is obvious in the lower troposphere. At 850 hPa level, the
breaking of the zonal high belt takes place around the Qinghai-Xizang

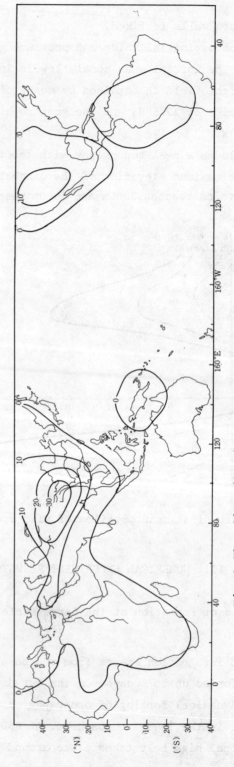

Fig.2 The distribution of topographj heights
(unit in 100m).

Plateau. The Iranian anticyclonic circulation with a center located at 27°N, 55°E develops to the south of the Iranian Plateau. However, the breaking of the zonal high belt can not be seen in the middle and upper tropospheres.

On the 2nd day, the perturbation produced by a purely dynamical forcing of topography extends up to the middle and upper tropospheres. At 200 hPa level (Fig.3a), three anticyclonic circulations appear to the south of the Iranian Plateau, Qinghai-Xizang Plateau and Rocky mountains respectively, with their centers located at 23°N, 45°E, 23°N, 106°E and 23°, 100°W respectively. These anticyclones just correspond to the observed counterparts on the climatological charts. At the same time, the Iranian high and Rocky high are fully developed at 500 hPa level (Fig.3b), with their centers located at 27°N, 40°E and 27°N, 106°W respectively, which are further south-west than climatological positions. The breaking of zonal high belt at 850 hPa (Fig.3c) is more obvious than 500 hPa.

After 5 days integration, the variation of flow pattern becomes much smaller. It seems to have reached a quasi-equilibrium state. Fig.4 indicates the simulated circulation on the 6th day. We can see that the Qinghai-Xizang high and Rocky high are quasi-stationary anticyclones and the Qinghai-Xizang high is the strongest anticyclonic system in the upper troposphere over the entire subtropical region in the Northern Hemisphere (see Fig.4a). This is in good agreement with the climatological patterns, though its centers are further southeast compared with the mean case. The center of the Rocky high is also further south compared with the observed one.

It is worthy to notice that the formation of other subtropical highs, the North Africa high, the Pacific high and the Atlantic high, had not been simulated at 850 hPa until the 5 day integration took place. They first appear in the lower troposphere and then extend upward. It is seen from Fig.4c that the domains of the Pacific high and the Atlantic high are very huge, which almost extends over the entire Pacific and Atlantic oceans. However the center of the simulated Pacific high is located further west. The Atlantic high is further north-east than climatic means.

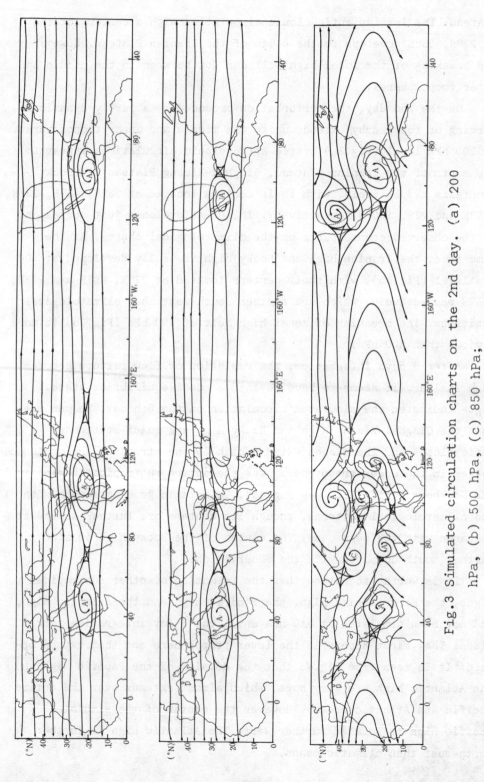

Fig.3 Simulated circulation charts on the 2nd day. (a) 200 hPa, (b) 500 hPa, (c) 850 hPa; A-Anticyclone, C-cyclone.

The Atlantic high is not formed at 500 hPa until the 5th day, while the North Africa high is not formed until the 6th day. It is seen from Fig.4 that the center of the simulated Pacific high is located at 27°N, 151°E, corresponding climatic mean center at 27°N, 152°E. The position of the simulated Atlantic high is also consistent with the observations, although the intensity of the simulated one is very weak. The center of the North African high is located at 27°N, 6°W, corresponding climaticmean center at 30°N 4°W. The numerical simulations at 500 hPa are in good agreement with the observations.

It is found from the above results that the six subtropical highs are simulated by purely dynamic effect, only the locations of the simulated high centers over the continents are further south than the observed climatic positions.

According to the difference between the formation process of these subtropical highs described as above, we can classify them into two types. One type is produced by the direct dynamical forcing of orography, such as the Qinghai-Xizang high, the Rocky high and Iranian high. We feel that these highs may be classified as direct topographic highs. The formation process of this type of subtropical high is very rapid. The Iranian high appears at low level on the 1st day and is located at Arabian Peninsular and it first moves westward, then turns eastward into the Iranian Plateau. This process is consistent with the observed facts pointed out by Zhu and Song (1984). The Iranian high is a moving system, its variations are much greater than the Qinghai-Xizang high and the Rocky high, which are quasi-stationary systems with little movements, since they are formed on the 2nd day.

2. Three-dimensional Structure of Subtropical Highs

We will discuss some features of the structures of the subtropical highs and compare the simulated structures with those on the climatic mean charts by Van de Boogaard. Fig.5 shows the vertical structures of highs along east-west direction.

The Iranian high is of a dynamic structure. The anticyclonic circulations are developed in the upper, middle and lower tropospheres. Their intensity weakens with height, the scale of anticyclonic circulation at lower and middle level is bigger than that at upper level. The

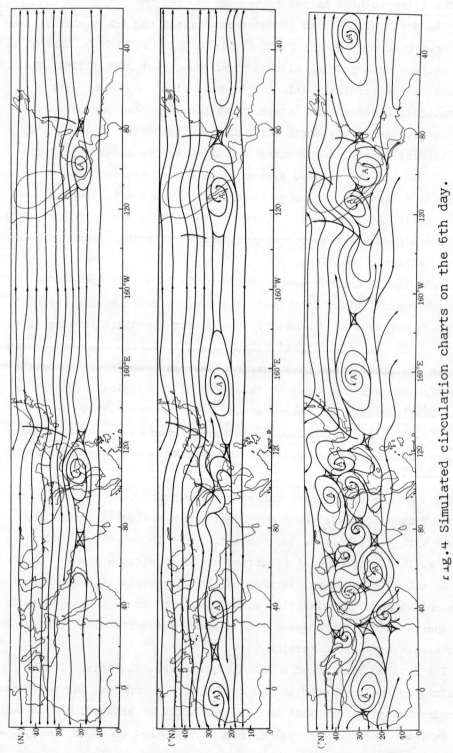

Fig.4 Simulated circulation charts on the 6th day.

(a) 200 hPa, (b) 500 hPa, (c) 850 hPa.

intensity of anticyclonic circulation weakens markedly at upper level
without an obvious center. The vertical distribution of the phases
shows a westward inclination with height (Fig.5). This is very similar
to the general climatic observations.

The Rocky high is of a thermal structure, with a obvious anticyclonic circulation in the middle and upper tropospheres, and a saddle
field between two anticyclones at 850 hPa. These are very similar to
the observed climatology. The axis of the high centers is quasi-
vertical (see Fig.5).

It is noteworthy that the Qinghai-Xizang high is also a thermal
high in nature. The vertical distribution of the phases in Fig.4 is
rather interesting with an obvious anticyclonic circulation at 200 hPa,
a low trough and cyclonic circulation at 500 hPa and 850 hPa respectively. This is very similar to the observed mean case.

The vertical structures of the North Africa high and highs over
the Atlantic and Pacific oceans, an obvious anticyclonic circulation
in the lower and middle tropospheres and a ridge at 200 hPa, are all
in agreement with the general observations.

Fig.4 shows that the vertical structures of all the six subtropical highs have a southward phase inclination with height. This is
also qualitatively consistent with climatic facts.

The above analysis reveals the major structures characteristic of
the subtropical highs. The vertical variations of the phases may be
divided into two kinds:

1) Thermal high structures. There are a number of subtropical
highs, such as Qinghai-Xizang high and the Rocky high. Their striking
features are that there is an intensified anticyclone prevailing in
the upper level and a cyclonic circulation or a transition region
between cyclonic and anticyclonic circulation in the lower troposphere.

2) Dynamical high structures. There are also a number of other
subtropical highs, such as the Iranian high, the North Africa high,
the Pacific high and the Atlantic high. Their structure features are
that an obvious anticyclonic circulation prevails in the middle and
lower tropospheres. The intensity of anticyclone weakens with heights.

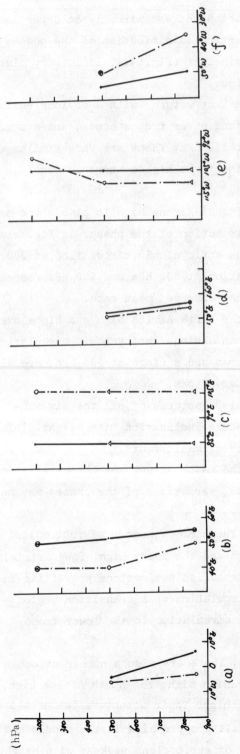

Fig.5 The vertical structure of subtropical highs. (a) denotes North Africa high, (b) denotes Iranian high, (c) denotes Qinghai-Xizang high, (d) denotes Pacific high, (e) denotes Rocky high, (f) denotes Atlantic high. Δ denotes saddle field or trough, ⊗ denotes anticycl c center which is not obvious, · denotes climatic positions. Point line shows simulated, solid line shows climatic case.

3. The Dynamic Effect of Large-scale Topography on the Formation of Thermal High

Krishnamurti et al. (1973) demonstrated that the formation of the Qinghai-Xizang high was associated with considerable heating, and the Mexican high and the North Africa high are produced by dynamical process. They formed as a consequence of the blocking influence of the thermal Qinghai-Xizang high on mean zonal flows. Ji et al. (1982) pointed out that the Qinghai-Xizang high could not be simulated by a purely dynamical forcing.

The result of this paper shows that the thermal Qinghai-Xizang high with the ultra-longwave scale may be simulated by a purely dynamical forcing of topography, though the position of high center is located further southeastward. The Qinghai-Xizang high, Rocky (Mexican) high, Iranian high can be produced by a direct dynamical forcing of orography, the first two of which are of thermal structures and the third is a dynamical one, while Africa high and the two highs over the oceans are also the dynamical high but take shape as a consequence of dynamical adjustment of atmospheric circulation caused by the above three orographic highs.

The characteristics of the structure of the Qinghai-Xizang high are broadly similar to the Rocky high, but they are different from each other. Under the Qinghai-Xizang high a low trough at 500 hPa remains, while for the Rocky high an anticyclonic circulation is seen (see Fig.4). These are in agreement with the observations. We feel that this difference may be caused partly by the difference in the terrain features of the Qinghai-Xizang Plateau and Rocky mountains.

The Qinghai-Xizang Plateau is a topography with its major axis oriented in the west-east direction and a sharp north-south slope, while Rocky mountains is a topography with the long axis in the north-south direction and an obvious west-east slope. When the air moves from the west to the Qinghai-Xizang Plateau, the airflows go around rather than go over the Plateau through the dynamical forcing of topography. Thus, the diffluence of the northern ridge and the southern trough is observed. So, a trough at 500 hPa may take shape. When the air moves from the west to the Rocky mountains, the flows go over rather than go around the mountains and the diffluence cannot take place, leading to the formation of an anticyclonic circulation at 500 hPa.

IV. CONCLUSION

The zonally uniform high pressure belt can be broken into some anticyclones in regions near the mountains by purely dynamical forcing of topography. The influence of dynamical effects of topography on basic zonal current is not confined to the lower troposphere. Its influence can reach the upper troposphere and bears a feature of planetary scale in the horizontal. It seems that not only the thermal Qinghai-Xizang high and Rocky high, but also the dynamical Iranian high may be produced by direct mechanic action of topography. Then the dynamical North-Africa high, Pacific high and Atlantic high are formed in the regions upstream and downstream of the above systems as a consequence of dynamical adjustement of atmospheric circulation in entire global belt. The six subtropical highs are simulated by purely dynamical effect. Only the locations of the centers of the simulated highs over the continents are further south than the observed, which may be attributed to the neglected heating effect in our model. Thus, we may conclude that the dynamic effect of large-scale topography of zonal asymmetry plays an important role in the formation of the thermal Qinghai-Xizang high.

REFERENCES

[1] Henry Van de Boogaard, The Mean Circulation of the Tropical and Subtropical Atmosphere—July, NCAR/TN—118+STR NCAR TECHNICAL NOTE, September 1977.

[2] Ji, L. R., Shen, R.J. and Chen, Y.X., A Numerical Experiment on the Dynamic Effect of the Qinghai-Xizang Plateau in Summer, Collected Papers of QXPMEX (2), Science Press, 1984.

[3] Zhu Baozhen, et al., Annual Report, Institute of Atmospheric

Physics, Academia Sinica, 1, 38-50, 1982.

[4] Krishnamurti, T.N., et al., Tibetan high and Upper Tropospheric
     Tropical Circulation During Northern Summer, Bull. Amer. Meteor.
     Soc., 54, 1234-1249, 1973.

[5] Zhu Baozhen and Song Zhengshan, The formation of the Qinghai-
     Xizang high and its quasi-periodic oscillations, Collected papers
     of QXPMEX (1), Science Press, 303-313, 1984.

SOME LABORATORY EXPERIMENTS RELATED TO TOPOGRAPHIC EFFECTS
ON ROTATING AND STRATIFIED FLOWS

Don L. Boyer
Department of Mechanical Engineering
University of Wyoming

I. INTRODUCTION

It is well known that large scale features of the earth's
orography can have an important influence on atmospheric motions. At
horizontal scales of the order of tens of kilometers one can cite the
marvelous "ship-wave" patterns or vortex streets noted in satellite
photographs in the lee of island mountains; e.g. Cheju, Guadalupe and
Jan Mayen. Figure 1 is a photograph of a vortex-street cloud pattern
in the lee of Guadalupe taken from the United States' National and
Aeronautics Administration's Skylab. Note the dominance of the cyclonic
vortices, a phenomenon which will be discussed with relation to the
laboratory experiments considered in this review. For larger scale
features, say the order of a thousand or more kilometers, one can also
note the important effects of mountain range complexes on major weather
patterns; e.g. the strong influence of the Alps on cyclogenesis in the
vicinity of the Gulf of Genoa.

While the increased availability of high speed computation has
allowed numerical modelers to address some of the problems related to
predicting such flows, the fact remains that our basic understanding
of the phenomena is lacking or at best incomplete. In the spirit of
obtaining an improved understanding of atmospheric (and oceanic)
phenomena, laboratory models can be used as an adjunct to field and

numerical studies. It is of course true that laboratory investigations cannot under any circumstances model all of the detailed physics (e.g. turbulence structure and details of the earth's orography) of the atmosphere. Nevertheless some of the principal forcing functions (e.g. rotation, including the beta effect, and stratification) can be simulated and much can be learned in the process.

In this review numerous laboratory facilities and experimental techniques are described which can be used to investigate the effects of rotation and stratification on flows past topography; each is considered relevant to some scale of motion in the atmosphere.

The experiments to be discussed are those in which the fluid motion is driven by mechanical means. In these studies the fluid medium is water and is either homogeneous or linearly stratified in the vertical, using salt as the stratifying agent. The characteristic dimensional parameters are the fluid speed, U; the horizontal and vertical dimensions of the topographic feature, R and h, respectively; the fluid depth, H; a characteristic horizontal dimension of the container, S_0 the background rotation, ω; the container slope, α, (to simulate the beta-effect); the kinematic viscosity, ν; and for the linearly stratified studies the density difference between the top and bottom of the container, $\Delta\rho$, mean density, ρ_0, and acceleration of gravity, g.

The experiments thus involve a large number of independent dimensionless parameters, as follows:

$R_0 = U/2\omega R$	Rossby number,
$Ek = \nu/2\omega R$	Ekman number,
$\beta = \alpha(R/H)/R_0$	beta parameter,
East or West	flow direction in beta plane experiments,
$Fr = U/(gH)^{1/2}$	Froude number,
$S = g\Delta\rho H/4\rho_0\omega^2 R^2$	stratification parameter,
$\Delta\rho/\rho_0$	stratification level,
h/R	topography aspect ratio,
R/H	topography width, fluid depth ratio, and
R/S_0	topography width, channel width ratio.

The first experiments to be considered were conducted in a small rotating water channel in which the upper and lower surfaces were inclined to simulate the beta-effect. In these studies the fluid is homogeneous and is in uniform motion past the topographic features under investigation; the topographies considered include a right circular cylinder extending from the lower to the upper bounding surfaces of the channel (axis parallel to the rotation axis), right circular disks and right circular depressions on the upper and lower bounding surfaces. Effects due to beta-plane eastward and westward flows are emphasized; the importance of the other dynamical and geometrical dimensionless parameters are also discussed. An electrolytic precipitation system is used for flow visualization and this technique is addressed in some detail.

The next set of experiments were performed in a rotating tow-tank In these studies the working fluid is either homogenous or linearly stratified and at the time of initiation of the experiments is at rest relative to a rotating observer. The false bottom of the tank consists of a flexible belt upon which the topographic feature under consideration is mounted. The belt (and obstacle) is then translated through th tank and observations of the resulting fluid motion made with respect to an observer translating (and rotating) with the obstacle. The electrolytic precipitation technique is again used for flow visualization. The effects of rotation and stratification on flow past conical and "cosine-squared" topographic features are presented.

The third system is a 2 (wide) by 10 (long) by 0.6 (deep) meter open-channel water facility mounted on a 14 meter diameter turntable; this facility is operated by the Institut de Mecanique de Grenoble, France. Turbulent flow past a right circular cylinder is presented to demonstrate the capabilities of the system. Flow visualization techniques utilized include dye tracers and a system whereby a large number of surface floats are simultaneously released into the flow stream. Using time lapse photography the latter method allows one to obtain a horizontal synoptic velocity field from which the distributio of the vertical vorticity component can be approximately determined.

II. ROTATING WATER CHANNEL-HOMOGENEOUS INCOMPRESSIBLE FLOW

The rotating water channel apparatus, given schematically in
Figure 2, is analagous to water and wind tunnel systems used for the
study of flow past obstacles in non-rotating systems. That is, the
obstacle or topographic feature under consideration is fixed in the
test section and the working fluid is forced through the system by a
pump or fan. In both the non-rotating and rotation cases it is, of
course, important to establish a well-defined uniform flow in the test
section. Details concerning the operation of the water channel, and
of the associated experimental techniques are given in Boyer and Davies
(1982).

The essential elements of the facility (see Figure 2) are the
water channel, C, whose upper and lower surfaces can be sloped to
simulate the beta-effect; inflow and outflow baffles B and B', respectively, designed to approximate the cross-stream pressure gradient in
the uniform geostrophic flow of the test section (without these baffles
a uniform flow in such a short channel cannot be realized); a pumping
system, P, to recirculate the fluid through the channel; and a flow
visualization system, FV. Other system components include a valve, V,
to vary the flow rate, a flow meter, F, to monitor the free stream
speed; and a rectangular tank, T, filled with water and surrounding the
channel to act as a constant temperature bath for the system.

The flow visualization system used was an electrolytic precipitation technique developed by Honji et al. (1980). In this method a small
potential difference (a few volts) is applied between a solder anode
and, say, a brass cathode in an electrically conducting liquid. The
liquid in the present experiments is water to which approximately one
cup (50 grams) of common salt and a few grams of sodium carbonate have
been added. A white colloidal cloud is observed to be released from the
solder electrode. Thus the anode can be used as a tracer release
position to establish a streakline in the flow field. In the present
experiments thin solder wires are used to release a dye tracer upstream
of the topographic feature and at various cross-channel locations in
the free-stream; see Figure 3 for examples. The tracer location can

also be placed on the topography itself by embedding solder into the model; see Figure 4 for an example. A horizontal slit of light from an ordinary slide projector is used to illuminate the tracer for photographic purposes.

In the Boyer and Davies (1982) study an extensive series of experiments on the flow past a full cylinder extending from the channel floor to the upper bounding surface was conducted. For details, the reader is referred to that article. Some of the principal findings, however, will be discussed below.

For f-plane flow ($\beta=0$; i.e., zero channel slope) and for sufficiently small Rossby and Ekman numbers the flow is approximately fully attached and represents potential flow past a cylinder. Figure 3(a), a photograph of the mid-channel streamlines using the electrolytic percipitation dye tracer technique, demonstrates this phenomenon. In this figure the flow is from left to right and the rotation is anticlockwise.

For beta-plane eastward motion (from west to east) the flow in the lee of the cylinder tends to jet; i.e., the characteristic speed downstream of the cylinder and in the vicinity of the streamwise centerline axis is larger than that of the free-stream. A damped Rossby-wave pattern is also set up in the lee of the cylinder. These observations are depicted in Figure 3(b), here the flow is from left (west) to right (east) with the rotation anticlockwise. The channel is sloped with the thin portion at the top (north).

For beta-plane westward motion (from east to west) the flow in the lee does not have a tendency to concentrate into a jet; see Figure 3(c) where the flow is from right (east) to left (west) with the rotation vertically upward. The channel slope is the same as that in 3(b). The tendency for the flow to separate from the cylinder is greatest for the beta-plane westward case, least for the beta-plane eastward case and intermediate for the f-plane case. This tendency for greater separation for beta-plane westward flows is clearly apparent in Figure 3, depicting experiments in which all parameters with the exception of β and the flow direction are approximately the same. Details of the separation phenomena as a function of system parameters are given in Boyer and Davies (1982).

The wake structure for all flow cases (i.e., f-plane and beta-plane eastward or westward) is not symmetric about the streamwise axis. Figure 4 depicts a sequence of photographs for a parameter set in which eddy-shedding occurs; the flow is from right to left and the rotation is anticlockwise. These photographs seem to indicate that the shed cyclones (eddies with the same sense as the background rotation) are stronger than the anticyclones (opposite sense of the background rotation). This matter will be discussed further below but note the similarity with Figure 1 in that the vortex street shed from Guadalupe is dominated by cyclonic eddies.

The second set of experiments concerns the flow past circular disks and depressions located symmetrically on the upper and lower surfaces of the channel; see Figure 5. The detailed results of this study are given in Boyer, Davies and Holland (1984). Some of the principal findings are discussed below.

Unlike the solid cylinder in which the flow is forced around the obstacle, the streaklines in the mid-plane of the channel can now pass over the topographic feature. From arguments of conservation of potential vorticity (see Pedlosky (1979)), the streaklines above the topography should develop anticyclonic vorticity for the disk and cyclonic vorticity for the depression. Such tendencies are demonstrated, respectively, in Figures 6 and 7. These figures are similar to Figure 3 in that the background rotation in each case is anticlockwise while the flow in the first figure (a) is for an f-plane, the second (b) is for beta-plane eastward, and the third (c) is for beta-plane westward; in each case north is upward.

Once again note in Figures 6(b) and 7(b) the tendency for jetting in the wake for beta-plane eastward flows. The wakes in these cases also indicate the presence of a damped Rossby wave; because of the fact that fluid passes over the topography the wake is not as symmetric with respect to the streamwise axis as was the case for the solid cylinder in Figure 3(b).

The principal features of these laboratory flows can be predicted quite accurately by numerically integrating the quasi-geostrophic potential vorticity equation; see Boyer Davies, and Holland (1984).

III. ROTATING TOW TANK

Investigations at relatively large Reynolds numbers (say Re>2500) and studies of topographic effects on rotating and stratified flows cannot be accomplished with a water channel of the size and character of are discussed in Section 2. A rotating tow-tank was thus developed in which topographic features could be translated through a fluid at rest relative to a rotating observer; see the schematic diagram in Figure 8.

The entire bottom of the tow-tank is covered by a flexible belt which can be translated in either direction at a speed, U. The topography to be investigated is mounted on the belt. The drive system for the belt also translates a towing carriage above the tank and at the same speed as the belt. Cameras and other experimental accessories are mounted on the carriage to facilitate flow visualization.

Boyer and Kmetz (1982) used this apparatus to investigate the character of vortex shedding from right circular cylinders in rotating systems; this study is not addressed further in the present review. The apparatus is presently being used to investigate the motion of a linearly stratified rotating fluid over topographic features. While this work is not as yet complete, some preliminary results are given here.

One series of studies concerns the flow past a conical topography of rather steep slope. The physical system is sketched in Figure 8. Figure 9 depicts photographs of the streakline fields obtained by moving a "rake" of vertically oriented solder wires ahead of the cone. The dye tracer system used in these photographs is again the electrolytic precipitation technique. Figures 9(a), 9(b) and 9(c) are photographs of the streakline fields obtained at the non-dimensional elevations $z/H = 0.86$, 0.62 and 0.37 respectively. The background rotation for all figures is anticlockwise. The base of the cone is indicated by the white circle on the right of each photograph.

The photographs are taken relative to an observer translating with the topographic feature. Such an observer thus notes the flow streaming from right to left. The shadows on the lower portion of 9(b) and 9(c) in the vicinity of the topography are caused by the blocking of the horizontal light beam originating at the top of each frame by the cone. While these results are preliminary one can onte the dramatic effect of stratification in damping disturbances forced by the obstacle. For example, Figure 9(a) is taken at approximately the elevation of the vertex of the cone; one can note that in spite of the large unsteady flow patterns at the lower levels, these disturbances are very weak in the vicinity of the top of the cone. This is to be contrasted with low Rossby number homogeneous flows in which disturbances tend to propagate throughout the entire fluid depth. Finally the photographs in 9(b) and 9(c) suggest that rotation effects again tend to develop stronger cyclones than anticyclones.

The tow tank has also been used to investigate linearly stratified flow over shallow topographies. Figure 10 depicts the lee wave structure for flow over a "cosine-squared" topographic feature obtained at various lateral locations, y, with y=0 representing the center of the topography. In each of the photographs the dye tracer is released at the non-dimensional vertical locations $z/H = 0.86$, 0.62, and 0.37 respectively. A vertical slit of light approximately 6 mm in width is directed onto the dye tracer along surfaces parallel to the x-z plane and at the y-locations noted. The diameter of the base of the topography is 10.0 cm with the center located at x=y=0. Thus each of the runs depicted are for y-locations for which the various streaklines pass over the topography.

Figure 11 depicts sketches of the various streaklines given in the photographs of Figure 10. These experiments thus nicely depict one type of lee wave pattern that might be expected in the lee of large-scale topographic features for which the Earth's rotation may be important. Note the asymmetry in the sense that the streaklines passing over the topography on the right side facing downstream have a larger amplitude than those on the left (e.g. compare Figures 10(b), 11(b) with 10(d), 11(d) respectively).

Figure 12 depicts photographs of the horizontal streamline patterns of the flow over the cosine-squared feature. Note the small deflections at all levels but with an almost imperceptible disturbance at the upper level. Such small deflection effects are again symptomatic of stratified flows since the steering by the feature would be substantially stronger for homogeneous flows.

IV. ROTATING OPEN-CHANNEL FLOW FACILITY

The writer is presently engaged in a cooperative research program with the Institut de Mecanique de Grenoble (IMG) to investigate large Reynolds number (i.e., 5000<Re<50,000) order unity Rossby number rotating open channel flow past topography. The flow facility is 2 meters wide, 10 meters long and 60 cm deep and is mounted on a 14 meter diameter turntable capable of rotation rates of up to 2 revolutions per minute. The channel and turntable are shown in the photograph of Figure 13 (the experiment on the floor of the turntable and in the foreground is a tidal model of the English channel).

A series of experiments on the flow past a right circular cylinder have been conducted by Boyer et al. (1984). The principal results of this study are discussed in the following. The experiments indicate that even at large Reynolds numbers and relatively large Rossby numbers, the asymmetry in the eddy shedding noted above is still present; i.e., the cyclones are stronger than the anticyclones. A qualitative example using dye tracer techniques (potassium permaganete solution) is given in Figure 14.

A tracer system using a large number of surface floats was developed to obtain a quantitative measure of this phenomenon. The floats were cylindrical in shape, 7 cm long and 7 mm in diameter at the top. The floats were painted white on the top and were weighted so as to float stably with about 0.5 cm of the float being above the water surface. The floats were released on a 10 cm grid. Using flash and streak photography the floats were then used to determine a synoptic velocity field in the wake of the cylinder. Figure 15 is an example of a photograph of the velocity field so obtained and depicts both an anticyclone

676

(center of photograph) and a cyclone (right of pnotograph), respectively. By subtracting the approximate advective speed of the eddies
from this vector field one can obtain the velocity distribution of the
eddies as seen by an observer translating with the vortices. Figure 16
depicts the results of this analysis for the cyclone in Figure 15.

By plotting the azimuthal velocity field, v_θ, as a function of
distance from the vortex center, r, and fitting this data to a vortex
model in which $v_\theta = Ar/(1+br^2)$ where A and B are constants to be determined from experiments, one can obtain an estimate of the maximum
vorticity in the vortex core; i.e., $\zeta_{max} = 2A$. Figure 17 is a plot of
the maximum relative vorticity as a function of the streamwise distance
from the cylinder center for various cyclones and anticyclones. The
data clearly show that for all observations at specified sownstream
coordinate locations, the cyclones are stronger. Furthermore these
eddies decay in the streamwise direction and as discussed in Boyer et
al. (1984), this decay is the Ekman suction (using a constant eddy
viscosity model).

ACKNOWLEDGEMENT

This work was sponsored by the United States National Science
Foundation under grants ATM 8018173, ATM 8218488 and INT 8116114.

REFERENCES

Boyer, D.L. and Davies, P.A., Flow past a circular cylinder on a
 beta-plane. Phil. Trans. R. Soc. Lond. A306, 533-556 (1982).

Boyer, D.L. and Kmetz, M.L., Vortex shedding in rotating flows. Geoph.
 and Astro. Fluid Dyn. 26, 51-83 (1983).

Boyer, D.L., Kmetz, M., Smathers, L., Chabert d'Hieres, G. and Didelle,
 H., Rotating open channel flow past right circular cylinders.
 Submitted to Geoph. and Astroph. Fluid Dyn.

Honji, H., Taneda, S. and Tatsuno, A., Some practical details of the
 electrolytic percipitation method of flow visualization. Res.
 Inst. for Appl. Mech., Kyushu Univ., Japan, 28, 83-89 (1980).

Pedlosky, J., Geophysical Fluid Dynamics. Berlin: Springer-Verlag
 (1979).

Pitts, D.E., Lee, J.T., Fein, J., Saski, Y., Wagner K. and Johnson, R.,
 Mesoscale cloud features observed from Skylab. Skylab Explores the
 Earth, NASASP-380, 479-501 (1977).

Figure 1. Vortex-street cloud pattern in the lee of Guadalupe Island photographed by Skylab (from Pitts et al. 1977). Note the dominance of the cyclonic vortices.

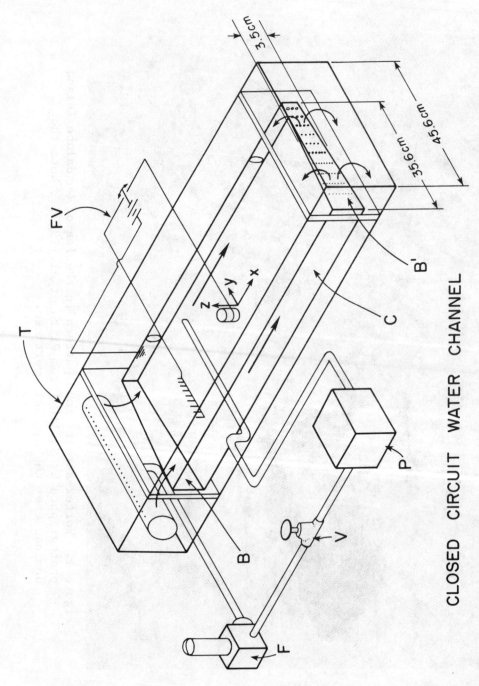

3.5 cm

45.6 cm

35.6 cm

FV

T

x
y
z

B'

C

B

V

F

P

CLOSED CIRCUIT WATER CHANNEL

Figure 2. Schematic representation of the closed circuit water channel. (See text for explanation of lettering.)

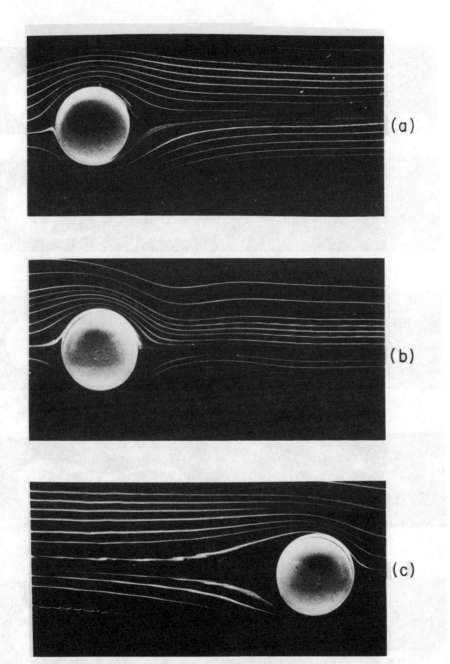

Figure 3. Mid-channel streaklines of flow past a full circular
    cylinder: Ro = 4.9 x 10^{-2}, R/H = 0.73; (a) f-plane, Ek = 6.7 x
    10^{-4}, Re = 73.0, β = 0.0; (b) beta-plane eastward, Ek = 7.2 x 10^{-4},
    Re = 68.3, β = 0.8; (c) beta-plane westward; Ek = 7.2 x 10^{-4},
    Re = 68.3, β = 0.8.

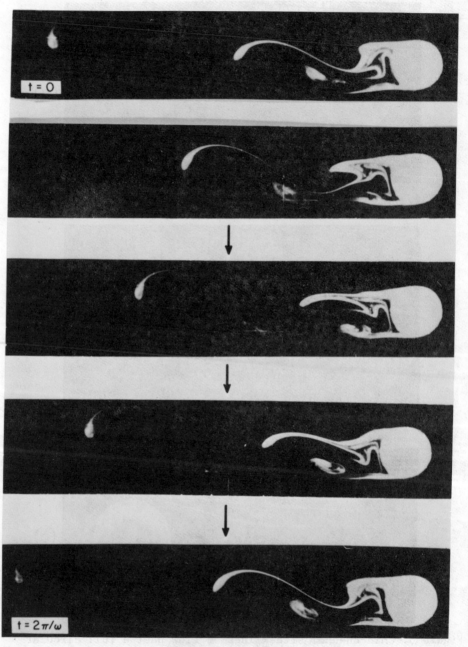

Figure 4. Time sequence showing vortex shedding for unsteady beta-plane westward flow past a cylinder; Ro = 0.82, Ek = 5.2 x 10^{-3} (Re = 158), β = 0.02, R/H = 0.36. The interval between successive photographs is one quarter of the background rotation period.

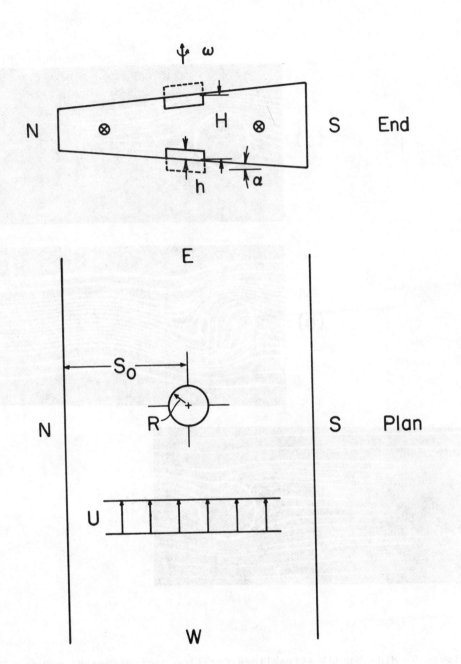

Figure 5. Physical system for flow past disks (———) and cylindrical
   depressions (----).

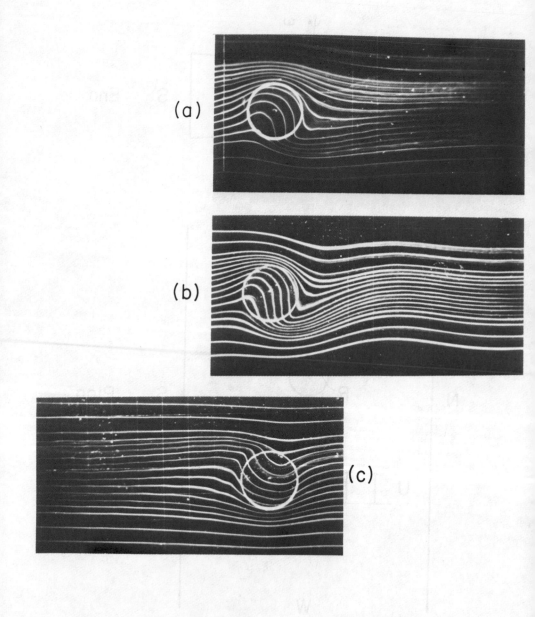

Figure 6. Mid-channel streaklines for flow past disks; $Ro = 9.8 \times 10^{-2}$, $Ek = 15.4 \times 10^{-4}$ ($Re = 64$), $2h/H = 0.25$, $R/H = 0.73$, $R/S_o = 0.14$; (a) f-plane (i.e. $\beta = 0$); (b) beta-plane eastward ($\beta = 0.40$); (c) beta-plane westward ($\beta = 0.40$).

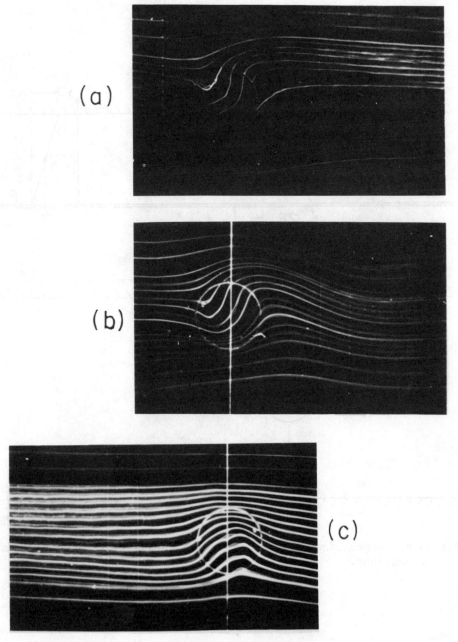

Figure 7. Mid-channel streaklines for flow past cylindrical depression
$Ro = 9.8 \times 10^{2}$, $Ek = 13.8 \times 10^{-4}$ $(Re = 71)$, $2h/H = 0.75$, $R/H = 0.73$, $R/S_0 = 0.14$; (a) f-plane ($\beta = 0$); (b) beta-plane eastward
($\beta = 0.40$); (c) beta-plane westward ($\beta = 0.40$).

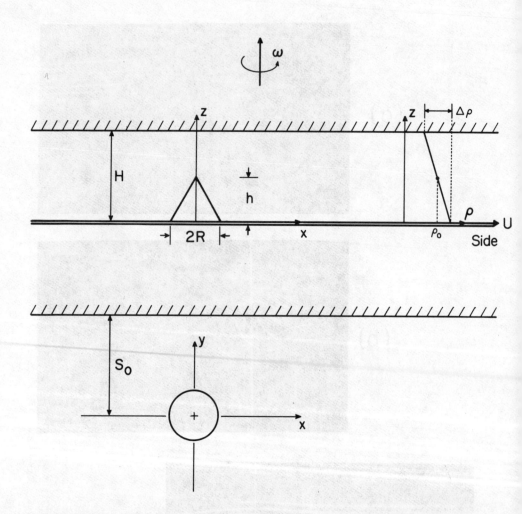

Figure 8.   Schematic diagram of tow-tank facility; flow past a conical
   topography.

686

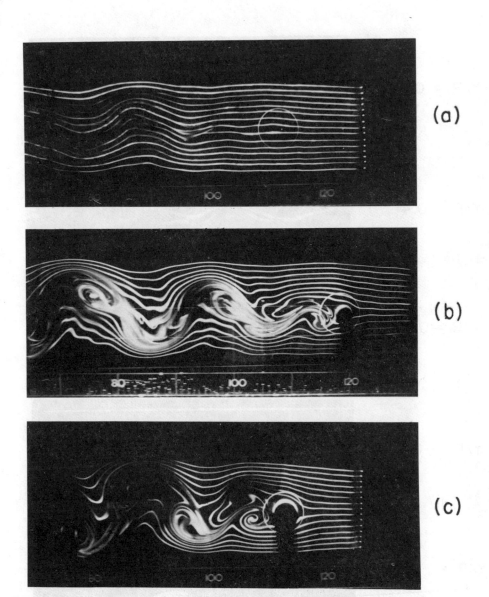

Figure 9. Streakline patterns in horizontal planes at various
elevations in linearly stratified rotating flow past a conical
obstacle; H = 8.1 cm, h = 6.6 cm, R = 3.8 cm, S_o = 15.6 cm, U =
0.50 cm/sec, ω = 0.25 sec^{-1}, Ro = 0.26, Ek = 1.4 x 10^{-3} (Re = 188),
$\Delta\rho/\rho_o$ = 5.6 x 10^{-3}, S = 12.3, h/H = 0.81, h/R = 1.74, R/S_o = 0.24;
(a) z = 7.0 cm (z/H = 0.86); (b) z = 5.0 cm; (z/H = 0.62); (c) z =
3.0 cm (z/H = 0.37).

FLO →

Y/R = 0.8   (a)

Y/R = 0.4   (b)

Y/R = 0.0   (c)

Y/R = -0.4   (d)

Y/R = -0.8   (e)

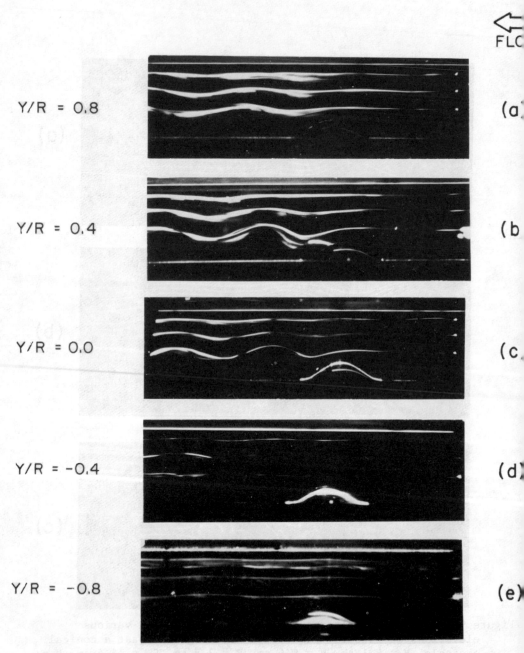

Figure 10. Lee wave pattern for cosine-squared topography of
    revolution: $H = 8.1$ cm, $h = 2.06$ cm, $R = 5.0$ cm, $S_o = 15.6$ cm, $U =$
    1.5 cm/sec, $\omega = 0.25$ sec^{-1}, $Ro = 0.60$, $Ek = 8.7 \times 10^{-4}$ (Re = 688),
    $\Delta\rho/\rho_o = 5.3 \times 10^{-3}$, $S = 6.7$, $h/H = 0.25$, $h/R = 0.41$, $R/S_o = 0.32$;
    (a) $y = 4.0$ cm $(y/R = 0.8)$, (b) $y = 2.0$ cm $(y/R = 0.4)$, (c) $y =$
    0.0 cm $(y/R = 0.0)$, (d) $y = -2.0$ cm $(y/R = -0.4)$, (e) $y = -4.0$ cm
    $(y/R = -0.8)$ (See Figure 8 for sign of y).

688

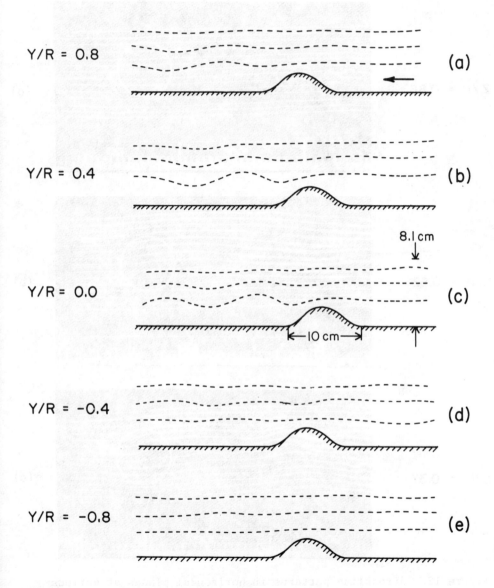

Figure 11. Sketches of lee wave patterns from the photographs of Figure 10; same parameters as Figure 10.

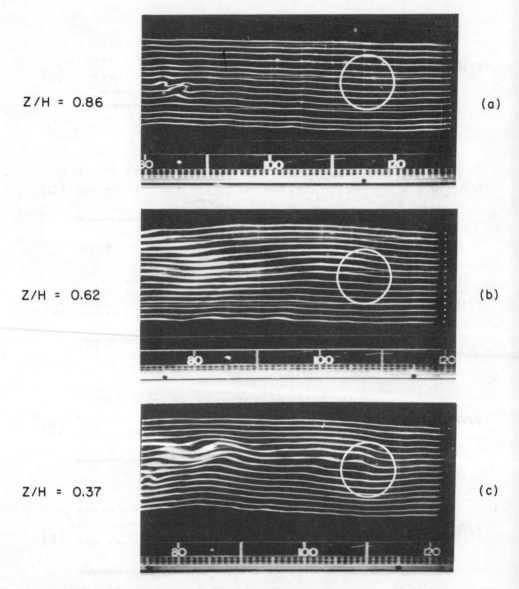

Z/H = 0.86   (a)

Z/H = 0.62   (b)

Z/H = 0.37   (c)

Figure 12. Streakline patterns in horizontal planes at various
    elevations in linearly stratified rotating flow past a shallow
    cosine-squared topography; H = 8.1 cm, h = 2.06 cm, R = 5.0 cm,
    S = 15.6 cm, U = 0.5 cm/sec, ω = 0.25 sec^{-1}, Ro = 0.20, Ek =
    8.7 x 10^{-4} (Re = 230), $\Delta\rho/\rho$ = 3.4 x 10^{-3}, S = 4.3, h/H = 0.25,
    h/R = 0.41, R/S = 0.32; (a) z = 7.0 cm (z/H = 0.86), (b) z = 5.0
    cm, (z/H = 0.62), (c) z = 3.0 cm (z/H = 0.37).

690

Figure 13. IMG rotating platform and open channel flow facility. (Note another experiment, the English Channel tidal model is in the foreground.)

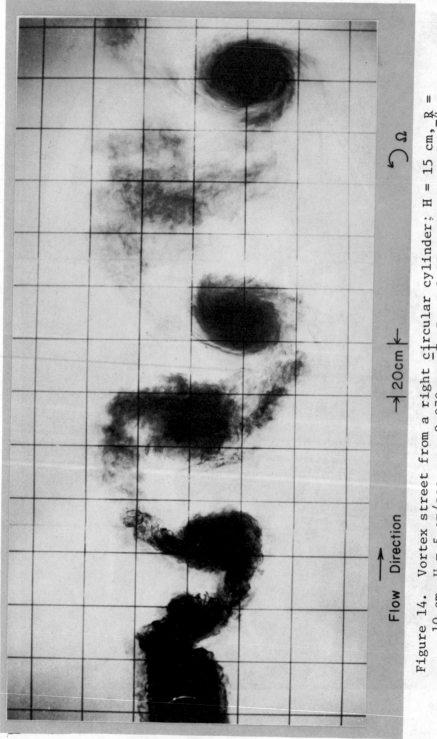

Figure 14. Vortex street from a right circular cylinder; H = 15 cm, R = 10 cm, U = 5 cm/sec, ω = 0.070 sec, Ro = 3.57, Ek = 7.14 x 10⁻⁴ (Re = 5,000), R/H = 0.67, and R/S₀ = 0.10.

Flow Direction

→ 20cm ←

Ω

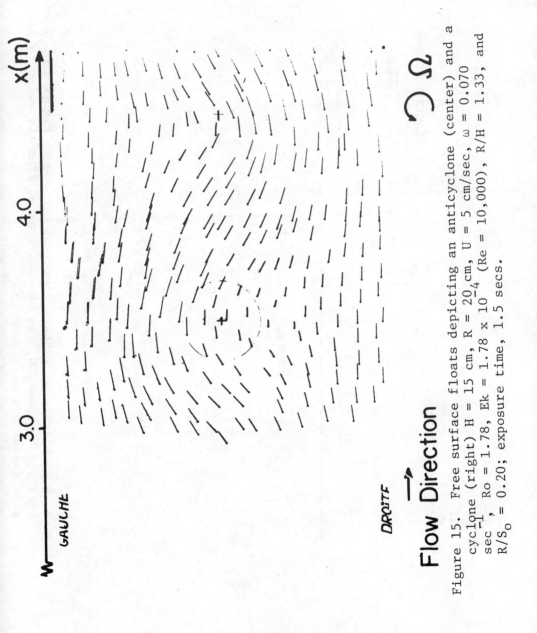

Figure 15. Free surface floats depicting an anticyclone (center) and a cyclone (right) H = 15 cm, R = 20 cm, U = 5 cm/sec, ω = 0.070 sec^{-1}, Ro = 1.78, Ek = 1.78 x 10^{-4} (Re = 10,000), R/H = 1.33, and R/S$_o$ = 0.20; exposure time, 1.5 secs.

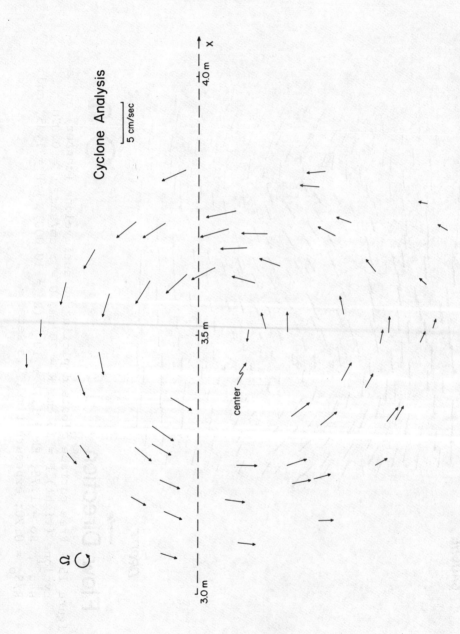

Figure 16. Analysis of surface floats of cyclone in Figure 15. The free stream speed, U, has been subtracted from the velocity field; same parameters as Figure 15.

694

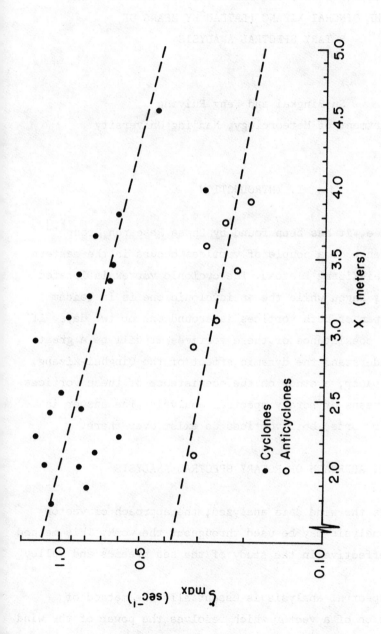

Figure 17. Maximum relative vorticity in vortex core, $|\zeta_{max}| = 2A$, as a function of the streamwise coordinate, x, for cyclones ($\bullet$) and anticyclones (o); Ro = 1.78, Ek = 1.78 x 10^{-4} (Re = 10,000), R/H = 1.33 and R/S$_o$ = 0.20.

AN INVESTIGATION INTO THE EXISTENCE OF VORTICES OVER THE QINGHAI-XIZANG PLATEAU BY MEANS OF ROTARY SPECTRAL ANALYSIS

Su Bingkai and Feng Ruiying

Department of Meteorology, Nanjing University

I. INTRODUCTION

Not long since, it has been found by Lhasa Research Group[1] that the co-existence of a couple of vortices occurs in the eastern part of the Qinghai-Xizang Plateau. The cyclonic vortex is located near the region of Nagqu while the anticyclonic one is in Qaidam Basin. The life span of both vortices is around one or two days. If there is indeed a coexistence of these vortices, it will be a great help for us to understand the dynamic effect of the Qinghai-Xizang Plateau. In this paper, a study on the coexistence of these vortices has been made by means of rotary spectral analysis. The answer is positive, in other words, both vortices do exist over there.

II. APPROACH OF ROTARY SPECTRAL ANALYSIS

Because it is the wind data analyzed, the approach of vector rotary spectral analysis may be used throughout the work. This method has been proven effective in the study of the sea breezes and valley wind etc [2-4].

The rotary spectral analysis is essentially the method of frequency resolution of a vector which resolves the power of the wind velocity at all frequence bands into the clockwise (negative frequency) and anti-clockwise (positive frequency) parts. If $\sigma > 0$, $\omega = -\sigma$ in the case of clockwise rotation; otherwise $\omega = \sigma$. Where σ is the angular frequency and ω is the angular velocity.

The vector of wind can be decomposited into u and v in x and y direction,respectively, i.e.

$$\vec{V}(t) = u(t) + iv(t) \tag{1}$$

where $i=\sqrt{-1}$. The Fourier transformation of $\hat{V}(t)$ is

$$\vec{V}_\omega = \frac{1}{T} \int_0^T \vec{V}(t)e^{-i\omega t} \, dt = |\vec{V}_\omega|e^{i\theta\omega} \tag{2}$$

where T is the period, $|\vec{V}_\omega|$ and θ_ω are the amplitude and phase corresponding to the frequency ω respectively. It is different from the scalar analysis in the ω, in this case it can be expressed as

$$|\omega| = |\frac{\partial n\pi}{T}| = \sigma, \quad n=0, \pm1, \pm2,\cdots \tag{3}$$

It is well known that the Fourier transformations of u and v for an angular frequency σ are

$$u_\sigma = a_{u\sigma}\cos\sigma t + b_{u\sigma}\sin\sigma t$$
$$v_\sigma = a_{v\sigma}\cos\sigma t + b_{v\sigma}\sin\sigma t \tag{4}$$

Thus, the Fourier transformation of $\vec{V}(t)$ can be written as

$$u_\sigma + iV_\sigma = \vec{V}_+ e^{i\sigma t} + \vec{V}_- e^{-i\sigma t} \tag{5}$$

where $\vec{V}_+$ and $\vec{V}_-$ are the components of wind vector with anticlockwise and clockwise rotation in the frequency space respectively. Therefore, $\vec{V}_+$ and $\vec{V}_-$ can be written as

$$\vec{V}_+ = \frac{1}{2}[(a_{u\sigma} + b_{v\sigma}) + i(a_{v\sigma} - b_{u\sigma})] = |\vec{V}_+|e^{i\theta}+$$
$$\vec{V}_- = \frac{1}{2}[(a_{u\sigma} + b_{v\sigma}) + i(a_{v\sigma} + b_{u\sigma})] = |\vec{V}_-|e^{i\theta}- \tag{6}$$

where $|\vec{V}|$ and θ are the amplitude and phase of $\vec{V}$ respectively. The subscript "+" denotes anti-cyclonic rotation and "-" cyclonic. Eq.(5) describes an ellipse on the plane of hodograph.

The mean kinetic energy S_t of wind vector can also be resolved into S_- and S_+, i.e.

$$S_t = S_- + S_+$$
$$S_- = <\vec{V}_-^*\vec{V}_->/2, \quad S_+ = <\vec{V}_+^*\vec{V}_+>/2 \tag{7}$$

where the symbol < > represents the ensemble average, the superscript "*" denotes the conjugate of V. The coefficient of rotation $C_{R\sigma}$ which is an invariant is as follows:

$$C_{R\sigma} = (S_- - S_+)/S_t \qquad (8)$$

It denotes the distribution of mean kinetic energy. The cases for $C_{R\sigma}=1$, -1, and 0 correspond to the motions in the clockwise, anticlockwise and non-rotary sense respectively.

III. RESULTS

Lhasa and Golmud are two stations selected on southern and northern slopes in the eastern part of the Qinghai-Xizang Plateau respectively. The wind data on 500hPa at 00z and 12z from Feb. 18 to Oct. 31, 1979 are used. Thus time intervals $\Delta T=12h$, total samples are 512. The missing data occurred during this period is interpolated by means of spline analysis[5].

The window of cosine slope is used for wind data before FFT. Its expression is as follows:

$$W_t = \begin{cases} \frac{1}{2}(1 - \cos\frac{\pi}{64}t) & t = \overline{1,64} \\ 1 & t = \overline{65,448} \\ \frac{1}{2}[1 - \cos\frac{\pi}{64}(512 - t)] & t = \overline{449,512} \end{cases} \qquad (9)$$

The spectral estimation calculated should be divided by $C=\frac{1}{512}\sum\limits_{t=1}^{512} W_t$ Then the spectral density of the clockwise and anti-clockwise rotation energy and the rotary coefficient of 500hPa wind vector for both stations are evaluated from Eq. (7) and Eq. (8).

The momentum equation can be written as

$$\frac{\partial \vec{V}}{\partial t} + if\vec{V} = \vec{\Gamma} \qquad (10)$$

where f is Coriolis parameter, Γ is the forcing function which includes the effects of pressure gradient force and friction. Except for the case of $\omega = -f$, we have

698

$$i(f + \omega)\vec{V}_\omega = \vec{\Gamma} \tag{11}$$

obviously, the forcing function Γ can be resolved into components of clockwise and anticlockwise rotation,

$$\vec{\Gamma} = \vec{\Gamma}_- + \vec{\Gamma}_+ \tag{12}$$

If the strength of clockwise rotation is equal to that of anticlockwise, i.e.

$$|\vec{\Gamma}_-| = |\vec{\Gamma}_+| \tag{13}$$

then the disturbance has no contribution to the rotation and the rotation is entirely due to the Coriolis force. Thus the rotary coefficient of Eq. (8) can be rewritten as

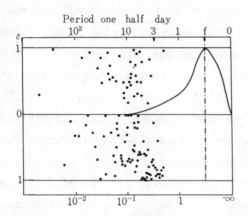

Period one half day

Fig.1 The Rotary Coefficients of Lhasa.

$$C_{R\sigma} = \frac{|\vec{V}_-|^2 - |\vec{V}_+|^2}{|\vec{V}_-|^2 + |\vec{V}_+|^2} = \frac{2\sigma f}{\sigma^2 + f^2} \tag{14}$$

and it is termed as the theoretical rotary coefficient.

The rotation coefficients at 500hPa of Lhasa and Golmud are given in Figs 1 and 2. The solid curves in both illustrations are theoretical rotary coefficient, the dots are the empirical rotary coefficients derived by the observed wind. The difference between these coefficients in both stations is rather significant. Theoretical rotary coefficient is approximately equal to zero for the higher frequency fluctuation with the period shorter than 36 hours. It implies

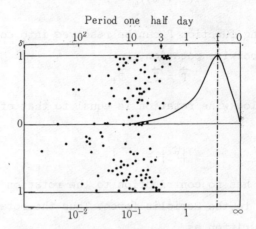

Period one half day

Fig.2 The Rotary Coefficients of Golmud.

that the motion is straight and undirectional. But the empirical value
for Lhasa in near -1, that means the motion is anticlockwise. It is
very interesting to note that the empirical value for Golmud is +1 as
expected. It corresponds to a clockwise motion. This is quite in agreement with the result based on synoptic analysis[1]. The result of
this paper provides another evidence to justify the existence of the
"twins" over the eastern part of the Qinghai-Xizang Plateau.

We believe that the effects of plateau will violate the assumption
of Eq. (13) that means the equality of strength of rotation in clockwise and anti-clockwise. In other words, the existence of the Qinghai-
Xizang plateau will strengthen the anti-clockwise component of rotation at Lhasa, thus there is a cyclonic vortex over there. On the contrary, the effect of the Qinghai-Xizang Plateau will strenghthen the
clockwise component of rotation over Golmud, thus anti-cyclonic vortex
appears.

REFERENCES

[1] Lhasa Research Group, a study on the low and shear at 500hPa over
    the Qinghai-Xizang plateau in summer, Science Press, 66-67, 1981
    (in Chinese).
[2] O'Brien, J. J. and Pillsbury, R. D., Rotary wind spectra in a sea

breeze region. J. Appl. Met. 13, 820-825, 1974.

[3] Hanafusa Tatsuo, The approach of spectrum analysis and spectrum of meteorological elements. Notes on meteorological research, 131, 16-17. 1977 (in Japanese).

[4] Joseph Gonella, A rotary-component method for analysing meteorological and oceanographic vector time series, Deep Sea Research, 19, 833-846, 1972.

[5] Academia Sinica,Shenyang Computing Technique Research Institute et al., A routine algorithm on electronic computer, Science Press, 35-40, 1976 (in Chinese).

THE DYNAMIC EFFECT OF THE QINGHAI-XIZANG PLATEAU
ON THE FORMATION OF LOW LEVEL JET IN EAST ASIA

Sun Shuqing and Ji Liren

Institute of Atmospheric Physics, Chinese Academy of Sciences

Lorenzo Dell'Osso

European Centre for Medium Range Weather

Forecast, Reading, England

Low-level jet, due to its close relationship with severe weather in summer, attracts broadly the interest and attention of meteorologists. In recent years a large number of researches have been done to the structure and characteristics of jets, the mechanism responsible for their formation and their roles in the formation of severe convective weather. Large-scale low-level jets, whether in East Asia or in the great plain of North America or East Africa, are located east of mountains or high lands and facing the oceans. This fact suggests that this particular geographic location might contribute to the formation of jets. An explanation has been given by Wexler[1] to the southly low-level jets over the mid-west of America which is located in the east of the Rockies. It is therefore suggested that there is a theory, which might be considered as an extension or modified version of the theory for oceanic current at the west boundary of the ocean, that the southly jets to the east of the Rockies are but the result from the interaction of the easterly from Bermuda high pressure with the N-S oriented mountain barrier. The same theory has been used to explain the low-level jet over East Africa. The works of Hahn and Manabe[2] and Washington et al.[3] show that when the mountains are removed from the model the cross-equatorial flows diverge and become

rather weak. When the highland is included, a jet with a scale similar to reality appears.

The monthly mean wind speed of large scale low-level jet in the eastern Asia may reach more than 12m/s in mid-summer. It also lies to the east of mountains, the Qinghai-Xizang Plateau, and the axis of the jet clearly experiences a seasonal change in position. What kinds of influence of the huge Plateau are there on the formation of low-level jet and how are these influences exerted? This problem has not been fully investigated yet. The present paper attempts to study the impacts of various version of orography on the jet by using the ECMWF limited area model for a real case of low-level jet.

II. THE MODEL

The ECMWF limited area model[4] is used for the experiments. The model uses the primitive equations on a regular latitude, longitude grid of 1.875×1.875 degrees. The vertical coordinate is well known as sigma coordinate. The vertical resolution is of 15 temperature and wind carrying levels (as shown in Fig.1). There are more levels (about 7 levels under 500mb) in the lower part of the troposphere suitable

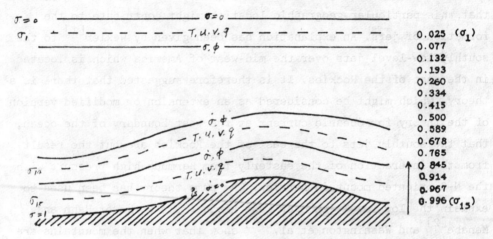

Fig.1 Vertical coordinate.

for describing systems like low level jets. The parameterization of physical effects includes a full hydrological cycle, involving the large-scale condensation and convective parameterization, and radiation turbulent transfer in the boundary layer and free atmosphere. The integration domain is bounded by 10°S-70°N and 60°E-160°E. The boundary values for every time step are provided by a linear interpolation of the analysis taken at 12 hour interval. The interior variables are smoothly adjusted towards their prescribed boundary values using relaxation technique suggested by Davies[5].

III. THE CASE AND FORECAST

The case selected was a strong low level jet in East China during 12-13 July, 1979. The axis with a heavy rain area to its left

side extended about 2000km along the east of the Qinghai-Xizang
Plateau. The strength of the jet reached its maximum around 00GMT
12th. Fig.2 gives the wind field charts of this case. The low level
jet with the maximum wind speed of 19m/s is located near 30°N, 110°E.
Looking at 700mb, it is very clear that the jet is closely related to
three systems: the westerlies trough in mid-latitude, subtropical
high over the west Pacific and the monsoon westerlies.

Comparing the three branches of southly currents coming from those
three systems, we find that the monsoon at lower layer of troposphere
is much stronger than others. The area with wind speed more than 10m/s
stretched to the Indo-China Peninsula. Making a cross section across
the monsoon current (Fig.3), we find that the centre with velocity
more than 10m/s was located under the level of 700mb, and the west
component of wind speed at 500mb in that area was nearly zero. On the
contrary, the currents of westerlies and subtropical high were
stronger at 500mb. The southly wind speed of subtropical high to the
west of 125°E at 850mb was less than 5m/s. So the low level jet is
much related to the SW monsoon so far as the lower layer is concerned.
This is different from the previous viewpoint which emphasized the
effect of the trough in mid-latitude and the subtropical high[6].

A 48hr forecast (J03) with initial field of 00GMT 10th July is
made as a control run. Fig.4(a,b) are the 48hr wind field charts of
850mb and 700mb for J03. The solid dashed lines are the observed
trough lines. It is obvious that the position of troughs in westerlies
are very close to the observed ones. Compared to Fig.2, the predicted
patterns are very similar to the observation. The position of trough
and ridge interested are in good agreement. The monsoon westerlies is
also well simulated except slightly weaker than that observed. Looking
into the detail of low-level jet we find that its position and
strength all agree well with those observed. The maximum velocity
centre is located at 30°–35°N, a little further north to the observed.

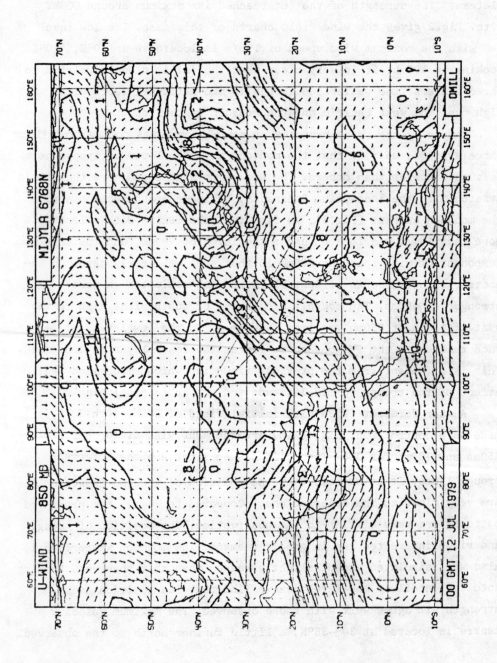

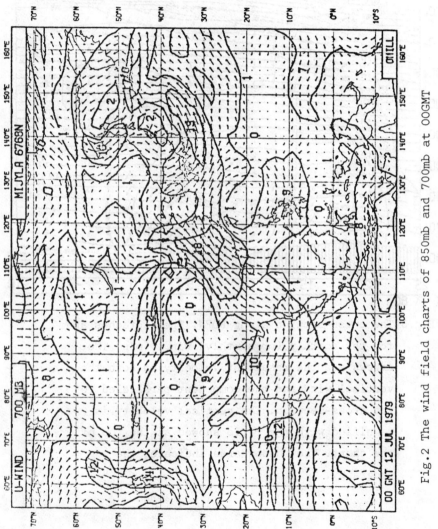

Fig.2 The wind field charts of 850mb and 700mb at 00GMT
12th, July 1979. The solid lines show the isotaches
with intervals of 5m/s. The heavy lines are the
orography trough line. The arrows are the wind vectors.

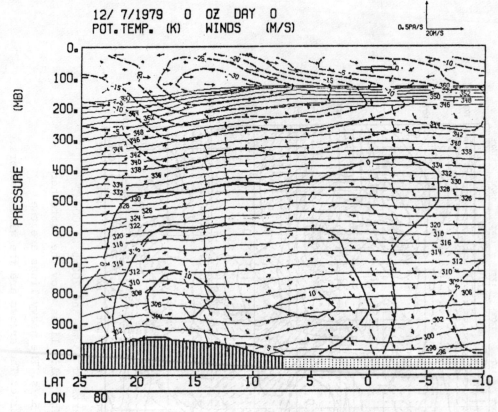

Fig.3 The cross-section along 80°N. The isotaches are
      for the northly (or southly) wind. (m/s) The
      arrows are the resultant velocity of both the
      vertical wind and the component paralleled to
      the cross-section. The vertical velocity is
      magnified with the ratio showed at the top of
      the chart. The shadow area indicates the
      orography.

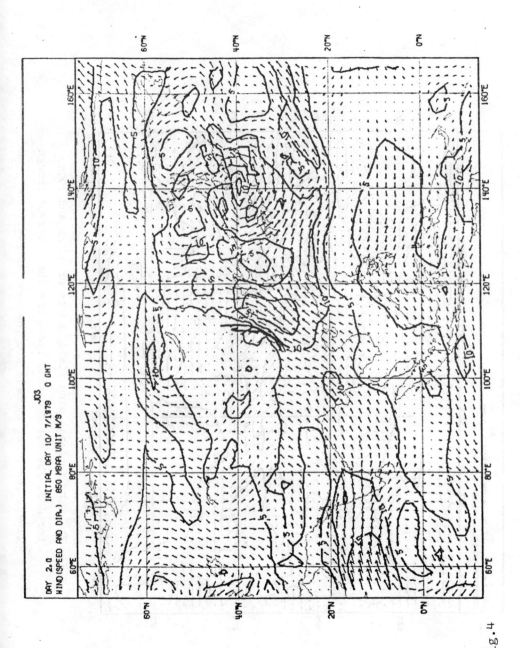

DAY 2.0   INITIAL DAY 10/ 7/1979 0 GMT
WIND(SPEED AND DIR.) 850 MBAR UNIT M/S

Fig.4

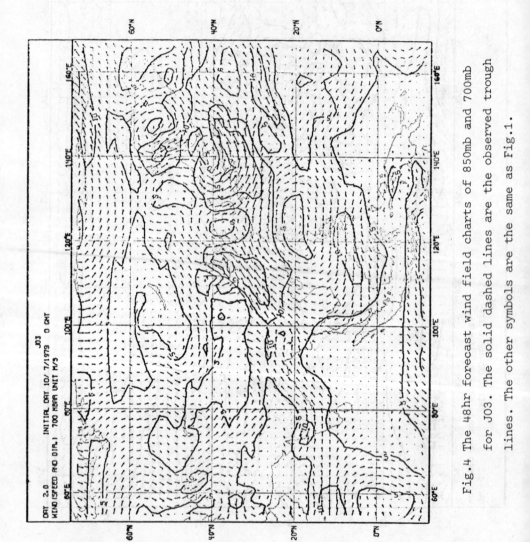

DAY 2.0   INITIAL DAY 10/ 7/1979   0 GMT   J03
WIND (SPEED AND DIR.) 700 MBAR UNIT M/S

Fig.4 The 48hr forecast wind field charts of 850mb and 700mb for J03. The solid dashed lines are the observed trough lines. The other symbols are the same as Fig.1.

Therefore, this run is suitable for comparison.

IV. THE IMPACT OF THE PLATEAU ON THE JET

Two experiments in which the orography has been modified or totally removed are carried out to examine the effect of orography. JO8 is the experiment where the orography has been totally removed. In summer time most systems on the retreated westerlies, under the influence of orography of Qinghai-Xizang Plateau, tend to turn around the north side of the Plateau and propagate southeastward in wave's form. It is rare that troughs or ridges would develop into planetary-scale ones. But once the Plateau is removed, the effect of westerlies, without any barrier, could get over the "Plateau" area. Fig.5 shows the 500mb height difference between JO8 and the control (JO3), the negative denoting the drop of height due to the removal of orography and the positive being the increase of height. One can see from the

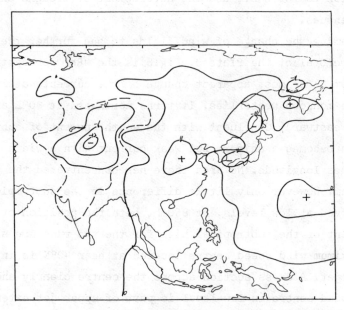

Fig.5 The difference map of 500mb geopotential height between JO8 and JO3.

map that this is a negative area to the west of 100°E. The maximum
may attain 12 dkm, equivalent to the total interval of three contours
in conventional analysis. This is the consequence of intensive development of a trough in westerlies. The influence may reach close to 10°N.
On the east side of the Plateau and in the eastern part of China there
are positive areas. This results from the fact that the ridge of
subtropical high extends northwestward after the Qinghai-Xizang Plateau
is removed. The maximum of positive area is located at where the
mountain trough lies in the control run. This means that for this run
(without mountain) the mountain trough disappears. Corresponding
developments can also be observed of the major trough on the eastern
coast of East Asia and two minor vortex at 140°E. Summarizing the
change of height field (over East Asia) to the east of 60°E leads to
the general impression that the effect of Qinghai-Xizang Plateau on
the flow fields is mainly of large-scale. Whether the major trough and
ridge develop or not depends much on the very existence of the Plateau.
The removal of the Plateau first leads to the development of the major
trough above the plateau. This in turn enhances the meridional extent
of circulation to the east, and the development of troughs and ridges
downstream ensues.

Now examine the change of wind fields to see further the effect
due to the removal of the Plateau. Fig.6 is the 48hr forecast of wind
field at 850mb. The most apparent change of the currents at lower
level is the monsoon westerlies. It turns northward at 80°E and cannot
keep moving eastward. Confluent with the south current of subtropical
high monsoon becomes a uniform but weak current with a width about
thirty degrees longitude. No area there has concentrated the south
wind momentum. However only little difference in the subtropical high
can be observed at low levels. At 850mb, both the position of the centre
and the extent of the subtropical high for the two runs are similar.

The maximum wind speed centre located at near 40°N is in fact not
a low level jet. A cross section across the centre clearly shows that
the strong wind centre in low-level is part of upper jet extending
downward. The gradient of velocity in the vertical direction is uniform.
The longitude sections along 28°N for both control run and no mountain

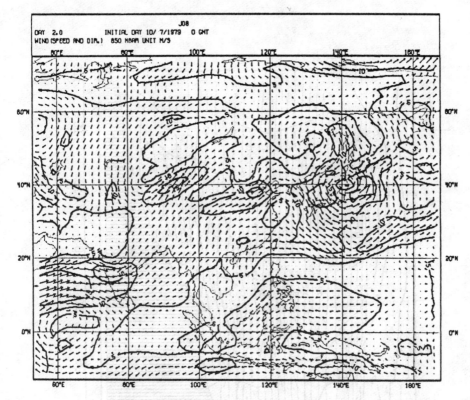

Fig.6 Same as Fig.4, except for J08.

run are given in Fig.7(a) and 7(b) respectively. There is a centre of
strong wind along the east side of the Plateau in the control run,
while for no mountain case there is not a centre under the mid-tropo-
sphere.

In a word, without the Plateau, there would be no low level jet
to its east, the current being weak and uniform. This result is in good
agreement with those studies about the East African Jet.

As stated above, low level jet seems to be closely connected with
summer monsoon current. Another experiment (J11) is designed to illustrate this connection and also to further study the influence of the
orography. We lift the orography at the Indo-China Peninsula to more
than 3000m with an attempt to set an artificial barrier in the monsoon
current. The new orography here is shown as Fig.8. There is a wall standing
along 100°E. It means that the east side of the Plateau is extended
southward to about 10°N. Fig.9 is the 48hr forecast of J11 for the

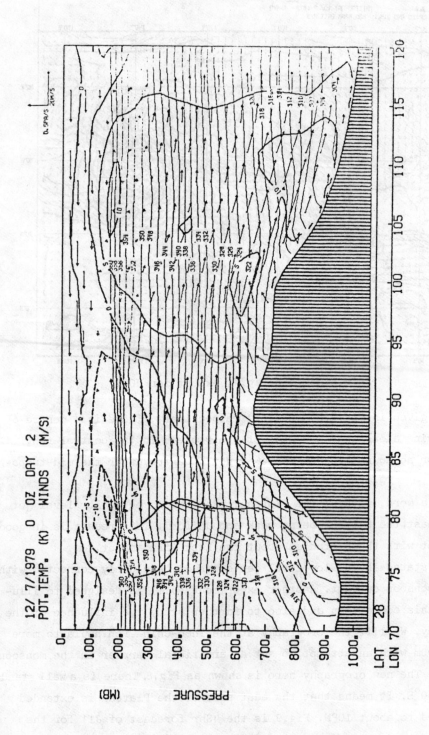

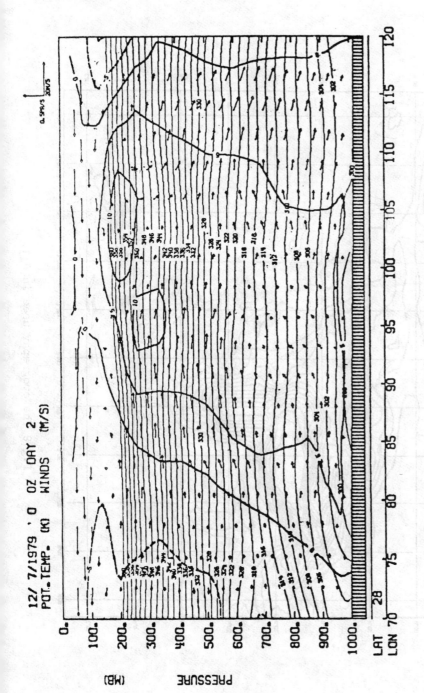

Fig.7 The cross-sections along 28°N. (a) J03, (b) J08. The symbols are same as Fig.3.

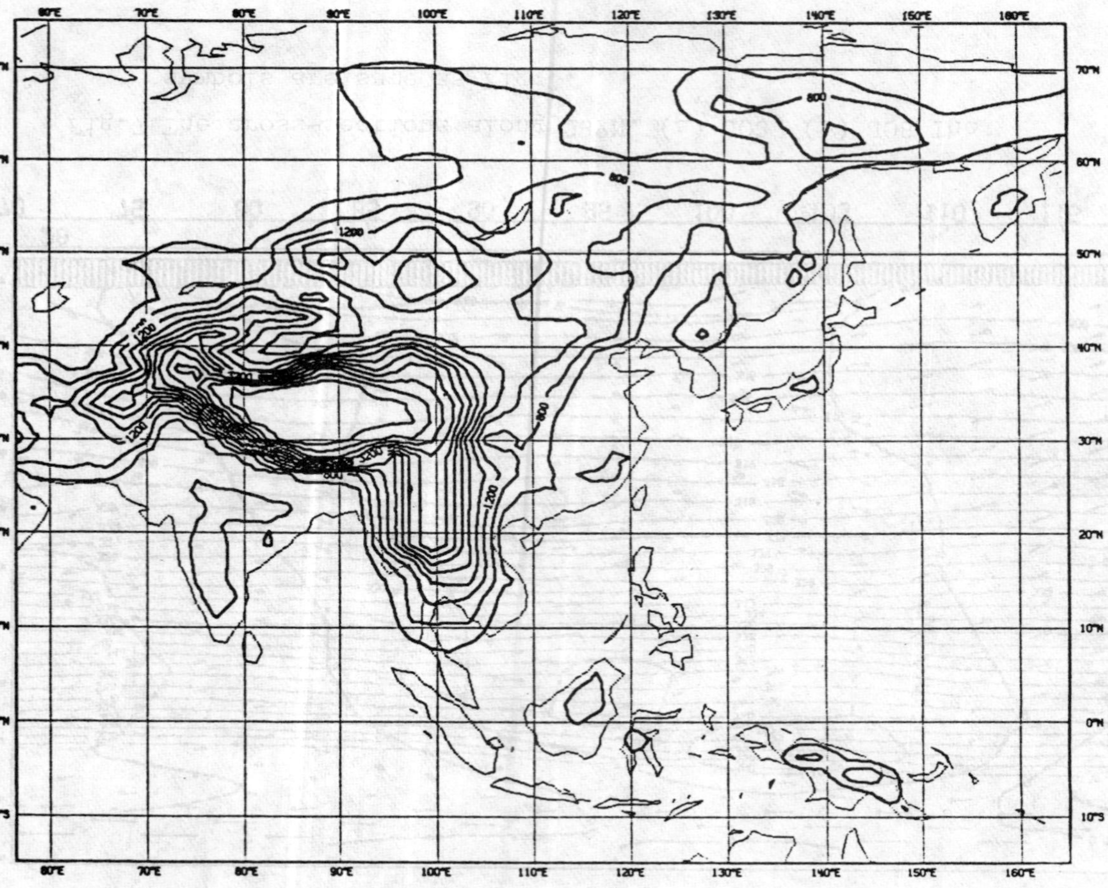

Fig.8 The orography used in J11.

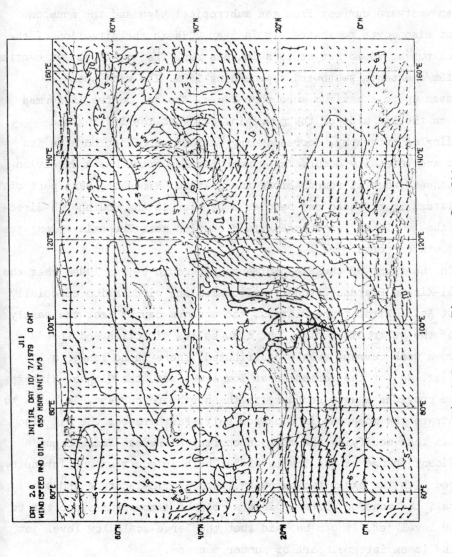

Fig.9 Same as Fig.4, except for J11.

wind field at 850mb. The monsoon current turning around the south tip
of the 'wall' does not slow down due to the block of the artificial
mountain barrier, but speed up instead because of confluence. It
becomes a narrow belt around the barrier. The latitude of confluence
between westward current from the subtropical high and the monsoon
current also moves southward. It is near 20°N in the situation of the
control run at 700mb but at 10°N in this run. The low level jet centre
with 15m/s extends southward approaching 15°N. The wind speed
increases up to a maximum more than 20m/s. The jet centre is formed
right on the lee side of the mountain. Owing to the lifting and
extending of orography, corresponding variations of jet both in its
height and intensity occur. The summer monsoon often has a direction
of southwest when it rounds along the plateau. Since the east part of
the Plateau is extended and becomes a barrier in the southnorth direction, the low level jet changes its direction too. The axis orientates
from south to north.

In the light of these experiments, it could be concluded that the
Qinghai-Xizang Plateau strongly influences the circulation especially
in East Asia. Without the Plateau, the monsoon cannot meet the eastly
flow of the subtropical high, but changes its direction and speeds
to form a weak current evenly distributed. There is no longer a low
level jet. So for the formation of low level jet, the Qinghai-Xizang
Plateau provides a west boundary to the northward current. It helps to
concentrate the momentum in a narrow belt. Furthermore, the J11 also
suggests that modifying the monsoon current alone can bring about a
significant change of the low level jet both in position and intensity.
Compared with the other two currents, both in mid-latitude and subtropical, the monsoon current plays a role of a momentum supporter to
the low level jet. It can be said that the large-scale low level jet
in China is an integral part of summer monsoon.

V. IMPACT OF THE PLATEAU ON THE LOW-LEVEL WIND
FIELDS ON ITS LEE SIDE

After the investigation of the influence of the Plateau on the

718

large-scale current to the east, it is worth while noting that the low-level jet over East Asia is located on the lee side of the Qinghai-Xizang Plateau relative to the currents of monsoon and westerlies. This is different from the relative positions of East African Highland to the cross-equatorial current, and of the Rocky mountains to the SE trade wind from the sea, both basic currents blowing towards the mountain barrier.

It has been pointed out in the above sections that in the case with the orography of the Plateau the lee side trough to the east is quite clear. In this section the formation of lee trough under conditions of different orographies and its relation to the low level jet will be examined in some detail.

Fig.2 clearly indicates that at 00E 12th July when the low level jet being most intense, orographically induced troughs appear at all levels at and below 500mb near 110°E. It lies in the vicinity of the Plateau, west of the centre of wind speed of the jet. The troughs look even sharper at lower levels (700mb and 850mb). The SW flow in the front of the lee trough, no doubt, enhances the local wind speed of the jet in that area. In 48h forecast map (Fig.4) for J03 where all orographies are retained, the lee trough is well predicted both its position and intensity. But for no-plateau case, as mentioned above, the current is uniform and weak. The lee trough near 110°E no longer exists, and southwestly flow dominates the East Asia continent region, which leads to an intensively developed major trough in the westerlies. However, it is of interest to investigate J11 where an artificial mountain barrier is added on the Indo-China Peninsula. In Fig.6, the 48h forecast map, the westly wind in the west changes into southly wind rapidly along the east side of the extended barrier, and the contours of geopotential at the southern end of the mountain also sharply turn clockwise.

Fig.10 are the maps of geopotential and temperature predicted by J11. To save space, only 12h and 48h forecasts are shown here. As is seen in Fig.10(a), a cyclonic curvature of geopotential makes its first apperence at the south tip of the extended mountain at 12h, forming a

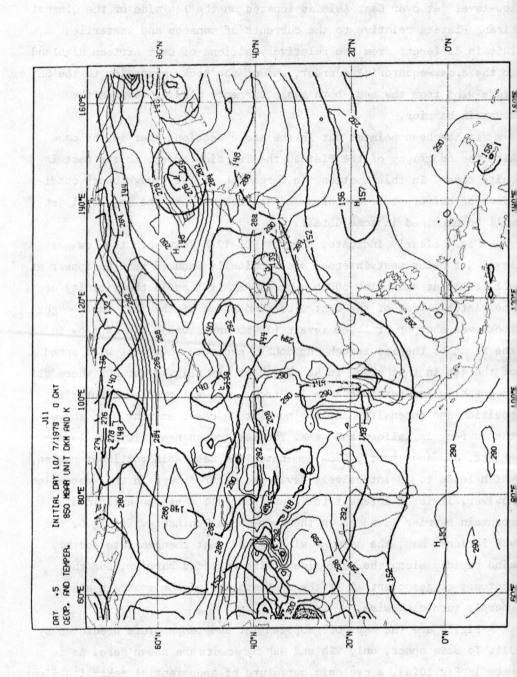

DAY .5   INITIAL DAY 10/ 7/1979 0 GMT
            J11
GEOP. AND TEMPER.   850 MBAR UNIT DKM AND K

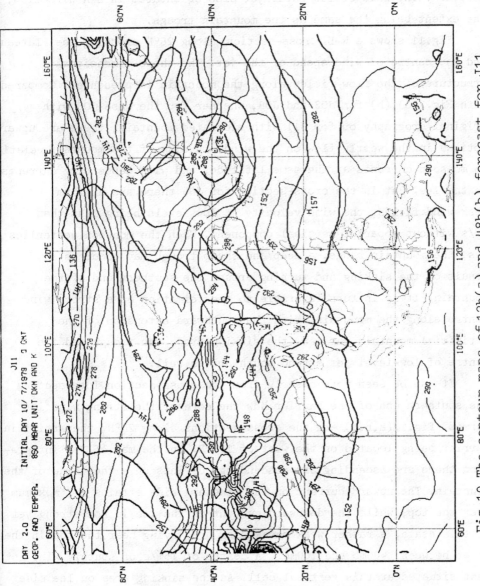

Fig.10 The contour maps of 12h(a) and 48h(b) forecast for J11.

clear-cut orographic trough at 24h (Fig. not shown here), it extended
northward, and at 48h it becomes a narrow and longated trough with a
S-N orientation, from 10°N to 35°N, parallel and close to the extended
mountain. The axis of low level jet also orientates in S-N direction,
the extent being the same as the mountain trough.

Fig.11 shows a W-E cross-section along 28°N crossing the Plateau
and the centre of wind speed of the jet. The three-dimensional
structure of the flow fields along the mountain is presented. Compared
with Fig.7(a),(b) for J03 and J08, whether for the case J03 with
original orography or for J11 with elongated mountain, there are upward
motions in the westly flow on the windward side of the mountain relativ
to monsoon currents at the same latitude, and with descending currents
on the lee. But in the cross-section for J08 there are uniform and
weak wind fields, the wind velocity at low levels being less than
5m/s without apparent vertical component. When the monsoon westerlies
pass the mountain range, a deep and sharp trough may occur as a
result of the sinking and turning around the southern end of the
mountain. It is in turn favorable to the formation of a strong wind
centre along the mountain. With the southward extension of the
artificial mountain barrier, the sinking area, the trough and the
centre of low-level jet extend southward as well.

It can be seen more clearly from the cross-section intersecting
the southern end of the barrier how the core of the strong wind is
formed. Fig.12(a,b,c) are the cross-sections along 20°N, the artificial
barrier being located on the west of 110°E. On the map of 24h integration there are ascending and descending currents on either side of the
mountain. The upward current on the windward side attains its maximum
over the top, while the downward current on the lee is not strong yet
at this stage, a rather weak descending area lying near the top in the
layer between 400mb and 700mb. A centre of 15m/s of southly wind is
just situated at this vertical cell. As the sinking area on lee side
expands and strengthens later on, the wind speed centre comes down
gradually. At 48h (Fig.12c) the centre descends below 800mb with a
strength of more than 20m/s. and strong southly wind prevails all over
the lee.

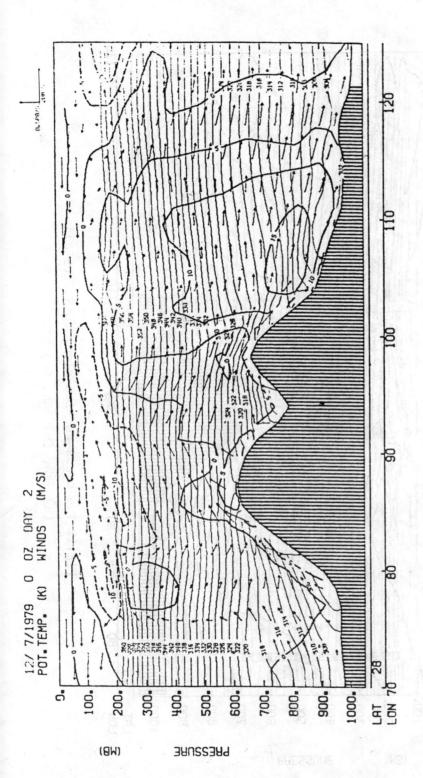

Fig.11 Same as Fig.7, except for J11.

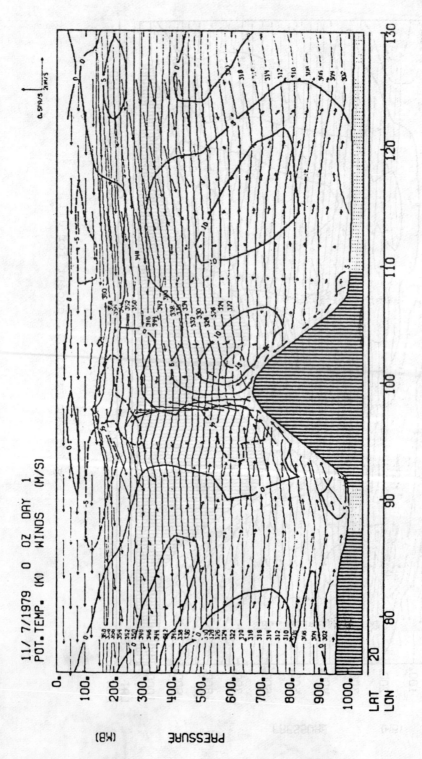

Fig.12 The cross-section along 20°N, for J11
(a) 24h, (b) 36h, (c) 48h.

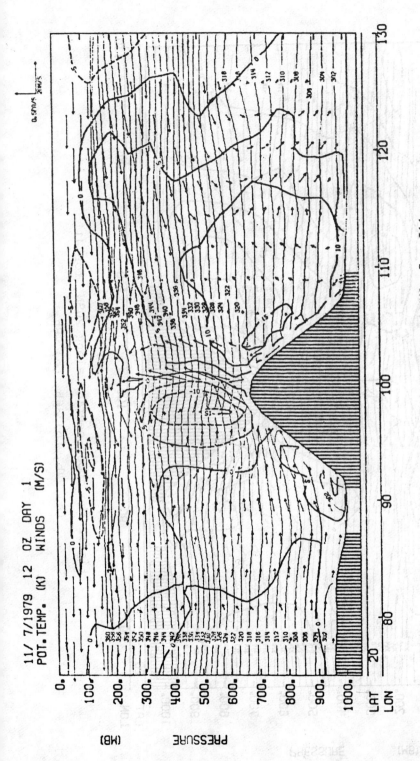

11/ 7/1979 12 0Z DAY 1
POT.TEMP. (K)   WINDS   (M/S)

Fig.12 The cross-section along 20°N, for J11.
(a) 24h, (b) 36h, (c) 48h.

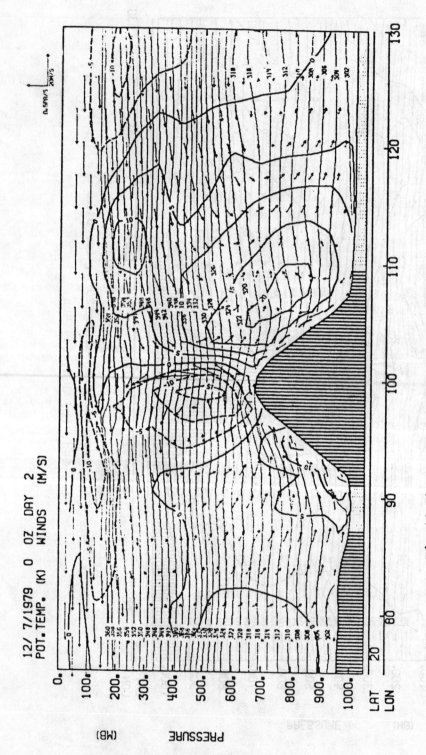

Fig.12 The cross-section along 20°N, for J11.

(a) 24h, (b) 36h, (c) 48h.

From what is mentioned above we may describe as follows the
process of the formation of lee-side strong wind centre:

First the currents round the southern end of the Plateau form
a cyclonic turning on the lee, and a mountain trough is resulted. A
sinking branch of the current passing over the top of the mountain
first appears, where southly current which is but the stretch of the
current round the mountain, attains its maximum, Later on, the mountain
trough develops and extends to the north. At the same time, the sinking
motion increases its strength and lowers its height. The wind speed
at the front of the lee trough is consistently increasing, and a centre
of strong wind is thus formed. This process precisely shows what
distinguishes the jets of East Asia from those of other regions in
the world.

From Fig.12 one may also see an interesting phenomenon that no
matter how the descending currents develop into lower layers, ascending
currents remain above the narrow area close to the mountain. The
current is probably part of the one turning around the mountain and
climbing up the mountain side.

VI. CONCLUSION

All these experiments confirm the extreme importance of the
Qinghai-Xizang Plateau for the onset of low level jet in East Asia.
If there were no mountain, the southly current consisting of three
flows would have been very weak and evenly distributed. The axis of
the low level jet extends southward when the orography is extended in
the same direction. It also shows that the effect of an "obstacle" in
the summer monsoon rapidly brings about the change of the low level
jet both in direction and intensity in its down-stream. The summer
monsoon may become main background circulation of the large-scale low
level jet. The descending air on the lee side of the Plateau is responsible for the formation of the orography induced trough and then for
the high wind speed centre.

REFERENCES

[1] Wexler, H. A boundary layer interpretation of the low level jet, Tellus, 13, 368-378, 1961.

[2] Hahn, D.G. and Manabe. S. The role of mountains in the South Asia monsoon circulation, J. A. S., 32, 1515-1541, 1975.

[3] Washington. W. and Daggupaty. S. Numerical simulation with the NCAR global circulation model of the mean conditions during the Asia-African summer monsoon, Mon. Wea. Rev., 103, 105-144, 1975.

[4] Tiedtke, M., J.-F. Geleyn, A.Hollingsworth and J-F, Louis, ECMWF Model-Parameterization of Sub-grid scale processes, ECMWF Technical Report No. 10, 1979.

[5] ECMWF Reseach Department, "ECMWF Forecast Model Documentation Manual". Edited by J-F Louis, 1981.

[6] Yuan Xin-Xuan A Synoptic study of the southwesterly low level jet stream south of the Changjiang River, Acta Meteorologica Sinica. Vol. 39, 245-251, 1981 (in Chinese).

Session 7: Observation and Analysis (5)

THE GENERAL CIRCULATION AND HEAT SOURCES OVER THE TIBETAN
PLATEAU AND SURROUNDING AREAS DURING THE ONSET
OF THE 1979 SUMMER MONSOON

Luo Huibang

Department of Meteorology

Zhongshan University, Guangzhou, People's Republic of China

Michio Yanai

Department of Atmospheric Sciences University of California

I. INTRODUCTION

The importance of the Tibetan Plateau (the Qinghai-Xizang Plateau) as an elevated heat source for the establishment and maintenance of the Asian summer monsoon circulation has been discussed by many authors (e.g., Flohn, 1957, 1968; Staff Members of Academia Sinica, 1958; Murakami, 1958; Yeh and Gao et al., 1979; Yeh, 1981, 1982) [see Luo and Yanai (1984) for a detailed review].

Since the First GARP Global Experiment (FGGE) of 1978-1979, quantitative knowledge of the tropospheric heat sources over the Tibetan Plateau and surrounding areas has begun to emerge. Wei et al. (1983) obtained the seasonal global distributions of the vertically averaged heating during the FGGE and showed that there is a heating maximum over the Plateau. Nitta (1983) presented the mean vertical profiles of the tropospheric heat source and moisture sink over several parts of the eastern Plateau for a 100-day summer period in 1979.

Luo and Yanai (1983) [hereafter referred to as LY Part I] presented a detailed analysis of the time evolution of the large-scale precipita-tion, low-level (850 hPa) wind, moisture and vertical motion fields over the Tibetan Plateau and surrounding areas during a 40-day period from late May to early July 1979, using the objectively analyzed FGGE Level II-b data. Subsequently, Luo and Yanai (1984) [hereafter referred to as LY Part II] examined the large-scale heat and moisture budgets of the same areas during the 40-day period.

The main objectives of the two-part paper were 1) to determine quantitatively the horizontal and vertical distributions of the tropospheric heat sources and moisture sinks over the Tibetan Plateau and surrounding areas in the early summer of 1979, and 2) to identify the heating mechanisms operating in the major heat source regions. It is the purpose of the present paper to give a concise summary of our work with a discussion of potential areas for further investigation.

II. DATA AND METHOD

The data used in this work are the daily precipitation amounts and the twice daily (0000 and 1200 GMT) wind, temperature, moisture and vertical motion fields at the ground surface and at standard pressure levels (850, 700, 500, 400, 300, 250, 200, 150 and 100 hPa) which have been analyzed objectively on a 2.5°×2.5° grid for the domain 10°-50°N, 60°-130°E (see Fig.1), from 26 May to 4 July 1979. The details of the objective analysis scheme were given in Appendix of LY Part I.

The apparent heat source Q_1 and the apparent moisture sink Q_2 (e.g. Yanai et al., 1973) are computed by

$$Q_1 = c_p \left[\frac{\partial T}{\partial t} + \underset{\sim}{v} \cdot \nabla T + \left(\frac{p}{p_0} \right)^\kappa \omega \frac{\partial \theta}{\partial p} \right] \tag{1}$$

and

$$Q_2 = - L \left(\frac{\partial q}{\partial t} + \underset{\sim}{v} \cdot \nabla q + \omega \frac{\partial q}{\partial p} \right) \tag{2}$$

where T is the temperature, θ the potential temperature, q the mixing ratio of water vapor, $\underset{\sim}{v}$ the horizontal wind, ω the vertical p-velocity, $\kappa = R/c_p$, R and c_p are the gas constant and the specific heat at constant pressure of dry air, L the latent heat of condensation, and $p_0 = 1000$ hPa.

The vertical p-velocity ω in (1) and (2) has been obtained kinematically by integrating the equation of mass continuity with the surface boundary condition which takes into account the sliding motion along the sloping topography. To reduce errors in the computation of Q_1 in the upper troposphere, we have imposed another constraint on the kinematically obtained vertical p-velocity by requiring

$$\omega = \left[\frac{\frac{\partial \theta}{\partial t} + \underset{\sim}{V} \cdot \nabla \theta - \frac{(p_0/p)^\kappa}{c_p} Q_R}{- \partial \theta / \partial p} \right] \qquad (3)$$

at p=125 hPa as suggested by Nitta (1977). In (3) Q_R is the radiative heating rate. (3) assumes that there are no heat sources other than the radiative heating, i.e., $Q_1 = Q_R$ in the uppermost layer between 100 and 150 hPa. We have also assumed that $Q_2=0$ above the 200 hPa level.

As discussed by Yanai et al. (1973), the apparent heat source and the apparent moisture sink are interpreted as

$$Q_1 = Q_R + L (c-e) - \frac{\partial}{\partial p} \overline{s'\omega'} \qquad (4)$$

$$Q_2 = L (c-e) + L \frac{\partial}{\partial p} \overline{q'\omega'} \qquad (5)$$

where $s \equiv c_p T + gz$ is the static energy, c the rate of condensation per unit mass of air, and e the rate of re-evaporation of cloud and rain water. The overbars denote running horizontal averages, and the primes denote the deviations from the large-scale values due to small-scale eddies such as cumulus convection and turbulence.

Integrating (4) from p_T (the pressure at which the eddy motion vanishes) to p_s (the pressure at the ground surface), we obtain

$$\frac{1}{g} \int_{p_T}^{p_s} (Q_1 - Q_R) dp = \frac{L}{g} \int_{p_T}^{p_s} (c-e) dp - \frac{1}{g} (\overline{s'\omega'})_{p=p_s}$$

$$\underset{\sim}{} LP + \rho_s c_p (\overline{T'w'})_{p=p_s}$$

$$= LP + S \qquad (6)$$

733

where ρ_s is the density of surface air, w the vertical velocity, P and S are the amount of precipitation and the sensible heat flux per unit area at the surface, respectively.

Similarly, from (5) we obtain

$$\frac{1}{g} \int_{p_T}^{p_s} Q_2 \, dp = \frac{1}{g} \int_{p_T}^{p_s} (c-e) \, dp + \frac{L}{g} (\overline{q'\omega'})_{p=p_s}$$

$$\approx LP - \rho_s L(\overline{q'w'})_{p=p_s}$$

$$= L(P-E) \tag{7}$$

where E is the eddy moisture flux (evaporation) per unit area at the surface.

In this study we do not attempt to determine individual components of Q_1 and Q_2 on the right side of (4) and (5). Instead we use (4), (5), (6) and (7) to infer the nature of heating processes from the horizontal and vertical distributions of Q_1 and Q_2.

Q_1 and Q_2 are calculated for the inner domain 15°-45°N, 65°-125°E (see Fig.1), and for the layer between the ground surface and the first standard pressure level above the surface then for each successive layer between the standard pressure levels. Because of horizontally varying terrain heights and lack of detailed data within the planetary boundary layer, separate computational schemes for the horizontal and vertical advection terms are used to reduce errors in the lowest layer (see Appendix A of LY Part II).

The heat and moisture budget calculations are made at every observation time (80 times). The budget results are often averaged for each of the 5-day sub-periods for brevity of discussion (Table 1). The major 850 hPa flow features and the distributions of organized precipitation for the eight sub-periods have been described in detail in LY Part I.

Table 1. Five-day sub-periods for description.

Period 1	26 May – 30 May	
2	31 May – 4 June	the pre-onset phase
3	5 June – 9 June	
4	10 June – 14 June	
5	15 June – 19 June	the onset phase
6	20 June – 24 June	
7	25 June – 29 June	the post-onset phase
8*	30 June – 4 July	

*A "break monsoon" period for most parts of India.

III. MEAN LARGE-SCALE CONDITIONS

The major events characterizing the distinct time change of the general circulation over East Asia around the time of monsoon onset such as the commencement of Mei-yu, the northward shift of the westerly jet stream and the development of the upper tropospheric Tibetan (or South Asia) anticyclone have been summarized ty Tao and Ding (1981) and in LY Part I. Although the development of the 1979 monsoon is considered to be subnormal in terms of the total precipitation over most parts of the Indian subcontinent (e.g., Fein and Kuettner, 1980), the analysis reveals that major events characterizing the onset of the Asian summer monsoon are well represented in the analyzed period.

a. The mean wind field

Figs. 2a-d show the 40-day mean resultant winds at 850, 700, 500 and 300 hPa, respectively. The mean 850 hPa wind field (Fig.2a) shows pronounced inflow toward the Plateau, especially along the southeastern periphery and on the east side of the mean Burma-India trough which intersects the Plateau at 85°E. The confluence zone associated with the Mei-yu front is the dominant feature of the mean flow on the east side of the Plateau. The mean 700 hPa wind field (Fig.2b) clearly shows the inflow along the southeastern, eastern and northern boundaries of the Plateau, and indicates the presence of a mean inflow center on the

735

eastern Plateau. The inflow exhibits a remarkable diurnal variation
(LY Part I) and it is a manifestation of the heat low on the Plateau
which has been discussed by Yeh and Gao et al. (1979), Gao et al.
(1981) and Yeh (1981).

At 500 hPa (Fig. 2c) the mean winds are generally westerly. The
wind speed of the 500 hPa flow above the Plateau is noticeably small,
and the westerlies on the upstream and downstream sides tend to split
into northern and southern branches. At 300 hPa (Fig. 2d) we begin to
see the low-level portion of the upper tropospheric South Asia anticyclone whose mean center is located over the southeastern periphery
of the Plateau. The intensity of the anticyclone during the analyzed
period was generally weak and its center often tended to diffuse (e.g.,
Krishnamurti et al., 1979, 1980).

b. The temperature field

Fig.3(a) shows the 40-day mean vertical distribution of the
temperature anomaly (the deviation of temperature from the horizontal
mean in the analysis domain) in the longitudinal plane along 32.5°N
(see Fig.1). There is a region of positive temperature anomaly (∿ 4-5°C)
extending from the Plateau surface to the 400 hPa level. We also
recognize another cneter of warm anomaly (∿ 4°C) in the upper troposphere (200-300 hPa) above the Plateau. The relation between the two
anomaly centers is more clearly revealed in the meridional vertical
cross section along 92.5°E (Fig.3b). The axis of the warm air tilts
southward with height. The center of the upper-level warm anomaly is
located at 20°N and it is closely associated with the upper tropospheric
anticyclone. A more detailed discussion of the warm centers will be
made in a future paper.

c. Mean horizontal transports of water vapor

Fig.4 shows the distributions of the 40-day mean horizontal tran-
sports of water vapor $\overline{q\mathbf{v}}$ at 850 and 300 hPa. In the figures the field
$\overline{q\mathbf{v}}$ is expressed by "streamlines" of $\overline{q\mathbf{v}}$ vectors and isolines of $|\overline{q\mathbf{v}}|$.
The $\overline{q\mathbf{v}}$ charts reveal the roles played by the Plateau in determining
the distribution of moisture sources and sinks in the domain. At 850
hPa (Fig.4a) the down-gradient moisture flow towards the southern
slope of the Plateau is the most pronounced, corresponding to the major

736

moisture sink (thus precipitation) over the Assam-Bengal region. Inflow of moisture towards the Plateau takes place also along the southeastern periphery. To the east of the Plateau, the down-gradient moisture flow toward the confluence zone is seen over the Changjiang River region.

An interesting effect of the Tibetan Plateau on the moisture distribution is seen at 300 hPa (Fig.4b). There is a conspicuous maximum of $|\overline{q\,\underline{v}}|$ above the eastern corner of the Plateau. At this level the Plateau acts as the major moisture source downstream of the westerlies i.e., the China Plain and Japan. The eastern Plateau acts as a huge chimney funneling water vapor from the lower to the upper troposphere. Previously, Flohn (1968) suggested a similar feature in the vertical transport of heat.

d. Total rainfall

The areas of organized precipitation are well related to synoptic systems in the low-level flow. The major synoptic systems related to the precipitation are 1) the quasi-stationary India-Burma trough and disturbances forming on the trough on the south side of the Plateau and 2) the confluence zone along the Mei-yu front, and a shear line on the east side of the Plateau which has been discussed by Tao and Ding (1981).

Fig.5 shows the distribution of total rainfall during the 40-day period. The heaviest rainfall (> 1000 mm) is located along the west coast of India. In addition, there are two major maxima of heavy rainfall, i.e., the Assam-Bengal region (> 560-640 mm) and the South China coast (> 400 mm). The northwestern part of the analysis domain is very dry and observed very little precipitation. The area of minimum precipitation (< 5 mm) extends from the western Plateau to Afganistan and desert areas of Turkestan.

There are considerable spatial variations in rainfall on the Tibetan Plateau. The 40-day total precipitation amount in the eastern part of the Plateau exceeds 80 mm. Rainfall amounts more than 160 mm are found along the southern and southeastern slopes of the Plateau. On the other hand, there is very little precipitation in the western part of the Plateau. A station located in the western Plateau (32.5°N,

80.1°E) observed only 2.0 mm of rainfall during the 40-day period.

IV. MEAN DISTRIBUTIONS OF HEAT SOURCES AND MOISTURE SINKS

a. Mean horizontal distributions of $<Q_1>$ and $<Q_2>$

Figs. 6a-b show the horizontal distributions of the vertically integrated heat source $<Q_1>$ and moisture sink $<Q_2>$ averaged for the 40-day period. With the observed mean precipitation rates shown in Fig.5, we may infer the following.

The pronounced heat source of ~ 300 Wm^{-2} over the Assam-Bengal region is consistent with the moisture sink of the same order of magnitude and is undoubtedly related to the heavy rains of ~ 10 mm d^{-1} in this region. Heat sources > 150 Wm^{-2} in a belt extending from Thailand to the South China Plain along the Mei-yu frontal zone also corresponds well to moisture sinks and precipitation (LP) of similar magnitudes.

We find heat sources of $\sim 100-150$ Wm^{-2} over the eastern Tibetan Plateau, which are accompanied by moisture sinks of $\sim 50-100$ Wm^{-2}. Wei et al. (1983) obtained the net heating of ~ 120 Wm^{-2} over the eastern Plateau for June-August 1979. Heat sources of $\sim 100-150$ Wm^{-2} are also found over the western Tibetan Plateau, the Western periphery of the Plateau near 37°N, 67.5°E, and the Takla Makan Desert. The heat sources in these regions are not accompanied by significant moisture sinks and this suggests intense sensible heating from the dry ground surface exceeding the net radiative cooling.

b. Mean vertical cross sections

Figs. 7a-b show the 40-day mean vertical distributions of the heating rate Q_1/c_p and the drying rate Q_2/c_p in the longitudinal plane along 32.5°N. Over the Tibetan Plateau between 82.5°E and 97.5°E, a deep layer of heating occupies the whole troposphere except near the ground surface (Fig. 7a). The heating over the Plateau appears to consist of two distinctly different regimes. The heat source over the eastern Plateau with peak values of ~ 5 K d^{-1} in the 300-400 hPa layer at 92.5°E is accompanied by a moisture sink with peak values of ~ 4 K d^{-1} in the 400-500 hPa layer (Fig. 7b). On the other hand, the heat source over the western Plateau with a peak value of 5.5 K d^{-1} in the

200-500 hPa layer is associated with a weak moisture source in the lower layer.

There are heat source regions over the eastern slope of the Plateau near 105°E and over the China Plain. The heat source over the slope has a maximum intensity of ~ 3 K d^{-1} at 450 hPa and it is associated with a moisture sink of nearly the same magnitude at the same level. This heat source appears to be related to the precipitation associated with synoptic systems along the low-level shear line (see Figs. 2a,b). The heat sources extending from the China Plain to the East China Sea are clearly the results of the precipitation in the Mei-yu frontal zone.

V. REGIONAL CHARACTERISTICS OF HEAT SOURCES AND MOISTURE SINKS

To identify the principal factors contributing to the mean heat budgets of different parts of the Tibetan Plateau and surrounding areas, we shall examine the areal mean values of Q_1 and Q_2 (denoted by $[Q_1]$ and $[Q_2]$) with the areal mean precipitation amounts [RR] in four selected regions shown in Fig.1. These are: Region I, a large domain including the western Tibetan Plateau; Region II, the eastern Tibetan Plateau; Region III, the middle and downstream region of the Changjiang River; and Region IV, the Assam-Bengal-Northeast India region. The four regions represent heat source areas with different characteristics in terms of underlying surfaces and prevailing meteorological conditions.

The western Tibetan Plateau is known to be semi-arid and the significance of sensible heat flux from its elevated surface has been emphasized by several authors (e.g., Flohn, 1968; Yeh and Gao et al., 1979; Yeh, 1982). As discussed in LY Part I we extended the area of analysis because of the scarcity of upper-air data on the western Plateau proper. Although the extended area (Region I) is affected by rains on the slopes of the Plateau after the monsoon onset, the time series of data for this region contains a "dry" period for which the effects of sensible heating may be isolated. On the other hand, the eastern Plateau is covered by a dense observation network (see Fig.1 of LY Part I). Thus, Region II is chosen to have the mean elevation of about 4,000m (600 hPa) as in the work of Yeh and Gao et al.

a. Vertical time sections of $[Q_1]$ and $[Q_2]$

1) REGION I (THE WESTERN PLATEAU AND ADJACENT AREAS)

In Fig.8, we show vertical time sections of $[Q_1]$ and $[Q_2]$ with daily values of ,[RR] for Region I. There is very little precipitation in this region for Periods 1—2. From Period 3 onward to the end of the analyzed period some amount of daily rainfall is observed. Substantial rainfall amounts are observed in later periods especially in Periods 7 and 8, which are contributed by the monsoonal rains on the southern slope of the Plateau (LY Part I). These are refelcted in the time section of $[Q_2]$ which shows small negative values (moisture source) near the surface for Periods 1-3 and intermittent positive values corresponding to precipitation peaks after Period 4.

On the other hand, $[Q_1]$ is generally positive in the whole layer and heating of $\sim$ 3—5 Kd^{-1} is observed even before the onset of rains. After Period 4 the peaks of $[Q_1]$ occur simultaneously with those of postive $[Q_2]$, suggesting the contributions from condensation heating. Because the net radiative heating rate Q_R is usually negative in the troposphere, we infer from Eq.(4) that the postive heat source appearing in the first two sub-periods is due to the vertical convergence of eddy sensible heat flux which is supplied from the ground surface and redistributed by turbulence or convection. However, the distribution of $[Q_1]$ during this dry phase is somewhat irregular and it may suggest inaccuracy of the heat budget calculated for this region.

2) REGION II (THE EASTERN PLATEAU)

Fig.9 shows the time sections of $[Q_1]$, $[Q_2]$ and [RR] for Region II. The analyses for this region are considered to be more reliable because of much denser observations. This region is generally more humid than Region I, but the values of $[Q_2]$ in the whole layer are negative for the first 10 days (Periods 1-2). The $[Q_2]$ values are positive when significant amounts of rainfall are observed in Periods 3—8.

Even in the first 10-day period there is substantial heating in the upper troposphere with peak values of ~ 4 Kd^{-1} above the 300 hPa level. The heating in this phase occurs with negative [Q_2] in the lower layer. These features are qualitatively similar to those seen in the pre-onset phase for Region I. The positive values of [Q_1] after Period 3 correspond well to positive values of [Q_2] and peaks of [RR]. The heating rate in the post-onset phase increases with time simultaneously with the increase of [Q_2] and attains a peak value of ~ 14 Kd^{-1} in the 250-300 hPa layer in Period 8.

3) REGION III (THE CHANGJIANG RIVER)

Fig.10 shows the time sections of [Q_1], [Q_2] and [RR] for Region III. This region contains most of the Central China Plain where a large amount of precipitation fell along the Mei-yu frontal zone (LY Part I). The time sections of [Q_1] and [Q_2] clearly reflect the time change of daily precipitation [RR] associated with the frontal zone. The time sequences of [Q_1] and [Q_2] show similar magnitudes and they are in phase, indicating that the primary source of heating in this region is the heat released by condensation.

During heavy rains intense heating rates of 6—12 Kd^{-1} occur in the middle troposphere with moisture sinks of the order of 4—8 Kd^{-1}. The level of maximum [Q_1] is located between 400 and 500 hPa. The level of maximum [Q_2] is also located in this layer or slightly below for Periods 1—4, but it becomes lower for Periods 5—8. These features suggest that the precipitation in this region is primarily from strati-form clouds at the beginning but it becomes more convective in later periods.

b. Mean vetical profiles

The mean vertical profiles of [ω], [Q_1] up to the 150 hPa level, and [Q_2] up to the 200 hPa level are shown for the four regions in Figs. 11a-d. They show distinct regional differences of heating proces-ses.

In Region I (Fig. 11a), the mean vertical motion is weakly upward and the heating (1-3 Kd^{-1}) in the whole layer is accompanied by very small values (~ 0.5 Kd^{-1}) of [Q_2]. The heating is more intense in the layer above the 500 hPa level than in the layer near the ground surface. In Region II (Fig. 11b) the mean vertical motion is distinctly upward

741

with a peak value at 400 hPa. The mean vertical profile of $[Q_1]$ shows strong heating (2–3 Kd^{-1}) in the deep tropospheric layer between 150 and 500 hPa and slight cooling (-0.1 Kd^{-1}) near the surface. The heating is accompanied by positive $[Q_2]$ of the order of 1 Kd^{-1} below the 300 hPa level. These results show that nearly half the heating in Region II is contributed by the latent heat release. Nitta (1983) obtained similar mean profiles of $[Q_1]$ and $[Q_2]$ for the eastern Plateau during a 100-day period from 30 May to 7 September 1979.

The mean vertical profiles of $[Q_1]$ and $[Q_2]$ for Region III (Fig. 11c) are remarkably similar to each other, showing that the heating in this region is mostly the result of frontal precipitation. On the other hand, the mean $[Q_1]$ and $[Q_2]$ profiles for Region IV show large differences in both their magnitudes and the levels of maximum values (Fig. 11d). The upward shift of the $[Q_1]$ profile from the $[Q_2]$ profile is a phenomenon typical for a cumulus convective atmosphere (e.g., Yanai et al., 1973; Thompson et al., 1979).

c. Vetically integrated values

The mean values of the vertical integrals $<[Q_1]>$, $<[Q_2]>$ and those of [LP] for the total period are summarized in Table 2 for the four regions. Using the relations (6) and (7), we can also estimate the mean sensible heat flux [S] and the mean evaporation rate [LE] as residuals using the mean values of $<[Q_1]>$ and $<[Q_2]>$ and climatological values of $<[Q_R]>$. We use July mean net radiative heating rates over the respective region estimated by Katayama (1967). The residuals are also listed in Table 2.

The indirectly obtained values of [S] and [LE] for the four regions are positive and quite reasonable. The estimated mean evaporation rates for Regions I, II, III and IV are 1.2, 0.9, 1.3 and 3.5 mm d^{-1} respectively. The estimates of [S] and [LE] for Regions I and II will be compared with the estimates obtained by other workers in the next subsection

d. Heat and moisture budgets above the Tibetan Plateau

The mean heat and moisture budgets results for Regions I and II are compared with the budget estimates given by Yeh and Gao et al.,

742

Table 2. The mean heat and moisture budgets for the total period for four regions (units: W m^{-2}).

Region	Heat balance				Moisture balance		
	$<[Q_1]>$	$<[Q_R]>^*$	[LP]	[S]	$<[Q_2]>$	[LP]	[LE]
I	107	-77	58	(126)	24	58	(34)
II	113	-62	71	(104)	44	71	(27)
III	161	-79	208	(32)	171	208	(37)
IV	250	-81	273	(58)	172	273	(101)

* From Katayama (1967). ( ) ... residuals.

(1979) for the western and eastern Plateau and those obtained recently by Nitta (1983) for the eastern Plateau. Yeh and Gao et al. list climatological monthly values of [S] and [LP] for the western and eastern Plateau separately and those of [LE] for the whole Plateau. These are based on long-term surface records obtained at stations with a mean elevation of 4000 m (600 hPa) and the border between two parts of the Plateau was placed at 85°E. They also list representative January and July values of the solar and long-wave radiation fluxes on the Plateau. Nitta (1983) estimated mean values of $<[Q_1]>$ and $<[Q_2]>$ on the eastern Plateau for a 100-day summer period in 1979 and obtained [S] and [LE] as residuals as in the present work.

Comparisons of the mean heat and moisture budget components for Region I with those for the western Plateau given by Yeh and Gao et al. (1979) are made in Table 3. Because our Region I is much larger than the western Plateau proper studied by Yeh and Gao et al. and includes the slopes where substantial rains fell from Period 3 onward, the mean values for the dry pre-onset phase (Periods 1-2) are listed for a closer comparison. As seen in Table 3, the estimated value of [S] for Region I during the pre-onset phase (169 W m^{-2}) is somewhat

743

smaller than the June mean value for the western Plateau proper
($219 \, W \, m^{-2}$) given by Yeh and Gao et al. The estimate of [LE] for Reg
I ($31 \, W \, m^{-2}$) is close to the June mean value for the whole Plateau
($39 \, W \, m^{-2}$) obtained by Yeh and Gao et al.

Table 3. Comparisons of the heat and moisture budgets for the western
Plateau (units: $W \, m^{-2}$).

	$<[Q_1]>$	$<[Q_R]>$	[LP]	[S]	$<[Q_2]>$	[LP]
Periods 1-2 mean	101	-77*	9	(169)	-22	9
Yeh and Gao (June)	(142)	-94**	17	219	(-22)	17

  * From Katayama (1967). ( ) --- residuals.
 ** Mean July value for the whole Plateau.
*** Mean June value for the whole Plateau.

Comparisons of the heat and moisture budget components for Regi
II with those for the eastern Plateau given by Yeh and Gao et al.
(1979) and Nitta (1983) are shown in Table 4. The estimated values o
[S] of three independent studies agree very well with each other. Th
estimated value of [LE] for Region II ($27 \, Wm^{-2}$) is closer to the Jun
mean value for the whole Plateau obtained by Yeh and Gao et al.

VI. HEATING MECHANISMS ABOVE THE TIBETAN PLATEAU

To identify the mechanisms which generate the observed troposphe
heating above the Plateau during the pre-onset and post-onset phases,
we shall closely examine the vertical heating profiles and the therma
stratification above the Plateau during each of the respective period
a. Mean heating profiles for the dry and rain periods
To clearly isolate the effects of the sensible heat flux from

Table **4·** Comparisons of the heat and moisture budgets for the eastern
Plateau (units: $W\,m^{-2}$).

	$<[Q_1]>$	$<[Q_R]>$	[LP]	[S]	$<[Q_2]>$	[LP]	[LE]
39-day mean	113	-62*	71	(104)	44	71	(27)
Yeh and Gao (June)	(94)	-94**	86	102	(47)	86	39***
Nitta (100 days)	120	-75	90	(105)	25	90	(65)

* From Katayama (1967).   ( ) --- residuals.
** Mean July value for the whole Plateau.
*** Mean June value for the whole Plateau.

those due to the condensation heating, we shall examine the mean vertical profiles of $[\omega]$, $[Q_1]$ and $[Q_2]$ during the first 10 days (26 May–4 June) for Regions I and II. Although the 10-day period is too short to obtain representative heating profiles for the pre-onset phase, we may find some clues suggesting the heating mechanism above the Plateau during the dry phase.

In Region I (Fig. 12) the mean vertical motion during this period is downward above the 300 hPa level and weakly upward below. We find positive values of $[Q_1]$ of 1–3 Kd^{-1} with no appreciable values of $[Q_2]$ in a deep upper tropospheric layer between 150 and 500 hPa. Below the 500 hPa level we observe weak heating and weak moistening which indicate vertical convergence of eddy heat and moisture fluxes. The mean vertical profiles for Region II during the same period are shown in Fig. 13. Here the mean upward motion below the 250 hPa level is stronger, and there is a deep layer of heating of $\sim$ 1–2 Kd^{-1} between 150 and 500 hPa. The deep heating layer in Region II is also associated with negative values of $[Q_2]$. These features of the $[Q_1]$ and $[Q_2]$ distributions over the Plateau are suggestive of dry (unsaturated) thermal convection rising from the heated surface and penetrating into the upper troposphere.

The heating profiles for the dry period may be compared with the

$[Q_1]$ and $[Q_2]$ profiles during the rain period for Region II which are shown in Fig.14. The mean is taken over all days with daily precipitation exceeding 2.5 mm d^{-1}. The intense upward motion has a maximum at 400 hPa and large heating and drying rates are observed in the whole troposphere. The maximum heating occurs in the 300-400 hPa layer during the rain period. The separation of the levels of maxima between the $[Q_1]$ and $[Q_2]$ profiles shows the cumulus-convective nature of the summer rains on the eastern Plateau, which has been documented by many authors (e.g., Flohn, 1968; Yeh and Gao et al., 1979; Yeh, 1981, 1982; Tao and Ding, 1981).

b. Thermal structure of the boundary layer on the Plateau

To explain the vertical heating profiles during the pre-onset phase over the Plateau shown in Figs. 12-13, which are characterized by large heating in the upper tropospheric layer with zero or negative moisture sink, we examine the possibility of dry thermal convection originating at the diurnally heated surface and reaching the upper troposphere. The diurnal temperature variation at the surface in summer is known to be of the order of 12°C over most parts of the Plateau (Yeh and Gao et al. 1979).

In Fig.15, the mean vertical distributions of potential temperature at 0000 GMT (0800 Beijing time) and 1200 GMT (2000 Beijing time) at Lhasa (55591, 29.7°N, 91.1°E) and Nagqu (55299, 31.5°N, 92.1°E) are illustrated. These stations are located on the eastern Plateau and their elevations are indicated in the figures. The vertical distributions of potential temperature show remarkable diurnal variations especially near the surface, and deep, nearly mixed layers are seen in the evening (1200 GMT). Inspection of daily soundings at these stations reveals that the mixed layer is most pronounced during the first 10-day dry period with higher mean potential temperatures, as shown by dash-dotted curves in Fig.15. The top of the nearly mixed layer reaches the 400 to 300 hPa levels during the dry period. Similar potential temperature profiles are commonly observed in the evening at other stations on the Plateau (LY Part II).

Fig.16 shows the horizontal distribution of the surface potential temperature at 1200 GMT averaged for the first 10 days. There is a warm center on the eastern Plateau with a maximum potential temperature of ~ 336 K. Because of a wide gap devoid of data, the estimated surface potential temperatures on the western Plateau between 75°E and 91°E are uncertain and are likely to be underestimated. The high surface potential temperatures on the Plateau have a profound effect on the static stability of the deep tropospheric layer above the Plateau. The elevated and diurnally heated surface generates very small or even negative static stability over a large area of the Plateau in the afternoon hours (LY Part II).

The center of high potential temperature observed on the eastern Plateau at 1200 GMT extends well above the surface and is a clearly recognizable feature at the 500 hPa level. Fig.17 shows the mean (0000 and 1200 GMT together) temperature and winds at 500 hPa for the first 10-day period. There is a well-defined warm center on the eastern Plateau to the west of which a pronounced temperature trough is located. Therefore the prevailing westerly flow leads to a strong cold advection above the Plateau at this level. The existence of intense heat source on the Plateau is evident because the warm center is maintained despite the advection of cold air from the west.

c. Plausible mechanisms of heating

In Fig.18 we show the mean distributions of the potential temperature θ and the equivalent potential temperature

$$\theta_e \approx \theta \, \exp(\frac{Lq}{c_p T_c}) \tag{8}$$

where T_c is the temperature at the lifting condensation level, observed at 1200 GMT at Lhasa for the first and the last 10-day periods.

During the first (pre-onset) 10-day period, the mean surface value of θ at 1200 GMT is 336 K and the top of the mixed layer reaches near the 400 hPa level. In Fig.18 we also show the observed maximum value of the surface potential temperature (339 K) during this period. The mean value of the daily maximum surface potential temperature for the 10-day period is higher than the mean 1200 GMT value for the same period by 2.1 K. Therefore the air parcel rising from the surface may reach well above the 400 hPa level in the afternoon and deposit the sensible heat in the upper troposphere. The distribution of θ_e for this period shows that the layer above the Plateau is conditionally unstable but the surface air is very dry as indicated by its high mean lifting condensation level (383 hPa).

On the other hand, the vertical distributions of θ and θ_e during the rainy 10-day period show a more stable stratification for dry convection but a more unstable condition for moist convection. Both the slight cooling of the surface air and the remarkable warming of the upper tropospheric air contribute to the increase of the dry static stability over the Plateau. However, the surface value of θ_e has increased from 344.4 K to 362.4 K and the mean lifting condensation level of the surface air has become much lower (517 hPa). The lifted surface air, after saturation, can reach the 200 hPa level during this period.

In summary, the tropospheric heating above the Plateau prior to the onset may be explained by dry thermal convection generated near the heated surface in the afternoon hours. After the onset of summer rains this mechanism may cease to operate in the eastern Plateau because of the stabilization of the tropospheric layer. However, the heat of condensation associated with cumulus convection assumes the principal role in the heating process during the post-onset phase.

VII. SUMMARY AND DISCUSSION

In this paper we have discussed the mean large-scale conditions and heat and moisture budgets over the Tibetan Plateau and surrounding areas during a 40-day period of early summer 1979, using the FGGE Level II-b data. During this period the general circulation and the precipitation pattern showed distinct time changes characterizing the onset of the Asian summer monsoon circulation.

The main findings of this work may be summarized as follows:

1) The presence of the Tibetan Plateau has profound effects on the large-scale circulation and the distributions of moisture and precipitation. There is a pronounced inflow towards the Plateau at lower levels, and a reduction of wind speed above the Plateau is seen at the 500 hPa level. There are two centers of positive temperature anomaly above

748

the plateau. The low-level center extends from the Plateau surface to the 400 hPa level, and the upper-tropospheric center is associated with the South Asia anticyclone. The eastern Plateau acts as a huge chimney funneling water vapor from the lower to the upper troposphere.

2) The analyses of the mean horizontal distributions of the vertically integrated heat source $<Q_1>$ and moisture sink $<Q_2>$ clearly locate major heat source regions and reveal their different degrees of association with precipitation. Intense heat sources of 150-300 Wm^{-2} with moisture sinks of nearly equal magnitude appear over the Assam-Bengal region and in a broad belt extending from Thailand to the China Plain along the Mei-yu frontal zone. The heat source of $\sim$ 100-150 Wm^{-2} over the eastern Plateau is accompanied by a moisture sink of about half that magnitude. The heat sources found over the western Plateau and the Takla Makan Desert are not accompanied by appreciable moisture sinks. The heating above the Plateau is pronounced in the upper tropospheric layer between 200 and 500 hPa with a mean heating rate of $\sim$ 3K d^{-1}.

3) Detailed comparisons of the vertical profiles of the areal mean heat source $[Q_1]$ and moisture sink $[Q_2]$ with the precipitation data show distinct regional differences in the heating process and its time change among the western Plateau and its vicinity (Region I), the eastern Plateau (Region II), the Changjiang River (Region III) and the Assam-Bengal region (Region IV). Heating in the upper tropospheric layer is observed prior to the onset of the summer rains in Regions I and II. After the onset the tropospheric heating in Region II intensifies with the addition of condensation heating. The heating in Region III is primarily due to frontal rains and that in Region IV is due to highly convective monsoonal rains.

4) The sensible heat flux from the ground surface is the dominant factor in the heat budget of the Tibetan Plateau especially for the western Plateau during the early summer of 1979. The contribution from condensation heating is also significant in the heat budget of the eastern Plateau after the onset of summer rains. These findings confirm the previous results of Flohn (1968) and Yeh and Gao et al. (1979).

5) The mean vertical profiles of $[Q_1]$ for Regions I and II during

749

the dry pre-onset period are characterized by a deep heating layer with almost zero or small negative values of $[Q_2]$. These profiles are distinctly different from the mean heating profile obtained for Region II during the rain period, which shows intense heating accompanied by large positive values of $[Q_2]$. The separation between the levels of maxima of $[Q_1]$ and $[Q_2]$ during the rain period indicates the presence of active cumulus convection.

6) There is a large diurnal change of the surface air temperature on the Plateau. A deep, nearly mixed boundary layer with very high potential temperature is observed in the evening (1200 GMT). The mixed layer is most pronounced during the dry pre-onset period. It is suggested that dry thermal convection originating near the heated surface in the afternoon hours may reach the upper troposphere and deposit the sensible heat there.

The unique heating mechanism operating above the Plateau is relate to the diurnally heated boundary layer on the elevated surface. This heated boundary layer is linked to the heat low with a warm center on the Plateau (Yeh and Gao et al., 1979; Gao et al., 1981; Yeh, 1981) whose presence is reflected in the diurnal variation of the low-level inflow towards the Plateau (LY Part I).

Because of the lack of temperature soundings over the western Plateau proper, the heat budget results for this region should be considered preliminary. It is also desirable to extend the analysis period to early spring to obtain more reliable vertical heating profiles during the pre-onset phase and to clarify the heating mechanism that establishes the warm center on the Plateau. More frequent soundings and detailed vertical resolution of data are needed for a detailed examination of the boundary layer on the Plateau.

The relationship between the Tibetan heat sources and the actually observed warming of the tropospheric air during the onset of the 1979 monsoon is not immediately obvious. The time change of the mean temperature of the upper tropospheric layer between 200 and 500 hPa during the 40-day period is shown in Fig.19. As discussed by Murakami and Ding (1982), the maximum warming took place over Afganistan—the western Tibetan Plateau region and over the east China Sea and Japan.

750

A detailed analysis of the time change of the temperature field in relation to the heat sources and to the horizontal and vertical thermal advection processes will be presented in a future paper.

Acknowledgments. The authors wish to thank many colleagues, especially Professor T. Murakami of the University of Hawaii and Dr. Tsuyoshi Nitta of the University of Tokyo for their continuous interest in this study and many useful comments. Thanks are extended to Mrs. Bianca Gola for typing the manuscript and Miss Mimi Archie for drafting the figures. This work was supported jointly by the National Science Foundation and the National Oceanic and Atmospheric Administration under Grant ATM 82-04086.

References

Fein, J.S., and J.P. Kuettner, 1980: Report on the summer MONEX field phase. Bull. Amer. Meteor. Soc., 61, 461—474.

Flohn, H., 1957: Large-scale aspects of the "summer monsoon" in South and East Asia. J. Meteor. Soc. Japan, 75th Anniversary Volume, 180—186.

Flohn, H., 1968: Contributions to a meteorology of the Tibetan Highlands Atmos. Sci. Paper No. 130, Colorado State University, 120 pp.

Gao, Y.-X., M.-C. Tang, S.-W. Luo, Z.-B. Shen and C. Li, 1981: Some aspects of recent research on the Qinghai-Xizang Plateau meteorology. Bull. Amer. Meteor. Soc., 62, 31—35.

Katayama, A., 1967: On the radiation budget of the troposphere over the northern hemisphere (III). Zonal cross-section and energy consideration. J. Meteor. Soc. Japan, 45, 26—39.

Krishnamurti, T.N., P. Ardanuy, Y. Ramanathan and R. Pasch, 1979: Quick look 'Summer MONEX Atlas', Part II, The onset phase. Department of Meteorology, Florida State University, Tallahassee, 205 pp.

Krishnamurti, T.N., Y. Ramanathan, P. Ardanuy, R. Pasch and P. Greiman, 1980: Quick look 'Summer MONEX Atlas', Part III. Monsoon depression phase. Department of Meteorology, Florida State University, Tallahassee, 135 pp.

Luo, H., and M. Yanai, 1983: The large-scale circulation and heat sources over the Tibetan Plateau and surrounding areas during the early summer of 1979. Part I. Precipitation and kinematic analyses. Mon. Wea. Rev.,111, 922—944.

Luo, H., and M. Yanai, 1984: The large-scale circulation and heat sources over the Tibetan Plateau and surrounding areas during the early summer of 1979. Part II: Heat and moisture budgets. Mon. Wea. Rev., 112, (in press).

Murakami, T., 1958: The sudden change of upper westerlies near the Tibetan Plateau at the beginning of summer season. J. Meteor. Soc. Japan, 36, 239—247. [in Japanese].

Murakami, T. and Y.-H. Ding, 1982: Wind and temperature changes over Eurasia during the early summer of 1979. J. Meteor. Soc. Japan, 60, 183—196.

Nitta, T., 1977: Response of cumulus updraft and downdraft to GATE A/B-scale motion systems. J. Atmos. Sci., 34, 1163—1186.

Nitta, T., 1983: Observational study of heat sources over the eastern Tibetan Plateau during the summer monsoon. J. Meteor. Soc. Japan, 61, 590—605.

Staff members of the Section of Synoptic and Dynamic Meteorology, Inst. of Geophys. and Meteor., Acad. Sin., 1958: On the general circulation over eastern Asia (II). Tellus, 10, 58—75.

Tao, S.-Y., and Y.-H. Ding, 1981: Observational evidence of the influence of the Qinghai-Xixang (Tibet) Plateau on the occurrence of heavy rain and severe convective storms in China. Bull. Amer. Meteor. Soc., 62, 23—30.

Thompson, R.M., Jr., S.W. Payne, E.E. Recker and R.J. Reed, 1979: Structure and properties of synoptic-scale wave disturbances in the intertropical convergence zone of the eastern Atlantic. J. Atmos. Sci., 36, 53—72.

Wei, M.-Y., D.R. Johnson and R.D. Townsend, 1983: Seasonal distributions of diabatic heating during the First GARP Global Experiment. Tellus, 35A, 241-255.

Yanai, M., S. Ebensen and J.-H. Chu, 1973: Determination of bulk properties of tropical cloud clusters from large-scale heat and moisture budgets. J. Atmos. Sci., 30, 611—627.

Yeh, T.-C. (Ye. D.), 1981: Some characteristics of the summer circula-

tion over the Qinghai-Xizang (Tibet) Plateau and its neighborhood.
Bull. Amer. Meteor. Soc., 62, 14—19.
Yeh, T.-C. 1982: Some aspects of the thermal influences of the Qinghai-
    Xizang (Tibetan) Plateau on the atmospheric circulation. Arch.
    Met. Geoph. Biokl., A31, 205—220.
Yeh, T.-C. and Y.-X. Gao et al.,1979: The Meteorology of the Qinghai-
    Xizang (Tibet) Plateau. Science Press, Beijing, 278 pp. [in
    Chinese].

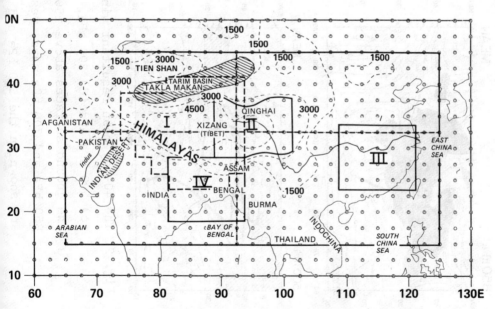

Fig. 1. The 2.5°×2.5° grid and smoothed topography (dashed contours
    in m). Vertical cross sections in Sections 3 and 4 are along
    32.5°N and along 92.5°E as shown by dash-dotted lines. Boxes
    I, II, III and IV represent four regions for detailed study
    (see Section 5).

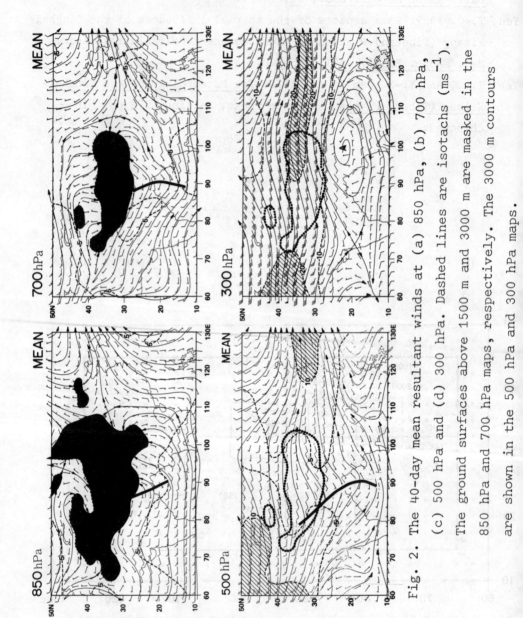

Fig. 2. The 40-day mean resultant winds at (a) 850 hPa, (b) 700 hPa, (c) 500 hPa and (d) 300 hPa. Dashed lines are isotachs (ms^{-1}). The ground surfaces above 1500 m and 3000 m are masked in the 850 hPa and 700 hPa maps, respectively. The 3000 m contours are shown in the 500 hPa and 300 hPa maps.

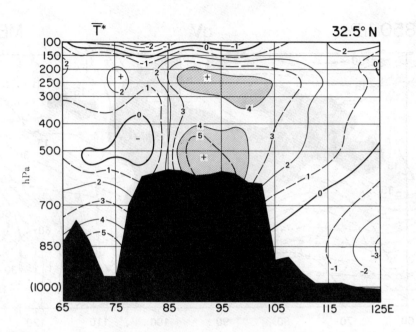

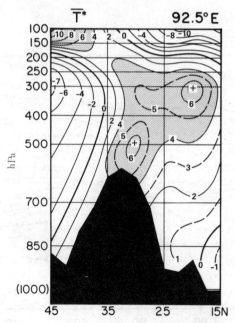

Fig. 3. (a) East-west vertical cross section showing the 40-day mean temperature anomaly (°C) along 32.5°N.

Fig. 3. (b) North-south vertical cross section showing the 40-day mean temperature anomaly (°C) along 92.5°E.

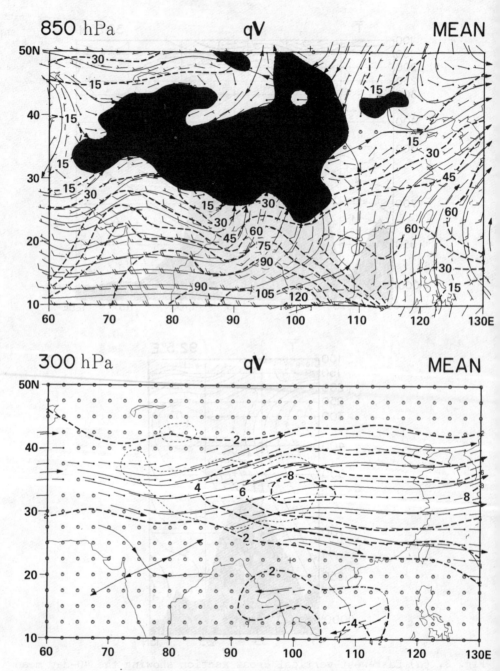

Fig. 4. Vector representations of 40-day mean horizontal transports of water vapor $\overline{q\underline{V}}$ at (a) 850 hPa (top) and (b) 300 hPa (bottom). Thick dashed lines are the isolines of $|\overline{q\underline{v}}|$ (in g kg^{-1} ms^{-1}).

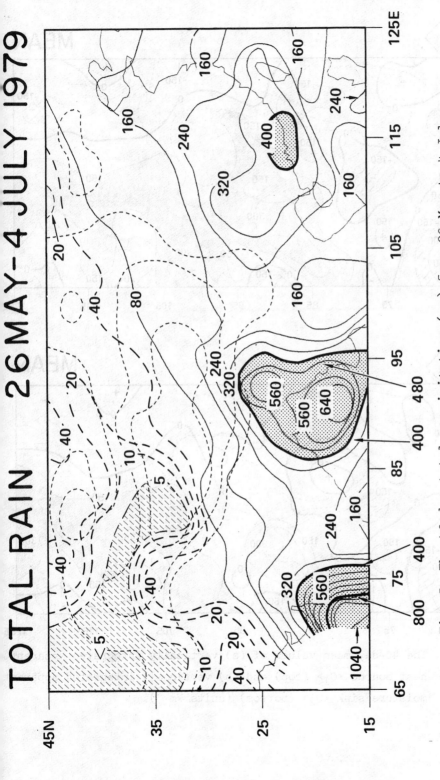

Fig. 5. The 40-day total precipitation (mm) from 26 May to 4 July 1979. Thin dashed lines are the 1500, 3000 and 4500 mm topographic contours.

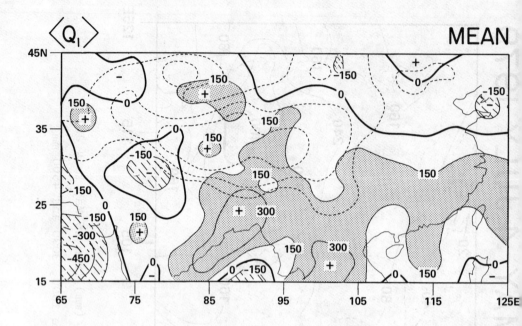

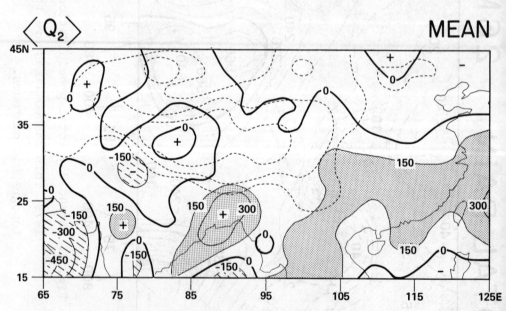

Fig. 6. The 40-day mean values of (a) vertically integrated apparent
heat source $\langle Q_1 \rangle$ (top) and (b) vertically integrated apparent
moisture sink $\langle Q_2 \rangle$ (bottom) (units Wm^{-2}).

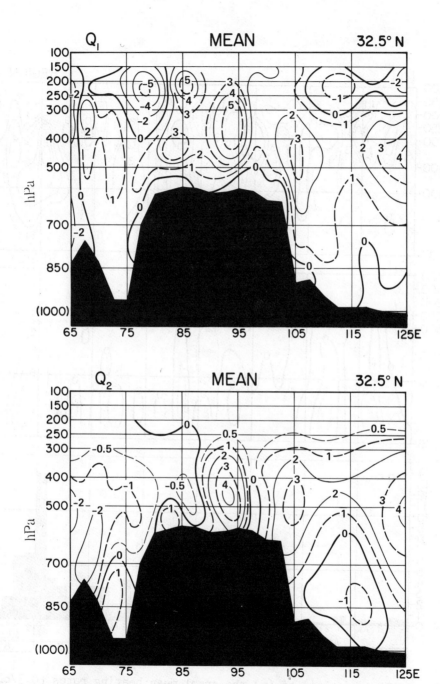

Fig. 7. East-west vertical cross sections showing (a) the 40-day
mean heating rate Q_1/c_p (K d^{-1}) and (b) the 40-day mean
drying rate Q_2/c_p (K d^{-1}) along 32.5°N.

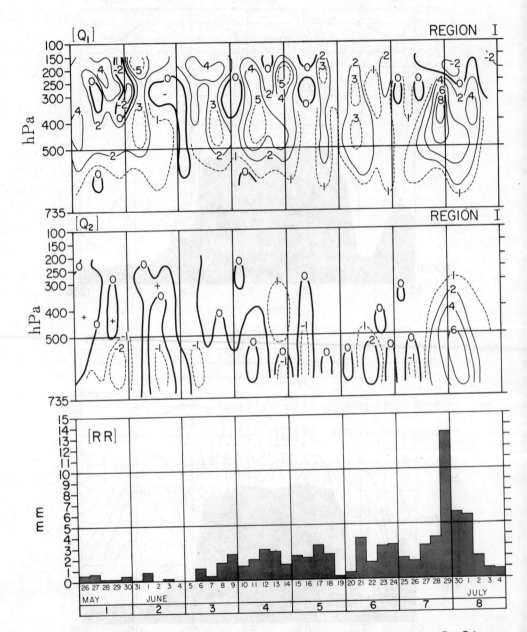

Fig. 8. Time sections of (a) the areal mean heating rates $[Q_1]/c_p$ $(K\ d^{-1})$, (b) the areal mean drying rate $[Q_2]/c_p$ $(K\ d^{-1})$ and (c) the areal mean daily precipitation $(mm\ d^{-1})$ for Region I.

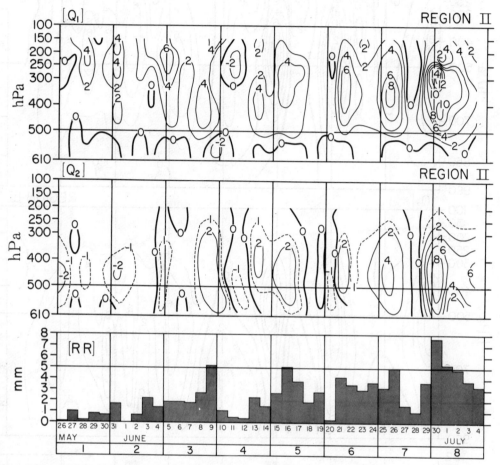

Fig. 9. As in Fig. 8 but for Region II.

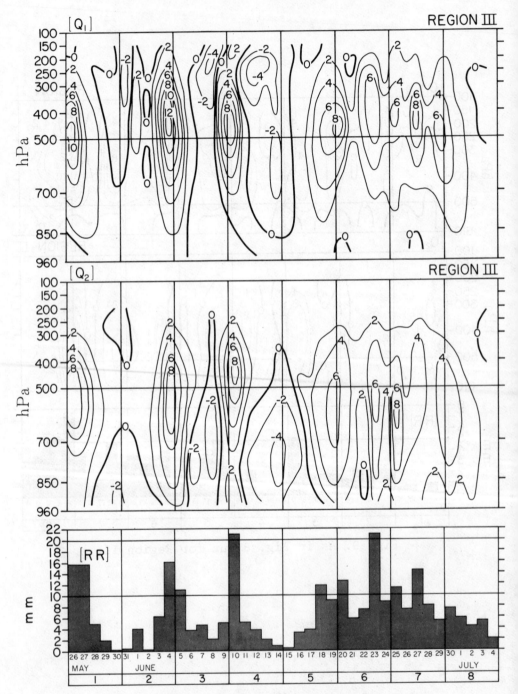

Fig. 10. As in Fig. 8 but for Region III.

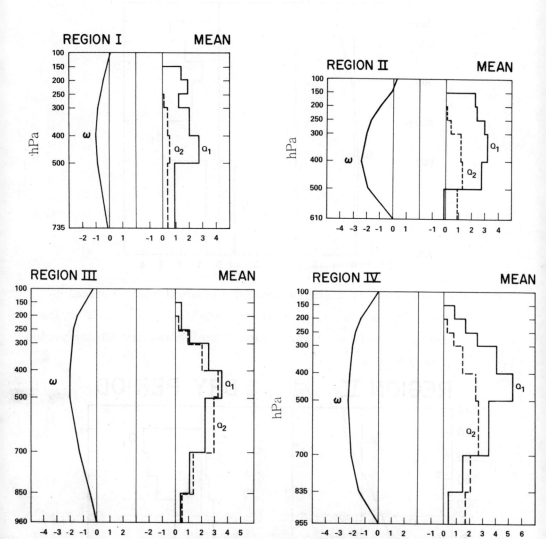

Fig. 11. The 40-day mean vertical distributions of the areal mean vertical p-velocity (hPa h^{-1}), heating rate $[Q_1]/c_p$ (K d^{-1}) and drying rate $[Q_2]/c_p$ (K d^{-1}) for (a) Region I. (b) Region II, (c) Region III and (d) Region IV.

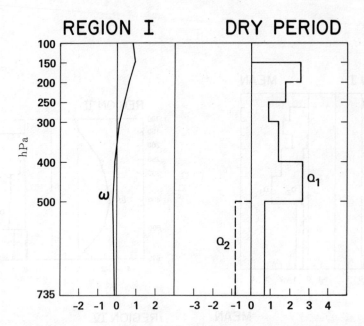

Fig. 12. Mean vertical distributions of the areal mean vertical p-velocity (hPa h^{-1}), heating rate $[Q_1]/c_p$ (K d^{-1}) and drying rate $[Q_2]/c_p$ (K d^{-1}) for Region I during the dry 10-day period (see text).

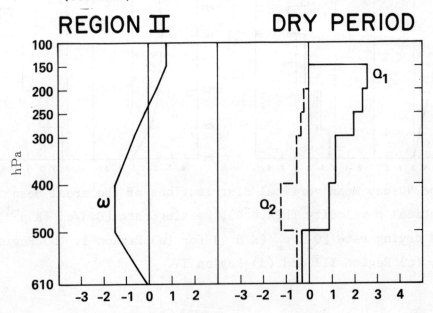

Fig. 13. As in Fig. 12 but for Region II during the dry 10-dar period (see text).

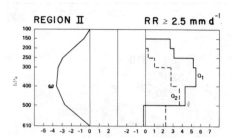

Fig. 14. As in Fig. 12 but for Region II during the rain period (see text).

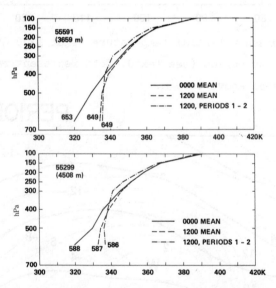

Fig. 15. Mean vertical distributions of potential temperature (K) at 0000 GMT (solid) and at 1200 GMT (dashed) for the total 40-day period, and at 1200 GMT during the first 10-day period (dash-dotted) for (a) Lhasa (29.7°N, 91.1°E) (top) and (b) Nagqu (31.5°N, 92.1°E) (bottom). The elevation of each station is indicated under the international station identification number. The mean surface pressure is shown at the lower end of each curve.

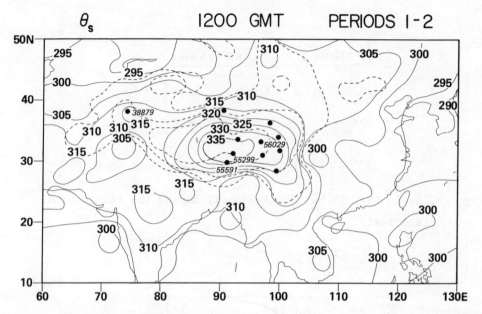

Fig. 16. The mean surface potential temperature θ_s at 1200 GMT during the first 10-day period (see text). Dots represent radiosonde stations located above 3000 m.

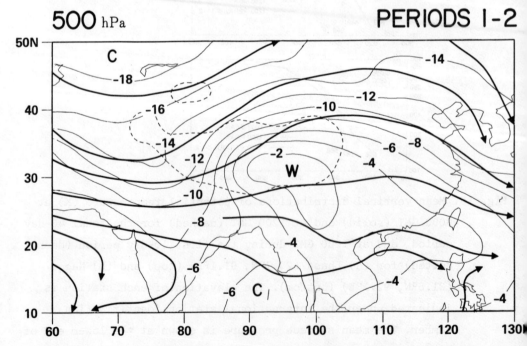

Fig. 17. The mean isotherms (°C) and streamlines at 500 mb during the first 10-day period (see text).

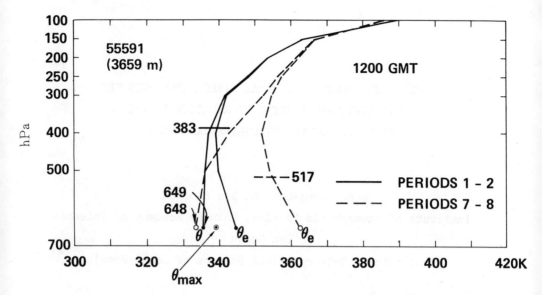

Fig. 18. Mean vertical distributions of the potential temperature
θ (K) and the equivalent potential temperature θ_e(K) at Lhasa
for the first 10-day period (solid) and for the last 10-day
period (dashed). The mean surface pressure and the mean lifting
condensation level pressure of the surface air are shown for the
respective periods. The observed maximum surface potential
temperature is shown by a circled dot.

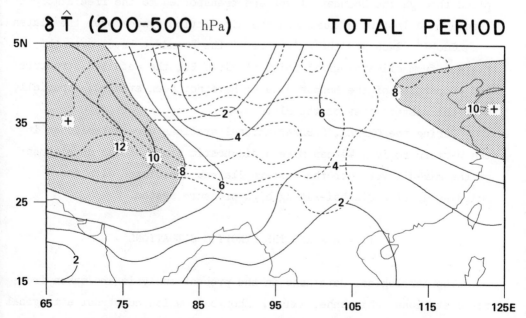

Fig. 19. The time change of the mean temperature of the upper tropospheric
(200—500 hPa) layer during the 40-day period (K).

THE DEVELOPMENTS OF THERMAL MIXED LAYER OVER THE QINGHAI-XIZANG PLATEAU IN RELATION TO THE FORMATION OF THE QINGHAI-XIZANG HIGH

Song Chengshan and Zhu Baozhen

Institute of Atmospheric Physics, Chinese Academy of Sciences

Sun Guowu

Institute of Meteorological Science of Gansu Province

I. INTRODUCTION

The Qinghai-Xizang Plateau is a huge land mass with an area of about 250 million km^2 and an average altitude of 4km forming a special boundary layer in the middle free atmosphere of the troposphere. As the Plateau is a heat source in summer, the heat energy may be supplied through the boundary layer and transported to the free atmosphere leading to a dynamic influence on the circulation system over the Plateau. Guo[1] discussed the existence of the boundary layer over the Plateau from climatological point of view. But the detailed structure and properties of the boundary layer have not been studied, especially over the western area of the Plateau.

During the QXPMEX, 3 observational stations were established in the western region. We can make a diagnostic study of the development of the mixed layer over the western Plateau associated with the building up of a Qinghai-Xizang high pressure system.

II. DATA AND THE SYNOPTIC SITUATIONS

The aerological data used in the present analysis are taken from the 3 stations (Shiquanhe, Gêrzê, Zhongba), which construct a trigonal region with an area of about 57 thousand km^2 and the length of the 3 sides are 410km(AB), 280km(BC) and 500km(AC) respectively (Fig.1).

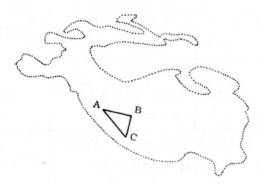

Fig.1 The calculation area.

The diagnostic analysis is made in the period from 1-15 July 1979 twice a day (00 and 12 GMT). In order to learn the detailed atmosphere structures, calculation was done on ten pressure levels (596, 550,···, 150 hPa) with the intervals about 50 hPa. The data of temperature and dew point was taken from radiosonde. Using the aerological wind data (from surface to 18km), the various dynamic parameters (divergence, vorticity and vertical motion) have also been calculated and interpolated to each pressure level.

The total daily rainfall for the three stations was used to calculate the areal mean precipitation.

During the first half of July 1979, a cold trough influenced the western Plateau at first, then the Iran high shifted eastward and the Qinghai-Xizang thermal high intensified in the upper troposphere. The main circulation features may be classified into three categories listed in table 1.

III. THE DEVELOPMENT OF MIXED LAYER OVER THE WESTERN
     PLATEAU

1. The vertical structure of the mixed layer

As well-known, the heating effect from surface can create convec-

Table 1 The main synoptic situations in the first half
of July 1979 over the western Plateau.

phases	data	synoptic feature	precipitation
Prevailing of a cold trough	2-4 July	cold trough with a cold and wet air	some
Invading of the Iran high system	5-7 July	high pressure system with a warm and dry air	few
Formation of the Qinghai-Xizang high	8-11 July	high system modified by warming and moistening	few

tive transport of heat energy due to the instability in the low layer.
A certain atmospheric quantity can be mixed in a deep layer resulting
in a mixed layer characterizing a uniform distribution of conservative
parameters.

The equivalent potential temperature (θ_e) is one of the conservative parameters for wet air. Fig.2 depicts its vertical distribution
at 00 and 12 GMT for three phases with following features:

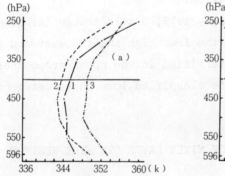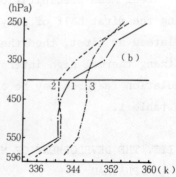

Fig.2 The vertical distribution of equivalent potential
temperature $\bar{\theta}_e$ for three phases. (a) 12GMT;
(b) 00GMT (unit: K).

(1) In the day-time (Fig.2a), an unstable layer below the 550 hPa is developed during the 1st phase and an unstable layer extends to 500 hPa during the 2nd phase, the instability is most unstable during the 3rd phase.

Under 550 hPa, the surface layer becomes stable in the early morning due to surface cooling (Fig.2b), showing a strong diurnal variation in the surface layer.

(2) The mixing process prevails between 400-500 hPa layer, in which the $\bar{\theta}_{\acute{e}}$ only has small changes. We can see a very uniform and fully mixed layer during the 2nd phase between 400-550 hPa (Fig.2b).

(3) Above 350-400 hPa, the $\bar{\theta}_e$ increases with height rapidly, displaying a stable stratification capped over the mixed layer. In the following we shall consider 400 hPa pressure level as the top of the mixed layer.

From Fig.2 we can see that the existence of thermal mixed layer is independent of the synoptic system. Therefore, we can say that the mixed layer is an inherent phenomenon over the Plateau.

2. The variation of the height of the mixed layer

The state of the mixed layer may be influenced by the synoptic situations. During the period of cold trough (phase 1), the height of uniform layer of $\bar{\theta}_e$ (i.e. the top of the mixed layer) is near 450 hPa, above which the $\bar{\theta}_e$ curve has inclined rapidly toward higher values, displaying that the stability increases especially above 350 hPa.

In comparison, when the high pressure system stagnated over this region (phase 3), the curve of $\bar{\theta}_e$ has a small inclination from 400 to 350 hPa level, showing the strengthening of mixed process. It seems that the effect of heating caused by the Plateau has reached the higher level of atmosphere (Fig.2b).

That the heights of mixed layer have an apparent diurnal variation is another feature. In the daytime, the uniform layer lifts to about 350 hPa due to the strong heating process.

IV. THE DYNAMICAL STRUCTURE OF THE MIXED LAYER IN RELATION TO THE FORMATION OF THE QINGHAI-XIZANG HIGH

1. The characteristic divergence field of mixed layer

At first, we calculate the areal mean divergence, vorticity and vertical motion using the data of wind at each level by the areal mean continuity equation of mass:

$$\overline{\nabla \cdot \mathbf{V}} + \overline{\frac{\partial \omega}{\partial p}} = 0 \tag{1}$$

Vertical motion can be obtained from (1) with the boundary condition:

$$p = p_0 = 596 \text{ hPa}, \quad \omega = \overline{\frac{\partial p_0}{\partial t}} + \overline{\mathbf{V} \cdot \nabla p_0}$$

$$p = 60 \text{ hPa}, \quad \omega = 0$$

The vertical distribution of divergence over the western Plateau has a characteristic feature. Its pattern for the three phases is given in Fig.3. There is a very thin convergence layer below 500 hPa

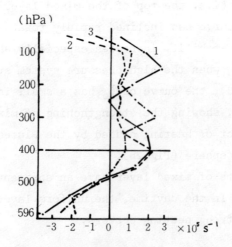

Fig.3 As Fig.2, but for divergence
(unit: 10^{-5}s^{-1}).

with a thick divergence layer in the middle and upper tropospheres.

It is very interesting to note that a nondivergence surface is at the level around 500 hPa (1.5km from surface). Though the whole mixed layer is quasi-nondivergent, the total atmosphere column over the western Plateau is divergent. On the average, there is a maximum of divergence about $2.2 \times 10^{-5} s^{-1}$ at 400 hPa level and the convergence value within the near surface layer has the value $-3.4 \times 10^{-5} s^{-1}$, which is independent of synoptic situations. It can be considered to be the inherent properties of the Plateau itself and able to give a dynamic effect on the circulation systems over the Plateau.

2. The surface sensible heat flux

The areal mean of the thermodynamical equation may be written:

$$\frac{\partial \bar{T}}{\partial t} + \overline{\nabla \cdot \mathbf{V}_T} + \frac{\partial \overline{\omega T}}{\partial p} - \frac{R}{c_p} \frac{\overline{\omega T}}{p} = \frac{L}{c_p} m^* + \frac{Q_R}{c_p} \tag{2}$$

The symbols are defined as follows:

'——': areal average of the quantity.

' ′ ': deviation from the areal average showing the role of the subgrid scale systems.

m^* : the rate of condensation and evaporation.

Q_R : the radiational heating.

In the mixed layer, the vertical convection becomes dominant. It is reasonable for us only to hold the vertical eddy term in the mean equation. Therefore the thermodynamic equation can be written as:

$$\frac{D\bar{T}}{Dt} + \frac{\partial}{\partial p} \overline{\omega'T'} = \frac{L}{c_p} m^* + \frac{Q_R}{c_p} \tag{3}$$

where

$$\frac{D\bar{T}}{Dt} = \frac{\partial \bar{T}}{\partial t} + \overline{\mathbf{V} \cdot \nabla_T} + \bar{\omega}(\frac{\partial \bar{T}}{\partial p} - \frac{R\bar{T}}{c_p p})$$

It can be obtained from the synoptic scale observational data.

Above the mixed layer, the vertical eddy flux of temperature may be negligible considering the stable stratifications. Therefore, integrating equation (3) from the surface to the top of the mixed layer (P_T=400hPa) we give the following expression of the heat flux on the ground.

The surface sensible heat flux:

$$F_S = - \frac{c_p}{g} \overline{\omega'T'}\Big|_{P_0} = \frac{1}{g} \int_{P_T}^{P_0} (\frac{D\bar{T}}{Dt} - L_m* - Q_R)dp$$

where the integrated value of m* is taken from the rainfall observation and Q_R is assumed as $-1.5°C \cdot d^{-1}$.

The value of F_S during phases 1, 2 and 3 are 102, 184 and 232 $W \cdot m^{-2}$ respectively. In general, they are larger than Yiao's results[2]

3. The transformation of the dynamic Iran high

The dynamic characteristic structure of mixed layer and the heat energy supplied from the Plateau exerts a great influence on the circulation system.

The dynamical characteristic distributions of horizontal divergence within the mixed layer have an inherent properties as shown in Fig.3. They are not dependent of different phases. They will affect the synoptic system passing over the Plateau. The convergence field below 500 hPa is (not) favorable to the maintenance of positive (negative) vorticity, while the divergence above 500 hPa is (not) favorable to the existence of negative (positive) vorticity. Therefore, when the Iran high shifts to the Plateau (as phase, 2, 5-7 July), the cyclonic vorticity rapidly changes to anticyclonic vorticity and the downward motion prevails in this period. Then the negative vorticity developed upward but the vertical motion changed from downward motion to upward motion (Fig.4,5). It means a dynamic Iran high had been transformed to a thermal Qinghai-Xizang high, owing to the strong heating supplied from the ground surface, which, in turn, would be favorable to the maintenance of the positive vorticity below 500 hPa. When the high system invades to the Plateau, it cannot stay below 500 hPa due to the strong convergence and the heating effect. But the compensation divergence from 500 to 300 hPa provides a favorable environment for the maintenance of the upper high system.

REFERENCES

[1] Yeh, T. C. and Gao, Y.X. et al., The meteorology of the Qinghai-Xizang Plateau, Science Press, 1979 (in Chinese).

[2] Yiao, L.C. et al., The characteristics of the monthly neating field distribution in the Qinghai-Xizang Plateau and its surrounding region in the summer of 1979, Collected papers of QXPMEX, Science Press, 1984 (in Chinese).

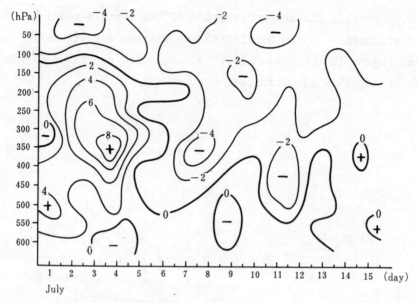

Fig.4 The time variation of vorticity in the
first half of July 1979 at different
levels (unit: $10^{-5}s^{-1}$).

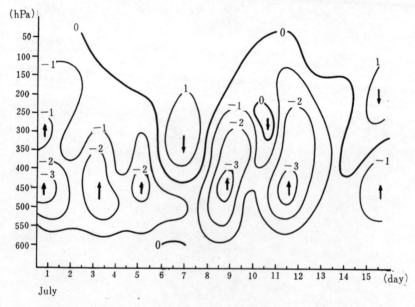

Fig.5 As Fig.4, but for vertical velocity
(unit: $10^{-3}hPa \cdot s^{-1}$).

THE TROPOSPHERIC ENERGY BUDGETS OF TWO HEAT SOURCE REGIONS WITHIN THE SOUTHWEST-EAST ASIAN SUMMER MONSOON SYSTEM: THE ARABIAN DESERT EMPTY QUARTER AND THE QINGHAI-XIZANG PLATEAU

Eric A. Smith

Department of Atmospheric Science

Colorado State University

1.0 INTRODUCTION

One of the fascinating aspects of the Southwest-East Asian Summer Monsoon is its multiplicity of component parts and scales of motion. Earlier papers by Krishnamurti and his co-workers have carefully high-lighted this aspect of monsoon dynamics; see e.g., Krishnamurti et al. (1973, 1977, 1981), Krishnamurti and Bhalme (1976), and Krishnamurti and Ardanuy (1980). In recent years, due in large part to the new data sets resulting from the 1979 First GARP Global Experiment (FGGE) and its embedded Summer Monsoon Experiment (SMONEX), it has been possible to examine in detail the various atmospheric phenomena making up the individual parts of the monsoon. An important focus of the 1979 Summer Monsoon Experiment, and of principal concern here, is the heat sources and sinks sub-programme. This sub-programme was designed to identify, quantify, and examine the development, maintenance, and role of the important heat source and sink regions within or in the proximity of the overall monsoon system.

This investigation focuses on two of these regions; the interior southern desert of the Arabian Peninsula (i.e., what the British designated as the Arabian Empty Quarter), and the extensive mountainous highland of the Qinghai-Xizang Plateau (Tibetan Plateau). It would be difficult to find two more sharply contrasted regions within the context

of the Summer Monsoon. In climatological terms, the Empty Quarter is
a vast, low elevation, sub-tropical sand desert, extremely dry towards
its interior, virtually free of vegetation, undergoing moderate seasonal
but extreme diurnal variations (both thermal and dynamic), and during
the monsoon season lacking any significant cloud or rainfall processes.
The Tibetan Plateau, on the other hand, is an elevated desert (≈ 4500
meters), highly influenced by and highly influencing both the mid-
latitude westerly and tropical-monsoonal circulations, undergoing both
seasonal and diurnal variation extremes, and during the Summer Monsoon
season exhibiting a wide variety of cloud and precipitation systems
which help support the extensive Tibetan summer upper tropospheric
anti-cyclone (the Qinghai-Xizang high).

Setting aside climatological differences, however, there are
basic similarities in how these regions manifest themselves energeti-
cally during the monsoon period. Foremost, both regions serve as important atmospheric heat sources. This feature of the summer Tibetan
Plateau has been understood for many years (i.e., its role as a summertime elevated heat source); however, it is only since the 1979
Summer MONEX experiment that we have acquired an understanding of the
heat source character of the thermal heat low region over the Arabian
Peninsula. This finding negates, in a sense, our classic notion of
desert heat lows as energy deficit regions.

Both regions maintain stationary surface heat lows. The low over
the Arabian Peninsula is locked to the center of the Empty Quarter, is
relatively shallow, and is driven primarily by surface sensible heating
and shortwave absorption into a dust-laden mixed layer; the low over
the Tibetan Highlands is generally tied to the southeastern corner of
the Plateau, is relatively deep, and is maintained by both sensible and
latent heating processes, as well as radiative convergence. Flohn
(1968) described this feature as a giant heat chimney serving to help
maintain the rather unique monsoon anticyclone in the high troposphere.
Both of these heat sources lie at the boundaries of major monsoon
features; i.e., the low level Arabian jet and the monsoon trough respectively. Therefore, both of these heat source regions are intrinsi-
cally tied to monsoon processes.

778

There are many remaining questions concerning the development, onset, and maintenance of these heat sources, and their exact role in monsoon dynamics. A great deal of uncertainty is due to the lack of a thorough data set describing top-of-atmosphere radiative exchange, surface energy budget, and the atmosphere water budget; these later two processes have been examined with existing data sets by Luo and Yanai (1983, 1984) and Yeh and Gao (1979). Of particular interest to many are the remaining uncertainties in the relative magnitudes of the three principal diabatic heating terms (sensible, latent, and radiative) driving the Tibetan Plateau heat exhaust; see Luo and Yanai (1984) and Nitta (1983).

This investigation focuses on the problem of determining the tropospheric energy budgets in these two regions. The methodology involves combining satellite retrievals of radiation and precipitation with surface energy budget measurements. Radiance data derived from polar orbiting and geosynchronous weather satellite instruments (AVHRR and VISSR) are used to estimate the top-of-atmosphere radiation budgets at a high time and space resolution scale. Lower resolution radiation budget measurements available from the Nimbus 7 Earth Radiation Budget (ERB) instrument are used to validate these estimates. Microwave brightness temperatures available from the Scanning Multi-channel Microwave Radiometer (SMMR), also flown on Nimbus 7, are used to provide an understanding of the precipitation process. Finally, specialized surface energy budget monitoring stations, which were developed at Colorado State University, are used to generate the surface energy budget parameters.

First, results from an already completed experiment concerning the Arabian heat low are presented in which the above measurement strategy has been employed. In this discussion a description of the principal measurement composites is provided. The vertical structure and possible role of the Arabian heat low is presented. Second, a description is given of an improved surface energy budget and boundary layer monitoring system, targeted for deployment on the Tibetan Plateau as part of a proposed new surface energy budget monitoring program. Included in this section is a discussion of various preliminary tests of the new system, taken at a mountain site in the Colorado Rocky Mountains. Finally, various satellite depictions of plateau processes are presented. Emphasis is given to the objective of coupling surface information with the satellite data in order to close the boundaries of tropospheric energetics.

2.0 THE ARABIAN PENINSULA EMPTY QUARTER EXPERIMENT

Based on a surface energy budget measurement system similar to the 'Radiation Station' which is discussed in Section 4.0, a fairly thorough characterization of the heat exchange process within the interior of the Arabian Empty Quarter during the pre-monsoon-monsoon onset phase of 1981 has been obtained; see Smith et al (1984). These measurements initiated a wider-scale Arabian desert monitoring program also discussed in Smith et al (1984). The salient features of the surface terms within the Arabian heat low region are illustrated in Figures 1-4. These figures demonstrate the rather dramatic diurnal directional reflectance characteristics (Fig. 1), the diurnal surface radiative process (Fig. 2), the diurnal sub-surface thermal waves (Fig. 3), and finally, the diurnal energy processes of net radiation (Q^*), storage (S), and sensible heat transfer (SH) (Fig. 4). By coupling these measurements with aircraft and GOES-1 satellite measurements obtained during the 1979 SMONEX via a multi-platform monitoring strategy (see Figures 5-7), a characterization of the principal radiative processes has been obtained (see Figures 8-9). These results and additional measurements obtained from the Nimbus 7 Earth Radiation Budget (ERB) Narrow-Field-of-View (NFOV) instrument (see Figures 10 and 11), along with the observational results of Blake et al (1983) describing the thermodynamic profiles and vertical motion fields, the calculations of Ackerman and Cox (1982) describing an enhanced shortwave aerosol absorption effect over the deep heat low mixed layer, and the modeling results of Freeman (1984), provide a composite picture of the vertical structure of the Arabian heat low (see Figure 12).

The important features of the heat low, in part depicted in

Figure 12, are as follows:

1) The heat low region is a tropospheric energy source-a finding in contrast to the classic view of desert heat lows.

2) This is mostly due to the enhanced shortwave aerosol absorption into the very deep mixed layer created by the low level convergent cyclonic circulation associated with intense daytime surface heating.

3) This enhanced solar absorption leads to a neutral net radiation condition seen at the top-of-atmosphere (see Figure 11), again in contrast to a widely held view of deserts as strongly deficit in net radiation; see Charney (1975).

4) The near-infrared surface albedo is over twice that of the visible spectrum albedo.

5) The shallow surface cyclone is maintained throughout the evening and into the night-not disappearing until early morning.

A possible outcome of this process, in terms of the role of the Arabian heat low within the Southwest Monsoon system, is that the excess tropospheric energy source is transported over the western Arabian Sea and serves to maintain the strong low level thermal inversion existing in this region (see Figure 13). This, in turn, serves to stabilize a region of the Arabian Sea through which the Somalian jet can then transport an important moisture source along a long Arabian Sea path into the Indian sub-continent. Indirect evidence of this effect is based on the structure of the Arabian Sea low level inversion; it is known to weaken both towards the equator (south) and towards the Indian sub-continent (east), that is, away from the horizontal heat exhaust region.

The next step in the process of understanding radiative exchange within other regions of the monsoon system is to compare and contrast the Arabian region with other regional components. This has been achieved by generating detailed radiative budget maps covering the complete monsoon system (see Figure 14). These data have been derived from TIROS-N AVHRR measurements which have been transformed into radiative fluxes based on a parameterization scheme developed by Smith (1984). Illustrations of the principal radiative terms prior to,

781

during, and after the 1979 monsoon onset period are given in Figures
15-17 for the latitude belt 0-35°N. It is noted here that the radiative process taking place over the Tibetan Plateau clearly illustrates
the multi-cellular pattern characteristic of the summer plateau circulation systems; see Yeh (1981).

Finally, Figures 17 and 18 illustrate three-dimensional isopleths
of the zonal and meridional averages of the albedo and infrared emittance over the monsoon region. Note how the zonal averages tend to dis-
guise the effect of monsoon disturbances in the radiative parameters;
the meridional averages tend to suggest that monsoon disturbances move
more in an east to west sense, than south to north, as the monsoon
develops.

3.0 A PROPOSED NEW MEASUREMENT PROGRAM ON THE TIBETAN PLATEAU

In order to depict the energetics of the Tibetan heat source along
the lines previously discussed for the Arabian heat low, a new measuring
program is in progress as a joint research effort between the Department of Atmospheric Science, at Colorado State University, and The
Academia Sinica and State Meteorological Administration of the People's
Republic of China.

The proposed measurement program on the Qinghai-Xizang Plateau is
directed at two separate mountain meteorology topics. The first concerns the impact of the plateau on the Southwest Indian-East Asian
Summer Monsoon and the large scale planetary wave structure; the
second concerns improving the knowledge of the surface energy budget of
the plateau and its impact on the overall tropospheric heat budget
profile. The 1984 expedition is in part planned as an information and
technology exchange with plans to collect an initial data set for preliminary analysis. One dual station, as illustrated in Figures 20 and
21, will be delivered to the PRC in March 1984. The objective, in the
following years, is to deploy various stations at different sites, in
recognition of the fact that there are large meridional and zonal
gradients of heat sources and moisture sinks, as well as multiple vertical circulation cells, over the vast Tibetan highland region.

The role of the Tibetan Plateau as both an orographic barrier and an elevated heat source on the large-scale circulation has been well established through the examination of synoptic data, annulus experiments, and numerical model experiments. Its downstream impact on weather systems in Northern and Eastern China and its important role in setting the timetable for the evolution, positioning, and maintenance of the Southwest Monsoon and interacting with the monsoon surges and breaks, and its controlling influence on replacing the subtropical jet with the tropical easterly jet in conjunction with the upper tropospheric plateau anticyclone during the monsoon season are now well accepted ideas. Although modeling results are not always realistic, it has required numerical model experiments and model frameworks to provide some of the essential detail in the vertical and horizontal structure of the dynamic and thermodynamic fields over and adjacent to the plateau, their seasonal variability, and their diurnal characteristics.

In recent years, based on the FGGE data sets and long-term Chinese data archives, some of the first detailed accounts of the atmospheric and surface energy budget processes over the plateau have been published; see Luo and Yanai (1984) and Yeh and Gao (1979). Nevertheless, there remain gaps in the data, particularly concerning radiative properties, radiative heating rates, precise estimates of sensible and latent heat exchange and subsurface heat storage. These data gaps provide the motiviation for our proposed experiments.

There is still much to be learned about plateau meteorology; the similarities and differences between the roles of the Qinghai-Xizang Plateau and the North American Rocky Mountains; the structure and role of the Summer Plateau heat low and its comparisons and contrasts to desert heat lows; the manner in which the surface radiation budget controls the sensible heat exchange process; and finally, the nature of the diurnal and seasonal cycles of deep soil temperature, their interannual variation and the impact of these processes on the planetary circulation and teleconnective feedback cycles.

An examination of the Tibetan Plateau from a satellite perspective alone provides some measure of the similarities and differences with respect to the Arabian heat low region. The time series of radiation budget (RADBUD) parameters (net flux, albedo, and day-night infrared emittances) over the course of the 1979 monsoon season, given in Figure 22, illustrate that the principal difference between the two regions is the positive net flux term (Q^*); the mean value over the monsoon season is on the order of 80 $W \cdot m^{-2}$ beginning in mid-May, decreasing to 30 $W \cdot m^{-2}$ at the end of August. As was illustrated in Figure 11, the Arabian heat low region remains radiatively neutral at the top-of-atmosphere throughout the course of the monsoon season. Thus the heat surplus nature of the plateau region is due, in part, to the radiative process, the bulk of which results from in-cloud absorption once the summertime convection process is initiated.

Another key feature seen in Figure 22 is the oscillatory nature of the RADBUD signals. In Figures 23a-c the periodograms of the albedo, infrared emittance, and net radiation time series (derived from the TIROS-N radiation budget estimates) are given. There is a clear dominant harmonic with a period of approximately 8 days ($\approx$ 1 week) which is about half the period of the low frequency monsoon cycles emphasized by Krishnamurti and Bhalme (1976), Sikka and Gadgil (1980), and Webster (1983). This periodicity is easily noted in the Nimbus 7 time series; it is also evident that the perturbation cycle begin prior to the beginning of June—well in advance of the monsoon onset. This is most likely an indication of the Burma/Assam-Bengel monsoon which preceeds the Southwest monsoon and is somehow tied to periodic short waves in the sub-tropical westerly jet, directly to the south of the South-eastern plateau. The same features are duplicated in the periodograms of albedo and infrared flux time series over the Arabian peninsula; however, the periodicity is not evident in the Q^* term. A clear difference in the Arabian RADBUD process is that the low frequency oscillations do not initiate until the beginning of July.

A final satellite rendition of the Tibetan Plateau is demonstrated by the use of Nimbus 7 Scanning Multichannel Microwave Radiometer (SMMR) data. Figures 24a-b provide histograms of the horizontal, vertical, and non-polarized brightness temperatures from Channel 5 (37 GHz), along with the histogram of the polarization factor, for the months of May and June, 1979, It is noted that the June analysis indicates a slight bulge and shift to the warmer temperatures ($\approx 10°$) as the East Asian monsoon commences. There is also a small redistribution of the permutations in the ordering of brightness temperatures for Channel 3(18 GHz), Channel 4(21 GHz), and Channel 5; these figures, given in percentage tables, are seen at the bottom of the histogram plots. Whether this effect can be deconvolved into rainfall statistics is a theoretical problem still under study.

4.0 A SURFACE ENERGY BUDGET MONITORING SYSTEM

The system developed for this project is configured in two parts-a radiation, rainfall, wind, state parameters, and subsurface heat and moisture monitoring station (referred to as the Radiation Station), and a four-level tower eddy flux monitoring station (referred to as the Tower Station). Schematic illustrations of the Radiation Station and the Tower Station are provided in Figures 20–21. All sensors on the two systems are interfaced with programmable, microprocessor driven data loggers which periodically record their memory contents onto conventional cassette tape recorders. The data logger and recording electronics are powered by rechargeable battery packs which are con-tinually repowered (during sunny conditions) by single-panel solar energy collector converter systems.

In the design of this system conventional sensor technology has been incorporated. The design effort has been more concerned with system reliability, portability, low maintenance, automation and cost, than it has with the use of sophisticated sensors. In this regard these systems have proved to be of great success.

In order to implement the measurement program, to be conducted in both Tibetan Plateau and Rocky Mountain regions, a full dual-station configuration, as illustrated in Figures 20–21, has recently been deployed (September, 1983) at a mountain site, west of Fort Collins, Colorado. The objectives of this preliminary mountain measurement program were to carry out trial preparations, shakedown tests, and sensor sampling and integration tests essential for the expedition to China taking place in March-April, 1984, in conjunction with the

International Symposium on the Qinghai-Xizang Plateau and Mountain
Meteorology. The eventual target site for an initial operational experiment will be near Lhasa, at approximately 30°N-91°E.

The mountain site presently being used is directly north of the
Mummy Range in Rocky Mountain National Park, at a Colorado State University Forestry campus called Pingree Park (40°N, 105°40'E). The elevation at this site is approximately 9,500 feet (2900 meters). The site
itself is located in a mountain valley just east of the Continental
Divide. The Radiation and Tower Stations were erected in a forest
clear-cut adjacent to a meadow area which lies in a major part of the
headwaters of the Poudre River drainage basin. The Pingree site will
be used along with a site on Storm Peak near Steamboat Springs, Colora-
ado, to carry out cooperative measurement programs in conjunction with
the experimental program in China. A significant aspect of the research
is to understand the analogous energetics roles of both plateau systems.

In the following, results are presented from two brief Tower
Station experiments that were essential before launching a measurement
program in China. The first experiment, called the 'Sampling Interval
Test', was designed to determine the effect of the data logger sampling
time on the derived mean quantities and eddy flux terms. The second
experiment, called the 'Integration Time Test', was designed to determine the effect of the averaging time period (eddy flux bar operator)
on the derived parameters. The first test was required to determine if
a sampling time of one second is sufficient to monitor vertical flux
structure. One second sampling is the upper allowable limit with the
eddy flux firmware unit in the Tower Station's data logger and also
close to the order of the response time of the tower's sensors. The
second test was needed to define a maximum integration time for characterizing the turbulent fluctuations. From an operations point of view,
the longer the specified integration time, the longer the station can
remain unattended because of tape recording space. It is emphasized
that these are relatively qualitative tests and yet they reveal much
about the impact of sampling and integration times on monitoring the
turbulent boundary layer parameters.

Before discussing the Tower Station test results, Figure 25 is presented as a means to characterize the data from the Radiation Station and to set the tone for the typical background meteorological conditions during mid-autumn at the Pingree site. In this figure, the state parameters, wind parameters, soil temperature and moisture parameters, and total solar radiation parameters are presented for the period September 24 to October 6, 1983. Data samples were taken every 15 minutes. It is clear from this figure that there are obvious diurnal periodicities in the thermal parameters (both above and below the surface), that the subsurface is undergoing a cooling trend, and that the winds tend to show very little diurnal preference in either magnitude or direction. The soil is drying out during this period (there was no rain throughout), except at the deepest level (40 cm) where the moisture remained relatively constant (the surface probe dried completely on October 3 and went off scale). It is apparent from the solar fluxes (these have been plotted in engineering units because the calibration constants were unavailable at the time) that the first five days underwent intermittent cloudiness, whereas the last three days were relatively cloud free. The cloudiness has a dramatic effect on destroying the uniformity of the diurnal waves in the air temperature and moisture parameters, as well as deamplifying the subsurface soil temperature waves.

The key results from the 'Sampling Interval Test' are shown in Figure 26. This experiment was conducted during midday on October 4, 1983, a relatively cold autumn day, with nearly neutral PBL conditions, and undergoing intermittent, light snow showers. Six different sampling intervals were incorporated in the runs; 1, 2, 3, 4, 5, and 10 second. Each run lasted 20 minutes. The integration remained at 1 minute throughout the two-hour test. In Figure 26 the vertical velocity-temperature correlations (proportional to eddy heat flux) are plotted. These results indicate that neither the time-scale of the turbulent plumes nor the resolution of their magnitude are highly impacted by degrading the sampling interval. This gives us reasonable confidence that we will be successful when operating at the 1-second sampling time interval.

The results from the 'Integration Time Tests' are shown in

787

Figure 27. These tests were carried out on October 27 over a 5 hour, 30 minute period on a relatively warm and slightly unstable autumn day. The integration time was initially set to 1 minute at 11:00 a.m., then degraded to 2 minutes at 11:40, to 4 minutes at 12:20, to 8 minutes at 13:00, and finally to 16 minutes at 15:00. Again the correlations proportional to eddy heat flux are shown. Whereas the high frequency plume structure can only be resolved from the 1-to 2-minute integration times, it is evident that the main features of the overall sensibl heat transport process are reasonably well depicted at an integration time scale of on the order of 10 minutes. Thus, for situations when the 'Tower Station' would go unattended for extended periods (30 days), a 10-minute integration time would be appropriate. For instances when an operator is available at the site, the 1-minute integration time would be preferable.

5.0 CONCLUSIONS

An energy budget monitoring program carried out in remote desert and mountain environments has been described. This program has been designed to characterize the tropospheric boundary terms in lowland and highland deserts known to interact with the Southwest-East Asian Summer Monsoon. In order to implement the surface measurement program, an energy budget system has been developed and tested. This system has demonstrated the capability of returning quantitative and detailed measurements of the surface energy exchange process. The systems have proved dependable, relatively maintenance free, and their automated microprocessing capability have eliminated much of the lag time between observation and analysis due to carrying out most of the calibration and data reduction procedures (statistical processing) in the field. They are relatively inexpensive, easy to operate, and will provide excellent ground truth needed for the development of satellite algorithms and parameterizations.

The systems have been deployed in the deserts of Saudi Arabia and the Colorado Rocky Mountains. A dual-system configuration is soon to be deployed on the Tibetan Plateau. The results of the Arabian

measurement program illustrate not only the daily and diurnal processes
that characterize the desert interior, they also illustrate some unique
radiative features characteristic of excessively dry desert regions
and have provided critical surface boundary information needed in the
description and understanding of the Arabian Desert heat low.

The tests at Pingree Park have demonstrated the feasibility of
using less expensive, but maintenance free sensors for monitoring
components of the PBL turbulent heat exchange process. The plans to
deploy a dual-station configuration in China have been described.
Finally, discussion of how the surface energy budget stations can be
used in conjunction with satellite data to complete the specification
of the boundary conditions essential in any thorough treatment of
atmospheric energetics. In this context, various examples and approaches
have been provided of how to implement this methodology for both the
Arabian Desert and the Tibetan Plateau.

6.0 ACKNOWLEDGMENTS

This research has been supported by National Science Foundation
Grants ATM 82-00808 and ATM 83-04813 and the Air Force Office of
Sponsored Research under Grant AFOSR 82-0162. Part of the computations
were performed at the National Center of Atmospheric Research, a
division of The National Science Foundation.

7.0 REFERENCES

Ackerman, S. A. and S. K. Cox, 1982: The Saudi Arabian heat low:
    Aerosol distribution and thermodynamic structure. J. Geophys. Res.,
    87, 8991-9002.
Blake, D. W., T. N. Krishnamurti, S. V. Low-Nam and J. S. Fein, 1983:
    Heat low over the Saudi Arabian desert during May 1979 (Summer
    MONEX). Mon. Wea. Rev., 111, 1759-1775.
Charney, J., 1975: Dynamics of deserts and drought in the Sahel. Quart.
    Jour. Roy. Meteor. Soc., 101, 193-202.
Flohn, H., 1968: Contributions to a meteorology of the Tibetan highlands.

Atmospheric Science paper No. 130, Colorado State University, Fort Collins, CO, 120 pp.

Freeman, L., 1984: Ph.D. Dissertation-In Progress. Dept. of Atm. Sci., Colorado State University, Fort Collins, CO.

Krishnamurti, T. N., S. M. Daggupaty, J. Fein, M. Kanamitsu and J. D. Lee, 1973: Tibetan high and upper tropospheric circulations during northern summer. Bull. Amer. Meteor. Soc., 54, 1234-1249.

Krishnamurti, T. N. and H. N. Bhalme, 1976: Oscillations of a monsoon system. Part I. Observational aspects. J. Atmos. Sci., 33, 1937-1954.

Krishnamurti, T. N., J. Molinari, H. L. Pan and V. Wong, 1977: Down-stream amplification and formation of monsoon disturbances. Mon. Wea. Rev., 105, 1281-1297.

Krishnamurti, T. N. and P. Ardanuy, 1980: The 10 to 20-day westward propagating mode and 'Breaks in the Monsoons'. Tellus, 32, 15-26.

Krishnamurti, T. N., P. Ardanuy, Y. Ramanthan and R. Pasch, 1981: On the onset vortex of the summer monsoon. Mon. Wea. Rev., 109, 344-363.

Luo, H. L. and M. Yanai, 1983: The large-scale circulation and heat sources over the Tibetan Plateau and surrounding areas during the early summer of 1979. Part I: Precipitation and kinematic analyses Mon. Wea. Rev., 111, 922-944.

Luo, H. L. and M. Yanai, 1984: The large-scale circulation and heat sources over the Tibetan Plateau and surrounding areas during the early summer of 1979. Accepted for publication by the Mon. Wea. Rev.

Nitta, T., 1983: Observational study of heat sources over the eastern Tibetan Plateau during the summer monsoon. J. Meteor. Soc. Japan, 61, 590-605.

Sikka, D. R. and S. Gadgil, 1980: On the maximum cloud zone and the ITCZ over Indian longitudes during the Southwest monsoon. Mon. Wea. Rev., 108, 1840-1853.

Smith, E. A., 1984: The estimation of radiation budget parameters from weather satellite spectral radiance measurements. Submitted to J. Atmos. Sci. for publication.

Smith, E. A., T. Henmi, C. Johnson-Pasqua, E. R. Reiter, J. Sheaffer, T. H. Vonder Haar and Y. X. Gao, 1984: Investigation of the surface energy budgets in remote desert and mountain environments. Submitted to the Bull. Amer. Meteor. Soc. for publication.

Webster, P. J., 1983: Mechanisms of monsoon low-frequency variability: surface hydrological effects. J. Atmos. Sci., 40, 2110-2124.

Yeh, T. C., 1981: Some characteristics of the summer circulation over the Qinghai-Xizang (Tibet) Plateau and its neighborhood. Bull. Amer. Meteor. Soc., 62, 14-19.

Yeh, T. C. and Y. X. Gao, 1979: The meteorology of the Qinghai-Xizang (Tibet) Plateau. Science Press, Beijing, 278 pp. (in Chinese).

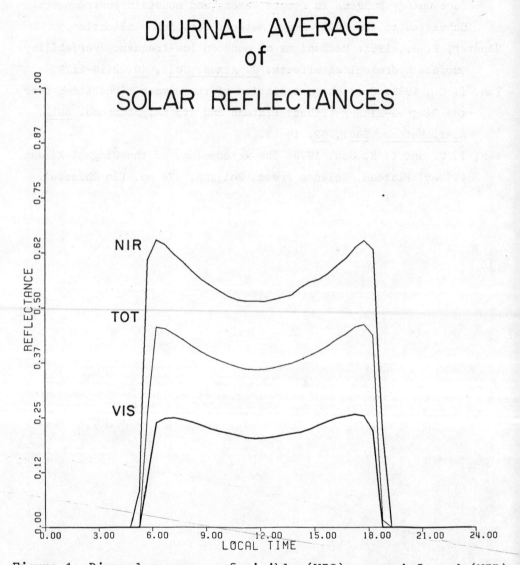

Figure 1: Diurnal averages of visible (VIS), near-infrared (NIR), and total solar (TOT) directional reflectances during June, 1981 at Sharouwrah, Saudi Arabia.

DIURNAL AVERAGE of RADIATIVE FLUXES

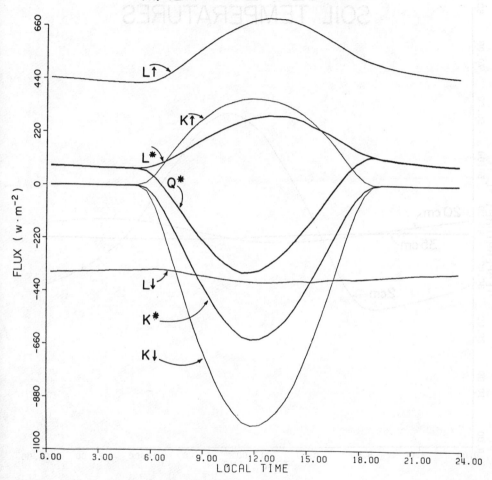

Figure 2: Same as Fig.1 for the principal terms of the surface radiation process.

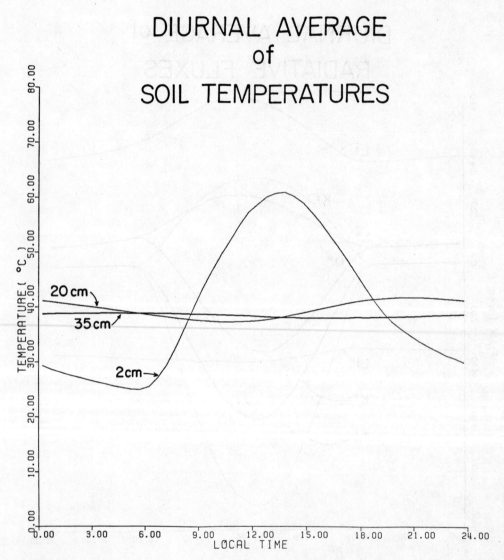

Figure 3: Same as Fig.1 for the sub-surface thermal waves at 2, 20 and 35 cm.

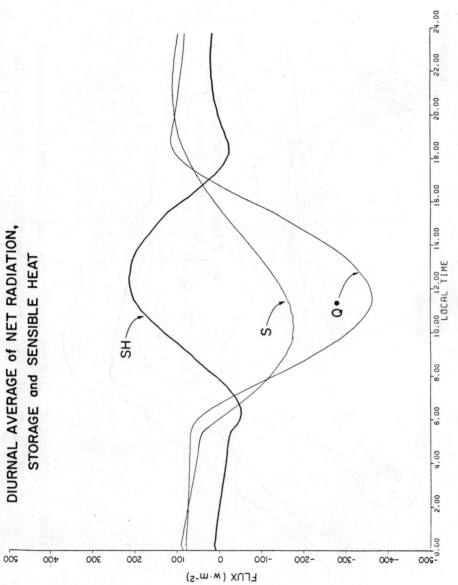

Figure 4: The diurnal surface energy budget; representative of June for the Arabian Empty Quarter.

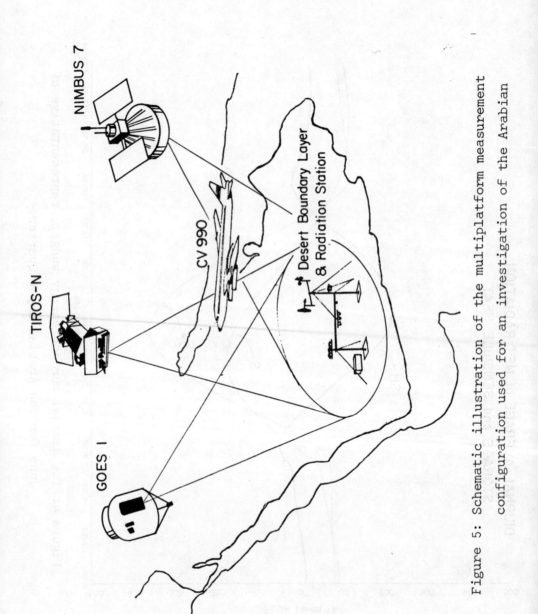

Figure 5: Schematic illustration of the multiplatform measurement configuration used for an investigation of the Arabian

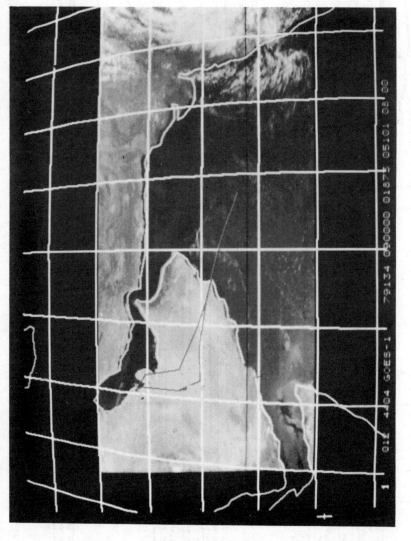

Figure 6: A GOES-1 visible satellite image with an overlay of the flight track of NASA CV-990 mission flown during the Saudi Arabian phase of the SMONEX. This was a differential heating mission taking place on May 14, 1979.

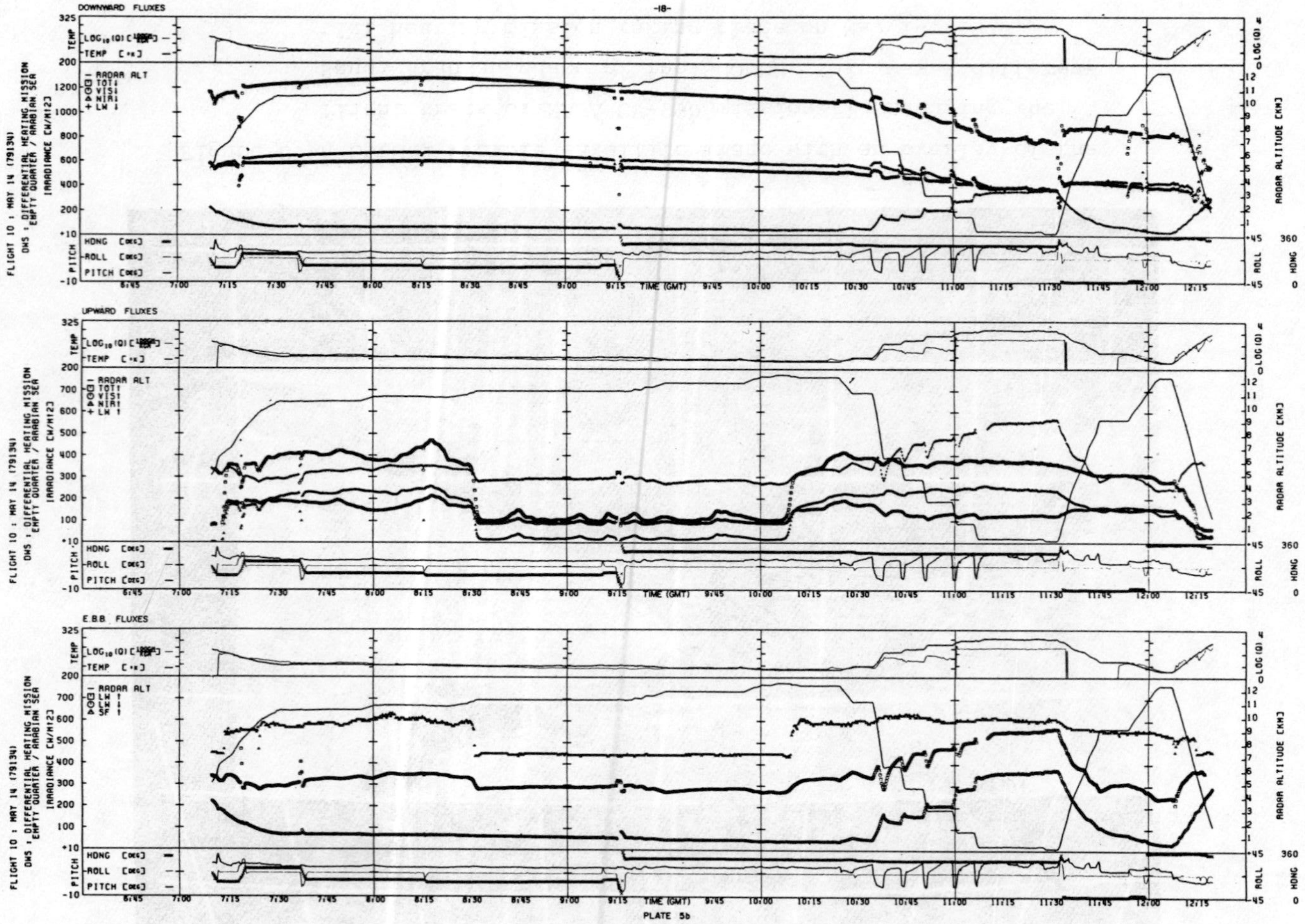

Figure 7: The aircraft radiation retrieved from the May 14 differential heating mission.

Figure 7: Continued.

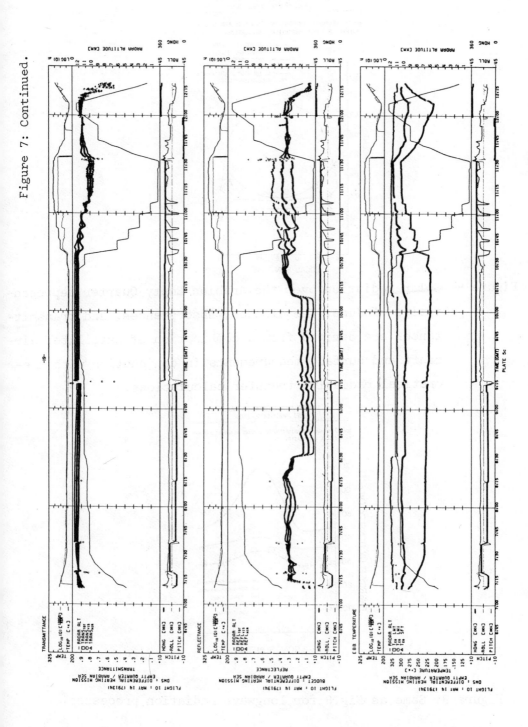

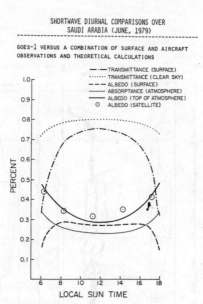

Figure 8: Solar radiation over the Arabian Empty Quarter representative of June. The albedos, absorptances, and transmit-
tances are derived from a combination of satellite, aircraft and surface measurements in conjunction with theoretical radiative transfer calculations.

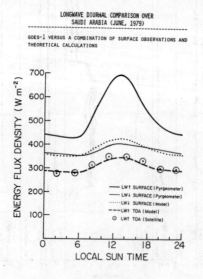

Figure 9: Same as Fig.8 for longwave radiation processes.

Figure 10: Net radiation map of the monsoon region for the period
May 19-24, 1979 derived from the Nimbus 7 ERB-NFOV data.
The color scheme is given below the figure.

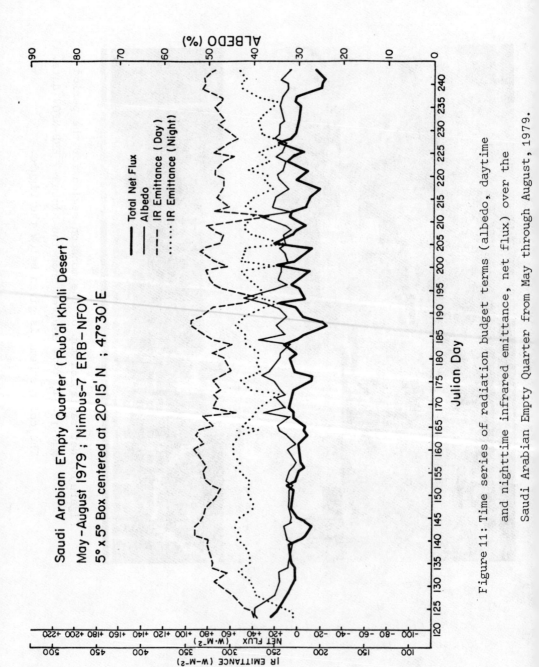

Figure 11: Time series of radiation budget terms (albedo, daytime and nighttime infrared emittance, net flux) over the Saudi Arabian Empty Quarter from May through August, 1979.

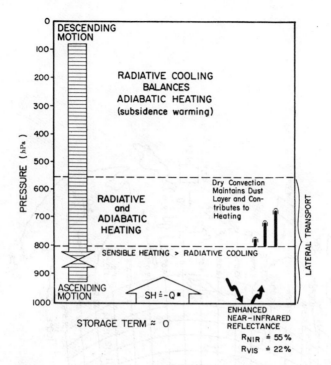

Figure 12: Vertical structure of the Arabian heat low representative
of June conditions.

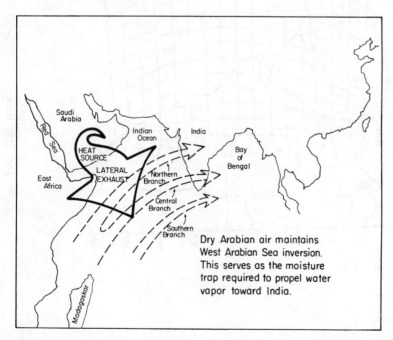

Figure 13: Role of Arabian tropospheric heat source on Southwest
Monsoon.

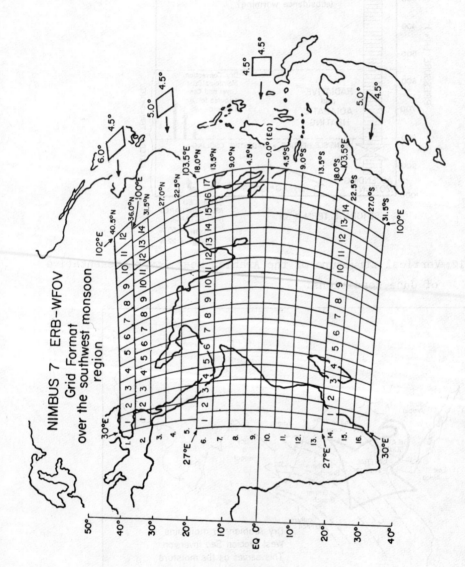

Figure 14: The Nimbus 7 ERB-NFOV grid format portraying the Southwest-East Asian monsoon region.

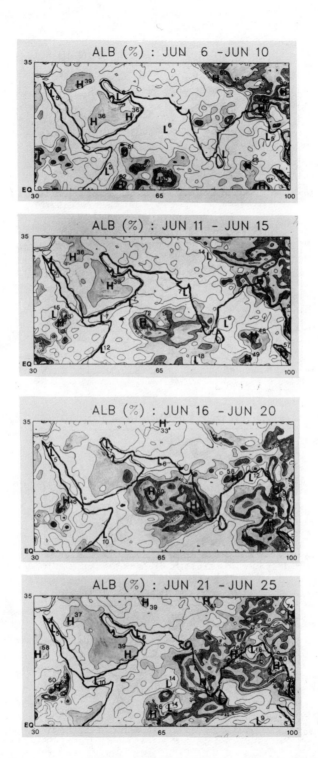

Figure 15: Five day average albedo fields (derived from TIROS-N AVHRR) from June 6 to June 25, 1979 over Southwest-East Asian monsoon region.

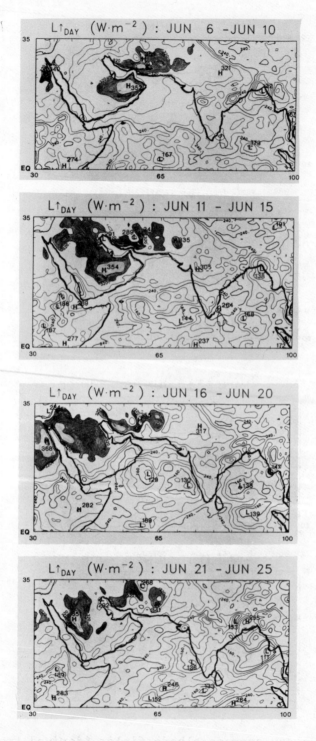

Figure 16: Same as Fig.15 for the emitted infrared flux.

Figure 17: Same as Fig.15 for the local-noon net flux.

ALB : ZONAL AVERAGE TIME SERIES

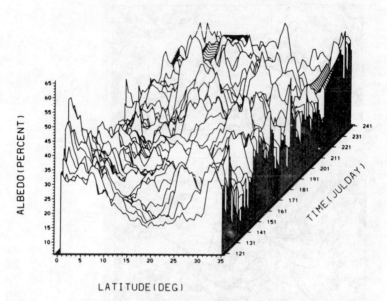

ALB : MERID AVERAGE TIME SERIES

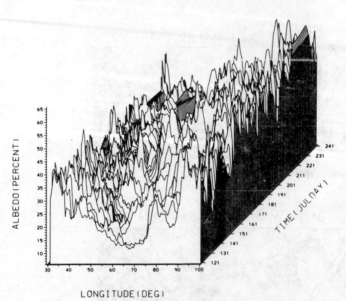

Figure 18(a): Three-dimensional isopleths of zonal average albedos over monsoon region (0-35°N; 30-100°E) from May 1 to August 30, 1979.

Figure 18(b): Same as Fig.18a for meridional averages.

L↑_DAY : ZONAL AVERAGE TIME SERIES

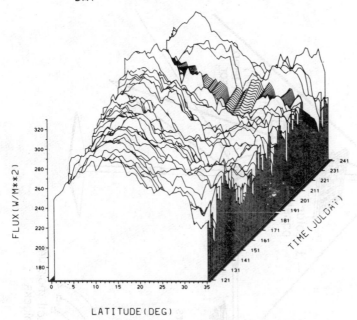

L↑_DAY : MERID AVERAGE TIME SERIES

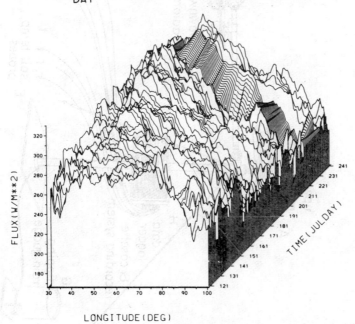

Figure 19(a): Three-dimensional isopleths of zonal average equivalent black body temperatures (broad band) over monsoon region (0-35°N; 30-100°E) from May 1 to August 30, 1979.

Figure 19(b): Same as Fig.19a for meridional averages.

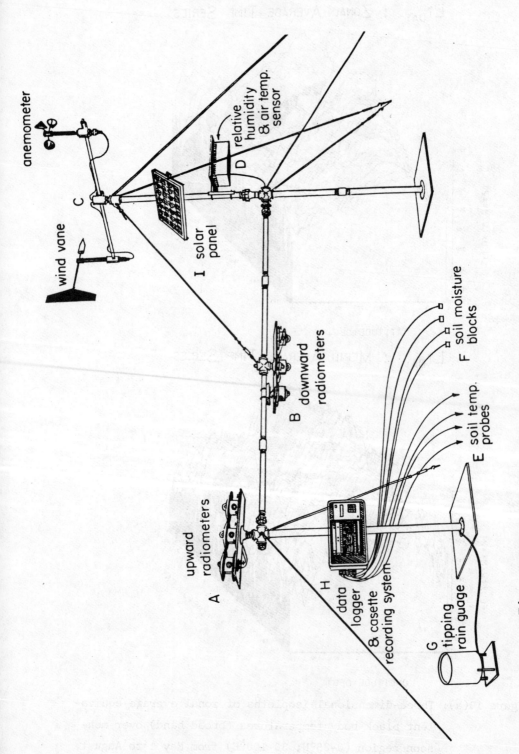

Figure 20: Schematic illustration of 'Radiation Station'.

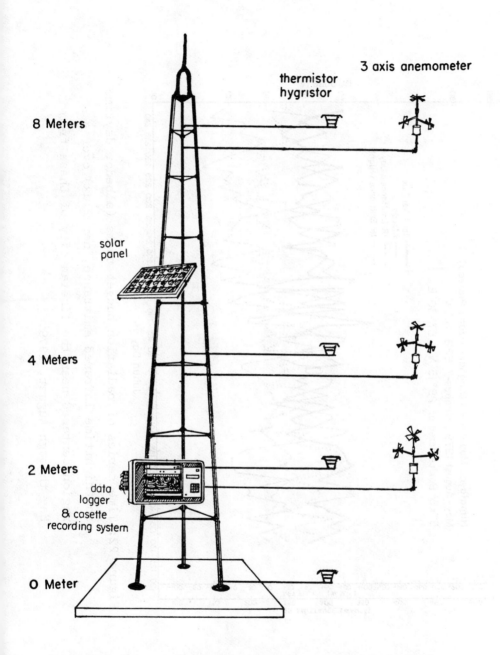

Figure 21: Schematic illustration of 'Tower Station'.

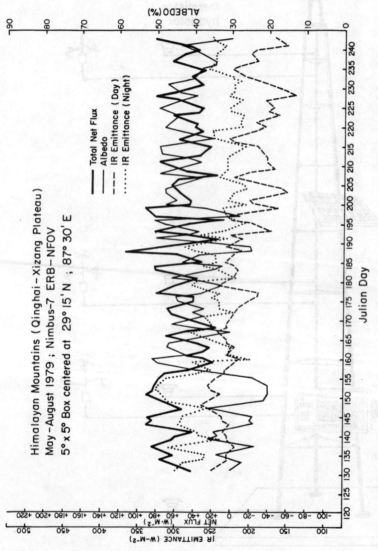

Figure 22: Time series of radiation budget terms (albedo, daytime and nighttime infrared emittance, net flux) over a region centered near the Tibetan city of Lhasa from May through August, 1979.

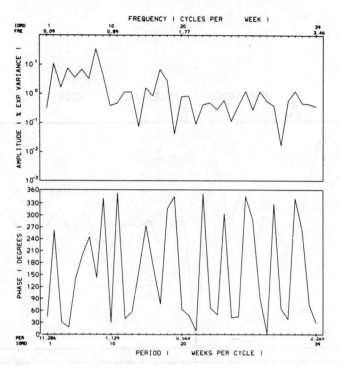

Figure 23(a): Periodogram of 80 day albedo time series (mid-May to
August 5, 1979) based on AVHRR RADBUD estimates over
Tibetan Plateau.

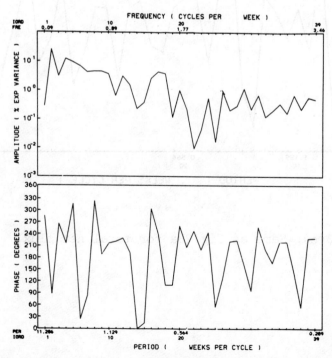

Figure 23(b): Same as Fig.23a for infrared emittance.

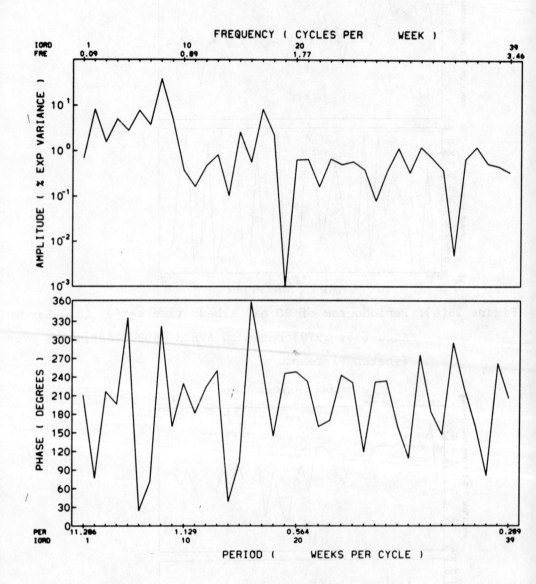

Figure 23(c): Same as Fig.23a for total net flux.

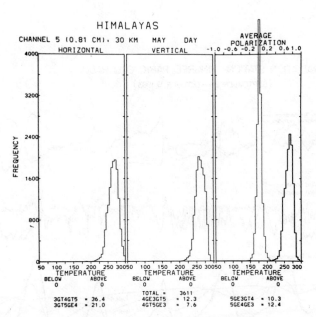

Figure 24(a): Histograms of vertical, horizontal, and non-polarized
brightness temperatures from SMMR channel 5 (37 GHz)
during May, 1979 over the Tibetan Plateau. The
histogram on the right also includes a histogram
of the polarization factor (P) where $P=(T_V+T_H)/T_{AVE}$.

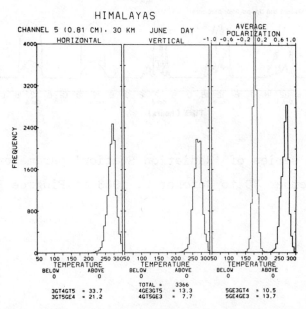

Figure 24(b): Same as Fig.24a for June, 1979.

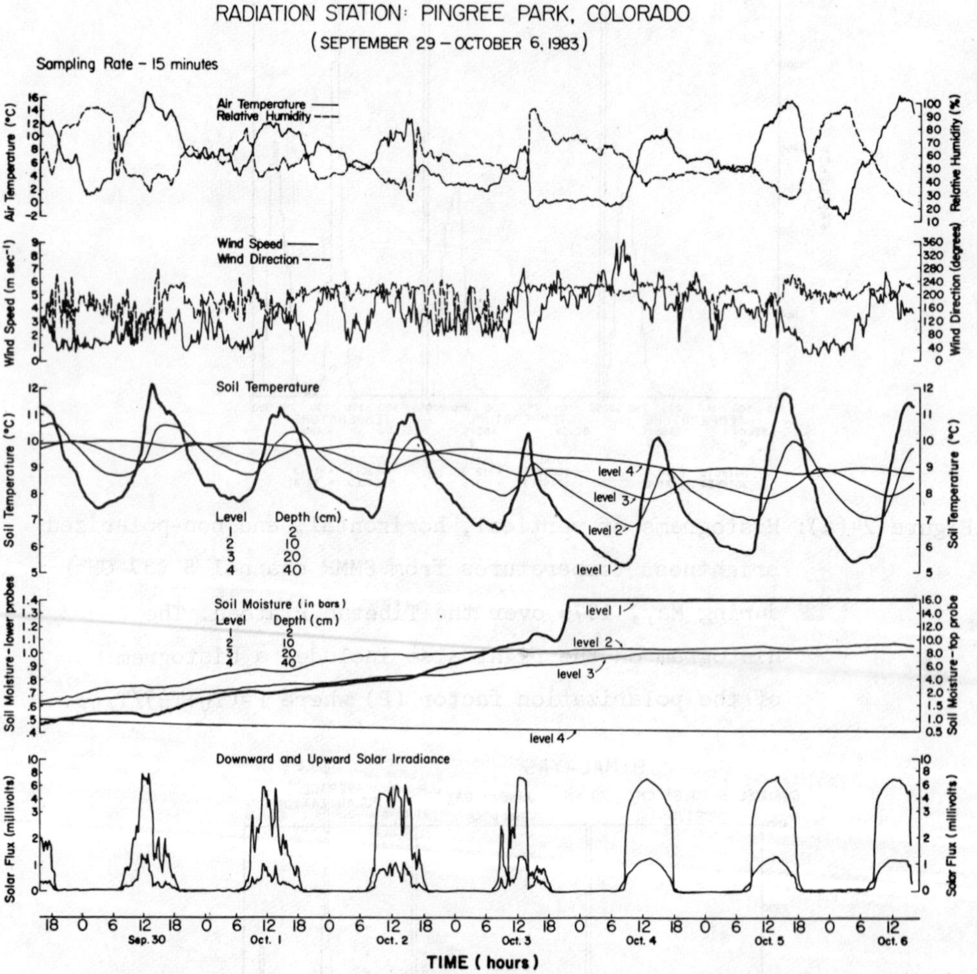

Figure 25: Time series of 'Radiation Station' parameters from
September 30 to October 6, 1983 at Pingree Park, CO.

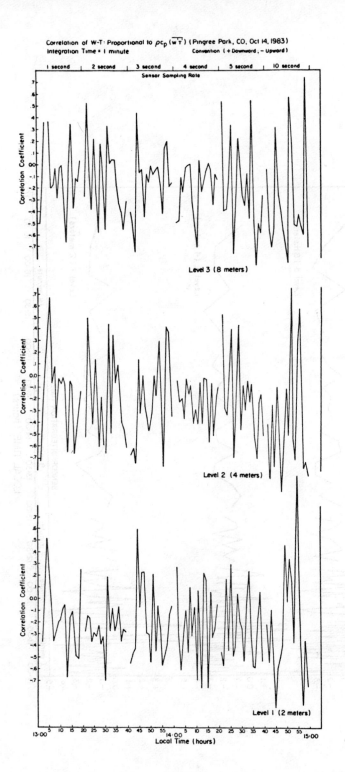

Figure 26: Correlation runs between vertical velocity and temperature from the 'Sampling Interval Test', held at Pingree Park on October 14, 1983.

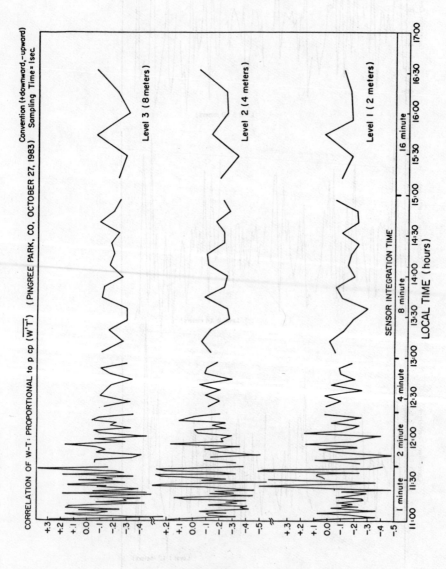

Figure 27: Correlation runs between vertical velocity and temperature from the 'Integration Time Test', held at Pingree
Park on October 27, 1983.

OBSERVATIONAL STUDIES OF THE AIRFLOW OVER AND AROUND THE ALPS

Reinhold Steinacker

Department of Meteorology and Geophysics

University of Innsbruck

Austria

1.INTRODUCTION

During the field experiment ALPEX a data set with much higher time and space resolution than usual was gathered, to be able to study the atmospheric processes in connexion with mountains in different scales. The Alps are small in size compared to other mountain massives (e.g. Himalaya, Rocky Mountains) but rather high, the main crest dividing roughly the lowest third of the atmospheres mass. Due to their frequent occurrence during ALPEX-SOP (Special-Observing-Period) the behaviour of coldfronts, coming from west to north, partly leading to cyclogenesis in the Gulf of Genoa is well suited to investigate.

2. ISENTROPIC TRAJECTORIES

To investigate the threedimensional motion in the vicinity of mountains, analyses on isentropic surfaces are advantageous and therefore often used (e.g. Buzzi and Tibaldi 1978). Due to considerable ageostrophic wind components the construction of Trajectories should not be carried out in the usual way (Pettersen, 1956; Danielsen, 1961) but rather by integration of the momentum equation (Petersen and Uccellini, 1979; Steinacker, 1984a). For this reason the Montgomery-potential, which represents the isentropic streamfunction has to be known to a high degree of accuracy. Fig. 1 shows an example of trajectories during cold-air advection from northwest against the Alps. In this case the blocking of the air by the Alps was strongly acting during a rather long time and reaching to a considerable height. It is clearly shown how the flow split leads to a northerly flow in the Rhone valley, which corresponds to the Mistral windsystem. The eastern branch shows a strong anticyclonic curvature so that the flow-direction becomes northeast, which finally leads to the Bora over the Yugoslavian coastal mountains.

3. APPLICATION OF VORTICITY DYNAMICS

To learn about the mechanism of the cyclogenesis in the Gulf of Genoa (Alpine cyclogenesis) the behaviour of the corresponding vorticity can be studied. Here again the isentropic vorticity recommends itself, being only time dependent on divergence, if adiabatic and frictionless flow can be assumed:

$$\frac{d\eta}{dt} = -\eta \, \nabla \cdot \vec{v} \tag{1}$$

Thus, following the vorticity along an isentropic trajectory, the divergence or the streching or shrinking of the vertical vortex-tube respectively can be obtained.

Two different cases of cyclogenesis in the Gulf of Genoa during ALPEX-SOP are now discussed. The first one discussed here took place on 20/21 March 1982 (Figs. 2-4). The initial situation was a north-westerly upper-tropospheric flow with an embedded well developed lowlevel cyclone. On the isentropic surface 290K on 20 March 0000 GMT an intense cyclonic vorticity center has reached the European continent (Fig. 2). 12 hours later (Fig. 3) this vorticity maximum has approached the Alps but at the same time a second new small-scale maximum has appeared in the Gulf of Genoa, strictly divided of the first by a zone of low absolute vorticity over the Alps. At 0000 GMT on 21 March (Fig. 4) the northern maximum has weakened while the Mediterranean has become very intensive. Finally the secondary low moved at about the same speed and direction as the primary did before, only a few hundred kilometers to the southwest.

For this case the change of vorticity of two parcels along their trajectories has been analyzed. Fig. 5 shows the western, Fig. 6 the eastern trajectory, for which the three-hourly positions are indicated in Fig. 2 by arrows.

Both trajectories show a decrease of vorticity (=Vortex shrinking, =divergence) beginning far upstream until the ridge of the Alps, followed by an increase (streching, convergence) in the lee. For the western parcel a total increase, for the eastern a total decrease of absolute vorticity occurred during the whole period.

To sum up, it can be said that this event of Genoa cyclogenesis was caused mainly by advection of vorticity, where the Alps have only acted in a deflecting way. The short-time vortex shrinking over the Alps causes a fictituous picture of a low "jumping" over the obstacle.

The second case (4/5 March 1982, Figs. 7-9) was associated with a west-south-westerly upper tropospheric flow. The isentropic analyses 290 K (Fig. 7) on 4 March 1200 GMT shows one vorticity center passing by, far north of the Alps. A second maximum appears in the Gulf of Biscay. The influence of the Alps on the mesoscale vorticity distribution comes out clearly at the eastern edge of the Alps. A strong dipole structure is caused by strong shear at the intersection of the isentropic surface with the Alps and downstream of that.

Six hours later (Fig. 8) the forming low in the Gulf of Genoa appears first, associated with a narrow vorticity maximum which is the counterpart to the above-mentioned minimum east of the Alps. Here the Mistral establishes a zone of strong cyclonic shear vorticity from the Cote d'Azur southward.

During the next 6 hours (Fig. 9) the cyclone developed explosively Again the narrow band of high vorticity is located south of the western edge of the Alps but also a widespread increase of absolute vorticity south of the Alps has taken place. Looking for the reason for that strong development it can be stated, that the advection in this case played only a minor role. The increase therefore must be due to convergence. This low level convergence is induced by heavy local pressure fall in the Gulf of Genoa, which again is caused by upper level divergence. In a schematic view this connections are demonstrated by Figs.10 and 11. In a case with a low level coldfront moving from northwest or west against the Alps the low level flow usually is blocked, leading to a deformation of the temperature field as shown in Fig. 10. If the upper level temperature gradient has the same direction as the low level, no (geostrophic) wind component across the ridge can be obtained, taking into account the thermal wind relation. The upper tropospheric flow therefore only shows a parallel component of the geostrophic flow over the mountains. Although the actual flow may be considerably ageostrophic, the sign of the actual relative vorticity should be equal to the geostrophic vorticity. Because of the stationarity of the vorticity distribution there must be a divergence above the Gulf of Genoa (Fig. 11). The wave pattern is depending on the shape of the mountains. In the case of the Alps it means that the arc, formed by the west-east oriented. East-Alps and the north-south oriented West-Alps is very important to generate a wave in the right way, the importance of which was demonstrated by Egger, 1972.

The boundary effect at the edges of the Alps is not included in Eq. 1. but should be taken into account. The generation of cyclonic vorticity by the Mistral may be estimated as follows: Assuming a windshear of 20 ms^{-1} existing across a distance of 100 km a cyclonic relative vorticity of 20×10^{-5} s^{-1} results. The area affected by the above increase after the time increment Δt is given by $A \simeq$ 100 km $\times$ $\bar{V} \times \Delta t$. Taking $\bar{V} \simeq$ [V(maximum)+V(minimum)]/2$\simeq$10 ms^{-1}, $A \simeq 2\times10^{4}$ km^{2}.

Comparing this with the observed increase $\frac{\partial \bar{\eta}}{\partial t} \simeq 4\times10^{-5}$s^{-1}/12 h within an area of 4×10^{5} km^{2}, the boundary effect is small, but not negligible for the mesoscale structure of the cyclone. The strong shear vorticity can be transformed into curvature vorticity by a mechanism discussed in Steinacker, 1984 b.

4. CONCLUSION

Although two cases are not sufficient to draw general conclusions, get it is evident and well known (Buzzi, Speranza, 1983) that there exist quite different mechanisms of Genoa cyclogenesis. For the advective type there exists already a distinct vorticity maximum upstream so that virtually there is no cyclogenesis, but the Alps give only the impression of that. The stationary type is characterized by a stronger blocking of the air by the Alps which can be effective higher than the crest. Here the Vorticity is created by low level convergence induced by heavy pressure fall which again is a result of the divergence aloft. The flow splitting in combination with the blocking finally produces very intense but small-scale vorticity extremes which help the explosive development of a Genoa cyclone in the first stage.

LITERATURE

Buzzi, A., S.Tibaldi (1978): Cyclogenesis in the Lee of the Alps;
       A Case Study. Q.J.R.M.S. 104, 271-288.

Buzzi, A., A.Speranza (1983): Cyclogenesis in the Lee of the Alps.
       In: D.K.Lilly and T. Gal-Chen (eds.), Mesoscale Meteorology-
       Theories, Observations and Models, 55-142. D.Reidel Publ.C.

Danielsen, E.F. (1961): Trajectories: Isobaric, Isentropic and Actual.
       J.Meteor. 18, 4, 479-486.

Egger, J. (1972): Numerical experiments on the cyclogenesis in the
       Gulf of Genoa. Beitr. Phys. Atmos., 45, 320-346.

Petersen, R.A., L.W. Uccellini (1979): The Computation of Isentropic
       Atmospheric Trajectories using a "Discrete Model" Formula-
       tion. Mon.Wea.Rev. 107, 566-574.

Pettersen, S. (1956): Weather Analysis and Forecasting. Mc Graw Hill,
       New York, 27-30.

Steinacker, R. (1984a): Airmass and Frontal movement around the Alps.
       Rivista Meteor. Aeronaut., in print.

Steinacker, R. (1984b): The Isentropic Vorticity and the Flow over
       and around the Alps. Rivista Meteor. Aeronaut., in print.

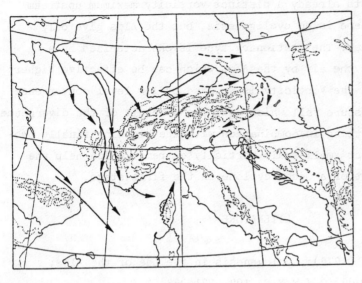

Fig.1. Isentropic tra-
jectories during a
situation with flow
blocking by the Alps.
Continuous: between 18
and 24 GMT, 4 March
1982.
Broken: between 12 and
18 GMT, 5 March 1982.

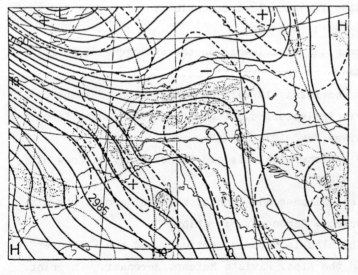

Fig.2. Isentropic
analysis of the surfac
290K on 20 March 1982
00GMT. the continuous
lines are lines of cor
tant Montgomery-poten-
tial in 1gpdam interva
broken are lines of
constant absolute vort
city in $4 \times 10^{-5} s^{-1}$ inte
val and dotted are
isobars in 100hPa in-
terval.

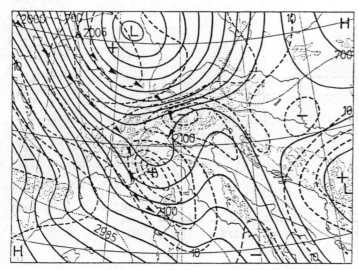

Fig.3. The same as Fig.2 but for 20 March 1982 12GMT. In addition two selected trajectories are indicated by 3-hourly positions of the parcels (arrows) between 20 March 00GMT (2000) (2006 respectively) and 21 March 1982 00GMT (2100).

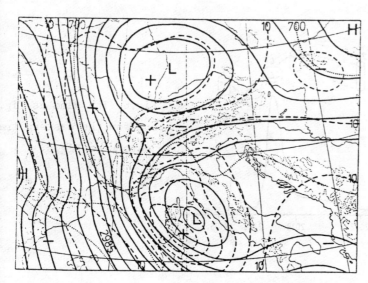

Fig.4. The same as Fig.2 but for 21 March 1982 00GMT.

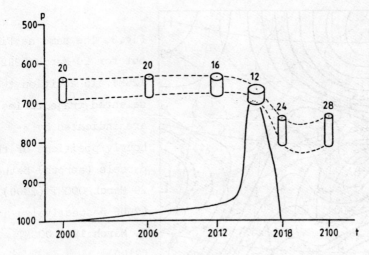

Fig.5. Schematic cros-
section of the absolut
vorticity $(10^{-5}s^{-1})$
along a trajectory cro
sing the Alps. The
three-hourly position
of the parcel is indic
ated in Fig.8, west.

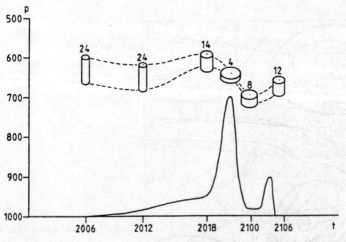

Fig.6. The same as Fig.
but for the eastern tra-
jectory of Fig.8.

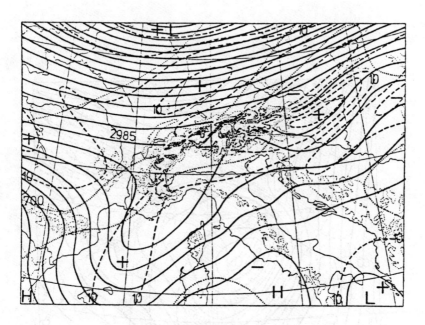

Fig.7. The same as Fig.2
but for 4 March 1982
12GMT.

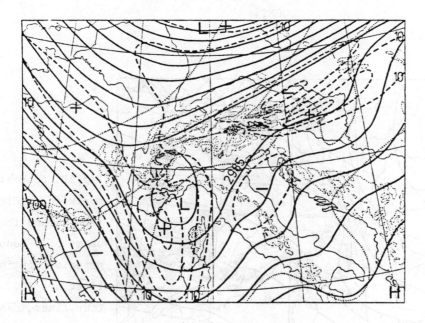

Fig.8. The same as Fig.2
but for 4 March 1982
18GMT.

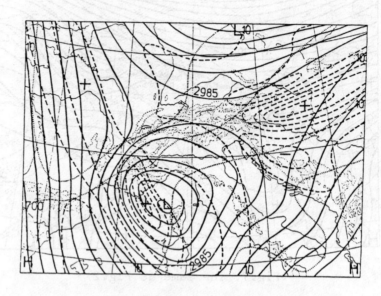

Fig.9. The same as Fig.2
but for 5 March 1982
00GMT.

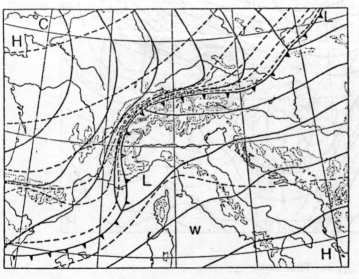

Fig.10. Schematic su
map with coldfront-p
tion during the init
phase of a Genoa cyc
genesis. Continuous
lines are 100 hPa co
tour lines, dashed a
1000 hPa isotherms.

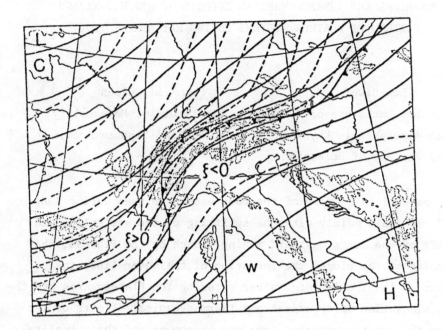

Fig. 11. Schematic upper troposheric contour lines (continuous)
corresponding to Fig. 10. The broken lines represent
the relative topography between the isobaric surfaces
of fig. 10 and Fig. 11. For comparison the surface
coldfront is plotted too.

DYNAMICAL AND THERMODYNAMICAL EFFECTS OF QINGHAI-XIZANG PLATEAU ON WEATHER SYSTEMS DURING MAY-AUGUST,1979

Luo Siwei
Lanzhou Institute of
Plateau Atmosphereic Physics,
Chinese Academy of Sciences

Li Guicen
Institute of Meteorology
Science of Sichuan Province

This paper is a summary of the four essays written during 1980-1982[1-4]. It deals mainly with the splitting of the westerlies and the cut-off of the large eastward moving trough by the Plateau, the strengthening of the ridge, the formation of the shear line, the development of the Iranian High after entering the Plateau area and the numerical simulation of dynamical and thermodynamical effects of the Plateau on the weather systems. Some new phenomena and their qualitative explanations have been found.

I. FORMATION OF TOPOGRAPHIC TROUGH AND VORTEX

The previous studies have suggested that the westerly trough is cut off by the Plateau, but none of them has described the process in detail yet. The analysis made in this paper shows that the trough mentioned can climb up the west slope of the Plateau but fails to pass over longitude 80°E instead of being cut off rapidly. It is seen from Fig.1 that there exists a long-wave trough from the Ural Mountains to Aral Sea which moves eastward during May 18 — 20 at a speed of 5 degrees of longitude per day, and which climbs up the west slope of the Plateau at 19 1200z. Then the northern part of the trough moves rapidly at a speed of 8 degrees of longitude per day and its southern part develops into a stationary "topographic vortex" over the northern part of Pakistan. The latter remains there for two days and then becomes weaker and transforms into a NW-SE small trough over New Delhi, moving eastward along the southern edge of the Plateau. It disappears on May 25 west of 90°E.

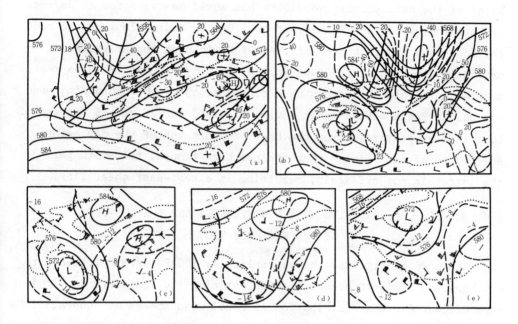

Fig. 1 500 hPa chart for 00 GMT May 19 (a), 21 (b),
22 (c), 23 (d) and 24 (e), 1979. The solid
line denotes geopotentials (in 10 m), the
dashed lines-isolines of vorticity (in Fig.1
(a)-(b)) or temperature (in Fig. 1(c)-(e)).
The heavy solid line-trough line, the numerals
—the dates. The small circles "0" in Fig.1(b)
denote the center of low pressure, their side
numerals denote the dates, their directions of
motion are shown by the arrows.

This "topographic trough" is closely related to the blocking effect of the Plateau. The NW flow south of the Plateau is very strong (for instance, the NW wind at 500hPa over New Delhi reached 25m·s^{-1} on May 23) and parallel to the orographic edge, while the air flows over the Plateau are all southerly winds with mean speeds as low as 4m·s^{-1}. The cyclonic shear vorticity of this flow field is very intense and as a result the "topographic trough" moves along the boundary of the Plateau.

The topography of the Plateau has an evident effect to strengthen the westerly ridge too. For instance, from 20-21 0000z the zero line ahead of the anticyclonic vorticity has moved eastward for 20 degrees of longitude from the vicinity of Aral Sea to South Xinjiang and made the long-wave trough broken off. At 22 0000z high center which was originally located at 21 0000z in the vicinity of Aral Sea disappeared and two anticyclonic centers have formed south of the South Xinjiang and Qaidam Basin with a speed moving eastward for 20 and 30 degrees of longitude respectively from the high pressure center the day before. As they moved much faster than the northern trough, the south part of the latter is cut off again, resulting in a west-east shear line.

II. THE SPLITTING OF THE WESTERLIES AND THE CUT-OFF OF THE WESTERLY TROUGH

As shown in Fig.1(a) the westerlies at 500hPa were splitted into southern and northern branches after entering the Plateau area. It is more evident for the distribution of zonal wind speeds at 500hPa,300hPa and 200hPa. Fig.2 is the profile along 65°E, 75°E and 82.5°E for 00 GMT May 19,1979. It is seen that the core of the westerly jet west of the Plateau at 65°E is located on 200hPa around 35°N and there is no splitting for either the westerly jet or the westerlies. Along 75°E just west of the Plateau, the original jet center is still located at 200hPa around 35°N, but the west wind below 200hPa has been split. Further east, above the main part of the Plateau (along 82.5°E), the splitting of the westerlies is even more evident and the jet core located previously above the Plateau is now divided into a northern and a southern ones .

The Qinghai-Xizang Plateau exerts a frequent effect on the cut-off of the westerly trough, not only on the long wave trough but also on the short wave trough, which may be clearly seen in the cloud chart (omitted). There are three topographic effects on the cut-off of trough:

1. Vorticity Advection

The dynamical effect of topography causes the west wind to be split, this effect destroys the original vorticity advection and its distribution.Because the wind speed is high by the jet core the vorticity advection transports fast and therefore the eastward movement of the trough is accelerated over south and north of the Plateau.

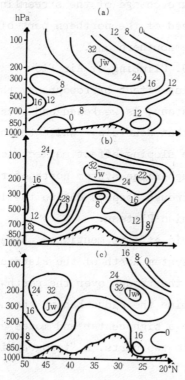

Fig.2 U-component profile along 65°E(a), 75°E(b) and 82.5°E(c) for 00 GMT May 19, 1979. The solid line denotes the isolines of zonal component of velocity U(m·s^{-1})

As shown in Fig.1(a,b) a pair of positive and negative vorticity centers in the neighbourhood of the westerly trough in the north of the Plateau has moved eastward about 20 degrees of longitude within two days, and the positive vorticity center in the south of the Plateau has moved eastward 15 degrees of longitude. Their moving speeds both are unusual and about the same as those of the westerly branches.

2. Shear Vorticity

Due to the topographic frictional effect and the splitting of
westerlies anticyclonic shear vorticities were produced north of the
Plateau and cyclonic shear vorticities south of it. Under certain proper conditions they might be transformed into curvature vorticities,
influencing the curvature change of the streamlines. This might enhance
the eastward moving speed of the northern part of the trough and decrease the speed of the southern one.

3. Climbing of Air Flow

Owing to the sloping effect of topography the trough, arriving at
the west end of the Plateau, is weakened and even more weakened on the
west slope of the Tianshan Mountain, while the high pressure is
strengthened. Thus, the sloping effect may accelerate further the eastward moving of the trough-ridge system north of the Plateau. In contrast,
the south part of the trough deepens due to the leeside effect after
it climbs over the Iranian Plateau.

All these orographic effects mentioned above result in the fact
that the trough-ridge system north of the Plateau moves more rapidly
than the south part of the trough over the northwest of India and
Pakistan. At the same time the ridge behind the north part of the trough is strengthened into a high center and moves eastward rapidly and
cuts off the trough into two parts. A "topographic vortex" was formed
west of the ridge then.

III. THE IRANIAN HIGH AT 500 hPa

Due to the limitation of data available in the past, little was
known about how the subtropical high developed at 500hPa after it moved
eastward from the Iranian Highland into the Qinghai-Xizang Plateau area.
On the basis of the experiment data in 1979, it is found that after the
Iranian high has moved eastward and entered into the west part of the
Qinghai-Xizang Plateau by way of South Xinjing, it is denatured and
weakened rapidly due to the heating effect of the underlying surface.
The downdraft in the high pressure center is changed into an updraft

below 400hPa and the high is moved south-eastward and finally disappears.

The following composite structure is obtained from the analysis of 7 cases of the high pressure, using the pressure center as the reference origin and the distance between two trough-lines along the east-to-west ridge line as one wave length. The averaged position of the pressure center at 500 hPa is shown at 34°N, 87.5°E and the high pressure center at 100 hPa is shifted northwestward for about 500 km. The longitudinal and latitudinal distances between the two grid points are 2.5° and 2.0° respectively. The vertical profile of east-west wind (u) and north-south wind (v) shows that the moving high pressure at 500 hPa exists below 400 hPa, and there is the Qinghai-Xizang high above 300 hPa level. Therefore, the vorticities near the high center are all negative from surface to 100 hPa (see Fig.3).

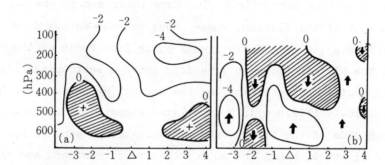

Fig.3 East-west vertical profile of vorticity field
(a) $(10^{-5}s^{-1})$ and vertical velocity (b)$(10^{-4}hPa\cdot s^{-1})$ along composite high pressure ridge-line and the sign"Δ" denotes the location of high center at the 500 hPa.

When the high is still located west of the Plateau, there is subsidence in the whole air column, as it enters into the west part of the Plateau there appears the convergent updraft below 400hPa in the vicinity of the high center, but subsidence above 400hPa is still dominant. This coincides with the temperature stratification below 400 hPa and a

sinking inversion between 300 and 400hPa. It means that the 500hPa
Iranian high has been denatured and weakened by surface heating after
entering into the Plateau area.

IV. NUMERICAL SIMULATION

To further understand the topographic effect of the Plateau upon
its nearby weather systems, we performed a simple numerical simulation,
in which a 2-layer P-σ primitive equation model, ideal topography
(ellipsoid shape, with its top height in 5.0 km), ideal zonal primitive
flow field and ideal stationary heating field were used. The calculation was performed at the grid point with resolution 5° × 5° longitudes
and latitudes covering an area of 40° — 150°E and 10° — 65°N. The size
and location of topography and the intensity of the heat source used
in the simulation were much similar to the reality of the Plateau.

Fig.4 shows the dynamic effect of the Plateau on the evolution of
500 hPa simulating streamfield. Two days later a high pressure center
appears west of the Plateau, whose ridge extends eastward and reaches
105°E (Fig.4 a); 3 days later a weak shear line forms on the eastern
side of the plateau (Fig.4 b); 4 days later a subtropical high over
the West Pacific appears also and the shear line between two highs moves
eastward slowly (Fig.4 c); 5 days later a small anticyclone forms on
the north side of the Plateau and a new shearline appears behind it
(Fig.4 d); at 132 hours the anticyclone moves eastward and the shearline behind it becomes stationary and a typical west-east one (Fig.4 e).

At the same time, a complete cyclontic circulation appears southeast of the Plateau on 700 hPa which is stationary (Fig.4f). With the
exception of this, we can see that there is another cyclonic circulation south of the Plateau (Fig.4c—f).

Fig.5 shows the evolution of 500 hPa simulating streamfield with
both the dynamical and thermodynamical effects of the Plateau. There are
two high centers, which are located west and east of the Plateau respectively, and a shearline on the Plateau at 48h (Fig.5a).Then the shearline strengthens continuously but its location does not change at all
(Fig.5b—d). At this time the vortex on 700 hPa south-east of the

836

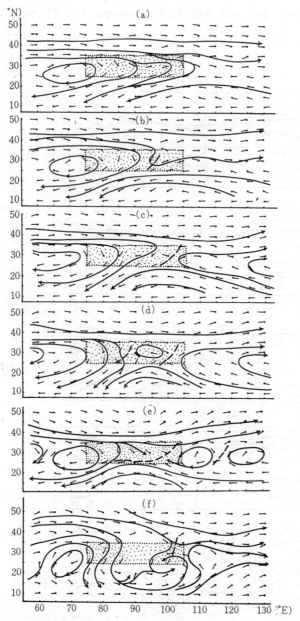

Fig. 4 The simulating streamfield with pure dynamic
effect of the Plateau: (a)-48h, 500 hPa; (b)-
72h, 500 hPa; (c)-96h, 500 hPa; (d)-120h,
500 hPa; (e)-132h, 500 hPa; and (f)-72h,
700 hPa.

Plateau does not appear and all air flows nearby the Plateau converge to it (Fig. 5e).

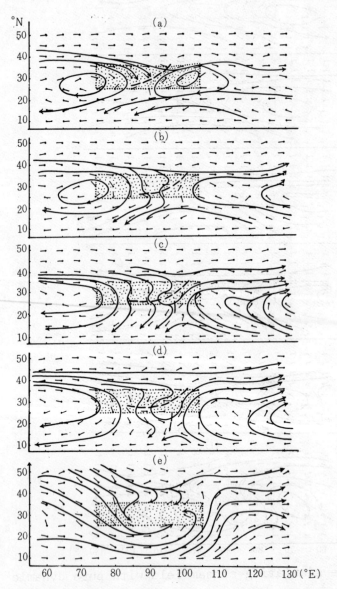

Fig. 5 The simulating streamfield with both the
dynamical and thermodynamical effects of
the Plateau: (a)-48h, 500 hPa; (b)-72h,
500 hPa; (c)-96h, 500 hPa; and (d)-120h,
500 hPa; (e)-120h, 700 hPa.

In addition, it is found from the simulation that when the west wind is weak over the Plateau, the formation of the Xizang high in the upper troposphere (100 hPa) depends mainly on the heating effect of the Plateau, and at this time, althongh the heat source south of the Plateau is five times more intensive than that on the Plateau, it makes no contribution at all.

From this numerical simulation we have got some preliminary results that the 500hPa shear line in the Plateau area in summer is caused mainly by the dynamic effect of the Plateau while its thermodynamical effect also plays a strenthening role. The Plateau's dynamic effect contributes a lot to the formation of the Iranian high upstream of the Plateau and the subtropical high over the West Pacific and to the formation and reinforcement of the low over the Bay of Bengal. The 700hPa SW vortex in China is formed mainly due to the dynamic effect of the Plateau while its thermodynamical effect has nothing to do with this formation.

REFREENCES

[1] Luo Siwei, Zhang Fuang, A case study of the effect of the Qinghai-Xizang Plateau on the westerly trough and ridge during their transits in May of 1979, Collected Papers of QXPMEX(3), Science Press, 1986 (in Chinese).

[2] Luo Siwei, Wei Li, The Synoptic-Dynamical Analysis on the Cut-off Process of a westerly trough in May of 1979 by the dynamic effect of the Qinghai-Xizang Plateau (to be published).

[3] Li Guicen, Analysis of the 500hPa moving high over the Qinghai-Xizang Plateau in the summer of 1979, Collected Papers of QXPMEX(3), Science Press, 1986 (in Chinese).

[4] Luo Siwei, Wang Anyu, Zhang Jugen Numerical experiment on the effect of the Qinghai-Xizang Plateau on its neighbouring flow field in summer, Plateau Meteorology, Vol. 4, No.2, 1984(in Chinese).

WINTER DISTURBANCES IN THE SOUTHERN
FRINGE OF THE TIBETAN PLATEAU-1981

SAN HLA THAW

National Weather Forecasting Division,

Department of Meteorology and Hydrology,

Rangoon

I. INTRODUCTION

The cool season rainfall of Burma during the month of January is
due to the migratory low pressure areas known as western disturbances
from North West India, which move along the foot hills of Himalaya.Yin
(1970) has stated that usually three to six western disturbances cross
North Burma in January. A Southward displacement of the westerly jet
axis is often found to be in association with these eastward moving
disturbances at the surface giving widespread but fairly light rain to
areas on the Southern side of the Tibetan Plateau.

It is not always possible to trace the origin of these eastward
moving disturbances though they appear to be associated with the major
troughs in the upper tropospheric westerlies. The disturbances attain
maximum intensity near Longitude 90°E and weaken as it proceeds further
towards the East.

Murakami (1981) showed the significant Orographic influence on
the structural wave features, changing the phase speeds and intensity
of the 200 hPa velocity potential field associated with the eastward
passage of the 12-20 days perturbations. Wester (1981) depicted the
eastward propagation of anomalous surface pressure waves in the neighbourhood of East Asia and Northwest Pacific with a phase speed of
10 ms^{-1} and concluded that the subtropical disturbances are not merely

the result of propagatory features which developed in situ over the ocean to the east of Asia.

Cheang and Krishnamurti. (1981) analysed eleven complete years of time longitude diagram of contour heights at 30°N and noted a very high frequency of occurrence of downstream amplification phenomena. Nearly all middle latitude major-trough/ridge systems are part of such pro-pagation. The dry and the wet spells of winter monsoon over Malaysia are found to be related with the positions of large amplitude 500 hPa troughs.

Sharma and Subramania (1983) showed a case of linkage of a western disturbance with a low pressure in low latitude tropospheric easterlies in the Arabian sea in winter causing extended rain along the trough line.

The aim of the present study is to bring out general features of subtropical disturbances moving along the Southern fringe of Tibetan Plateau, which might move into the North Burma areas. The data set for January 1981, published by the Hydrometeorological Division of USSR are used.

II. VARIATION OF GEOPOTENTIAL HEIGHTS AND TEMPERATURE

Daily variation of geopotemtial heights of the 500, 300, and 200 hPa are shown in figure 1 Trough / ridge systems are most pronounced between 75°E and 90°E particularly at the middle tropospheric level. The geopotential gradient is larger on the western side than that on the eastern side of the Southern fringe of Tibetan Plateau.

The trough / ridge systems are found to move generally eastwards from the area west of 60°E and attain maximum intensity near 80°E and then weakened as they moved east. But there were occasions when the trough / ridge system moved from area east of 90°E towards west for a distance of about 10° longitudes and intensified near 75°E. These phenomena can be found on the 7th and the 22nd of the month.

The temperature variation along latitude 30°N showed that the troposphere was cooler on the western side than the eastern side of the Tibetan Plateau. The average temperature near 75°E and that near 100°E are -20.4°C, and -12.7°C at 500 hPa, -47°C and -34°C at 300 hPa,

841

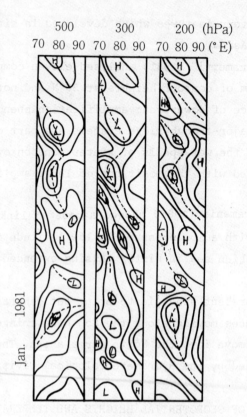

500　　　300　　　200 (hPa)
70　80　90　70　80　90　70　80　90 (°E)

Fig. 1 Time—longitude diagram.

and -55°C and -47°C at 200 hPa respectively. Cold pools are formed in association with the geopotential lows and the warm dome with the geopotential heights. The temperatures under different synoptic situations are shown in table (1). The temperature difference between the eastern and the western part of the Plateau is as large as 15°C.

MOVING DISTURBANCES

In winter, westerlies prevail in the subtropical region. Disturbanes embedded in the westerlies are steered eastwards. During the mid-winter season, they are expected to move eastwards along with the prevailing westerly currents of the troposphere. In January 1981, same subtropical lows are noted to move eastwards while the others such as those found on 7th and the 22nd. moved westwards along the southern

Table 1 Variation of temperature
under different synoptic situations.

Element	Level (hPa)	West of the Plateau	East of the Plateau
Geopotential lows	500	-25°C	-18°C
	300	-49°C	-41°C
	200	-62°C	-57°C
Geopotential Highs	500	-16°C	-10°C
	300	-38°C	-37°C
	200	-54°C	-49°C

fringe of Tibetan Plateau. These disturbances are found to have wave
number of 6, and the average zonal wind speed were less than the critical Rossby speed. See table 2.

III. SYNOPTIC FEATURES RESPONSIBLE FOR THE WESTWARD MOVING DISTURBANCES

1. Siberian Anticyclone

In January, the sea level anticyclone over the central/North Asia
was most pronounced. The intensity which simply refers to the highest
pressure value of the anticyclone is fluctuating from day to day while
its location was changing. The highest pressure value observed was
1075 hPa and the lowest pressure 1045 hPa. At times there appeared two
anticyclones: one over central and the other over North Asia, during
days when the development is less pronounced. During the period of
formation of the eastward moving disturbances, the intensities of the
anticyclones were not large. The highest pressure value of the anticy-

Table 2 Wave parameters in January 1981 (500 hPa.).

No.	Period in Jan 1981	Wavelength (Average) °Long	Direction of movement	Wave Number	Zonal speed Average ms^{-1}	Rossby critical speed ms^{-1}	Remark
1	1-2	45	East	8	14	7	
2	3-6	30	East	12	10	3	
3	7-10	62	West	6	10	12	
4	11-14	40	East	9	10	5	
5	15-19	48	East	7	10	8	
6	21-26	57	West	6	10	12	
7	27-31	40	East	9	18	5	

clone was 1055 hPa but during the period of the formation of the Westward moving disturbances, the highest pressure value of the anticyclone became large. See figure 2.

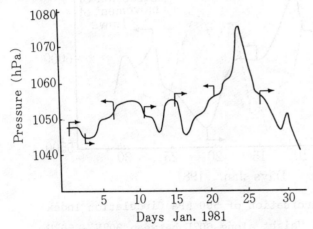

Fig. 2 Daily variation of Intensity (pressure) of the
Siberian Anticyclone.
Arrow indicates the direction of movement of
subtropical low.

The pressure tendency of the anticyclone was even more sensitive towards the direction of the movement of the disturbances as follows:

1) the eastward movement of the disturbance was followed by the weakening of the anticyclone;

2) the westward movement of the disturbance was followed by the strengthening of the same anticyclone.

2. Circulation Index

The index of circulation at 500 hPa level between 30°N and 60°N is analysed for the meridian of the western most fringe of the Tibetan Plateau which is 60°E. High/low and rising / falling of the geopotential field do not show that they have any relationship with the eastward / westward moving disturbance. See figure 3.

The circulation index reveals the following interesting features:

1) Disturbances, having moved into the area of interest which have

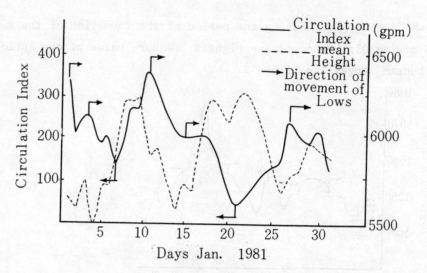

Fig. 3 Daily fluctuation of 500 hPa Circulation Index
and Mean Height along 60°E between 30°N - 60°N.

either a rising or falling tendency of the circulation index, can change from one state to another.

2) Periods with falling tendency of the circulation index coincide with the eastward moving disturbances and those with rising tendency coincide with the westward moving ones.

3) Subtropical disturbances exist together with both instances of high and low circulation indices.

4) Change of tendency of circulation index is very sensitive even to the shallow trough that passes through the area of interest.

3. Blocking High

It is a common synoptic feature to note that the prevailing cyclone over the mediterranean sea induced a trough at 500 hPa level. But there was an occasion when a ridge or a high from the area east of Ural mountains moved Southwest into the area Northeast of the Caspian Sea on the 21st, where it intensified into an anticyclone and anchored there until the 25th. See figure 4.

The period was found to coincide with the westward moving troughs and lows along the southern fringe of the Tibetan. Plateau Synoptic map showing the blocking high and the generated lows on the equatorward

846

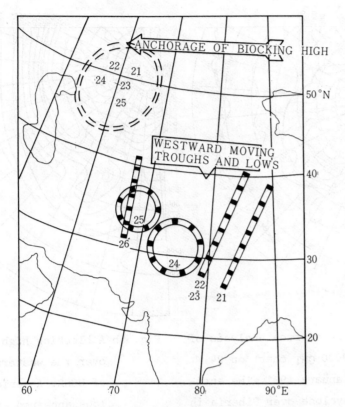

Fig. 4 Positions of westward moving troughs and lows
over the Tibetan Plateau on 500 hPa surface in
January 1981. The blocking high anchored over
the area North of Caspian Sea in the neighbourhood
of 50°N and 60°E.

presented in figure 5.

Sometimes, the sinusoidal oscillation becomes so large that the amplitude is extended towards north and south and covering a large meridianal interval the wave becomes unstable and cut-off low formed on the equatorward side followed by the formation of blocking high on the poleward.

On the 7th, January preexisting low on the southern fringe of Tibetan Plateau centred near 85°E had moved west until 10th due to the formation of this high as explained above. The westward movement ceased with the weakening of the departing anticyclone.

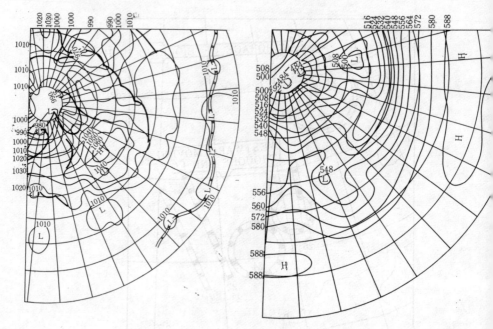

Fig. 5a The surface analysis of 0000 GMT chart on 24 January 1981. The anti-cyclone over Siberia is intensifying. The west-ward moving Subtropical disturbance may be observed over Northwest India.

Fig. 5b A blocking high anchored over the western Kazakh of USSR. A series of lows appeared at the southeast of the block-ing high; one over Afghanistan and another over Northwest India. The low over the North-west India further moved west and converged with the low over the Afghan-istan on the next day.

The intrusion of a blocking high into the Kazakstan, which is the windward side of the Plateau shows no effect upon the eastward moving subtropical disturbance along the southern fringe of the Tibetan Plateau.

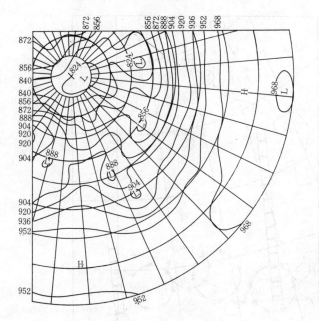

Fig. 5c   Same as Fig. 5-b except on 300 hPa.

In Figure 6, on the 15th, January, a trough from Kazakstan, extended into the North Arabian Sea where it developed into a low, while a blocking high from North Siberia moved Southwestward near latitudes 65°N and 55°E. The further movement of the blocking high seemed to have no effect on the eastward moving subtropical disturbance.

IV. DISCUSSION

The occurrences of eastward moving subtropical disturbances (trough/ridge system) are very common in winter. But westward moving disturbances are also observed. These westward moving phonomena are not just exceptionally rare features. It may be proposed that they are the interactions between the cyclonic cells and the blocking highs. There are similarities as well as differences between the eastward moving and the westward moving disturbances as summerized in table 3.

Meteorological parameters such as the circulation index, the synoptic situations that precede the disturbances, such as the blocking high anchoring over Kazakstan in the Northeast of Caspian Sea at 500 hPa

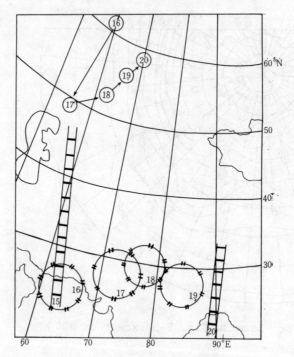

Fig. 6 Intrusion of blocking high into the Kazakstan and the
eastward moving subtropical disturbance.
Notice that trough formed in the Kazakstan prior to the
intrusion of the blocking high.

and the simultaneous fluctuation of the Siberian anticyclone are related to the westward moving subtropical disturbances on the Southern
foothills of the Tibetan Plateau. In one way or another, the phenomena
probably envisage part of the disruption of the circulation primarily
dominant over the Euro-Asiatic continent.

Schematic presentation of the eastward moving subtropical disturbances and the westward moving one is shown in Fig. 7. Eastward moving
subtropical disturbances are formed due to eastward propagation of
trough in the 500 hPa westerlies, while the westward moving disturbances
are formed due to the blocking high over the Kazakstan in the area
Northeast of the Caspian Sea which induces the low or which is induced
while the low exists. These cut-off cold lows are probably entrapped
by the tropography of the Tibetan Plateau.

Table 3 Physical features accompanied with the moving
trough / Ridge system of the subtropical
disturbances in the Southern fringe of the
Tibetan Plateau in January 1981.

Meteorological Element	movement	
	Eastward	Westward
1. Originates	West of 60°E	East of 90°E
2. Location of maximum Intensity	75° – 85°E	75° – 85°E
3. Mean rate of movement	5° / day	4° / day
4. Temperature	Cold low warm high	Cole low warm high
5. Direction of movement	Evident through out the troposhere	Evident at 500 hPa and Lower levels only. It is stationary at Upper levels.
6. Formation	Due to eastward moving trough over Caspian Sea	Due to blocking high over Kazakstan of USSR.
7. Siberia anticyclone	Moderate intensity. Pressure value 1055 hPa or less weakening on the following days.	Strong intensity. Pressure value more than 1055 hPa strengthening on the following days.
8. The circulation index at 500 hPa	Does not form in very low index condition	Does not form in very high index condition.
9. Nature of circulation index at 500 hPa	Falling on the following days.	Rising on the following days.
10. Effect of the mean contour height value of 500 hPa	NIL.	NIL.

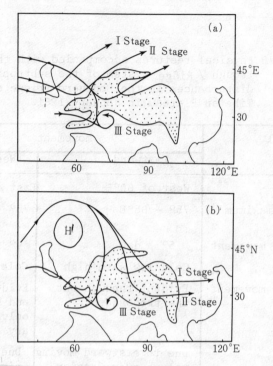

Fig.7 Schematic presentation of the waves in the
subtropical westerlies due to
(a)-Usual meandering of westerlies.
(b)-Stationary blocking high over Kazakstan.

V. CONCLUSION

The period covered in this study is very short but evidence shows
two possible types of motion exist under two distinct synoptic situations at 500 hPa level. The eastward / westward movement of the disturbance is closely related to the characteristics of circulation index
at 500 hPa at the entrance of the westerlies to the Tibetan Plateau and
the fluctuation of the Siberian anticyclone. It is possible that the
nature of eastward / westward movement of these disturbances along the
southern fringe of the Tibetan plateau could be a clue which explains
part of the general circulation during the winter season of the Euro-
Asiatic region. The knowledge about these moving disturbances may be
useful in operational weather forecasting.

REFERENCES.

1. Cheang B.K. and Krishnamurti, T.N. (1981), Middle lattitude inter
   actions During the winter Monsoon. Inter. Conf. on Early Results
   of FGGE and Large-scale Aspects of Monsoon Experiments. Tallahasse
   Florida. U.S.A. 4.50-4.37.
2. Murakami, T.(1981), Geographic influence of the Tibetan Plateau on
   the Asiatic winter Monsoon Circulation. Inter. Conf. on Early Results of FGGE and large-scale Aspects of Monsoon Experiments.
   Talahasse, Florida. U.S.A. 4.4-4.11.
3. Sharma, R.V. and subramanian D.V. (1983), The Western Disturbance of
   22nd. December 1980. A case study, Mausam, 34, 1, 117-120.
4. Webster, P.J.(1981), Mechanism Determining the mean and Transient
   Structure of the Large-scale Winter Monsoon, Cold surges, Inter.
   conf. on Early Results of FGGE and Large-scale Aspects of Monsoon
   Experiments. Tallahasse, Florida, USA 4.20-4.29.
5. Yin, M.T.(1970),The Weather and Climate of Burma.
6. 1962 January Rainfall of Burma, Burma Meteorological Department
   (1962), January Rainfall of Burma.

STUDIES OF STRUCTURE OF OROGRAPHIC LEE WAVES

G.S.Golitsyn, A.N.Gruzdev, N.E.Elansky,
N.N.Pertsev and N.N.Shefov
Institute of Atmospheric Physics of the Academy
of Sciences of the USSR

In the Institute there is a program of measurements of orographic lee waves by various methods. The measurements are carried out from an aircraft on flights across mountain ridges. At flights in the Northern Ural region simultaneous measurements were made of temperature variations at height of 3 km of the rotational temperature of the hydroxyl emission (the altitude of the emission layer is near 90 km). When winds blew across the ridge there was an increase of the rotational temperature by some 20K and no increase when the wind was along the ridge. This can be explained by penetration of orographic disturbances seen in the temperature record at 3 km into the upper atmosphere which break there and heat up the lower thermosphere.

Orographic lee waves were also observed at flights across ridges in Armenia and Southern Kazakhstan while measuring chemically the ozone concentration together with the temperature. The ozone measurements seem to supply an effective method of tracing the mesoscale dynamic phenomena in the atmosphere.

I. INTRODUCTION

The flow of stably stratified air over mountain ridges is accompanied by appearance of lee waves. The characteristics of the waves are connected to the structure of the flow, namely, to its velocity and vertical stratification. Theoretical studies of the waves began with works by Dorodnitsyn (1938), Lyra (1943), Queney (1948), Scorer (1949) and are vigorously carried out up to the present. A very comprehensive review of the studies was prepared by Smith (1979). The orographic waves are of interest because of their influence on the local weather and on flight conditions over mountain regions. By their essense the waves are internal gravity waves in the atmosphere caused by mountains in the air flow (Gossard and Hooke, 1975).

In the Institute of Atmospheric Physics of the USSR Academy of Sciences two different observational programs were carried out in 1980-1982 to study the waves from an aircraft. In both programs temperature sensors were used which measure the temperature pulsations with the precision of 0.1K with time resolution of about 5 sec at the flight trajectory. In the first program the intensity of the hydroxyl emission was also measured which is related to the atmospheric temperature at altitudes near 90 km. An increase of the emission intensity was observed when winds were across the ridge. This is connected with the penetration of the waves into the upper layers of the atmosphere and heating the layers by breaking waves.

The other program included measurements of the ozone local concentration. The latter is found to be a good indicator of atmospheric vertical motions and in favourable conditions is revealing well a spatial wavy structure.

II. OPTICAL PROGRAM OF OBSERVATIONS

Krassovsky (1972) was the first who observed internal gravity waves in the atmosphere by optical means. Further results and development of the observational method can be found in the paper by Krassovsky and Shefov (1976). The waves in the upper atmosphere usually have the origin from synoptic formations in the lower troposphere, from fronts, cyclones and jet streams. The wave period and the horizontal component of the wave vector can be determined from observations. Then using the dispersion relation the vertical component may be found and therefore the complete wave vector restored. After that neglecting the wind velocity in comparison with the wave propagation

velocity the trajectory of the wave ray can be restored and the place
of its origin determined. It is by this method that the synoptic
disturbances in the troposphere were found as sources of the gravity
waves observed in the upper atmosphere.

The flights to observe orographic perturbations of the upper
atmosphere were carried out from 14 to 21 March 1980 and from 26 January
to 8 February 1981. The flights were performed over Northern Ural
along 64°N at the altitude of near 3 km. During the flights at night
the hydroxyl emission spectra were registered. In 1981 the temperature
on the flight path was also measured (Semenov et al., 1981; Shefov et
al., 1983). After exclusion of flights in unfavourable weather conditions (cloudiness, polar auroras) the number of flights when the quality
hydroxyl spectra were obtained was 15; the number of flights when the
ambient temperature was also measured amounted to 7.

The method of determining the upper atmosphere temperature is
based on the existence of the relationship between the ratio of intensities of different lines in the hydroxyl spectrum and the rotational
temperature of OH (Kvifte, 1961). Assuming that this temperature
corresponds to the ambient temperature (Krassovsky, 1977) one can
determine the temperature of a layer responsible for the hydroxyl
emission. The height of the layer is about 90 km and its width by a
half of the emission intensity is from 8 to 15 km. For registration
of the rotational temperature of OH bands the spectrographic method
was used with application of electronic image transformer (Semenov
et al., 1981; Shefov et al., 1983).

Fig 1. presents averaged variations of the temperature increase
at height near 90 km in dependence on the distance from the middle of
the mountain ridge for two cases of the wind direction: across and
along the ridge. It is clear that the temperature increase ΔT is
observed only in the first case (Shefov et al., 1983).

To measure the aircraft ambient temperature a resistance thermometer was used with a precision of about 0.1K and the time constant
of 5 sec. The measurements were also carried out at the ascend and
descend of the aircraft during 40 minutes which gave vertical temperature profiles. The horizontal temperature gradients as the treatment

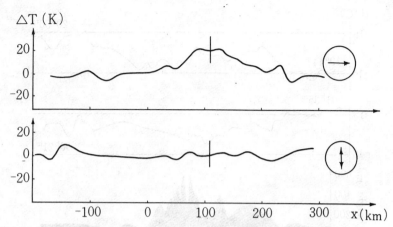

Fig.1 Averaged increase of the hydroxyl rotational temperature
in dependence on the distance from the axis, 0 km, of
the Ural ridge along 64°N for zonal (above, 7 cases) and
meridional (below, 5 cases) wind directions. Vertical
lines show the temperature standard deviation. For zonal
wind after the ridge there is a clear increase of the
upper atmosphere temperature.

of the measurements showed did not introduce, and as a rule, trended
significantly into the vertical temperature gradients.

The temperature records were treated by the harmonic analysis.
For various flights a large number of wave packets were found with
wave lengths from 5 to 50 km with amplitudes from 0.05 to 1K. Fig.2
presents temperature records of the two flights on the same path to
and fro with the same speed of 70 m.s^{-1} The time interval between the
records was about 4 hours. The similarity between the records is
apparent. To study the temporal variability of the wave patterns in
the lee of the ridge, we compared the results of harmonic analysis for
the same parts of the path flown first in the eastward direction and
then in the reverse westward direction. In both cases 11 wave components
were found with amplitudes from 0.05K and a good correlation was
revealed in frequencies and amplitudes of the components. This is an
evidence of the stationarity of the patterns. In other cases the
stationarity was conserved with lesser accuracy.

An analysis of the measurement results shows that at the increase

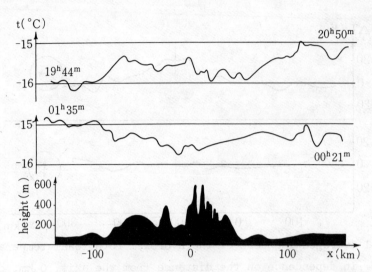

Fig.2 Two temperature records at the altitude
3 km during the flights across Ural
ridge on February 7-8, 1981. The wind
11-17 km is South-West. Below is the
ridge profile.

of the temperature wave harmonics there is a tendency for the growth
of the hydroxyl rotational temperature. Also is apparent the relationship between the value of ΔT_{OH} and the angle θ between the wind
direction at the altitude of 3 km and the mountain ridge. This is
illustrated by Fig.3: for $\theta > 25°$ there never was any increase of the OH
temperature while for $\theta < 60°$ the effect in the lee zone was always
observed.

The vertical component of the energy flux may be estimated as

$$F_z = -(g^2\rho\omega^3 k_z / 2k_x^2 N^4)(A_T/\bar{T})^2 \qquad (1)$$

where g is the gravity acceleration, $\rho, \bar{T}$ the density and temperature
at z=3 km, ω the frequency, N the Brunt-Väisälä frequency, A_T the
amplitude of the temperature perturbation. The formula gives F_z of
order of 0.01 to 0.02 Wm^{-2}. Such a flux if absorbed near or just above
the mesopause in the layer of 10-15 km thick by e.g. the mechanism of
non-linear wave breaking can provide a heating of the layer by about

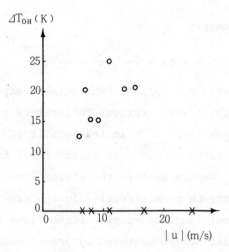

Fig.3 Dependence of the individual rotational hydroxyl temperature increase on the absolute value of the wind velocity. Crosses are for angles θ between wind direction and the ridge axis less than 25°, circles for θ>60°.

20K which is observed (see Fig.1).

The reported results are supported by Gavrilov and Shved (1982) who analyzed observations of the night glow in Ashkhabad and showed that a part of the glow variations can be explained by orographic perturbation in the air flow propagating into the upper atmosphere.

III. OZONE OBSERVATIONS

Concentration of a trace passive admixture in the atmosphere is determined by atmospheric dynamics and photochemical reactions in which the admixture is involved. If the admixture is conservative, i.e. the characteristic of its photochemical reaction is large compared to a dynamic time which in a stably stratified atmosphere is determined by the Brunt-Väisälä frequency N then the deviation of the concentration n from its value in the non-perturbed flow n_0 is connected to the stream line perturbation η in the adiabatic case by the following relationship

$$n' = n - n_0 = (\frac{\gamma_a}{\kappa-1} \frac{n_0}{T_0} - \frac{\partial n_0}{\partial z})\eta \qquad (2)$$

where $\eta = (T-T_0)/(\gamma_a-\gamma)$, $\kappa = C_p/C_v$ is the adiabat exponent, $\gamma_a=-g/c_p$ the adiabatic and γ the actual vertical temperature gradients. Simple estimates show that the first term in brackets in (2) is, as a rule, small in comparison to the second one. In any case, it is clear that by measuring n and n_0 we can restore the stream line pattern. Such a method of restoring may be more effective in the case of large vertical gradients of n and small amplitudes of perturbations than by temperature records. This will be demonstrated by ozone measurements in the region of orographic lee waves.

Ozone is convenient because it can be measured locally with a high precision, it is conservative enough and its vertical distributions are often characterized by large gradients with a sufficient homogeneity over horizontal.

The measurements were carried out aboard an aircraft using electrochemical gas analyzer with the time constant of few seconds and a high relative precision. Ambient air was pumped into the analyzer with a constant rate. Temperature measurements were done with a system similar to the one described in the previous section.

The horizontal parts of the flights were in isobaric surfaces with deviations less than 10-20 m. The flights were performed in August-September 1982 over Gegam ridge near lake Sevan in Armenia and over various ridges in Middle Asia. Gegam ridge is mainly meridional and at this time westerlies are prevailing there, which was checked each time by results of aerological sounding at Yerevan. Mean ozone concentrations and vertical temperature profiles were obtained during the aircraft descend and partly during its ascend.

In Fig.4 we present examples of the ozone concentration and temperature records which have a clear wavy structure. The oscillations of O_3 are in counterphase with the temperature ones. It was connected to a small temperature inversion above which the ozone concentration increased sharply with $\partial n/\partial z = 30$ mcgm $m^{-3}km^{-1}$.

If we know mean O_3 and T profiles we can determine from (2) the stream line displacement. The largest value of η reaching 80÷100 m were observed for stream lines between peaks of the ridge. Up to five waves with the length of about 10 km can be seen in Fig.4 (the flight speed here was about 100 m.s^{-1}). Note that using the well-known

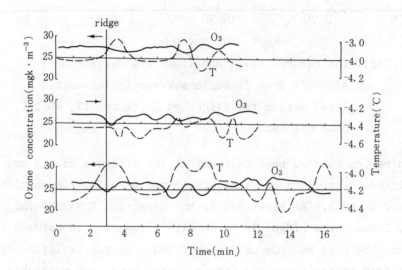

Fig.4 Results of measurements of ozone concentration and temperature at flights on August 8,
1982, over Gegam ridge in Armenia at altitude
5.1 km. The arrows show the flight direction
(West to East). The distance between paths
is about 10 km. The flight speed is 6.5
km/min.

Scorer's formula λ≈N/u the wave length may be estimated as 9 km for the wind velocity 12 m.s^{-1} observed at the heights of the ridge.

Note that during the flights in cloudy conditions when the temperature vertical gradient was close to wet adiabatic one the wavy perturbations were not observed as well as during the flights when wind was along the ridge. In other cases not described here in any detail complicated patterns of non-stationary or three-dimensional flow over mountains were observed.

Fig.5 presents the measurements on September 17, 1982 over

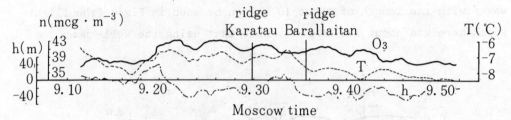

Fig.5 The records of O_3 and T and the deviation of
aircraft from isoboric surface (point-dotted
line) during the flight on September 17, 1982
near Chimkent.

mountain region Karatau near Chimkent at the altitude of 4.2 km. The
region is of 50 km width and consists of two parallel ridges with
heights of 1 and 1.8 km stretched in the direction NNW-SSE. The
distance between the ridges is 25 km with a lowering in between down
to 0.8 km. The time records of O_3 and T and also the deviation of the
aircraft from isobaric surface (point-dotted line) are presented in
Fig.5.

At the altitude of 1.8 km the wind was 10 m.s^{-1} in the direction
of 60° from the North, at the altitude 4.2 km it had 30 m.s^{-1} and 70°.
The waves were observed not only in the lee of the ridges but also
windward of them. Because there were no substantial hills south-west
of the region the waves before the ridges should have some other
origin. As a possible cause one may point out to a proximity of a cold
front. We may also note during flights across the frontal zones or
near them we often observed an oscillation structure in temperature
and ozone concentration. This also agrees with results of observations
of internal gravity waves in the upper atmopshere by optical means
when the ray tracing technique (see Section 2) points out to fronts as
sources of the waves (Krassovsky et al., 1977).

Therefore the ozone may serve as a good indicator of mesoscale
motions within the atmosphere. Large possibilities may supply here
lidar methods (Pelon and Magie, 1982). The use of lidars gives a pos-

sibility by remote sounding from aircrafts or from the ground to
restore the ozone distribution and to visualize actually the structure
of air flow of practically any scale.

IV. CONCLUSION

In this paper we tried to present material which shows possibilities of new methods of lee wave studies by optical and chemical
means. The possibility of the first method is rather large for studying
the upper atmosphere. Here, e.g., we have demonstrated the influence
of orographic waves on the temperature regime of the lower thermosphere.
The use of ozone for detection of orographic perturbations has been
demonstrated here for the first time. One may foresee that this or that
modification of the second method, the use of lidars or direct concentration measurements may have a future.

REFERENCES

Dorodnitsyn A.A. 1938. Perturbations of air flow caused by inhomo-
geneities of the Earth's surface. Trudy of the Main Geophys.
Observ., N° 23 (6) (in Russian).

Gavrilov N.M., Shved G.M. 1982. Studies of internal gravity waves in
the lower thermosphere using isophotos of night air glow.
Izvestia-Atmosph. Oceanic Physics, 18, N°1, 8-17.

Gossard E.E., Hooke W.H. 1975. Waves in the atmosphere. Elsevier Sci.
Publ. Co., Amsterdam-Oxford-New-York.

Krassovsky V.I. 1-72. Infrasonic variations of OH emission in the
upper atmosphere. Ann. Géophys. 28, N°4, 739-746.

Krassovsky V.I., Shefov N.N. 1976. The intensities, Doppler and
rotational temperatures of the upper atmospheric emission and
gravity waves. Ann. Géophys. 32. N°1, 43-46.

Krassovsky V.I., Potapov B.P., Semenov A.I., Shagayev M.V., Shefov N.,
Sobolev V.G. 1977. On the equilibrium nature of the rotational
temperature of hydroxyl airglow. Planet. Space Sci. 25, N°6,
596-597.

Kvifte G. 1961. Temperature measurements from OH bands. Planet. Space
    Sci. 5, N°2, 153-157.

Lyra G. 1943. Theory der stationaren Leewellenströhmung in freier
    Atmosphäre. Z.Angew Math. Mech. 23 (1), 1-28.

Pelon J., Megie G. 1982. Ozone vertical distribution and total content
    using a ground-based active remote sensing technique, Nature 299,
    N° 5879, 137-139.

Queney P. 1948. The problem of airflow over mountains. A summary of
    theoretical studies. Bull. Amer. Met Soc. 29, 16-26.

Scorer R.S. 1949. Theory of waves in the lee of mountains. Q.J. Roy.
    Met.Soc. 75, 41-56.

Semenov A.I., Shagayev M.V., Shefov N.N. 1981. On the influnce of
    orographic waves on the upper atmosphere. Izvestia-Atmos. Oceanic
    Physics 17, N°9, 982-983.

Shefov N.N., Pertsev N.N., Shagayev M.V., Yarov V.N. 1983. Oro-
    graphically caused variations of the upper atmospheric emissions.
    Izvestia-Atmos.Oceanic Physics 19, N°9, 920-926.

Smith R.B. 1979, The influence of mountains on the atmosphere. Adv.
    Geophys. 21, 87-230.

INTERACTION OF OROGRAPHY AND HEATING FOR PLANETARY SCALES

Julia N. Paegle[1] and Gu Hongdao[2]

I. INTRODUCTION

There has been an upsurge of investigations in the last 15 years describing the large scale quasi-stationary patterns of the Northern Hemisphere (i.e. Bjerknes, 1969, Sawyer, 1970, Van Loon and Rogers, 1978, Wallace and Gutzler, 1981, Namias, 1981, Horel and Wallace, 1981, Blackmon et al., 1983a, b). Some of the observed features have been interpreted as resulting from localized forcings in the equatorial Pacific or tropical Atlantic. This interpretation finds theoretical support in the results obtained from some linear and non-linear models.

Linear models which simulate atmospheric response to tropical heating have been developed by several investigators (Egger, 1977, Hoskins and Karoly, 1981, Simmons, 1982). General circulation models of various degrees of complexity have also been used to study the atmospheric response (i.e. Rowntree, 1972, Keshavamurty, 1982, Hanna et al., 1983). Some of these numerical models appear to simulate the train of waves which seemingly emanates from heating sources as in certain linear analyses and in some observations. However, others require a dipole heating structure in the tropics and mid-latitude for similar patterns. Simmons et al., (1983) note that observed long wave patterns of the northern hemisphere resemble those corresponding to baro-

[1]Department of Meteorology, University of Utah

[2]Lanzhou Institute of Plateau Atmospheric Physics, Chinese Academy of Sciences

tropically unstable mode of the 300 hPa zonally varying climatological
state. This interpretation relies on barotropically unstable modes to
efficiently extract energy from the mean flow and to therefore prevail
in the resulting configuration. This idealization does not allow for
the non-linear feedback between the perturbation and the basic state
as the disturbance reaches finite amplitude. However, the non-linear
integrations performed by Simmons et al., 1983, indicate that the long
term means exhibit similar patterns to those found in linear solutions.
These results differ from those reported by Paegle et al., 1983. In
their global integrations of non-linear barotropic models they found
substantial differences in the response of linear and non-linear models
forced by topography in the vicinity of North America while the response
over the Asian region was well represented by linear models. The
applicability of linear models to interpret long-term characteristics
of non-linear solutions appears to be regionally dependent.

In the present paper, differences between linear and non-linear
equilibrium solution of barotropic flows are studied for simple forcing
configurations. Paegle et al., 1983, have shown that forcing of a
global barotropic model with a divergent source of wavenumber 1 which
maximizes at the equator at 150°E has pronounced effects in the
circulation patterns of the northern hemisphere (such as changing the
latitude of the subtropical jet and creating a block over Greenland).
In the present paper we investigate the extent of the differences
between linear and non-linear equillibria, the geographical distribution of the response to specified stationary forcing and whether
regional differences between linear and non-linear solutions persist
when only the planetary scales are explicitly resolved. A spherical
harmonic expansion is chosen as the means to restrict the flow to the
largest atmospheric scales since in this representation the two-dimensional wavenumber vector is associated with a given length scale. This
is not true of other decompositions commonly used (such as Fourier
analyses) where the wavenumber vector represents different length
scales for different latitudes. Spherical harmonics are solutions of
the non-divergent barotropic vorticity equations and are a natural
choice when using barotropic models.

II. SPECTRAL MODEL

Modifications due to heating cannot be directly applied in barotropic models and are usually imposed as vorticity sources or divergence fields. The latter is applied in the present investigation. The
triangular truncation of order 4 (10 wave modes and four longitudinally
symmetric modes) is chosen as a compromise between high truncation and
a representation of earth orography that reproduces its main features
(Fig. 1).

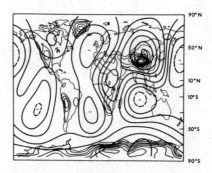

Fig.1 Earth orography (solid, thin lines, analysis interval
of 500 m) and that used by truncated model (0 and
positive values represented by thick solid lines,
negative values dashed lines, analysis interval
of 300 m)

The spectral model is based on the barotropic vorticity equation
as in Paegle et al., 1983:

$$\frac{\partial}{\partial t}\nabla^2\psi + J(\psi, \ \nabla^2\psi + f) = -(f + \nabla^2\psi)\nabla\cdot\vec{V} - k(\nabla^2\psi \ \nabla^2\psi^*) \tag{1}$$

where
$$\nabla \cdot \vec{V} = - \frac{\omega_{surface}}{700 \ hPa} + \nabla \cdot \vec{V^*}$$

and
$$\omega_{surface} = -pg \ \vec{K} \times \nabla \ \psi \cdot \nabla z$$

$\nabla \ V^*$ = divergent forcing excluding orographic effects.

In spherical coordinates this equation can be rewritten as follows:

$$\frac{\partial \zeta}{\partial t} = F - \frac{2\Omega}{r^2} \frac{\partial \psi}{\partial \lambda} + \alpha T - D - 2\Omega\mu \nabla \cdot \vec{V}* - k(\nabla^2\psi - \nabla^2\psi*) \qquad (2)$$

Here:

$$F = -J(\psi,\zeta) = \frac{1}{r^2} \left(\frac{\partial \psi}{\partial \mu} \frac{\partial \zeta}{\partial \lambda} - \frac{\partial \psi}{\partial \lambda} \frac{\partial \zeta}{\partial \mu}\right)$$

$$T = \frac{\mu}{r^2} \left(\frac{\partial \psi}{\partial \mu} \frac{\partial Z}{\partial \lambda} - \frac{\partial \psi}{\partial \lambda} \frac{\partial Z}{\partial \mu}\right) \qquad (3)$$

$$D = \zeta \nabla \cdot \vec{V}*$$

and

$$\zeta = \nabla^2 \psi$$

$$\nabla^2 = \frac{1}{r^2} \frac{\partial}{\partial \mu} \left[(1-\mu^2)\frac{\partial}{\partial \mu}\right] + \frac{1}{r^2(1-\mu^2)} \frac{\partial^2}{\partial \lambda^2}$$

$$\mu = \cos\theta$$

$$\alpha = \frac{2\Omega\rho g}{700 \text{ hPa}} \qquad (4)$$

$$\theta = \text{co-latitude}, \ \lambda = \text{longitude}$$

$$r = \text{earth's radius}$$

$$k = 1/15 \text{ days}$$

Variables are expanded in spherical harmonics as follows:

$$\begin{Bmatrix} \frac{\psi}{r^2} \\ z \\ \nabla \cdot \vec{V}* \\ \frac{\psi*}{r^2} \end{Bmatrix} = \underset{mn}{\Sigma\Sigma} \begin{Bmatrix} \frac{\psi^c_{nm}}{n(n+1)} \\ z^c_{nm} \\ \delta^c_{nm} \\ -\frac{G^c_{nm}}{n(n+1)} \end{Bmatrix} \bar{Y}^c_{nm} + \begin{Bmatrix} \frac{\psi^c_{nm}}{n(n+1)} \\ z^s_{nm} \\ \delta^s_{nm} \\ -\frac{G^c_{nm}}{n(n+1)} \end{Bmatrix} \bar{Y}^s_{nm} \qquad (5)$$

868

and a triangular truncation of order 4 is assumed (Fig. 2).
Here:

$$\frac{\psi_{nm}^c}{n(n+1)} = \frac{G_{nm}^c}{n(n+1)} = 0 \quad (m=n=0)$$

$$\bar{Y}_{nm}^c = [\frac{2n+1}{2\pi\varepsilon_m}\frac{(n-m)!}{(n+m)!}]^{\frac{1}{2}} P_{nm}(\mu) \cos m\lambda \quad (n \geqslant m \geqslant 0)$$

$$\bar{Y}_{nm}^s = [\frac{2n+1}{2\pi}\frac{(n-m)!}{(n+m)!}]^{\frac{1}{2}} P_{Hm}(\mu) \sin m\lambda \quad (n \geqslant m \geqslant 1)$$

$$P_{nm}(\mu) = (1-\mu^2)^{\frac{m}{2}} \frac{d^m}{d\mu^m} P_n(\mu)$$

$$P_n(\mu) = \frac{1}{2^n n!} \frac{d^n}{d\mu^n} (\mu^2-1)^n$$

$$\varepsilon_m = \begin{cases} 2 & m=0 \\ 1 & m \neq 0 \end{cases}$$

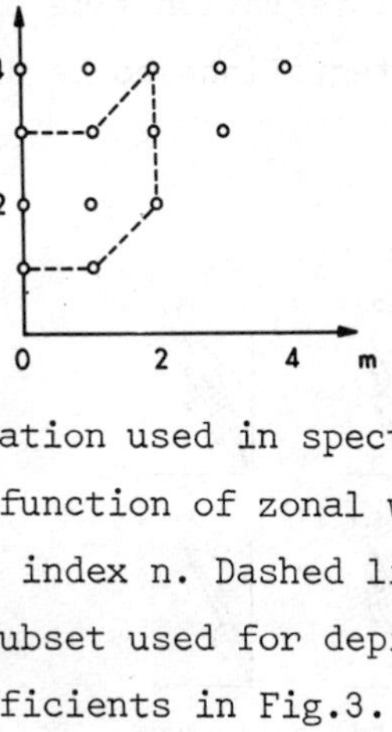

(6)

Fig.2 Triangular truncation used in spectral model
given by o as a function of zonal wavenumber
m and meridional index n. Dashed line indicates
the wavenumber subset used for depiction of
interaction coefficients in Fig.3.

Equation (2) is converted to a set of ordinary differential
equations in time using the expansion given in (5) and the orthogona-
lity properties of the spherical harmonics.

The resulting equations give the time changes of the streamfunction coefficients ($\dot{\psi}_{nm}$) as a function of advection of relative
vorticity (F_{nm}), advection of earth vorticity (Ω nm; for $n \neq m$), topographic forcing (T_{nm}); divergent forcing (other than topography)
associated with relative vorticity (D_{nm}) and earth rotation ($\delta*_{nm}$);

frictional dissipation $(k\psi\,nm)$ and forcing by a zonally invariant
streamfunction $(k\psi^*_{no})$.

The non-linear terms F_{nm}, T_{nm} and D_{nm} couple different zonal
wavenumbers and meridional indices. The δ^*_{nm} term gives the projection
of the divergent forcing multiplied by the Coriolis parameter (which
equals $2\Omega\mu$) to the Y_{nm} mode, linking different meridional indices for
the same zonal wavenumber. The beta terms allow for wave propagation.

Figure 3 depicts the non-linear interactions obtained from terms
D_{nm} and the dominant component of T_{nm} from $f\mathbf{V}\cdot\mathbf{V}$ term in (1) for a
rhomboidal subset of the spherical harmonics as indicated in Fig.2.
This figure shows that this component of the topographic forcing for
wavenumber 0 does not affect the zonally invariant components (Tno)
while the divergent forcing does (Dno and δ^*no). Integrations performed
with divergent forcing in modes m.0 did not lead to a steady state
within the maximum integration time of 6 weeks and were not included
in the numerical integrations to be discussed next.

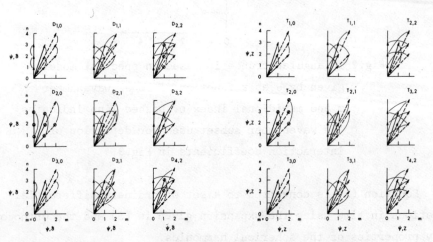

Fig.3 Graphical representation of non-linear interactions
for the rhomboidal subset indicated in Fig.2 due to
non-orographic (left) and topographic forcing (right)
as discussed in text. Circles denote self-interactions.

The forcing functions (Dnm; δ^*nm for m$\neq$0 and ψ^*no) are obtained

from projection of the 200 hPa velocity potential and streamfunction
during the Special Observing Period 1 of FGGE (January and February,
1979), using the Goddard Laboratory for Atmospheric Sciences level
III-b data. We selected the weekly averages for the second week in
January and last week in February (case 1 and 2) for input to the model.
Figure 4 depicts the divergence with a triangular 4 truncation. These
two cases resemble patterns observed during different phases of the
Southern Oscillatin (SO). Case 1, with the divergent outflow located
east of the dateline is reminiscent of the "warm" phase of the SO
(Horel and Wallace, 1981) when warm sea surface temperature anomalies
extend over the equatorial Pacific while Case 2 has similarities to the
opposite phase of the SO with cool surface temperature east of the
dateline and low precipitation values in the central and eastern
equatorial Pacific.

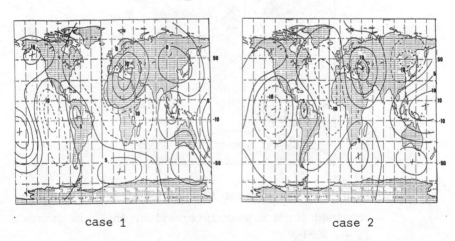

case 1   case 2

Fig.4 Divergence obtained from a triangular 4 truncation
      from averages for the 2nd and 8th week of the
      Special Observing Period I of FGGE. Analysis
      interval $5 \times 10^{-7} s^{-1}$.

Although the SO acts on a very different time scale, the choice
of these two particular weeks was motivated by these similarities, in
order to gain some insight in the effect that longitudinal shifts ór

tropical divergent outflows have upon stationary solutions which include only planetary scales. We take these divergence fields as a rough measure of all the divergent processes that influence the global scales that are not included in the barotropic model (such as diabatic heating in the tropics and quasi-geostrophic divergence in higher latitudes).

III. MODEL RESULTS

The non-linear model is integrated for six weeks with only topographic forcing. The relative vorticity for the flow adjusted to the orography of Figure 1 is shown in Figure 5. The stationary re-

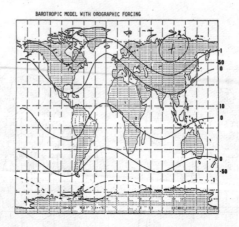

BAROTROPIC MODEL WITH OROGRAPHIC FORCING

Fig.5 Relative vorticity of flow adjusted to topography
obtained for a 6 week integration. Analysis interval
is $10^{-5}s^{-1}$.

sponse displays troughs over the higher elevations of the Northern Hemisphere. For these planetary scales the vorticity equation indicates a balance between the beta and divergence term. That is: $\beta_v = -f_0 \vec{\nabla} \cdot \vec{V}$ with divergence inducing equatorward flow and convergence indicating poleward flows.

This state is used as the initial conditions for integrations with the divergent forcing of cases 1 and 2. The equilibrium configurations

are given in Figure 6. The introduction of the divergent forcing intensifies the trough and ridge configuration excited by the orography. The
net effect in the streamfunction is to increase the amplitude of the
highs centered over the Pacific Ocean and Mediterranean Seas and the
troughs over the American and Asian continents.

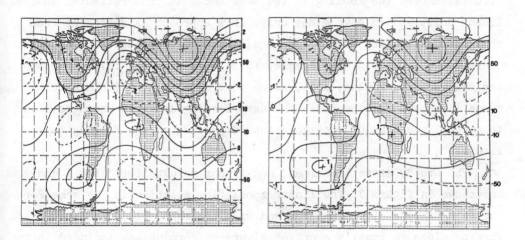

Fig.6 Relative vorticity as in Fig. 5 for cases with divergent forcing.

Comparison of case 1 and 2 shows the amplitude of the wave patterns created by the orography in the Northern Hemisphere to be most
intensified for the case of the "warm" tropical heating. Inspection of
the amplitude of the spherical harmonic coefficients helps elucidate
the nature of this amplification. This is done in the next section where
the stationary solutions from the non-linear integrations are compared
with those of a linearized model.

IV. LINEAR MODEL

The linear model assumes a) the zonally invariant modes to be
stronger than the zonally varying modes; b) the zonally invariant modes
must be solutions to the given equations in the absence of wave modes;
c) products of zonally varying modes can be neglected.

The steady state solutions are obtained by setting the time deriva-

tives equal to zero; determining first the zonally averaged state
which is in equilibrium with the zonally averaged forcing (in agreement
with (b) above) and then solving the steady state solutions for the
wave modes. This approach precludes wave modes in the forcing from influencing the steady state of the zonal flow in agreement with assumption (a) above. The forcing in the wave modes is thus reflected in the
linear solutions in wave modes only. This approach is different than
that used in studies of multiple equilibria in barotropic models (i.e.
Charney and De Vore, 1979; Paegle, 1979; Wiin-Nielsen, 1979; Trevisan
and Buzzi, 1980; Kallen, 1981, among others). The steady state solutions in these cases included the wave forcing and were shown to lead
to several equilibrium configurations depending on the forcing amplitude.

The steady state linear and non-linear results for cases 1 and 2
are shown in Figure 7. The linear zonal flow solution is the same for
the integrations with orography and with orography plus divergent
forcing since no zonally averaged divergent forcing was included in the
present calculations. Figure 7 shows a more pronounced strengthening
of the mean flow in the non-linear calculations for case 1. The non-
linear solutions for both cases display pronounced increases in the
zonal wavenumber 2 components when divergence forcing is included as
compared to those obtained from the non-linear solution with orography
only. The excitation of this mode depends on the structure and phasing
of the divergent forcing with respect to those induced by the orography
and can be best examined by considering divergent forcing in single
wave modes. This is discussed next.

V. SINGLE MODE DIVERGENT FORCING CASES

In this section we discuss the impact that divergence forcing in
a single mode has upon the linear and non-linear steady state solutions.
The phase of the modes is chosen equal to that of case 1 and 2 unless
otherwise stated and the amplitude is increased so that the forcing by
a single mode is comparable to that derived from observations for the
planetary waves (with a triangular truncation of 4). All non-linear

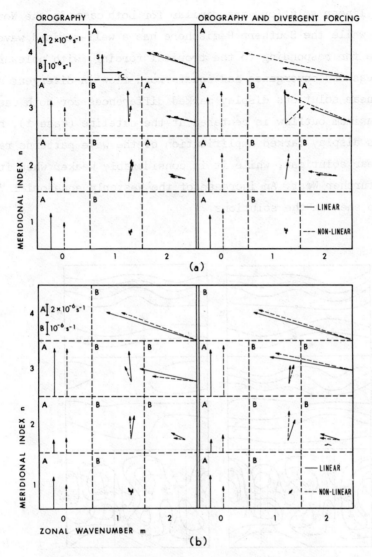

Fig.7 Comparison of spherical harmonic coefficients
obtained from linear and non-linear steady state
solutions for case 1 (a) and case 2 (b) and these
obtained without divergent forcing.

integrations are started with initial conditions given by the linearized
solutions.

Figure 8 shows the linear and non-linear steady state streamfunction and divergent forcing corresponding to the amplitude in mode

(1,1). The linear solutions are similar for both cases in the Northern Hemisphere while the Southern Hemisphere has a well-defined wavenumber 1 structure for responding to the tropical forcing, with poleward flow in areas of convergence and equatorward flow in divergent areas. The non-linear solutions display marked differences for both cases. When the maximum outflow is centered at the dateline (Case 1), both hemispheres display marked amplification of the wave patterns resolved by the linear solutions, while it is considerably weaker when it is displayed further West. An increase of the Westerly equatorial "Pacific" jet is also seen in the solutions.

Fig.8 Linear and non-linear steady state streamfunctions (top and middle panels) analyzed every $5 \times 10^{-7} m^2 s^{-1}$ for the divergent forcing (thin solid and dashed thin lines in units of $10^{-6} s^{-1}$) and orography (thick solid lines in meters) given in the lower panel. The phase of the divergence forcing is that of case 1 (left diagram) and case 2 (right diagram).

The sensitivity of the response to mid-latitude forcing phase is displayed in Figure 9 which shows the results for forcing from mode (2,1).

Fig.9 Same as Fig. 8 for divergence forcing in
mode (n=2, m=1).

The different phasing of this mode for both cases has strong impact both in the linear and non-linear solutions. The non-linear solutions reflect the asymmetric characteristics of the forcing in the two hemispheres with a high index situation whenever the divergence forcing is in the Pacific and a disturbed "blocked" situation when it is in the Atlantic. The blocked situations have a wavenumber 1 structure resulting from the relative phasing of the divergence forcing and orography and the projection into this mode due to non-linear interactions shown in Figure 3.

An example of a forcing by a mode that maximizes at the tropics and mid-latitudes is shown in Figure 10. The phasing of this mode corresponds to that of Case 2, with the equatorial forcing maximizing west of the dateline and over the northern and Southern latitudes of the Pacific Ocean. The Asian and Antartic trough are considerably

amplified in this case. It is of interest that forcing with the phase
and amplitude of the second week lead to solutions which amplified
continuously through the six weeks integration times. It appears that
the atmospheric response to this forcing structure may induce noticeable variability depending on the amplitude and strength of the forcing.

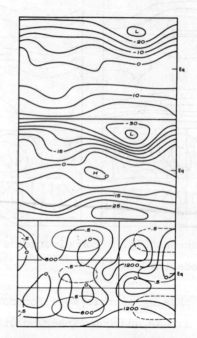

Fig.10 Same as Fig. 8 for divergence forcing (n=3, m=1).

We have emphasized wavenumber 1 in the discussion above since
this mode structure has received considerable attention recently
(Madden, 1978; Paegle et al., 1983). Since topographic effects produce
a dominant wavenumber 2 structure in planetary waves, it is of interest
to discuss the response to forcing by this wavenumber.

Figure 11 shows the impact of three different meridional forcing.
Comparison of linear and non-linear solutions show intensification of
the zonal flow in the Northern Hemisphere and wave patterns around the
dateline and mid-Pacific (Figure 11a). Decrease in the intensity of the
zonal flow is observed with other configurations (Figure 11b and c).

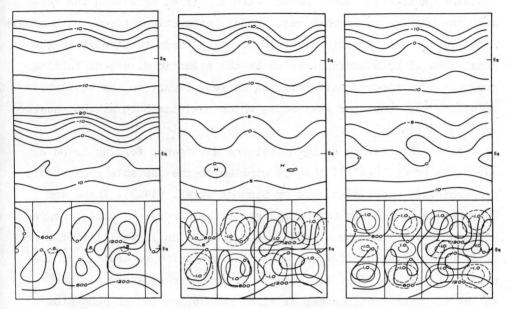

Fig.11 Same as Fig. 8 for m=2 for meridional indices
n=1, 2, 3.

VI. CONCLUSIONS

Many of the observed features of planetary waves have been interpreted by previous investigators in terms of the linear response of
barotropic models to topography and forcing. The present comparison of
linear and non-linear model integrations for planetary waves shows
marked difference in the results of linear and non-linear analyses.
The high truncation of the model was chosen to preclude interaction
with synoptic scale waves which have been shown to affect the structure
of planetary scales (Gall et al., 1979; Reinhold and Pierrehumbert,
1982), except as that effect may be parameterized in the observed large
scale divergence. Non-linear effects are global in nature in this model
and don't show the regional sensitivity of the results as in less
truncated models (Paegle et al., 1983).

The phasing and meridional structure of the forcing have a

pronounced effect on the barotropic response. It is shown that when the divergent forcing resembles that of the "warm" episodes of the Southern Oscillation wave patterns are amplified more than the case when the heating is further west. Model integrations forced with single modes reveal that Pacific divergent forcing tends to strengthen the zonal flow of both hemispheres while the Atlantic divergent forcing produces more blocked patterns. The largest impact on the equilibria solutions was found for wave modes which maximize both at the equator and mid-latitudes.

It is interesting to note that the divergence forcing is on a magnitude (0 (10^{-6} s^{-1})) that is similar to the unstable growth rate of barotropic modes discussed by Simmons et al. (1983). These are distinct mechanisms in that a change of the divergent forcing represents a re-distribution of the forcing, while an instability of a pre-existing state should trigger a similar evolution regardless of the details of the forcing. It would be of interest to ascertain the extent to which these rather distinct mechanisms impact upon slow transition. We are pursuing further work toward this understanding.

ACKNOWLEDGMENTS

This material is based upon work supported jointly by the National Science Foundation, the National Oceanic and Atmospheric Administration, and the National Aeronautics and Space Administration under Grant Numbers ATM 8219198, ATM 8018158 and NAG 5-127.

BIBLIOGRAPHY

Bjerknes, J., 1969: Atmospheric teleconnections from equatorial Pacific. Mon. Wea. Rev., 97, 163-172.

Blackmon, M.L., Y.-H. Lee and J.M. Wallace, 1983a: Horizontal structure of 500 mb height fluctuations with long, medium and short periods. Submitted to J. Atmos. Sci.

Blackmon, M.L., Y.-H. Lee and H.-H. Hsu, 1983b: Time evolutions of 500 mb height fluctuations with long, medium and short periods.

Submitted to J. Atmos. Sci.

Charney, J.G. and J.G. De Vore, 1979: Mulitple flow equilibria in the atmosphere and blocking. J. Atmos. Sci., 36, 1205-1216.

Egger, J., 1977: On the linear theory of the atmospheric response to sea surface temperature anomalies. J. Atmos. Sci., 34, 603-614.

Gall, R., R. Blakeslie and R.C.J. Somerville, 1979: Cyclonescale forcing of ultralong waves. J. Atmos. Sci., 34, 1040-1053.

Hanna, A.F., D.E. Stevens and E.R. Reiter, 1984: Short-term climatic fluctuations forced by thermal anomalies. J. Atmos. Sci., 41, 122-141.

Horel, J.D. and J.M. Wallace, 1981: Planetary scale atmospheric phenomena associated with the Southern Oscillation. Mon. Wea. Rev., 109, 813-829.

Hoskins, B.J. and D.J. Karoly, 1981: The steady linear response of a spherical atmoshere to thermal and orographic forcing. J. Atmos. Sci., 38, 1179-1196.

Kallen, E., 1981: The non-linear effects of orographic and momentum forcing in a low-order barotropic model. J. Atmos. Sci., 38, 2150-2163.

Keshavamurty, R.N., 1982: Response of the atmosphere to sea surface temperature anomalies over the equatorial Pacific and the teleconnections of the southern oscillation. J. Atmos. Sci., 39, 1241-1259.

Madden R., 1978: Further evidence of traveling planetary waves. J. Atmos. Sci., 35, 1605-1618.

Namias, J., 1981: Teleconnections of 700 mb height anomalies for the Northern Hemisphere. California Oceanic Fisheries Investigations (CALCOFI) Atlas No. 29, Marine Life Research Group, Scripps Institution of Oceanography, 265 pp.

Paegle, J.N., 1979: The effect of topography on a Rossby wave. J. Atmos. Sci., 36, 2267-2271.

Paegle, J., J.N. Paegle and H. Yan, 1983: The role of barotropic oscillations within atmospheres of highly variable refractive index. J. Atmos. Sci., 40, 2251-2265.

Reinhold, B.B. and R.T. Pierrehumbert, 1982: Dynamics of Weather regimes: quasi-stationary waves and blocking. Mon. Wea. Rev. Rev., 110, 1105-1145.

Rowntree, P.R., 1972: The influence of tropical East Pacific Ocean temperatures on the atmosphere. Quart. J. Roy. Meteo. Soc., 98, 290-321.

Sawyer, J.S., 1970: Observational characteristics of atmospheric fluctuations with a time scale of a month. Quart. J. Roy. Meteor. Soc., 96, 610-625.

Simmons, A.J., 1982: The forcing of stationary wave motion by tropical diabatic heating. Quart. J. Roy. Meteor. Soc., 108, 503-534.

Simmons, A.J., J.M Wallace and G.W. Branstator, 1983: Barotropic wave propagation and instability and atmospheric teleconnection patterns. J. Atmos. Sci., 40, 1363-1392.

Trevisan, A., and A. Buzzi, 1980: Stationary response of barotropic weakly nonlinear Rossby waves to quasi-resonant orographic forcing. J. Atmos. Sci., 37, 947-957.

van Loon, H., and J.C. Rogers, 1978: The seesaw in winter temperatures between Greenland and Northern Europe. Part I: General Description. Mon. Wea. Rev., 106, 296-310.

Wallace, J.M., and D.S. Gutzler, 1981: Teleconnections in the geopotential height field during the Northern Hemisphere winter. Mon. Wea. Rev., 109, 785-812.

Wiin-Nielson, A., 1979: Steady states and stability properties of a low order barotropic system with forcing and dissipation. Tellus, 31, 375-386.

Session 8: Modeling and Theory (3)

JET STREAK DYNAMICS AND GEOSTROPHIC ADJUSTMENT PROCESSES
DURING THE INITIAL STAGES OF LEE CYCLOGENESIS

Rainer Bleck and Craig Mattocks
Rosenstiel School of Marine and Atmospheric Science
University of Miami

I. INTRODUCTION

In the absence of friction and diabatic effects, the temporal
evolution of atmospheric flow is governed by the conservation of two
quantities: potential temperature and potential vorticity. This is
particularly easy to demonstrate in the case of quasi-geostrophic flow
where the potential vorticity equation reduces to a three-dimensional
elliptic equation for the mass field tendency. In isentropic coordinates, the variable representing the mass field is the Montgomery potential $M \equiv gz + c_p T$, and the elliptic equation for the M field tendency
$M_t \equiv (\partial M/\partial t)_\theta$ is

$$\left[A \, \nabla^2 + \frac{\partial}{\partial\theta} B \frac{\partial}{\partial\theta}\right] M_t = -\underline{x}\cdot\nabla Q \tag{1}$$

The coefficients A and B in (1) are positive-valued functions of x,y,θ
(rendering the differential operator on the left-hand side elliptic),
and

$$Q = (-\frac{\partial\bar{p}}{\partial\theta})(\frac{1}{f_0}\nabla^2 M + f) + f_0\frac{\partial p}{\partial\theta}$$

is the quasi-geostrophic potential vorticity. A derivation of (1) is
given in Appendix A.

The conservation law for potential temperature, $d\theta/dt=0$, is im-
plied in (1) by evaluating $-\underline{x}\cdot\nabla Q$ along isentropic surfaces, but it also

885

enters explicitly through the lower boundary condition

$$[M_t - \theta \frac{\partial}{\partial \theta} M_t] = -(\theta \frac{\partial^2 M}{\partial \theta^2}) \ (\chi \cdot \nabla \theta)_{ground} \tag{2}$$

This boundary condition, which according to the standard classification of elliptic boundary value problems is of the "mixed" type, is
based on an expression for the time tendency of Montgomery potential
at the ground (Bleck, 1974).

Equations (1) and (2) concisely express the fact that, in the
absence of friction and diabatic effects, changes in the three-dimensional mass field (including changes in surface pressure) are brought
about by the advection of potential vorticity aloft and/or advection
of potential temperature along the ground. We wish to emphasize that
within the quasi-geostrophic framework these results are formally
identical to those obtained in isobaric coordinates (e.g., Charney,
1973, section VII), and that our work may be regarded as an extension
of Kleinschmidt's (1950) pioneering work on the influence of large-
scale potential vorticity intrusions or "implants" on the ambient
pressure field.

Classical quasi-geostrophic theory views extratropical cyclogenesis as being the result of upper-tropospheric potential vorticity
overrunning a low-level air mass boundary. Ordinarily, the upper-level
potential vorticity and the surface cold front are constrained to
advance at more or less the same rate. However, if the low-level cold
air flow is retarded by an orographic barrier, the vertical alignment
between the upper-level trough advecting high potential vorticity and
the cold air flow at low levels is temporarily disturbed.

Lee cyclogenesis is most likely to occur in this situation if the
upper-level vorticity maximum is able to cross the barrier in spite of
the distortion of the three-dimensional mass field caused by the low-
level blocking. This is probably most easily accomplished during the
initial stage of upper-level trough development when the vorticity
exists mostly in the form of shear vorticity on the upwind side of the
trough. In that situation, the vorticity is bonded to highly energetic
flow (the "jet streak") which has significant self-advection capabi-

lities and may react rather slowly (in relative terms) to a change in the ambient pressure field.

An advancing jet streak (i.e., a jet stream of finite length) always triggers a geostrophic adjustment process. Since the isotachs forming the jet streak generally move more slowly than the air itself, the flow ahead of the velocity maximum is supergeostrophic. It thus veers to the right and causes pressure to fall in the forward left quadrant of the jet streak. Low-level convergence and upward motion set in, bringing about a cooling of the air column and a pressure fall aloft which in turn reduces the degree of supergeostrophy. Under quasi-geostrophic conditions, this adjustment process is described quantitatively by the omega equation or the mass field tendency equation. However, quasi-geostrophic diagnostic tools can be used even under less than ideal conditions to explain convergence and vertical motion patterns in a qualitative sense.

In the absence of an orographic obstacle, the amount of lifting required to cool the air column in the left forward quadrant of the jet streak is reduced by horizontal cold air advection. Thus, if a mountain range inhibits the influx of cold air, the vertical "pumping" action may be expected to be more vigorous than under flat bottom conditions. Furthermore, if the area is shielded against cold air advection on more than one side (Radinovic, 1965), that is, if the cold air must make a significant detour or must wait until it is piled high enough to spill over the ridge, the geostrophic adjustment forces causing low-level convergence and uplifting can be expected to persist in the same location for a significant amount of time, perhaps 12 hours or more. The result will be vigorous low-level vortex spin-up in the immediate lee of the barrier and significant release of latent heat in the ascending, warm air at a stage when the cold air has not yet rounded the barrier or swept over it.

It is hypothesized that the geostrophic adjustment process sketched above incorporates the essence of the first stage (often referred to as the "trigger" phase, see Buzzi and Tibaldi, 1978) of cyclogenesis in the lee of narrow, meso-scale mountain ranges like the Alps.

II. A NUMERICAL EXPERIMENT

The following experiment, inspired in part by the experiments of Tibaldi et al. (1980), was carried out to study the effect of upper level potential vorticity advection (i.e., jet streak intensity) on lee cyclogenesis. In a beta plane channel consisting of 20 grid points in east-west direction and 30 points in north-south direction the primitive equations were integrated at 12 isentropic levels spanning the atmosphere between the ground and approximately 100hPa. Grid points were spaced 200km apart. Initial conditions were chosen to produce a growing baroclinic disturbance of zonal wave number one in the northern half of the channel. The initial fields were fine-tuned to assure that no wave disturbance would form along the cold front in the absence of orographic or enhanced jet streak forcing.

After approximately 1.5 days of integration the flow field and boundary conditions were modified as follows:

Experiment 1: A hook-shaped, 2000m high mountain barrier resembling the Alps was inserted a short distance ahead of the position of the advancing surface cold front, and potential vorticity was enhanced in a small region behind the cold front. The location of this potential vorticity "implant", which extended from a level of about 300K to 350K but reached maximum intensity near 320K, was chosen in such a way that it would be steered toward the barrier.

Experiment 2: The barrier was introduced as in Experiment 1, but the potential vorticity field remained unaltered.

Experiment 3: The potential vorticity field was altered as in Experiment 1, but no mountain range was introduced.

Potential vorticity in a multi-level primitive equation model can be altered by modifying either the thermal stratification or the velocity field. Either process creates unacceptably large geostrophic imbalances. In order to circumvent this problem we chose to modify not the actual potential vorticity field, but the geostrophic potential vorticity, which by virtue of (A-2) is related to the horizontal and vertical derivatives of M through

888

$$Q_g = (\nabla^2 M + f)/(\partial p/\partial \theta) = (\nabla^2 M + f)/\left[\frac{c}{R}p\ (M_\theta/c_p)^{(1-k)/k}(M_{\theta\theta}/c_p)\right] \quad (3)$$

Note that this expression serves not only as a definition of Q_g, but also as a three-dimensional elliptic equation for M, given Q_g (Bleck, 1973). Our approach was to compute Q_g from the M field using (3) and, after introducing the implant, use (3) again to compute M from the modified Q_g field.

The wind field for the subsequent integration of the primitive equation model was determined geostrophically. In an effort to reduce the shock caused by altering the potential vorticity field and inserting the mountain barrier, quasi-geostrophic divergent velocity components deduced from the mass field tendency were added to the geostrophic velocity field. Details of this procedure, which requires the solution of (1), can be found in Bleck (1974).

Three types of output fields will be shown to illustrate the results of our numerical integrations: pressure at sea level, Montgomery potential on an upper-tropospheric isentropic surface, and "Q vector" divergence on a lower-tropospheric isentropic surface. The Q vector, $\underset{\sim}{Q} = (\nabla \underset{\sim}{v}_g) \cdot (\nabla \theta)$, was originally introduced by Hoskins et al. (1978) for the purpose of combining the vorticity and thermal advection terms in the quasi-geostrophic omega equation into a single forcing term which thereby assumes the form $-2\nabla \cdot Q$. In other words, the Q vector divergence is proportional to the three-dimensional Laplacian of the quasi-geostrophic omega field. As mentioned in the Introduction, we do not intend to extract quantitative information from this term, considering the fact that the Rossby number for Alpine lee cyclogenesis is typically of order one, and in particular see no need to actually solve the omega equation. For the same reason we may re-define the Q vector in a dimensionally different form which is particularly convenient for evaluation on isentropic surfaces (Draghici, pers.comm.):

$$\underset{\sim}{Q} = (\nabla_\theta \underset{\sim}{v}_g) \cdot (\nabla_\theta c_p p^k)$$

Here, the subscript θ indicates differentiation along an isentropic surface.

Due to the continual excitation of gravity waves by the steep
mountain barrier we have chosen to display fields pertaining to low-
level atmospheric conditions in time-averaged form only. Since the
results of the three experiments were saved at hourly intervals, the
reader should be aware that an n-hour time average actually represents
an average of only n instantaneous fields.

III. RESULTS

The surface pressure field at the start of Experiments 1—3 is
shown in Fig.1(a). The Montgomery potential field, together with the
potential vorticity at θ = 320K before and after enhancement, is shown
in Fig.1(b) and 1(c) respectively. Note that the potential vorticity
implant alters the orientation of the trough axis in a way which
synoptic meteorologists normally associate with increased potential
for cyclonic development.

After 21 hrs of time integration in the presence of the mountain
barrier (Experiment 1) the pattern shown in Fig.1(c) has evolved into
the one shown in Fig.2(a). The surface potential temperature field
valid at 21 hrs (Fig.2b) shows that no cold air has been able to reach
the lee of the mountain barrier during that time span.

As seen from Fig.2(a-c), a strong lee cyclone forms during the
first 30 hrs of integration in Experiment 1. The pressure in the lee
of the barrier actually does not start to fall appreciably before t=16
hrs. This is approximately the time when the forward edge of the surface cold front and the upper level potential vorticity maximum arrive
at the upwind side of the barrier. The intensification rate of the lee
cyclone thereafter actually increases with time right up to the moment
when the potential vorticity maximum is crossing the barrier. This
trend is well borne out by the Q vector divergence on the 292.5 isentropic surface (Fig.4(a-c) which gives the maximum amount of forcing
of the quasi-geostrophic omega field during the final 6-hr period.
While we did not carry out the time integration past t=30 hrs, a visual
inspection of the instantaneous Q vector divergence fields during the

final hours of integration (not shown here) suggests that the quasi-geostrophic forcing begins to fall off after 26 hrs.

Patterns corresponding to those shown in Figs.3 and 4, but pertaining to Experiment 2 (no jet streak enhancement) are shown in Figs.5 and 6 respectively. Results from Experiment 3 (no barrier, but enhanced jet streak) are shown in Figs.7 and 8.

The results from Experiment 2 (Figs.5 and 6) indicate that the intensity of lee cyclogenesis in this particular case is indeed very sensitive to the strength of the upper-level potential vorticity advection. Note, however, that these results have a slight defect which will have to be removed in the future to make the argument more convincing. The problem is that the boundary conditions for M used in solving (3) do not permit specification of the total mass in the system. In particular, any net increase of Q_g resulting from adding the potential vorticity "implant" will cause a drop in mean surface pressure. Since the mass field variables along the northern and southern wall remain unchanged during the enhancement procedure, the overall mass loss manifests itself in a zonal pressure trough in mid-channel. Thus, the easterlies in the northern domain and the westerlies in the southern domain are slightly stronger in Experiments 1 and 3 than in Experiment 2. While this is not likely to have a strong effect on the overall baroclinity in the channel, we should attempt to alter the potential vorticity field in the future in such a way that total mass in the grid domain is the same at the start of all three experiments.

The ability of a mountain barrier to focus the jet streak-related geostrophic adjustment process on the immediate vicinity of the barrier is amply demonstrated in Figs.4 and 8. While the enhanced jet streak in Experiment 3 is able to produce a wave disturbance along the cold front over flat terrain (Fig.7), this disturbance obviously does not resemble the intense "meso-low" obtained in Experiment 1 (Fig.3).

Acknowledgements: This research was supported by the National Science Foundation under grants No. ATM 80-22264 and ATM 83-06597. Computations were performed at the National Center for Atmospheric Research (NCAR) in Boulder, Colorado. NCAR is sponsored by the National Science Foundation.

APPENDIX A

The mass field tendency equation in isentropic coordinates.

The quasi-geostrophic potential vorticity equation in isentropic coordinates is derived by eliminating the divergence $\nabla \cdot \underline{v}$ between the quasi-geostrophically approximated vorticity and continuity equations,

$$\left(\frac{\partial}{\partial t} + \underline{v}_g \cdot \nabla\right)\left(\frac{1}{f_0}\nabla^2 M + f\right) + f_0 \nabla \cdot \underline{v} = 0$$

$$\left(\frac{\partial}{\partial t} + \underline{v}_g \cdot \nabla\right)\frac{\partial p}{\partial \theta} + \frac{\partial \bar{p}}{\partial \theta}\nabla \cdot \underline{v} = 0$$

where f_0 is a representative constant Coriolis parameter, $-\partial \bar{p}/\partial \theta$ is an area average of isentropic layer thickness, and $\underline{v}_g = f^{-1}\underline{k}\times\nabla M$ is the geostrophic velocity vector. The resulting equation is

$$\left(\frac{\partial}{\partial t} + \underline{v}_g \cdot \nabla\right)\left[\left(-\frac{\partial \bar{p}}{\partial \theta}\right)\left(\frac{1}{f_0}\nabla^2 M + f\right) + f_0 \frac{\partial p}{\partial \theta}\right] = 0 \qquad (A-1)$$

By virtue of the hydrostatic equation, we have

$$p = \left(\frac{1}{c_p}\frac{\partial M}{\partial \theta}\right)^{1/k} \qquad (A-2)$$

where $k = R/c_p$. Therefore,

$$\frac{\partial}{\partial t}\left(\frac{\partial p}{\partial \theta}\right) = \frac{\partial}{\partial \theta}\left(\frac{\partial p}{\partial t}\right) = \frac{1}{k}\frac{\partial}{\partial \theta}\left[\left(\frac{1}{c_p}\frac{\partial M}{\partial \theta}\right)^{(1-k)/k}\frac{1}{c_p}\frac{\partial}{\partial \theta}\left(\frac{\partial M}{\partial t}\right)\right]$$

This permits us to write (A-1) in the form shown in (1),

$$\left[A\nabla^2 + \frac{\partial}{\partial \theta}B\frac{\partial}{\partial \theta}\right]M_t = -\underline{v}_g \cdot \nabla\left[\left(-\frac{\partial \bar{p}}{\partial \theta}\right)\left(\frac{1}{f_0}\nabla^2 M + f\right) + f_0 \frac{\partial p}{\partial \theta}\right]$$

where

$$A = \frac{1}{f_0}\left(-\frac{\partial \bar{p}}{\partial \theta}\right)$$

$$B = \frac{f_0}{R}\left(\frac{1}{c_p}\frac{\partial M}{\partial \theta}\right)^{(1-k)/k}$$

REFERENCES

Bleck, R., 1973: Numerical forecasting experiments based on the conservation of potential vorticity on isentropic surfaces. J.Appl. Meteorol., 12, 737-752.

Bleck, R., 1974: Short-range prediction in isentropic coordinates with
    filtered and unfiltered numerical models. Mon.Wea.Rev., 102,
    813-829.

Buzzi, A., and S. Tibaldi, 1978: Cyclogenesis in the lee of the Alps:
    a case study. Quart.J.Roy.Met.Soc., 104, 271-287.

Charney, J., 1973: Planetary fluid dynamics. Dynamic Meteorology,
    P.Morel, Ed., D.Reidel, Dordrecht/Boston, 97-351.

Hoskins, B.J., I. Draghici, and H.C. Davies, 1978: A new look at the
    omega equation. Quart.J.Roy.Met.Soc., 104, 31-38

Kleinschmidt, E., 1950: Über Aufbau und Enstehung von Zyklonen (1.Teil),
    Meteor.Rundsch., 3, 1—6.

Radinovic, D., 1965: On forecasting of cyclogenesis in the West Mediter-
    ranean and other areas bounded by mountain ranges by baroclinic
    model. Arch.Met.Geoph.Biokl.,A 14, 279-299.

Tibaldi, S., A. Buzzi, and P. Malguzzi, 1980: Orographically induced
    cyclogenesis: analysis of numerical experiments. Mon,Wea.Rev.,
    108, 1302-1314.

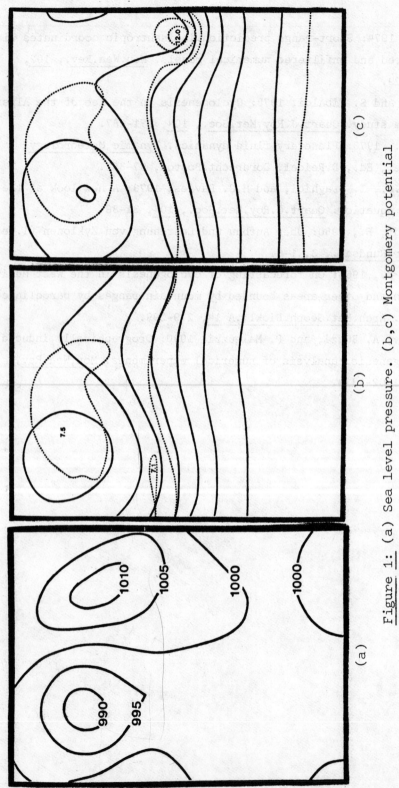

Figure 1: (a) Sea level pressure. (b,c) Montgomery potential (solid) with potential vorticity (dashed) on 320K isentropic surface before and after jet streak enhancement.

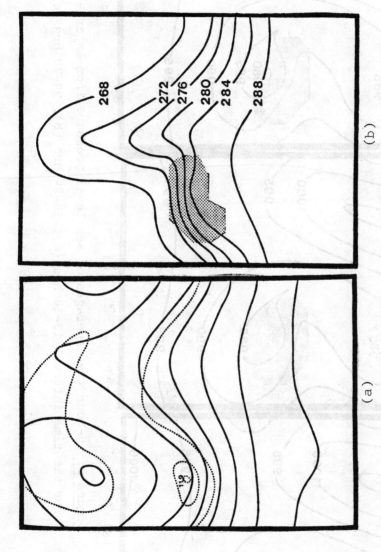

(a)

(b)

Figure 2: Montgomery potential and potential vorticity on 320K
isentropic surface (a) and surface potential temperature (b) from
Experiment 1 after 21 hrs of integration.

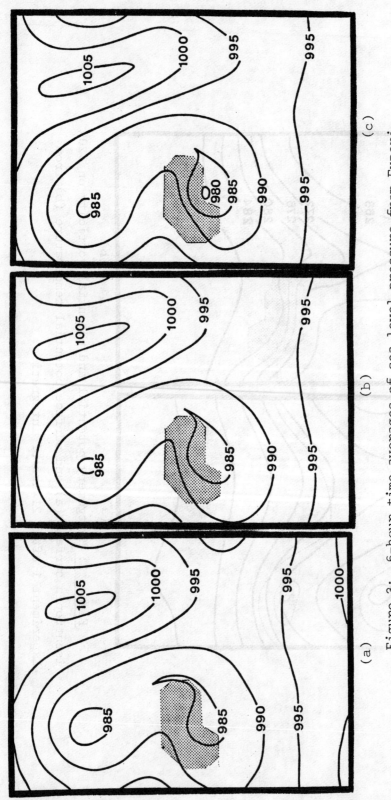

(a)   (b)   (c)

Figure 3: 6-hour time averages of sea level pressure from Experiment 1 for the time interval 12-18 hrs (a); 18-24 hrs (b); 24-30 hrs (c).

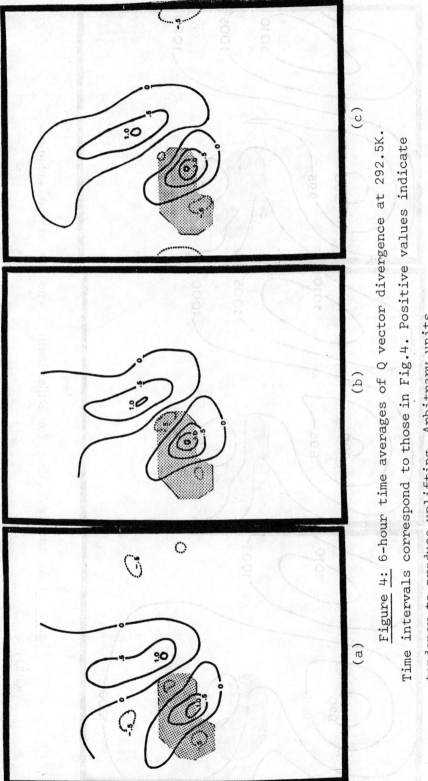

Figure 4: 6-hour time averages of Q vector divergence at 292.5K. Time intervals correspond to those in Fig.4. Positive values indicate tendency to produce uplifting. Arbitrary units.

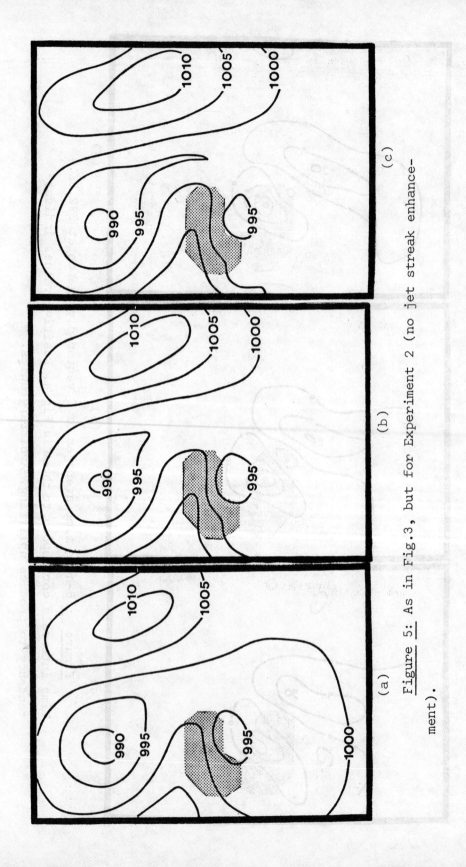

(a)

(b)

(c)

Figure 5: As in Fig.3, but for Experiment 2 (no jet streak enhancement).

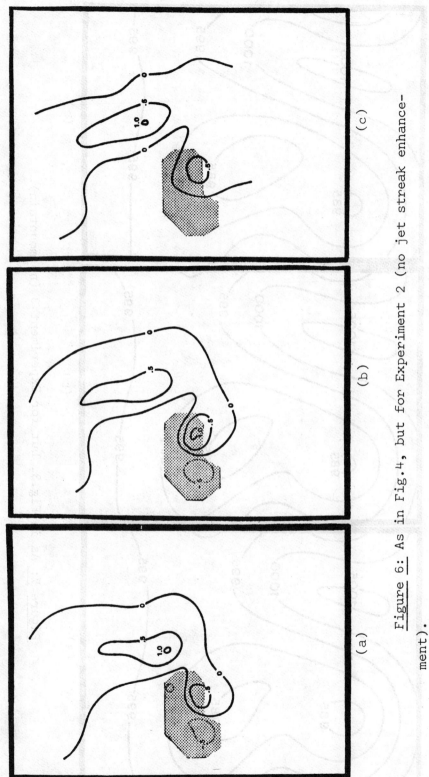

Figure 6: As in Fig. 4, but for Experiment 2 (no jet streak enhance-ment).

(a)   (b)   (c)

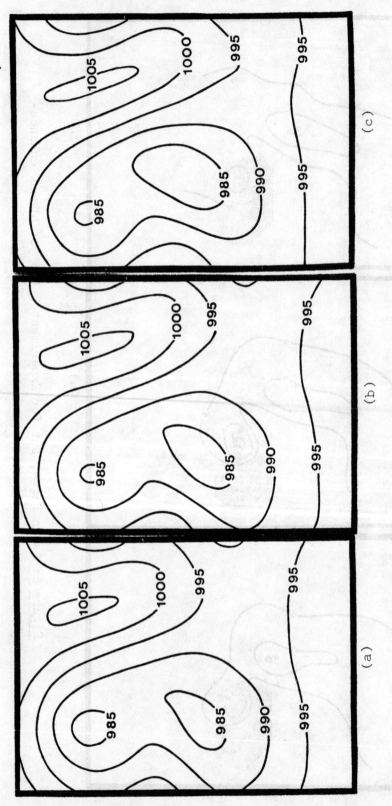

Figure 7: As in Fig.3, but for Experiment 3 (no mountain).

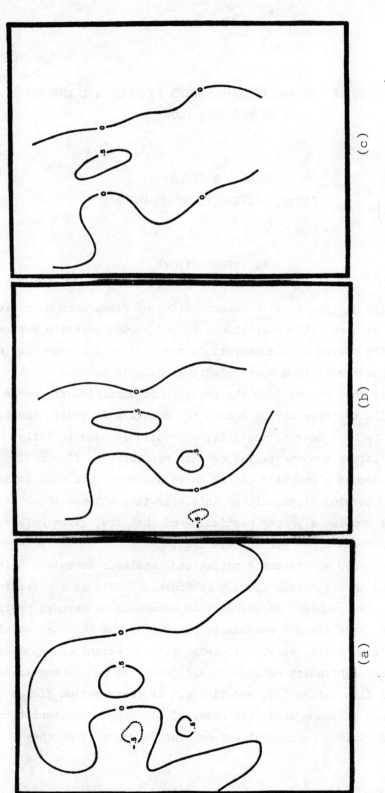

Figure 8: As in Fig. 4, but for Experiment 3 (no mountain).

APPLICATION OF THE VARIATIONAL METHOD IN OBJECTIVE ANALYSIS
OF MOUNTAIN FLOW

JOHN A. MCGINLEY
CIMMS, UNIVERSITY OF OKLAHOMA

I. INTRODUCTION

The major difficulty with determination of flows near mountains
is the lack of upper air observations in and around mountain regions.
When spacially sparse wind observations are used to diagnose flow near
mountains, the result from most objective analysis schemes is that the
amount of air that passes over the mountain is significantly overes-
timated, while the flow moving around the mountain is underestimated.
There seems to be a general inability of available data to fully
resolve the latter component. The dynamic responses to flow moving
over versus around a mountain can be quite different and thus it is
important to develop diagnostic or initialization schemes which can
consistently provide a proper partioning of the flow, given only a few
scattered upstream upper air observations.

In this paper we present a variational analysis technique which
has been used in diagnostic studies (McGinley, 1982), as a possible
approach to this problem. We assume that observations through preliminary analysis, have given a reasonable picture of the balanced quasi-
geostrophic flow field. We wish to learn to what extent can application of a mass continuity constraint influence the derived horizontal
and vertical flow deflection, and find out if these motion fields can
duplicate what is known about the observed blocking characteristics of
mountains of limited horizontal extent. The development of the

variational formalism will be shown, with examples of solutions for constant flow interacting with a crescent shaped (Alpine-like) mountain barrier. The characteristics of a user defined constant of proportionality will be examined with respect to solutions of the above problem. Objective methods to select the value of this constant will be discussed, including scaling and theoretical considerations. This will be validated by comparison with flow blocking computations using ALPEX data sets. Finally, the simulated and actual wind fields will be used to examine the frontogenetical characteristics of the crescent-shaped mountains and Alps, respectively.

II. FORMALISM

The variational formalism for this problem is developed assuming that the input winds fields have been preanalyzed to reflect the large scale. We obtain these fields as follows. The rawinsonde observations are first interpolated to a three dimensional grid using a cubic spline techinque similar to Fritsch, 1971. A second stage couples the mass and momentum fields to minimize geostrophic error. An input vertical motion is computed from these fields using the diagnostic quasigeostrophic omega equation. The formalism I, written in discrete form for a type C staggered grid in x,y and σ, requires that mass continuity be satisfied everywhere, but that RMS geostrophic error be minimum over the domain.

$$I = \sum_{i=1}^{I} \sum_{j=1}^{J} \sum_{k=1}^{K} (u_{i-1/2jk} - \tilde{u}_{i-1/2jk})^2 + (v_{ij-1/2k} - \tilde{v}_{ij-1/2k})^2$$

$$+ \tau(\dot{\sigma}_{ijk-1/2} - \tilde{\dot{\sigma}}_{ijk-1/2})^2 + \lambda_{ijk}[(u_{i+1/2jk} - u_{i-1/2jk})/\Delta x$$

$$+ (v_{ij+1/2k} - v_{ij-1/2k})/\Delta y$$

$$+ (\dot{\sigma}_{ijk-1/2} - \dot{\sigma}_{ijk+1/2})/\Delta\sigma$$

$$+ (\frac{\partial \tilde{p}_s}{\partial t})_{ijk}] \tag{1}$$

where $u=u'\Delta p$, $v=v'\Delta p$, $\dot{\sigma}=\dot{\sigma}'\Delta p$, with $\Delta p = p_s - p_{top}$, ($p_{top}=50hPa$). p_s is the surface pressure, u' and v' the horizontal velocities, and $\dot{\sigma}'=d\sigma/dt$, where σ is given by $\sigma=(p-p_{top})/(p_s-p_{top})$. The terms ($\tilde{}$) refer to the input winds. $\tilde{\dot{\sigma}}$ is derived from $\tilde{\dot{\sigma}}=\tilde{\omega}_g - \sigma\dot{p}_s$, where $\tilde{\omega}_g$ is the quasi-geostrophic vertical motion. λ is a Lagrange multiplier which insures that the equation of continuity (the [ ] term in Equation (1)), is satisfied at every point in the interior. The pressure tendency term in the equation is evaluated from surface observations. The term τ (units of m^2) serves as a constant of proportionality between horizontal and vertical motion.

At the stationary point of I where $\delta I=0$, solutions are determined for $u_{i-1/2jk}$, $v_{ij-1/2k}$, $\dot{\sigma}_{ijk-1/2}$, and λ_{ijk}, Providing the variations of these quantities, $\delta u_{i-1/2jk}$, $\delta v_{ij-1/2k}$, $\delta\sigma_{ijk-1/2}$, and $\delta\lambda_{ijk}$ are arbitrary in the interior. The adjustment equations that result are

$$u_{i-1/2jk} = \tilde{u}_{i-1/2jk} + \frac{(\lambda_{i+1jk} - \lambda_{ijk})}{2\Delta x} \tag{2}$$

$$v_{ij-1/2k} = \tilde{v}_{ij-1/2k} + \frac{(\lambda_{ij+1k} - \lambda_{ijk})}{2\Delta y} \tag{3}$$

$$\dot{\sigma}_{ijk-1/2} = \tilde{\dot{\sigma}}_{ijk-1/2} - \frac{(\lambda_{ijk+1} - \lambda_{ijk})}{2\tau\Delta\sigma} \tag{4}$$

$$\frac{(\lambda_{i+1jk} - 2\lambda_{ijk} + \lambda_{i-1jk})}{\Delta x^2} + \frac{(\lambda_{ij+1k} - 2\lambda_{ijk} + \lambda_{ij-1k})}{\Delta y^2}$$

$$+ \frac{1}{\tau}\frac{(\lambda_{ijk+1} - 2\lambda_{ijk} + \lambda_{ijk-1})}{\Delta\sigma^2} =$$

$$-2\left[\frac{(\tilde{u}_{i+1/2jk} - \tilde{u}_{i-1/2jk})}{\Delta x} + \frac{(\tilde{v}_{ij+1/2k} - \tilde{v}_{ij-1/2k})}{\Delta y}\right.$$

$$\left. + \frac{(\tilde{\dot{\sigma}}_{ijk-1/2} - \tilde{\dot{\sigma}}_{ijk+1/2})}{\Delta\sigma} + \left(\frac{\partial\tilde{p}_s}{\partial t}\right)_{ijk}\right] \tag{5}$$

With boundary conditions,

$$\lambda_{1jk}{}^{\delta}u_{1/2jk} = \lambda_{Ijk}{}^{\delta}u_{I+1/2jk} = \lambda_{i1k}{}^{\delta}v_{i1/2k} = \lambda_{iJk}{}^{\delta}v_{iJ+1/2k} =$$

$$\lambda_{ij1}{}^{\delta}\dot{\sigma}_{ij1/2} = \lambda_{ijK}{}^{\delta}\dot{\sigma}_{ijK+1/2} = 0 \tag{6}$$

The physical boundary conditions on $\dot{\sigma}$ are $\dot{\sigma}_{ij1/2}=0$ and $\dot{\sigma}_{ijk+1/2}=0$. Since in general, $\tilde{\sigma}_{ij1/2} \neq 0$, the variations cannot vanish (since we need to adjust $\dot{\sigma}$ at the boundary), and thus, $\lambda_{ij1}=\lambda_{ijK}=0$. is the proper choice. This, of course, eliminates the adjustment of the horizontal motions at the upper and lower levels. At the side boundaries, selection of $\lambda_{1jk}=\lambda_{I1k}=\lambda_{1Jk}=\lambda_{iJk}=0$, satisfies Equation (6). Equation (5) is solved by relaxation.

III. TESTS WITH SIMULATED FLOW

Given a surface pressure distribution which simulates an isolated, crescent-shaped mountain barrier (Figure 1), we will test the effect of varying τ given the following conditions:

1. A steady 10m s^{-1} westerly flow
2. An assumed input vertical motion of 0, ($\tilde{\omega}_g=0$ so $\tilde{\dot{\sigma}}=\sigma\dot{p}_s$).
3. A surface pressure tendency of zero

The resulting Figures (2a,b;3a,b;4a,b) illustrate how, for this case, τ controls the depth of penetration of terrain induced vertical motion. This in turn has significant effects on the configuration of the flow near the mountain, with respect to deflection and splitting. Increasing τ forces more flow around versus over the mountain. Interpreted another way τ controls the transition depth between the boundary effects and the free atmosphere. We will examine the implications of this in the next section.

IV. ON SELECTING A VALUE FOR τ

If the formalism I, (Equation 1) is non-dimensionalized, we can see what value of τ may be appropriate for the given input fields. The non dimensional form of I is written (with subscript notation eliminated for brevity)

$$\hat{I} = \Sigma \ \Sigma \ \Sigma (\hat{u}-\tilde{\hat{u}})^2 + (\hat{v}-\tilde{\hat{v}})^2 + \tau\frac{\Omega^2}{U^2}(\hat{\dot{\sigma}}-\tilde{\hat{\dot{\sigma}}})^2 \ + \ \hat{\lambda} \ \ldots \qquad (7)$$

where U is a representative velocity and Ω is a representative $\dot{\sigma}$
(recall $\dot{\sigma}=\Delta p \ d\sigma/dt$). ($\hat{\ }$) indicates a non dimensional quantity.

The condition that all RMS errors are of the same order of
magnitude in the non dimensional case $[(\hat{u}-\tilde{\hat{u}})^2 \approx (\hat{v}-\hat{v})^2 \approx (\hat{\dot{\sigma}} \ \tilde{\hat{\dot{\sigma}}})^2]$, is met
if $\tau\Omega^2/U^2=1$. This implies that accuracy of all input fields is likely
equal. Given the highly filtered nature and near geostrophy of the
input winds and vertical motions, this assumption seems reasonable.
Thus, the mountain induced perturbation of the quasigeostrophic flow
effects the horizontal and vertical motions equally in a non-dimensional
sense if $\tau=U^2/\Omega^2$. This brings up the problem of finding a suitable
estimate for Ω. We assumed that the perturbations of the isolated
mountain were nearly unresolved with respect to the data network. Over
most of the domain the terrain is relatively flat and the average in-
put vertical motion $\tilde{\dot{\sigma}}$ is approximately $\tilde{\omega}_g$. From basic scale analysis
a characteristic vertical motion can be estimated (Haltiner, 1971) for
large scale by

$$\Omega = \rho_0 g U^2 f/N^2 h\overline{\Delta p}^2 \qquad (8)$$

where N^2 is the Brunt-Väisäla frequency, h is a characteristic depth,
ρ_0 is mean density and g and f are gravity and the Coriolis parameter.

The scaled value Ω represents the characteristic value of vertical
motion for a particular atmosphere described by U,N, and h. Motions
larger than this value would tend to be suppressed in the free atmosphere at this scale. It should be noted that the kinematic scaling
estimate, $\Omega = U\rho_0 gh/L\overline{\Delta p}$ (where L is the mountain length scale), would
not be truly representative of this atmosphere owing to the limited
dimension of the terrain. Such a vertical motion would only be valid
for a few points near the terrain slope. Thus τ can be written as

$$\tau = \overline{\Delta p}^4 N^4 h^2/\rho_0^2 g^2 U^2 f^2 \qquad (9)$$

τ is large for stable atmospheres, low-speed horizontal flows, and for atmospheres of deep characteristic depth. For the sub-mountaintop layer this depth is the terrain height. In essence what the scheme does for large values of τ is to insure that vertical motion is as consistent as possible with the allowable large scale estimate. If this estimate is small, the terrain induced motion is suppressed, and the mass compensation manifests itself as deflected flow in the horizontal components.

To check the validity of such scaling we can compute actual flow blocking for two days with different upstream conditions (wind speed and stability), where aircraft data was collected during ALPEX. The model terrain (Figure 5) is designed to conserve the maximum height of the Alps. Two experiments were run:

A. Determine the blocking as accurately as possible using the combined data set (aircraft and rawinsonde observations)

B. Determine the characteristics of τ in controlling blocking, given only the geostrophic fields derived from the sparse rawinsonde observations

The blocking ratio B, is the ratio of mass passing around the mountain to the mass approaching the mountain upstream. This is illustrated schematically in Figure 6. By knowing the local density change in the volume (this is estimated using 24 hours of data centered on the time in question), and the vertical motion on the top surface of the volume, B can be evaluated. When B is 1 all flow is blocked; when B is 0 all flow passes over the mountain. It is evident that any scheme which can limit the upward penetration of vertical motion can control blocking. This is shown in Figure 7 for the test cases in section III. Consider now Figure 8 which shows the computed blocking from A. above. The dotted lines show the blocking values for 2 March and 9 April 82. 2 March was a low stability/high wind speed day, while 9 April was a high stability/low wind speed day. The solid lines illustrate the results of B. above. Blocking increases as τ increases.

From Figure 8 we can get some feeling for the sense and accuracy of our scaling procedure. There would appear to be an empirically correct estimate of τ (with liberal allowance for error in the computation for B), where the two lines intersect. This value, say τ_e, can be compared to the scaled estimate, τ_s, for conditions (U,N) on each case day. The finding from these comparisons is that there is a general correspondence between days which have intense observed blocking and large values of τ; and weak observed blocking and small values of τ. However, many more data sets are needed to substantiate this result.

V. FRONTOGENESIS STUDIES

An interesting question is the role of this terrain generated adjustment in frontogenetical processes. We evaluate the frontogenetic tendencies in both the simulated cases (assuming a constant background baroclinic zone), and four ALPEX case days. Numerous studies have shown that the frontogenetic tendencies acting at large scale (Radinovic, 1965; McGinley 1982; and Buzzi and Speranza, 1983), and at subsynoptic scale (Buzzi and Tibaldi, 1978), can lead to cyclogenesis in mountain regions, typically in the lee.

The frontogenesis seen at large scale is present even in the quasi-geostrophic flow; that is, the mountain can create deformation even in the geostrophic wind. What we see here is, in a sense, the ageostrophic contribution induced by windward side deformation and tilting at the mountain ridge.

We first look at computations of the simulated data from section III. We assume a baroclinic zone of constant magnitude exists over the domain, with a gradient directed along the flow. The change in the intensity of the front can be computed from the frontogenesis equation (Miller, 1948) written in pressure coordinates.

$$F = \frac{1}{|\nabla\theta|}[\underbrace{\nabla\theta\cdot\nabla\dot{\theta}}_{\text{Diabatic}} - \underbrace{(|\nabla\theta|^2\frac{D}{2})}_{\text{Divergence}} + \underbrace{(\theta_x^2 - \theta_y^2)\frac{S}{2} + \theta_x\theta_y R}_{\text{Deformation}} + \underbrace{\theta_p(\theta_x\omega_x + \theta_y\omega_y)}_{\text{Tilting}}] \quad (10)$$

where θ is potential temperature, S is stretching deformation, $S=u_x-v_y$; R is shearing deformation, $R=u_y+v_x$; D is horizontal divergence, $D=u_x+v_y$. The subscripts imply derivative quantities.

908

This equation is transformed to the x,y,σ system and evaluated using the output winds. The resulting fields are then interpolated to pressure coordinates for display.

Consider the three wind fields given by figures 2-4. We assume a standard constant value for θ_x and θ_p over the domain. Table 1 shows the maximum frontogenesis averaged vertically over the pressure height of the mountain, within 100km of the mountain surface. We have combined the divergence term with the deformation term in this Table.

TABLE 1: Frontogenesis ($K\,m^{-1}\,s^{-1}$) for simulated flow

$\tau(m^2)$	TOTAL	DEF	TILT	PRESSURE DEPTH ABOVE MOUNTAIN
10^9	$.21\times10^{-7}$	$.048\times10^{-7}$	$.163\times10^{-7}$	200 hPa
10^{11}	$.280\times10^{-7}$	$.058\times10^{-7}$	$.223\times10^{-7}$	160 hPa
10^{13}	$.377\times10^{-7}$	$.085\times10^{-7}$	$.282\times10^{-7}$	100 hPa

Note that computed frontogenesis increases as τ (blocking) increases. However, positive frontogenesis extends to a greater atmospheric depth for low values of τ (and weak blocking). Figure 9 shows a sample horizontal profile of frontogenesis for τ equal to 10^{12} m^2. Note how the frontogenesis is stronger on the windward side than the leeward side owing to contributions from the deformation term. The vertical extent of positive frontogenesis is determined by the tilting term.

The same procedure can be done for available case study days from the ALPEX data set. The data is processed as outlined in section II, and resulting mean fields of temperature and winds used to find an appropriate τ value (Equation 9) for each analysis period. These values ranged from 3×10^{12} to 3×10^{13} m^2. The resulting wind fields were used in Equation 10 to compute frontogenesis.

The results are shown for the 850 hPa level (Figures 10-13, a and b) with total frontogensis shown in solid contours, limits of the baroclinic zone shown in dashed contours, and terrain as indicated. The accompanying wind field is also shown. Two of these cases produced significant cyclones (20 March, 24 April), the third a weak cyclone (2 March), and the fourth, no cyclone (9 April). The studies cited have shown that cyclogenesis is preceded by frontogenesis in the mountain region 3 to 12 hours prior to rapid surface pressure falls. Unfortunately the results here are not too revealing.

The cases involving the most intense cyclogenesis show that, in the low levels at least, that mountain induced frontogenesis is weak. The strongest case of frontogenesis (2 March) had only weak cyclone development. The case with no cyclogenesis had by far the most intense low level baroclinic zone, and relatively moderate frontogenetical tendencies. It is strongly suspected and verified in other studies (Buzzi and Speranza, 1983) that the most important processes may be occurring at upper levels particularly during the initial stages of cyclone development when the surface system is subsynoptic in scale. One interesting aspect of this Alpine study is that the deformation term in the frontogenesis equation accounts for 30 to 40 percent of the total frontogenesis while for the simulated fields, this amount was less than 20 percent.

VI. CONCLUSIONS

This study has attempted to describe a technique to consistently analyze flow near mountains. Given input fields which describe only the geostrophic or quasigeostrophic flow, the scheme employs variational analysis to impose mass continuity over the gridded domain. It was shown that a constant of proportionality, τ, has the capability of controlling the vertical penetration of air, and hence the blocking. Limiting the vertical penetration produces deflection of the horizontal components around the mountain. Choice of an appropriate value for τ can be obtained by using a ratio of estimates of the mean horizontal motion and a quasigeostrophically-scaled estimate of vertical motion. Thus, τ is a function of vertical stability, mountain height, and mean wind speed. Verification of this estimate from direct blocking calculations indicated that the sense of such estimates is correct, but that the data sample was too small for any substantial conclusions. Further cases will be sought.

The terrain perturbed flow fields were used to examine the frontogenetical tendencies. The case of simulated constant flow produced stronger frontogenesis as blocking increased. The total frontogenesis was approximately 80 percent attributable to the tilting term and 20 percent attributable to the deformation term. The vertical extent of the frontogenetic zone increased as blocking decreased. The deformation frontogenesis occurred only on the windward side of the mountain.

Experiments with actual data were less conclusive. Low level frontogenetical tendencies did not correlate well with cyclogenesis nor did cyclogenesis seem related to the intensity of the low level baroclinic zone. This was not unexpected owing to evidence in other works that the upper level processes (the balance between upper level shear and mid-level baroclinicity) may be the triggering element in initiation of surface cyclones. In this study, the Alps had a definite effect in strengthening fronts that moved into the region. It may be that the major role of the Alps in the cyclogenetic process, is in providing conditions suitable for cyclogenesis (i.e. abundant available potential energy), which can be realized once a triggering mechanism arrives (most likely an upper level jet streak).

REFERENCES

Buzzi, A. and A. Speranza, 1983: Cyclogenesis in the lee of the Alps, Meso Meteorology: Theories and Observations, p. 55-142, D. Reidel Publishing Co. (Lilly and Gal-Chen, Editors).

Buzzi A. and A. Tibaldi, 1978: Cyclogenesis in the lee of the Alps: A case study. Quart. J. Roy. Meteor. Soc. 103 135-150.

Fritsch, J., 1971: Objective analysis of a two-dimensional data field by the cubic spline technique. Mon. Wea. Rev., 99, 179-186.

Haltiner, C.J., 1971: Numerical Weather Prediction John Wiley and Sons New York. 317 pp.

McGinley, J.A., 1982: A Diagnosis of Alpine Lee Cyclogenesis. Mon. Wea. Rev. 110, 1271-1287.

Miller, J.E., 1948: On the concept of frontogenesis, J. Meteor. 5, 169-171.

Radinovic, D., 1965: On forecasting of cyclogenesis in the Mediterranean and other areas bounded by baroclinic model Arch. Meteorol. Ser. A14 279-299.

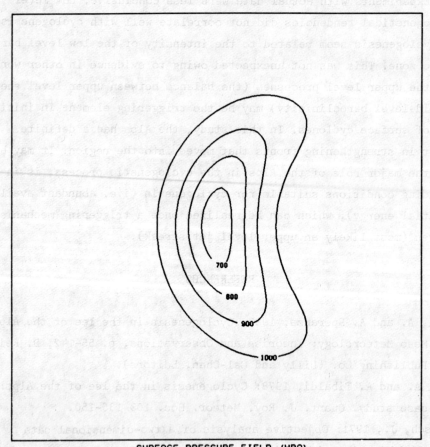

SURFACE PRESSURE FIELD (HPA)

Figure 1: Surface pressure field for simulated flow experiments, in hPa.

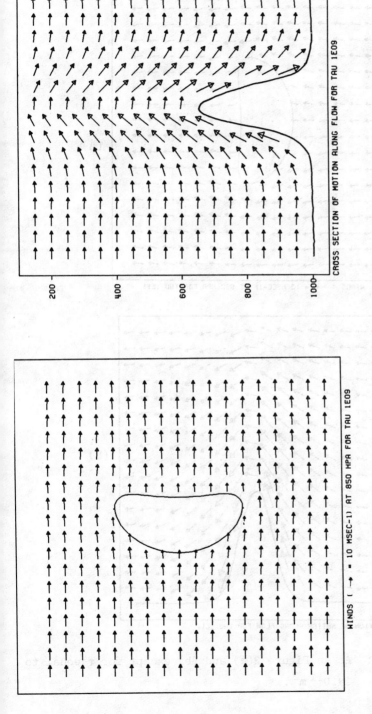

WINDS (→ = 10 MSEC-1) AT 850 HPA FOR TAU 1E09

CROSS SECTION OF MOTION ALONG FLOW FOR TAU 1E09

Figure 2: (a) (top) Horizontal winds (scale as indicated) at 850 hPa for simulated case where τ equal to 1×10⁹ m². Note minimal horizontal deflection of wind.

(b) (bottom) Cross-section along flow through highest part of mountain for τ as in (a). Vertical motion is scaled as in a) except vector represents 0.5 Pa sec⁻¹.

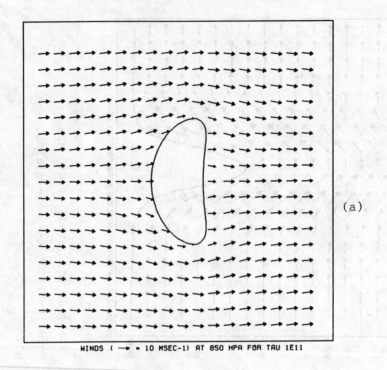

(a)

WINDS (→ = 10 MSEC-1) AT 850 HPA FOR TAU 1E11

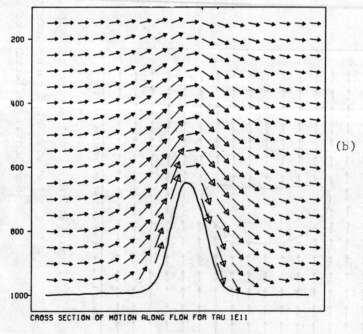

(b)

CROSS SECTION OF MOTION ALONG FLOW FOR TAU 1E11

Figure 3(a) and (b):   As in Figure 2(a) and (b) except for τ equal to $1\times10^{11}\,m^2$.

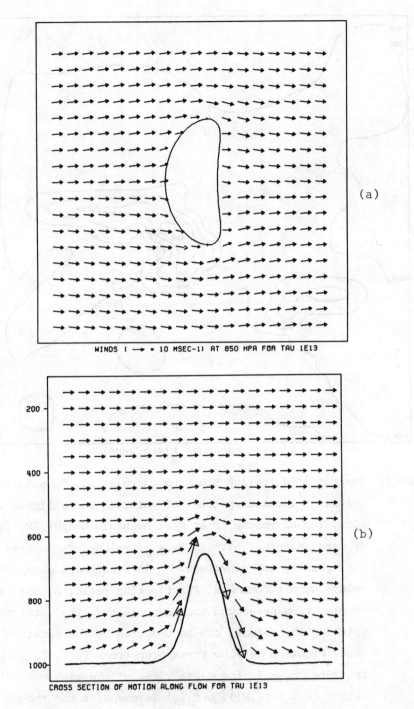

WINDS (→ = 10 MSEC-1) AT 850 HPA FOR TAU 1E13

(a)

CROSS SECTION OF MOTION ALONG FLOW FOR TAU 1E13

(b)

Figure 4(a) and (b):   As in Figure 2(a) and (b) except for τ equal to
$1 \times 10^{13} \, \mathrm{m}^2$. Note significant horizontal deflection.

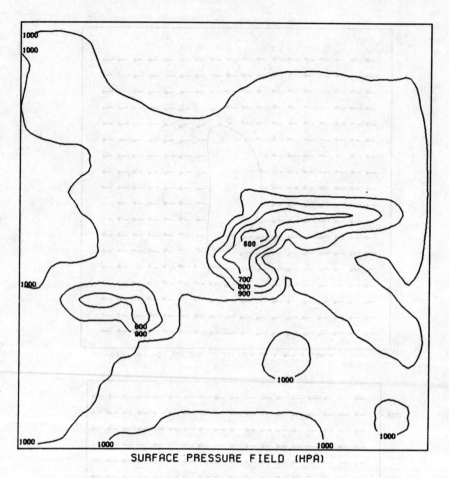

SURFACE PRESSURE FIELD (HPA)

Figure 5: Diagram illustrating blocking calculation. Computation
volume is located upstream of mountain and alligned along
flow. Top of volume is at mean mountain height. $\dot{M}_I$ is rate
of mass entering upstream face of volume, $\dot{M}_0$ is rate of
mass leaving top of volume. M_L is local mass change due to
volume density changes. The blocking ratio, B, the ratio
of mass flowing around the mountain over the total mass
entering the volume, can be computed by the equation at
lower right. U is flow through upstream face, P_S is surface
pressure, σ_{mtn} is the sigma value at the mean mountain
height ($\sigma_{mtn}=f(R,S)$). $\omega(\sigma_{mtn})$ is omega at the top of the
volume, g is gravity. Typically R and S are about 500 km
for the Alps. (The blocking computation is not sensitive to
these choices.)

916

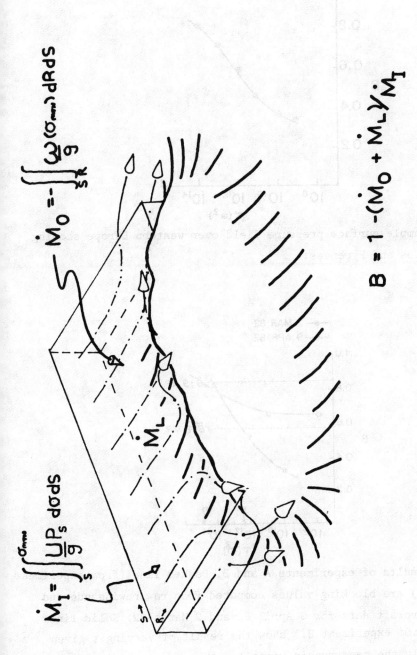

$$\dot{M}_O = - \iint\limits_{SR} \frac{\omega}{g}(\sigma_{max})dRdS$$

$$\dot{M}_L = \iint\limits_{S} \int\limits_{\sigma_{max}} \frac{U P_s}{g} d\sigma dS$$

$$B = 1 - (\dot{M}_O + \dot{M}_L)/\dot{M}_I$$

Figure 6: Blocking (B) computation for simulated flow against a crescent-shaped mountain, plotted as a function of τ. Dimension of τ is appropriate for pressure weighted wind components.

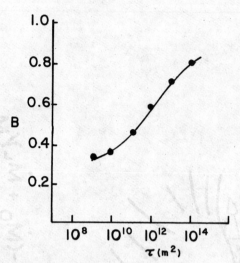

Figure 7: Sample surface pressure field over western Europe showing Alps and Pyrenees.

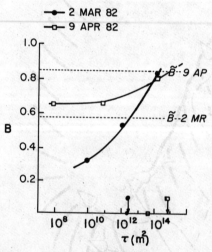

Figure 8: Results of experiments A and B. Dashed line (from experiment A.) are blocking values computed from raw rawinsonde and aircraft data for 9 April 82 and 2 March 82. Solid lines (from experiment B.) show the result of varying τ given only the geostrophic winds for the above case days. Arrow plots on τ axis show empirical estimate, τ_e (↓,↓) for each case day, while symbols directly on the τ axis (●,□) show scaling estimate, τ_s.

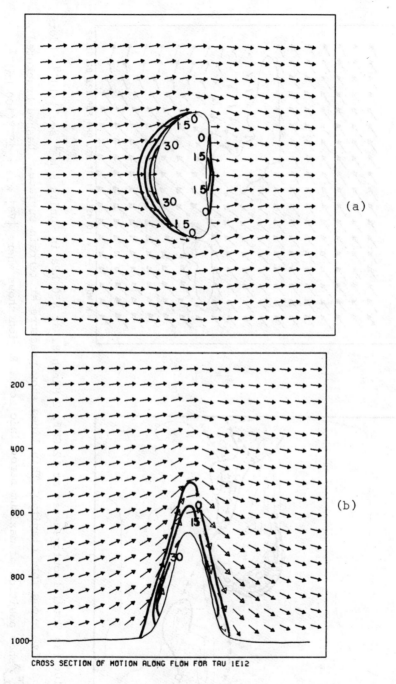

CROSS SECTION OF MOTION ALONG FLOW FOR TAU 1E12

Figure 9: Total frontogenesis at (a) 850 hPa and (b) along cross-section through highest part of mountain for τ equal to 1×10^{12}.
Units are K m^{-1} $sec^{-1} \times 10^{-9}$. Note how windward side frontogenesis extends about 140 hPa above mountain.

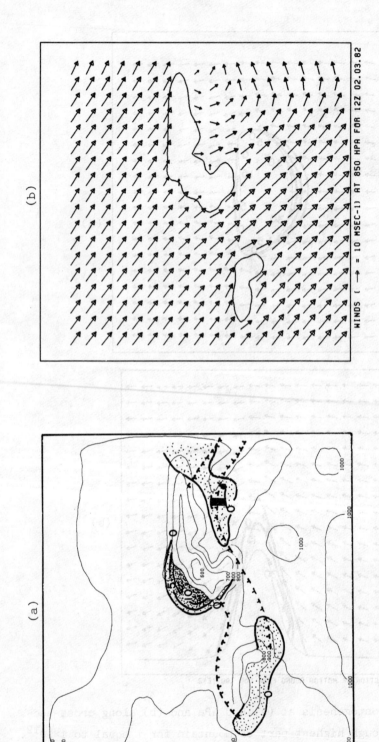

Figure 10: (a) Total frontogenesis at 12 GMT, 2 March 82 at 850 hPa in units K m^{-1} sec^{-1}×10^{-9}, (dots shade positive areas for clarity). Thin lines show pressure at terrain surface. Dashed bracketed areas ⟩$_\tau^\tau$⟨ show position of maximum baroclinic zone. Region shows where $|\nabla\theta|$ was > 1° K(100 km)$^{-1}$. Symbol (**L**) shows where surface low was developing.

(b) 850 hPa wind field for 12 GMT, 2 March 82. Value of τ for this day was 3.59×10^{12} m^2 from Equation 9.

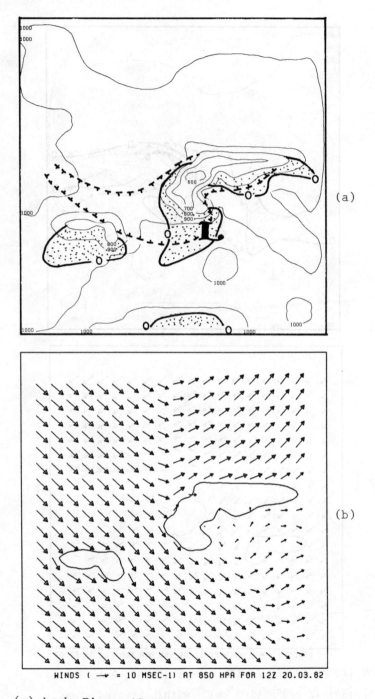

(a)

(b)

WINDS (⟶ = 10 MSEC-1) AT 850 HPA FOR 12Z 20.03.82

Figure 11: (a) As in Figure 10a except for 12 GMT, 20 March 82
          (b) As in Figure 10b except for 12 GMT, 20 March 82
              Value of τ was 1.27×10^{13} m^2.

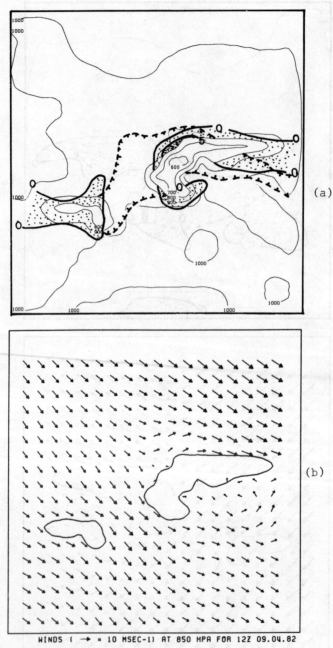

(a)

(b)

WINDS (→ = 10 MSEC-1) AT 850 HPA FOR 12Z 09.04.82

Figure 12: (a) As in Figure 10a except 12 GMT, 9 April 82.
$|\nabla\theta|$ is > 3°K (100 km)$^{-1}$ within bracked areas.
No surface cyclone developed.
(b) As in Figure 10b except 12 GMT, 9 April 82.
Value of τ was 3.0×10^{13} m^2.

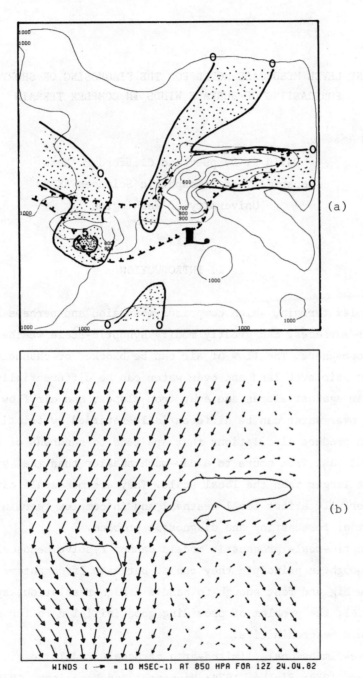

WINDS (→ = 10 MSEC-1) AT 850 HPA FOR 12Z 24.04.82

Figure 13: (a) As in Figure 10a except for 12 GMT, 24 April 82.
$|\nabla\theta|$ is > 2°K (100 km)$^{-1}$ within bracketed areas
(b) As in Figure 10b except for 24 April 82
Value of τ was 2.11×10^{13} m^2

A ONE-LEVEL MESOSCALE MODEL FOR THE DIAGNOSING OR SHORT-TERM FORECASTING OF SURFACE WINDS IN COMPLEX TERRAIN

David P. Dempsey and Clifford F. Mass
Department of Atmospheric Sciences AK-40
University of Washington

I. INTRODUCTION

Complex terrain, which comprises mountains and perhaps both land and water surfaces, can greatly modify synoptic-scale weather in the lower troposphere. The flow of air can be blocked or channeled by mountains; air over land and over water can be differentially heated, as can air against slopes and away from slopes; and air flowing over land and over water can be differentially retarded by friction. These processes produce circulations that are tens to hundreds of kilometers across and last from hours to a day—much smaller than the synoptic scale but larger than the local scale. They are mesoscale circulations. They powerfully affect local weather, and in complex terrain they pose an important forecasting and diagnostic problem.

Synoptic-scale prognostic models cannot resolve mesoscale circulations, though in principle they could. But even if computers existed that were big and fast enough to handle a high-resolution, synoptic-scale model, the problem of providing good initial data for such a model would be prohibitive.

Three-dimensional, limited-area mesoscale models (e.g., Anthes and Warner, 1978; Pielke, 1974; Nickerson and Magaziner, 1976) can resolve mesoscale circulations, but they are computationally expensive and require initial and boundary data that are rarely available from complex terrain. They are potent research tools but are hard to manage.

To diagnose and forecast mesoscale circulations in complex terrain, three types of relatively simple, manageable models have been
developed: (1) mass conservation models; (2) one-layer, vertically-
integrated primitive equation models that assume that the boundary
layer is well mixed; (3) and one-level, sigma-coordinate models that
solve the primitive equations without the continuity equation and
parameterize the vertical temperature structure.

The mass conservation models (e.g., Anderson, 1971; Fosberg et.
al., 1976; Dickerson, 1978; Sherman, 1978) are kinematic rather than
dynamic. They calculate a three-dimensional wind field within a
specified layer of topographic influence. The wind field is consistent
with the mass continuity equation, but there is no reference to non-
linear advection, the Coriolis force, adiabatically or diabatically
induced pressure gradient forces, or other dynamic forcing (though the
effects of some forces, such as friction, may be parameterized). Mass
conservation models are reasonably efficient and require little input
data, but their very simple physics limit their usefulness.

A one-layer, vertically-integrated primitive equation model was
introduced by Lavoie (1972) and later modified by Lavoie (1974),
Overland et al. (1979), and Keyser and Anthes (1977). In these models
the lower atmosphere is divided into four layers: a surface layer of
constant stress, a well-mixed boundary layer, an inversion layer, and
the free atmosphere above. Primitive equations in which the wind vector,
the potential temperature, and the height of the mixed layer are the
dependent variables, are vertically integrated through the mixed layer.
The dependent variables are thus averaged over the mixed layer and no
longer depend on height. The equations account for nonlinear advection,
Coriolis acceleration, pressure gradient accelerations, surface friction and boundary layer entrainment, and are integrated in time. The
only initial data required by the models are large-scale geostrophic
winds, mixed-layer heights along inflow boundaries, and the temperature
jump across a capping inversion. These models appear able to reproduce
some features of boundary layer flow in complex terain, but they are
limited to well-mixed conditions, are sensitive to boundary conditions,
and have trouble with large topographic relief.

A one-level, sigma-coordinate model was introduced by Danard (1977) and improved by Mass (1981b). These models parameterize the potential temperature tendency profile and solve explicit equations for the pressure, potential temperature, and momentum at one level—the surface. The continuity equation is not used. Diabatic heating and cooling are parameterized and the equations are integrated in time. The only input data required are the surface and 850 hPa geostrophic winds and a lower-tropospheric lapse rate.

The model described in this paper both refines and departs from the models of Danard (1977) and Mass (1981b). It has the following characteristics:

(1) The pressure-sigma coordinate, $\sigma = p/p_s$, is used in the vertical. At the surface, $\sigma = 1$ and $\dot{\sigma} = 0$.

(2) The model solves momentum and temperature tendency equations at one level only the surface. The continuity equation is not used.

(3) The model is heavily parameterized. The vertical temperature profile, friction, subgrid-scale horizontal temperature and momentum diffusions, and diabatic heating and cooling are all parameterized. Temperature advection, momentum advection, Coriolis acceleration, and the surface temperature and momentum tendencies are explicitly represented. The surface pressure gradient force term is specially treated before its components are finite-differenced, to remove the hydrostatic variation of the surface pressure that dominates the term in sigma coordinates.

(4) The model is hydrostatic. Because the temperature profile is parameterized, the hydrostatic approximation can be used to eliminate all explicit references to the surface pressure, the surface pressure tendency, and surface pressure gradients.

(5) The model is diagnostic or pseudo-prognostic. It calculates steady-state surface wind and temperature fields when given upper level synoptic-scale height and temperature fields (the synoptic-scale forcing). The synoptic-scale forcing may come from NMC analyses or from prognostic model output. A time-dependent diabatic term may be added and the model integrated for an arbitrarily long period.

(6) The model requires relatively little input: upper level (e.g.

850 hPa) height and temperature fields; a representative lapse rate
from the upper level; and terrain data.

(7) The model is relatively efficient and can be run on a mini-computer.

In three cases presented here, representing different surface wind
regimes from the Pacific Northwest, the model reproduces most of the
main features of the observed flow patterns remarkably well. Particularly interesting are long lines of surface convergence that the
model calculates downwind of large obstacles to the flow, such as the
Olympic Mountains. These convergence lines appear to be confirmed by
the observations.

II. MODEL DESCRIPTION

a. Basic equations

This one-level model integrates in time the dependent variables,
T_s and $\vec{V}_s$, at the surface (or $\sigma=1$) level (see Fig.1). The model assumes
hydrostatic balance, a restriction valid for all but the smallest
scales (less than a few km) or for strong flow rapidly accelerated by
highly curved slopes. The model's horizontal momentum equation in
sigma coordinates at the surface is:

$$\frac{\partial \vec{V}_s}{\partial t} = -\vec{V}_s \cdot \nabla_\sigma \vec{V}_s - f\vec{k}x\vec{V}_s - (g\vec{\nabla}_\sigma z_s + RT_s\vec{\nabla}_\sigma \ell n p_s) - \vec{F} + K_M \nabla_\sigma \vec{V}_s \qquad (1)$$

where $\vec{V}_s$, T_s, p_s and z_s are the wind vector, temperature (K), pressure
and height at the surface; f is the coriolis parameter, g is the
gravitational acceleration, R is the ideal gas constant, $\vec{F}$ is the
frictional force and K_m is the horizontal momentum diffusion coefficient. Equation (1) indicates that the wind vector at a point on the
surface can be altered by advection, coriolis acceleration, the pressure gradient force, frictional drag and horizontal diffusion. The
diffusion term prevents nonlinear computational instability and also
parameterizes subgrid-scale horizontal mixing that cannot be resolved
by the model. The friction and horizontal diffusion parameterizations
are described in Appendices I and II.

Starting with the first law of thermodynamics, and adding a horizontal diffusion term, a tendency equation for the surface temperature in sigma coordinates can be derived:

$$\frac{\partial T_s}{\partial t} = -\vec{V}_s \cdot \vec{\nabla}_\sigma T_\sigma + \frac{RT_s}{c_p} \left(\frac{\partial \ell np_s}{\partial t} + \vec{V}_s \cdot \nabla_\sigma \ell np_s\right) + \frac{Q}{c_p} + K_T \nabla_\sigma^2 T_s \qquad (2)$$

where c_p is the heat capacity of dry air, Q is the diabatic heating rate and K_T is a diffusion coefficient. Equation (2) says that the local surface temperature can change in response to temperature advection, local changes in surface pressure, advection of surface pressure, diabatic heating or cooling, and horizontal diffusion.

The final unknown, the surface pressure p_s, can be found by integrating the hydrostatic equation between the surface and a constant-pressure reference level, the height of which is known. This reference level height, however, cannot be influenced by the underlying terrain, since the model has no way of calculating the height if it does vary. Hence we assume that the topography affects the flow only through a finite depth, H, which is a constant of approximately 2 km. This assumption is open to question, but after examining several empirical studies (e.g., Reed, 1981; Marwitz, 1983), we accepted this value as a reasonable first guess under a wide variety of conditions. For our Pacific Northwest model runs, 850 hPa is the reference level; higher levels would be appropriate in areas with particularly high terrain.

Combining the hydrostatic equation with the perfect gas law and integrating between the surface and the reference level gives

$$\ell np_s = \ell np_R + \left(\frac{g}{R}\right) \int_{z_s}^{z_R} \frac{1}{T(z)} \, dz \qquad (3)$$

where p_R is the pressure at the constant-pressure reference level and z_R is the height of the reference level. If we assume that there is a linear lapse rate, γ, between the reference level (z_R) and the top of the layer of topographic influence (z_H), and further assume that the lapse rate, γ_2, within the layer of topographic influence (z_s to z_H) is also linear, then $T(z)$ is specified below the reference level and the integral in (3) can be explicitly evaluated. Equation (3) then takes the form

$$\ell np_s = F(T_S, T_H, T_R)\tag{4}$$

and the surface pressure varies only with the temperatures at the surface, at the top of the layer of topographic influence, and at the reference level. Of these, by hypothesis, only the surface temperature is affected by events at the surface and is therefore the only time-dependent variable on the right-hand side of (4).

The surface-temperature tendency equation (2) requires $\partial\ell np_s/\partial t$ and $\vec{\nabla}_\sigma\ell np_s$, which can be calculated by taking the local derivative and horizontal gradient, respectively, of (4):

$$\frac{\partial\ell np_s}{\partial t} = (\frac{\partial F}{\partial T_S})\,\frac{\partial T_S}{\partial t}\tag{5}$$

$$\vec{\nabla}_\sigma\ell np_s = (\frac{\partial F}{\partial T_S})\vec{\nabla}_\sigma T_S + (\frac{\partial F}{\partial T_H})\vec{\nabla}_\sigma T_H + (\frac{\partial F}{\partial T_R})\vec{\nabla}_\sigma T_R\tag{6}$$

Substituting (5) and (6) into (2) and solving for $\partial T_S/\partial t$ gives the surface-temperature tendency equation used by the model:

$$\frac{\partial T_S}{\partial t} = -\vec{V}_s\cdot\vec{\nabla}_\sigma T_S - \vec{V}_s\cdot\vec{\nabla}_\sigma T_H(A_2/A_1) - \vec{V}_s\cdot\vec{\nabla}_\sigma T_R(A_3/A_1)$$

$$+ Q/(A_1 c_p) + (K_T/A_1)\nabla_\sigma^2 T_S\tag{7}$$

where A_1, A_2, and A_3 are all functions of the temperatures at the surface, at the top of the layer of topographic influence, and at the reference level.

It is tempting to substitute (6) directly into the pressure gradient force term of the momentum equation (1). After this substitution, however, the pressure gradient force term would be a small difference between two large terms of opposite sign, namely $-g\vec{\nabla}_\sigma z_s$ and $-RT_S\vec{\nabla}_\sigma\ell np_s$. These terms largely cancel because most of the horizontal variation of surface pressure is hydrostatic; the surface pressure varies inversely with the surface elevation. Because both of the component terms of the pressure gradient force term must be approximated by finite differences, which possess some truncation error, the small difference between them would be a poor estimate of the pressure gradient force. It is therefore useful to calculate the horizontal pressure gradient force so that the hydrostatic variations are removed before the finite difference approximations are made. This can be done by differentiating (3) before the integral is explicitly evaluated, using the General Leibnitz Rule for differentiating integrals:

$$\vec{\nabla}_\sigma \ell n p_S = (\frac{g}{R})[T_R^{-1}\vec{\nabla}_\sigma z_R - T_S^{-1}\vec{\nabla}_\sigma z_S - \int_{z_S}^{z_R} T^{-2}(z)\vec{\nabla}_\sigma T(z)dz] \qquad (8)$$

Substituting (8) into the pressure gradient force term of the momentum equation gives

$$g\vec{\nabla}_\sigma z_S + RT_S\vec{\nabla}_\sigma \ell n p_S = g[(T_S/T_R)\vec{\nabla}_\sigma z_R - T_S\int_{z_S}^{z_R} T^{-2}(z)\vec{\nabla}_\sigma T(z)dz] \qquad (9)$$

Note that during the manipulation to produce (9) the large hydrostatic term $g\vec{\nabla}_\sigma z_S$ is explicitly cancelled so that the hydrostatic variation of pressure along the $\sigma=1$ surface has been removed.

Since the vertical temperature structure in the model is specified and therefore known, the second term of (9) can be explicitly integrated. Thus,

$$g\vec{\nabla}_\sigma z_S + RT_S\vec{\nabla}_\sigma \ell n p_S = f_1\vec{\nabla}_\sigma T_S + f_2\vec{\nabla}_\sigma T_R + f_3\vec{\nabla}_\sigma z_S + f_4\vec{\nabla}_\sigma z_R \qquad (10)$$

where f_1, f_2, f_3, and f_4 are all functions of T_S, T_H, and T_R.

Substituting the pressure gradient force term (10) into the momentum equation (1) eliminates the last explicit reference to surface pressure. Equation (1), with the substitution of (10), and (7) now form a closed pair of equations in the variables $\vec{V}_S$ and T_S. If needed, the surface pressure can be diagnosed from (4) once T_S is known.

b. Diabatic effects

Diabatic forcing can produce strong mesoscale circulations in complex terrain. In this model, diabatic forcing is parameterized by adding a constant term Q/c_p to the surface-temperature tendency equation (2). The term is positive for heating and negative for cooling. It has a greater magnitude over land than over water but otherwise does not vary spacially or temporally. Its magnitude is chosen to reflect representative rates of temperature change observed in the case of interest. This crude parameterization is sufficient to simulate land and sea breezes, and slope winds or mountain and valley winds.

Sea breezes are produced because air over land warms up more than air over water does. In the model, the temperature profile within the 2 km-deep layer of topographic influence is tied linearly to the surface temperature. Hence when the temperature at the surface increases more over land than it does over water, the same thing happens to the temperatures throughout the layer of topographic influence. Under the hydrostatic assumption, the surface pressure must then drop more over land than over water, and an onshore pressure gradient develops, producing a sea breeze. The opposite occurs at night when greater cooling over land produces an offshore flow, or land breeze.

Figure 2 shows how the model can produce upslope winds even when the surface is uniformly heated everywhere. Above a layer of constant depth H, the free atmosphere lapse rate is not affected by surface heating. Within the layer, the temperature profile is linear and the lapse rate is increasing because the profile is tied to the increasing surface temperature. Points A and B in Figure 2 are at the same height, but none of the air above point A is heated while some air above point B is heated. Under the hydrostatic assumption, the pressure falls at B but not at A. If the pressures at points A and B are hydrostatically reduced, respectively, to points C and D at the surface, and the purely hydrostatic pressure difference between these two points is removed, then there will be a net pressure gradient force directed from C to D, upslope. Similarly, downslope or drainage winds occur when the surface is cooled.

c. Numerical solution of the equations

For the cases presented in this paper, the reference pressure level is taken to be 850 hPa. At this level, temperatures and geopotential heights are interpolated to the model grid from the appropriate National Meteorological Center (NMC) analysis. From the closest upstream radiosonde sounding, a free atmospheric lapse rate is found. This lapse rate is used with the 850 hPa temperature and height data to establish a surface temperature field. The initial wind field is then computed by balancing the initial surface pressure-gradient, Coriolis and frictional forces. Appendix III gives details of the initialization.

The model variables are positioned on an Arakawa "C" staggered grid (Mesinger and Arakawa, 1976) as shown in Fig.3. Such a grid is convenient for finite differencing and reduces truncation error. All of the runs presented in this paper are done on a 74 by 75 point grid with a resolution of approximately 7.5 km. The domain is a square, approximately 555 km (5 degrees of latitude) on each side.

After testing several time integration schemes it was found that a modified second-order Adam's-Bashforth scheme is most stable. Specifically, we used:

$$\phi^{(n+1)} = \phi^n + \frac{\Delta t}{2} [3 \frac{\partial \phi}{\partial t}^{(n)} - \frac{\partial \phi}{\partial t}^{(n-1)}]$$

where ϕ is any dependent variable, Δt is the time step, and n+1, n and n-1 represent the next, current and previous times. For all model runs a time step of 180 s was used.

Once the initial wind and temperature fields are established, the equations are integrated as follows:

(1) Integrate the surface-temperature tendency equation one time step with no diabatic term and calculate a new surface temperature field.

(2) Parameterize new temperature prifiles and deduce new surface pressure gradients.

(3) Integrate the momentum equations one time step and calculate a new surface wind field.

(4) Test to see if the domain-averaged tendencies of the wind components have become sufficiently small (e.g., less than .000001 m/s/s). If not, then return to step (1) and repeat the sequence. If so, then the wind field is virtually steady state and diabatic effects can now be included.

(5) Integrate the surface-temperature tendency equation one time step with a diabatic term included and calculate a new surface temperature field.

(6) Same as (2) above.

(7) Same as (3) above.

(8) If more diabatic heating or cooling is desired, then return to

step (5) and repeat the sequence. If diabatic heating or cooling has gone on long enough, then stop and plot all desired fields.

The model with a 74 by 75 point grid, run to 500 time steps (enough in some cases to reach steady state and then run for six hours with diabatic effects), required 30 seconds of CRAY-1 CPU time. By reducing the size of the domain, decreasing the resolution of the grid, or relaxing the steady-state convergence criterion, the CPU time can be reduced to 6 seconds or less.

Surface temperatures are fixed at inflow boundaries but are allowed to vary at outflow boundaries. Wind components are allowed to vary on all boundaries. To compute $\partial \vec{V}_S / \partial t$ and $\partial T_S / \partial t$ on the boundaries it is necessary to assume values of T_S, u_S and v_S just outside the model boundary. Tests of several possibilities indicated that the best assumption is

$$u_S(B-1) = u_S(B+1)$$

$$v_S(B-1) = v_S(B+1)$$

$$T_S(B-1) = T_S(B+1)$$

where B+1 and B-1 signify one grid length inside and outside a boundary, respectively.

The advection terms in the momentum equations are finite differenced in space according to Gerrity et al. (1971) while the momentum diffusion terms are differenced following Danard (1971). Temperature advection was center-differenced and the temperature diffusion terms followed Danard (1971) except that temperatures surrounding each point were first adjusted to the same elevation using their respective local lapse rates (see Appendix II).

III. MODEL SIMULATIONS

All three of the model runs presented in this paper come from an area of the Pacific Northwest-hereafter referred to as "the domain" that encompasses southwestern British Columbia, western Washington State and northwestern Oregon (Fig. 4). This domain is nearly ideal for testing mesoscale complex terrain models. It is mountainous, it has lots of land, water and coastlines, and it contains a good network of routine surface observation stations with which to verify models, at least at lower elevations. Furthermore, several recent studies (e.g. Mass, 1981, 1982; Overland and Walter, 1981; Walter and Overland, 1982; Reed 1981, 1982) provide much observational knowledge of the mesoscale low-level flow of the region.

The Cascade Mountains range along the entire eastern edge of the domain, reaching elevations exceeding 2000 m and substantially blocking low-level flow. Along the coast to the west, separated by low gaps, several series of mountains-including Vancouver Island and the relatively symmetric Olympic Mountains-parallel the Cascades. The Strait of Georgia and the Puget Sound lowlands separate the Cascades from the coastal mountains. The Pacific Ocean lies to the west. Figure 5 shows the smoothed topography used for the model runs presented in this paper.

The following sections will describe three model runs that include synoptic-scale flows from various directions. Two are nighttime cases and one is a daytime case.

a. December 22, 1983 at 12 GMT: East-northeasterly large scale
   flow and moderately strong diabatic cooling

This case occurred near the beginning of a period during which extremely cold temperatures prevailed over the domain—record lows for individual days were recorded at many stations. North and east of the domain, a pressure ridge at the surface and at 850 hPa (Fig. 6a), oriented southeast-northwest, built explosively and record high mean sea level pressures (up to 1064 hPa) were recorded in Montana, east of the domain. Southeasterly to northeasterly flow predominated throughout the period at the surface and at 850 hPa.

Figure 6(b) shows the surface wind observations for this case. Note the strong northeasterly winds in southern Strait of Georgia, except at the mouth of the Fraser River where the wind is weak and southeasterly. In southern Puget Sound the wind is weak easterly, calm, and even weak westerly. The coastal winds suggest some deflection around the Olympic Mountains.

More intriguing is the infrared satellite photograph taken at this

934

time (Fig. 6c). The extremely cold air flowing offshore over the warmer Pacific Ocean was unstable, and convective rolls and patches formed, aligned with the low-level wind. One particularly large patch of convection formed approximately downwind of the Olympic Mountains. This region of enhanced convection persisted in the same place for several days. The satellite image in Fig. 6(e), taken 30 hours later, shows the patch most clearly.

The model output for the case, run with moderately strong cooling (5 degrees C in 6 hours over land, 1 degree C in 6 hours over water), is shown in Fig. 6(d). Many of the observed details of the wind field inland also appear in the model. Relatively strong northeasterly winds dominate southern Strait of Georgia, but at the mouth of the Fraser River the wind is weak and easterly. In southern Puget Sound the wind is weakly northeasterly, almost calm, and even northeasterly, very much like the observed field. But most interesting is the line of convergence that forms downwind of the Olympics over the Pacific Ocean, a little north but very close to the region of enhanced convection visible in Fig. 6(c). This line of convergence is characteristic in the model downwind of any large obstacle like the Olympics; the convergence would persist there as long as the large-scale flow persisted. The longevity of the elongated patch of convection downwind of the Olympics suggests that it is topographically forced, and the model supports that hypothesis. Conversely, the existence of this persistent, apparently topographically forced feature in the low-level wind field supports the model's tendency to produce long lines of surface convergence downwind of large obstacles like the Olympic Mountains.

b. May 3, 1978 at 00 GMT: Northwesterly large scale flow and
   strong heating

During the 24 hours preceding this case, a cold front had moved out of the Pacific Ocean, across the domain, and into eastern Washington State. The surface winds at well-exposed coastal stations veered from south-southwest to west-northwest when the front passed. Similarly, the 850 hPa winds on the coast veered from west-southwest at 00 GMT on May 2, to northwest at 00 GMT (5:00 PM local time) on May 3 (Fig. 7a). At Quillayute on the Washington coast, the radiosonde sounding showed a superadiabatic layer immediately above the heated surface, superposed by a weakly stable layer and topped by an inversion at 860 hPa. Above 860 hPa the air was quite stable.

Figure 7(b) shows the surface wind observations in the domain. The wind appears to deflect around the Olympic Mnts. and then converge in Puget Sound, southeast of the Olympics (note the calm wind in southern Puget Sound between the northerly and southerly flows). This phenomenon, called the Puget Sound Convergence Zone, is characteristic of the area and is often accompanied by a band of cloudiness and precipitation across Puget Sound (Mass, 1981) (see Fig. 7e).

The model, run to steady state without any diabatic surface heating or cooling, produced the surface wind field shown in Fig. 7(c). Note that northwesterly winds off the coast deflect north and south around the Olympics and then converge in northern Puget Sound. This line or zone of confluence and convergence looks similar to the one calculated by the model in the December 22, 1983 case previously described. This should not be surprising because the Olympic Mountains are relatively symmetric; the line of convergence downwind of the Olympics should not depend on the direction of the large-scale forcing.

The convergence zone in the unheated model run is substantially north of the actual location shown in Fig. 7(b). But previous studies (Staley, 1956; Mass, 1982) have shown that diurnal circulations are an important component of the low-level wind field in this region, particularly during the warm months from April through October. For example, daytime heating over land pulls air through the Strait of Juan de Fuca and southward into the Puget Sound basin. To include this and other diabatic effects, a diabatic surface heating rate of 9°C per 6 h over land and 2°C per 6 h over water was added to the model, which was then run for an additional 6 hours. The resulting surface wind field is shown in Fig. 7(d). As in the real world, diabatic heating in the model increases the flow into the Strait of Juan de Fuca and enhances the northerlies in Puget Sound. In addition,

Note that the plots of the model output show only alternate vectors in the domain, rather than all vectors, to make the plot easier to look at.

upslope and onshore flow increase throughout the domain. Significantly, the line of surface convergence in the model moves southward to a position just a few km to the north of the observed location. Thus, it appears that given the proper synoptic scale and diabatic forcings, the model can correctly diagnose the existence and position of this important mesoscale feature.

c. May 9, 1983 at 12 GMT: Westerly large scale surface flow with moderate diabatic cooling

At this time there was a large-scale surface trough north of Washington State, oriented southeast-northwest with lowest pressures to the east. A ridge was building to the southwest. The pattern was similar at 850 hPa (Fig. 8a). At Quillayute on the Washington coast, the radiosonde sounding showed a shallow radiative surface inversion superposed by a weakly stable layer to 805 hPa.

Figure 8(b) shows the surface wind observations. Along the coast, a weak westerly wind from the Pacific Ocean meets a weak easterly drainage flow and land breeze driven by the greater nighttime cooling over land. The wind in the western part of the Strait of Juan de Fuca is calm, and westerlies predominate in much of the Strait of Georgia and the eastern part of the Strait of Juan de Fuca. Northeast of the Olympic Mountains, southerly winds from Puget Sound meet westerly winds from the Strait of Juan de Fuca to form a line or zone of confluence and convergence (note the calm wind observation between the southerly and weterly flows).

The results of the model, run for 6 hours with 3°C of cooling over land and 1°C of cooling over water, are shown in Fig. 8(c). As in the observed windfield, westerlies from the Pacific Ocean meet easterlies forced by cooling over land along the Washington coast. Before cooling was added to this simulation, westerlies completely dominated the coastal region. At the western end of the Strait of Juan de Fuca, in a region oriented southwest-northeast, the model indicates calm winds along the confluence of drainage flows and land breezes between Vancouver Island and the Olympic Penninsula. This feature is amazingly similar to the observed configuration. Equally striking is the distinct line of confluence and convergence northeast of the Olympic Mountains, between southerly winds from Puget Sound and westerly winds from the Strait of Juan de Fuca, very close to the observed location. This feature is very similar to those observed and modeled in the previous two cases, which is again not surprising. The Olympics are symmetric and the physics in the model respond to the mountains in the same way in each case. The direction of the large scale forcing is the only really significant difference. An apparent failing of the model is the excessively northerly winds in the northern Strait of Georgia, close to the model boundary.

IV. SUMMARY AND CONCLUSIONS

This paper has described a relatively simple, one-level, sigma-coordinate mesoscale model that calculates surface wind fields in complex terrain. The three cases presented above as well as several additional cases not described in this paper indicate that this model can diagnose many details of the mesoscale flow in complex terrain. The model's success suggests that it may posses much of the essential physics that determine surface flow patterns in such terrain. This is surprising at first, considering the model's many simple parameterizations, its lack of the continuity equation, and its explicit representation of only one level. The latter seems especially crippling, because the model cannot explicitly account for effects on the flow away from the surface that are forced by the topography.

Why does the model do so well? The model can simulate blocking and channeling of air around mountains, and that is its great strength. It can simulate topographic deflection because of the simple way the temperature profile is parameterized. Consider a model run in which stably stratified air moves directly towards an isolated range of mountains. Initially the flow moves up and over the topography with little deflection. But in a stable atmosphere, air at a fixed point on the windward side will be replaced by air from below that has ascended and cooled adiabatically to a temperature lower than the initial temperature. Because the temperature throughout the hypothesized, 2 km-deep layer of topographic influence is tied linearly to the surface temperature, the temperature within 2 km of the atmosphere above the surface also drops. In a hydrostatic atmosphere, this implies that the surface pressure must rise, though the model has not worried at all about where the extra mass comes from to fill the cooling column above. So the pressure rises on windward slopes, creating a pressure gradient downslope that slows and deflects the approaching wind until advection approximately balances it. Nearly the opposite occurs on the leeward side, where descending air warms adiabatically, mass disappears from the layer of topographic influence, and the surface pressure drops. The adiabatically warmed air advects downwind, creating a line of decreased pressure that extends two or more mountain diameters, until diffusion smooths the temperature (and therefore the pressure) field. The line of decreased pressure downwind of a mountain top is what causes the convergence lines computed in the three cases presented in this paper.

Does the atmosphere really behave like the model does, or does the model simulate blocking and channeling by coincidence? Because the model lacks the continuity equation and the ability to explicitly model the flow away from the surface, the answer must in part be that the model contains a lucky parameterizaton. But the model's results also very strongly suggest that adiabatic warming and cooling of descending and ascending air in a layer near the surface may be the dominant mechanism by which mountains block and channel the wind. Unfortunately the model does not make obvious to us the role that mass continuity plays in this process, nor does it directly suggest what happens to the flow away from the surface over complex terrain. The model does raise questions about flow over complex terrain that we had not thought to ask before, and so it has already been useful.

There are several ways in which this model might be improved without adding more levels. These include:

a) Parameterizing the height of the layer of topographic influence, H, more physically. For example, H could be made a function of wind speed, surface roughness or vertical stability of the lower troposphere.

b) Trying to account for the adiabatic warming and cooling of

sinking or rising air that is forced in the layer of topographic influence by divergence or convergence at the surface. This would be an
attempt to parameterize one effect that mass continuity would sup-
posedly have, but it is not clear whether this is a significant effect
or even a real one.

c) Improving the model's surface-friction parameterization.
Improvements could account, for example, for the effects of variations
in surface roughness or variations in the stability near the ground.

d) Allowing the unchanging "free atmosphere" lapse rate to vary
in space. Currently, the model uses a single lapse rate at all grid
points that is taken from the sounding nearest to and upwind of the
domain. The free atmohere lapse rate at each grid point could be
interpolated from several surrounding radiosonde observations.

e) Improving the parameterization of diabatic effects. For example,
heating and cooling could affect the temperature profile to depths
different from each other and different from the depth of topographic
influence, and the diabatic term could be made time-and slope-dependent.

We believe that the above modifications could further improve the
model's verification in most cases. It should be noted that there are
some situations in which this simple model would be inadequate for
diagnosing the low-level wind field, such as during a frontal passage,
when the vertical structure aloft is complex and makes an important
contribution to the surface pressure field. However, it appears that
this model does have the potential for diagnosing the important details
of the low-level wind field most of the time and so should prove a
useful analysis and forecasting tool in regions of complex terrain.

ACKNOWLEDGMENTS

This work was funded by the Naval Environmental Prediction Research
Facility, Monterey, CA (under Program Element G3207N; Project 7W0513
"Automated Environmental Prediction System.") and the Atmospheric
Research Section, National Science Foundation, under Grant ATM82-05390.
Computer resources for some of the research were supplied by the
National Center for Atmospheric Research, which is sponsored by the

National Science Foundation. Ms. Kathryn Stout and Ms. Charlotte Arthur assisted in preparing the manuscript.

REFERENCES

Anderson, G. E., 1971: Mesoscale influences on wind fields. J. Appl. Meteor., 10, 377-386.

Anthes, R. A., and T. T. Warner, 1978: Development of hydrodynamic models suitable for air pollution and other mesometeorological studies. Mon. Wea. Rev., 106, 1045-1078.

Cressman, G. P., 1959: An operative objective analysis scheme. Mon. Wea. Rev., 87, 367-374.

Danard, M., 1971: Numerical study of the effects of longwave radiation and surface friction on cyclone development. Mon Wea. Rev., 99, 831-839.

Danard, M., 1977: A simple model for mesoscale effects of topography on surface winds. Mon. Wea. Rev., 105, 572-580.

Dickerson, M. H., 1978: MASCON-A mass consistent atmospheric flux model for regions of complex terrain. J. Appl. Meteor., 17, 241-253.

Fosberg, M. A., W. E. Marlatt, and L. Krupnak, 1976: Estimating airflow patterns over complex terrain. USDA Forest Service research paper RM-162, Rocky Mountain Forest and Range Experimental Station, Ft. Collins, CO.

Gerrity, J. P., R. D. McPherson and P. D. Polger, 1972: On the efficient reduction of truncation error in numerical prediction models. Mon. Wea. Rev., 100, 637-643.

Keyser, D., and R. A. Anthes, 1977: The applicability of a mixed-layer model of the planetary boundary layer to real-data forecasting. Mon. Wea. Rev., 105, 1351-1371.

Lavoie, R. L., 1972: A mesoscale model of lake-effect storms. J. Atmos. Sci., 29, 1025-1040.

Lavoie, R. L., 1974: A numerical model of the trade wind weather on Oahu. Mon. Wea. Rev., 102, 630-637.

Marwitz, J. D., 1983: The kinematics of orographic airflow during Sierra storms. J. Atmos. Sci., 40, 1218-1227.

Mass, C., 1981. Topographically forced convergence in western Washington State. Mon. Wea. Rev., 109, 1335-1347.

Mass, C., 1981: A single-level numerical model suitable for complex terrain. Proceedings of the Fifth Conference on Numerical Weather Prediction, American Meteorological Society, Boston, MA 02108, 316-319.

Mass, C., 1982: The topographically forced diurnal circulations of western Washington State and their influence on precipitation. Mon. Wea. Rev., 110, 170-183.

Mesinger, F., and A. Arakawa, 1976: Numerical methods used in atmospheric models, Vol. I. GARP Publications Series No. 17.

Nickerson, E. C. and E. L. Magaziner, 1976: A three-dimensional simulation of winds and non-precipitating orographic clouds over Hawaii. NOAA Technical Report ERL 377-APCL 39.

Overland, J. E., M. H. Hitchman and Y. J. Han, 1979: A regional surface wind model for mountainous coastal areas. NOAA Technical Report, ERL 407-PMEL 32.

Overland, J. E. and B. E. Walter, 1981: Gap winds in the Strait of Juan de Fuca. Mon. Wea. Rev., 109, 2221-2233.

Pielke, R., 1974: A comparison of three-dimensional and two-dimensional numerical predictions of sea breezes. J. Atmos. Sci., 31, 1577-1585.

Reed, R. J., 1980: Destructive winds caused by an orographically induced mesoscale cyclone. Bull. Amer. Meteor. Soc., 61, 1346-1355.

Reed, R. J., 1981: A case study of a bora-like windstorm in western Washington. Mon. Wea. Rev., 109, 2384-2393.

Sherman, C. A., 1978: A mass-consistent model for wind fields over complex terrain. J. Appl. Meteor., 17, 312-319.

Staley, D. O., 1957: The low-level sea breeze of northwest Washington, J. of Meteor., 14, 458-470.

Walter, B. A. and J. E. Overland, 1982: Response of stratified flow in the ice of the Olympic Mountains. Mon. Wea. Rev., 110, 1458-1473.

APPENDIX I

Parameterization of Surface Friction

The frictional force in the boundary layer $(\vec{F})$ can be expressed as the vertical divergence of the shearing stress:

$$\vec{F} = -\frac{1}{\rho}\frac{\partial \vec{S}}{\partial Z} \qquad\qquad 1.1$$

where ρ is density and $\vec{S}$ is the shearing stress. At the surface the stress can be parameterized by a drag law

$$\vec{S}_s = \rho C_D \vec{V}_s |\vec{V}_s| \qquad\qquad 1.2$$

where C_D is a drag coefficient and $\vec{V}_s$ is the surface wind vector; above the boundary layer $\vec{S}$ can be considered to be negligible. If one assumes a linear stress profile that vanishes at the top of the boundary layer (or in our case at H, the top of the layer of topographic influence) the mean frictional force in the boundary layer can be estimated as:

$$\vec{F} = -\frac{1}{\rho}\frac{(\vec{S}_H - \vec{S}_s)}{H} = -\frac{C_D \vec{V}_s |\vec{V}_s|}{H} \qquad\qquad 1.3$$

Deardorff (1972) suggests that under neutral or stable conditions the stress vanishes at a height lower than the inversion or boundary layer height and accordingly the stress should be increased by a factor c of 2.8. In addition, because the stress profile under stable and neutral conditions is not linear but is more steeply sloped near the surface, the stress divergence at the surface (i.e., at 10 m in our case) is greater than the mean stress divergence for the layer as a whole (by approximately a factor of two in Deardorff, 1972). Furthermore, frictional boundary layers are typically lower (e.g. 1 km) than the layer of topographic influence hypothesized in our model (2 km). For these reasons, we suggest that the surface frictional force is about 4 times the estimate given in (3), so that

$$\vec{F}_s = -\frac{4 C_D \vec{V}_s |\vec{V}_s|}{H} \qquad\qquad 1.4$$

where $\vec{F}_s$, the surface frictional force, is assumed to be directed in the opposite direction of the surface wind. In our model we used values of C_D of 2×10^{-2} over land and 1.4×10^{-3} over water. At each model grid point we determined the percentage of water and land in the surrounding 7.5 km square and scaled the drag coefficient proportionately. This land/water percentage was also used in the diabatic heating parameterization.

APPENDIX II

The Horizontal Diffusion Terms

The momentum (1) and thermodynamic energy (4) equations of this model possess horizontal diffusion terms to control computational instability and to represent the effects of horizontal subgrid-scale mixing. The momentum diffusion term is of the form $K_m \nabla_\sigma^2 \vec{V}_s$, where K_m is the momentum diffusion coefficient and the Laplacian ∇_σ^2, taken at the surface or $\sigma = 1$ level, is finite differenced using the method described in Danard (1971).

The temperature diffusion term in the thermodynamic energy equation (4) is of the form $K_T \nabla_H^2 T$, where K_T is the temperature diffusion coefficient and the Laplacian ∇_H^2 is taken on a horizontal plane rather than along the surface. To understand why a different Laplacian is used for this term, consider a hypothetical situation in which there are no large scale pressure (or height) or temperature gradients, no diabatic forcing, and a constant lapse rate everywhere. In such a case no surface flow should be produced by the model. However, if the temperature diffusion term uses a Laplacian evaluated at the surface, the Laplacian will generally be non-zero since surface temperatures usually vary non-linearly along slopes. The result is a non-zero diffusion term that forces spurious winds in the absence of large scale or diabatic forcing. Evaluating the Laplacian on a horizontal plane eliminates this problem. The finite difference form of this Laplacian is evaluated at the level of the center point with the temperatures at the surrounding points being vertically extrapolated from the surface

944

using the known local lapse rates.

The runs presented in this paper used diffusion coefficients (K_m and K_T) of either 2.5 or 3×10^4 m^2 s^{-1}.

APPENDIX III

Model Initialization

To begin an integration the model requires a lapse rate representative of the lower free atmosphere as well as the geopotential height and temperature fields at an undisturbed pressure level, in our case 850 hPa. The lower tropospheric lapse rate is taken from the radiosonde sounding nearest the large scale inflow into the domain. The 850 hPa heights and temperatures at each grid point are found by first subjectively interpolating the fields analyzed on the appropriate National Meteorologial Center (NMC) operational 850 hPa chart to a 4 by 4 grid covering the model domain. Then an iterative Cressman (1959) scheme is used to calculate interpolated values on the model grid.

The initial surface temperature field is calculated by using the given lower tropospheric lapse rate and 850 hPa fields:

$$T_s = T_{850} + \gamma(Z_{850} - Z_s) \qquad 3.1$$

where Z_s and Z_{850} are the heights of the surface and 850 hPa levels at a point and γ is the lower tropospheric lapse rate.

To compute the initial surface wind field we assume a balance between the surface pressure gradient, Coriolis and frictional forces. Because the temperature lapse rate in the layer of topographic influence is initially the same as the free atmosphere lapse rate above, the expression for the surface pressure gradient force $\vec{P}$ reduces to (see Eq. 8)

$$\vec{P} = g\vec{\nabla}_\sigma Z_{850} - \frac{g(Z_{850} - Z_s)}{T_{850}} \vec{\nabla}_\sigma T_{850} \qquad 3.2$$

Note that this initial force is only dependent on height and temperature variations at 850 hPa.

Using the frictional parameterization described in Appendix I and
the standard expression for the Coriolis force, $\vec{C} = -f\vec{k}\times\vec{V}_s$, where f is
the Coriolis parameter and $\vec{k}$ is the unit normal vector at the suface,
we can set up an expression for the balance between the pressure
gradient, Coriolis and frictional forces in which the only unknown is
$\vec{V}_s$. Solving for $\vec{V}_s$ and using the T_s field calculated above, the model
is ready for integration.

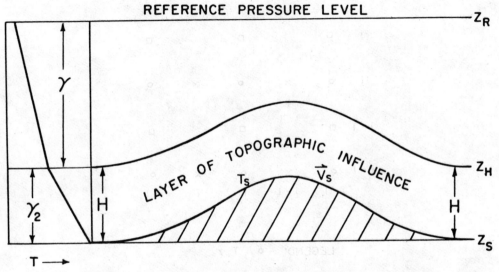

Figure 1   Vertical structure of the model

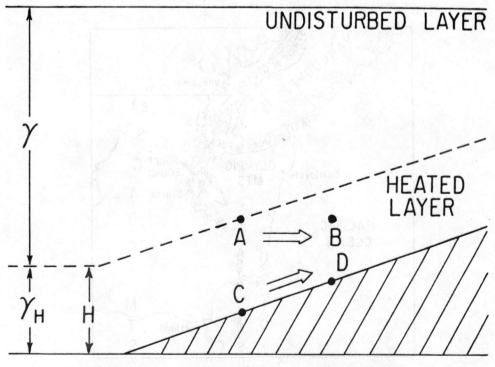

Figure 2   Schematic of a heated slope

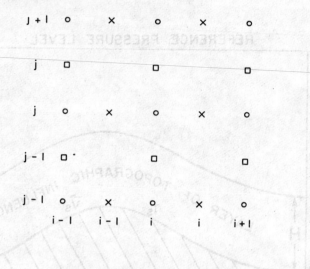

LEGEND: o T, z

x U

□ V

Figure 3   Model grid structure

Figure 4   Major geographical features of the model domain

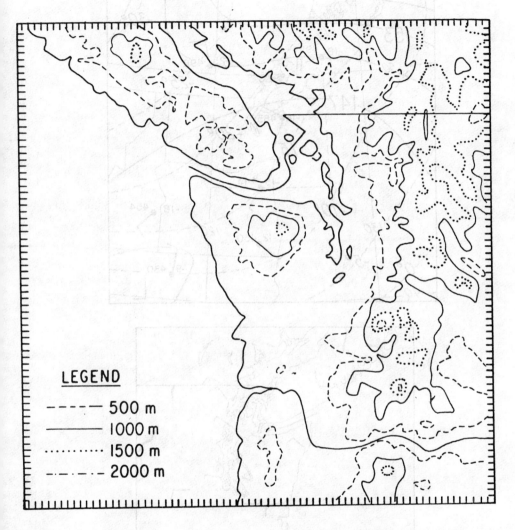

LEGEND

----- 500 m
——— 1000 m
········ 1500 m
—·—·— 2000 m

Figure 5   Topography used in the model runs

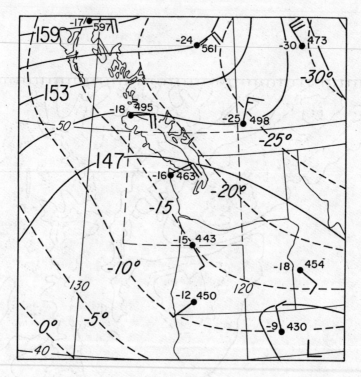

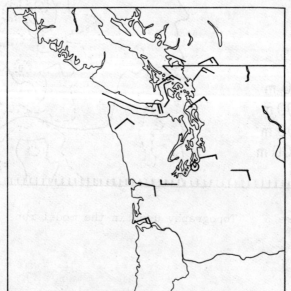

Figure 6   Run for December 22, 1983 at 12 GMT

      (a) 850 hPa chart

      (b) Observed surface winds

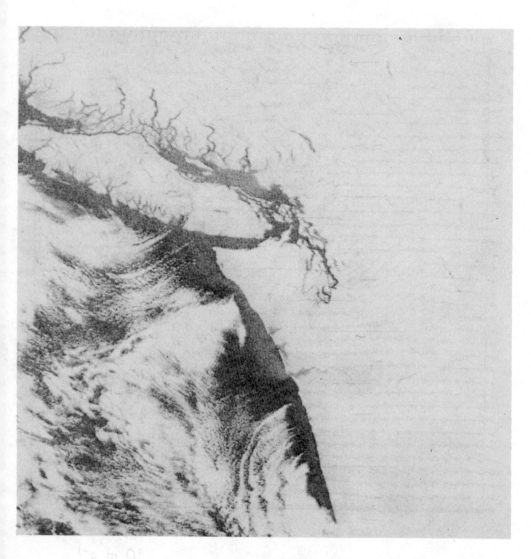

Figure 6: (c) Satellite photo taken at 1234 GMT

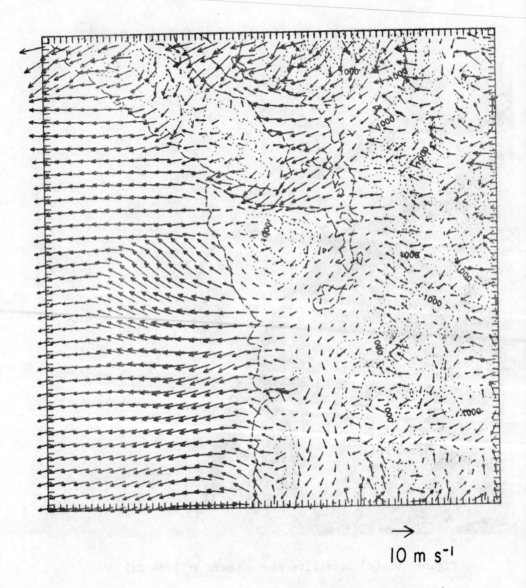

10 m s⁻¹

Figure 6: (d) Model surface winds with diabatic cooling

Figure 6 : (e) Satellite photo taken December 23, 1983 at 1816 GMT

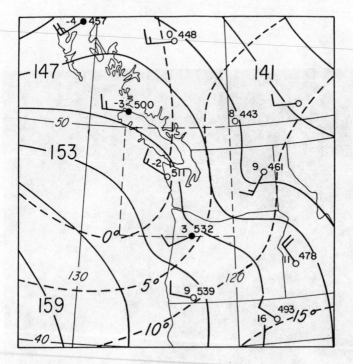

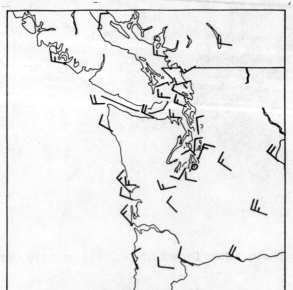

Figure 7   Run for May 3, 1978 at 00 GMT

          (a) 850 hPa chart

          (b) Observed surface winds

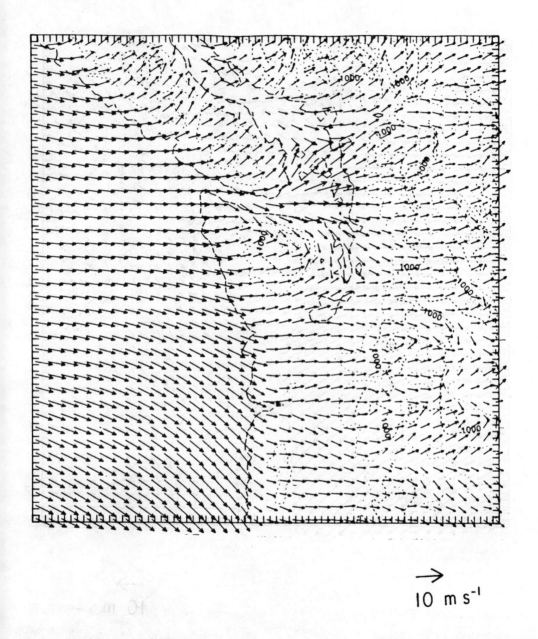

$$\overrightarrow{}$$
10 m s^{-1}

Figure 7 (c) Model surface winds without diabatic heating

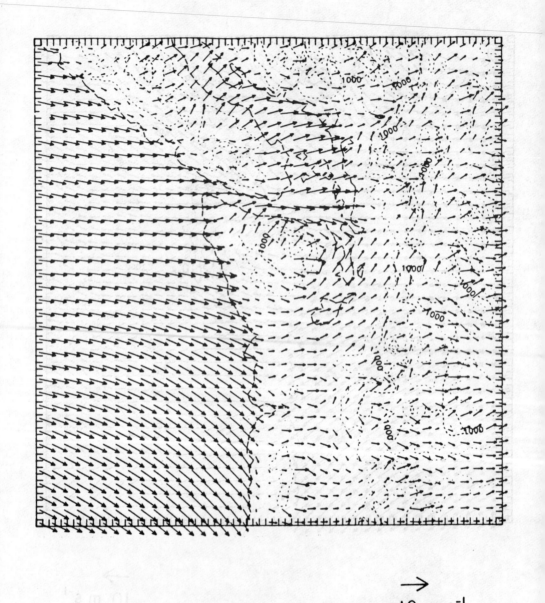

$$\xrightarrow{\hspace{2cm}}$$
10 m s⁻¹

Figure 7 (d) Model surface winds with diabatic heating

Figure 7 (e) Satellite photo taken May 2, 1978 at 2015 GMT

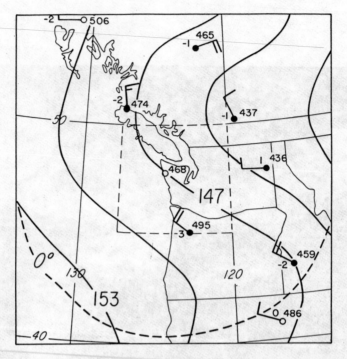

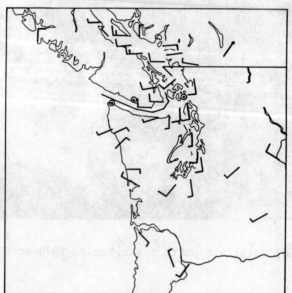

Figure 8   Run for May 9, 1983 at 12 GMT

(a) 850 hPa chart

(b) Observed surface winds

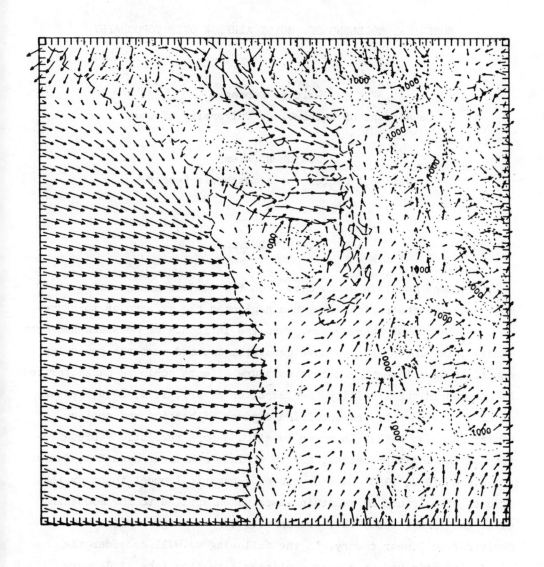

10 m s⁻¹

Figure 8 (c) Model surface winds with diabatic cooling

NUMERICAL MODELING OF MOIST AIRFLOW OVER TOPOGRAPHY

Dale R. Durran

Department of Meteorology University of Utah

Joseph B. Klemp

National Center for Atmospheric Research

I. INTRODUCTION

Although mountain waves have been the subject of extensive study, most previous theoretical work has neglected the effects of moisture on the dynamics of these waves. When the influence of moisture has been studied, the investigators have generally assumed that the governing equations were linear, and that moisture effects could be parameterized by changes in the stability (see Barcilon et al., 1979, 1980, Fraser et al., 1973), or by pre-specified regions of heating and cooling (Smith and Lin, 1982). In order to allow a more realistic representation of the moist processes, and to include nonlinear dynamics, Durran and Klemp (1982, 1983) have developed a numerical model to study the effects of moisture on mountain waves. When applied to low mountains (i.e. the linear case) they obtain results which agree with those derived from linear theory. In the following we will consider the effects of moisture in a very nonlinear situation (the 11 January 1972 windstorm in Boulder, Colorado). In addition, we will discuss the formulation of an appropriate upper boundary condition for the successful numerical simulation of highly nonlinear mountain waves.

II. A NUMERICAL CONSIDERATION

The numerical model used in these experiments solves the two-dimen-

960

sional, nonlinear, nonhydrostatic, compressible equations of motion.
Terrain is incorporated through a transformation of the vertical
coordinate; moist processes are included through a Kessler parameterization. (For a detailed description of the numerical model, see Durran
and Klemp, 1983). The model has been extensively tested and found to
reproduce analytic solutions with reasonable fidelity. During the
testing, the calculated flow did exhibit a strong sensitivity to the
location of the upper boundary whenever highly nonlinear, vertically
propagating waves were present.

The radiation boundary condition, which requires that all energy
transport be directed out of the domain, is approximated at the upper
boundary. This condition is crucial for the successful simulation of
vertically propagating mountain waves. There are, however, physical
situations in which downward propagating waves reflect from sharp
gradients in the atmospheric structure or regions of wave overturning
and breakdown and have a significant impact on the wave dynamics
below. In such instances the correct solution can be obtained only by
applying this boundary condition above the reflecting layers.

The radiation boundary condition is approximated by adding an
absorbing layer to the top of the domain[1]. The effective mean viscosity
in the absorber is chosen so that waves entering from below have
negligible amplitude when they arrive at the top of the absorbing
layer where the actual boundary condition is w=0. Reflections, which
might otherwise be produced by vertical variations in the viscosity,
are minimized by ensuring that the strength of the absorber increases
gradually with height. These requirements impose a constraint on the
minimum depth of an effective wave absorbing layer. Klemp and Lilly
(1978) have suggested that, for linear hydrostatic waves, this minimum
depth is approximately one vertical wavelength.

Both viscous and Rayleigh damping have been used in absorbing
layers (Clark, 1977; Klemp and Lilly, 1978). Rayleigh damping has
been chosen for this model because the second derivatives required for
viscous damping have a complicated finite difference structure in the
presence of the coordinate transformation. In the absorbing layer,

1)See Footnote on Page 974.

only the perturbations of a variable from its upstream value are
damped. The damping terms, which are added to the right-hand sides of
the u, w and θ equations, are

$$
\left.\begin{array}{l}
R_u = \tau(z)(u - \bar{u}) \\[2mm]
R_w = \tau(z)w \\[2mm]
R_\theta = \tau(z)(\theta - \bar{\theta})
\end{array}\right\} \tag{1}
$$

The damping coefficient has the structure

$$
\tau(z) = \left\{
\begin{array}{ll}
0, & \text{for } z \leq z_D \\[3mm]
-\dfrac{\alpha}{2}\left[1-\cos\dfrac{z-z_D}{z_T-z_D}\pi\right], & \text{for } 0 \leq \dfrac{z-z_D}{z_T-z_D} \leq 1/2 \\[4mm]
-\dfrac{\alpha}{2}\left[1+\left(\dfrac{z-z_D}{z_T-z_D}-\dfrac{1}{2}\right)\pi\right], & \text{for } \dfrac{1}{2} \leq \dfrac{z-z_D}{z_T-z_D} \leq 1
\end{array}\right. \tag{2}
$$

where Z_D is the height of the bottom of the absorbing layer and π is
3.1416. Klemp and Lilly (1978) have shown that for a single, linear
hydrostatic wave, an absorbing layer with a sinusoidal vertical
viscosity profile will be most effective when α satisfies $2 < \alpha/ku < 5$
where k is the horizontal wavenumber. In actual simulations, α is
chosen so that the dominant horizontal wavenumbers are absorbed most
efficiently.

The numerical solution is not strongly sensitive to the strength
of the damping in the wave-absorbing layer, but it can be very sensitive to changes in the height Z_D at which the absorbing layer begins,
i.e., the effective height of the upper boundary. This sensitivity increases as the amplitude of the waves entering the bottom of the
absorbing layer increases. Numerical simulations of linear, weakly non-
linear, and trapped waves, which have little amplitude at the effective
upper boundary, are not sensitive to changes in the height of that
boundary.

Figure 1 illustrates the difference in the behavior of the surface
pressure drag in two simulations with effective upper boundaries at
1.25 and 1.5 vertical wavelengths ($\lambda_z = 2\pi\bar{u}/N$), when the wave amplitude
is moderately nonlinear (Fig. 1a) and strongly nonlinear (Fig. 1b).

These simulations were conducted with the Boussinesq form of the model using 90 horizontal grid points and either 66 or 72 vertical grid levels. The wave absorbing layer occupied the top 36 levels (having a depth of 1.5λ). The grid resolution was Δx = 2 km and Δz = $\lambda_z/24$ = 250 m; the large and small time steps were 25 and 5 s. The wind speed and stability were constant, with $\bar{u}$ = 10 m s^{-1} and N = 0.01047 s^{-1}. The mountain contour was specified as

$$Z_s(x) = \frac{ha^2}{x^2 + a^2} \tag{2}$$

with a = 10 km; h was chosen so that Nh/$\bar{u}$ is 0.5 in Fig. 1(a) and 0.8 in Fig. 1(b).

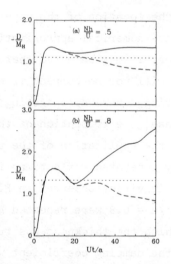

Fig.1 Surface pressure drag, normalized by its linear
hydrostatic value, as a function of time obtained
from numerical simulations in which the bottom of
a 1.5λ_z thick wave-absorbing layer is located at
1.25λ (solid line) or 1.5λ_z (dashed line) and
Nh/$\bar{u}$ is (a) 0.5, or (b) 0.8.

The steady state hydrostatic solution was also calculated from Long's equation (Lilly and Klemp, 1979); the surface pressure drag associated with that solution is plotted in Fig.1(a),(b) for reference.

Note that for this mountain shape, the steady Long's solution contains overturned streamlines (breaking waves) for $Nh/\bar{u} > 0.85$, so the situation simulated in Fig. 1(b) is indeed strongly nonlinear. As shown in both Figs. 1(a) and 1(b), the solutions with differing domain depths are almost identical until a nondimensional time $\bar{u}t/a = 20$ (the model is initialized over the first five nondimensional time units). After $\bar{u}t/a = 20$, the shallow-domain solutions amplify, while the deep solutions decay. At $\bar{u}t/a = 60$, the solutions are still relatively similar in the moderately nonlinear case ($Nh/\bar{u} = 0.5$, Fig. 1a), but in the highly nonlinear case ($Nh/\bar{u} = 0.8$, Fig. 1b) they are very different. The shallow simulation contains a breaking wave and a feature resembling a hydraulic jump which propagates downstream from the mountain, whereas the deep simulation is qualitatively similar to Long's solution.

This sensitivity to the location of the upper boundary seems to be produced by an inadequate numerical approximation to the radiation boundary condition. However, it may be noted that the radiation boundary condition itself will not be correct if nonsteady, nonlinear mountain waves are generating downwardpropagating waves through wave interactions. In such a case, the truncation of the numerical domain at any finite height with the application of the radiation condition would eliminate internally generated reflections from above that level and might produce the sensitivities observed in Fig. 1a,b.

The simulations of $Nh/\bar{u} = 0.8$ were repeated by use of a very deep wave-absorbing layer in which the thickness was tripled to $4.5\lambda_z$, and the vertical gradient in the damping coefficient was correspondingly reduced by a factor of 3. The simulations are otherwise identical to those shown in Fig. 1(b), except that the vertical grid resolution was halved to save computer time. As shown in Fig.2, the sensitivity of the surface pressure drag to the location of the upper boundary has been drastically reduced (note that these simulations are run out longer in time than those in the shallow domain). Similar experiments, in which the depth of the domain is increased (to $4.5\lambda_z$) but the thickness of the wave absorbing layer is held constant (at $1.5\lambda_z$), also show some decrease in sensitivity. However, in the latter case, this is primarily due to an increase in the time required for errors at the upper boundary

to propagate back down and degrade the solution. Although the sensitivity of the deep damping layer solutions is weak, as shown in Fig. 2,
the surface pressure drag differs from that determined from Long's
solution by 10-15%. The exact source of this error has not been determined but the lateral boundaries are the most likely candidates. Lilly
and Klemp (1979) have obtained numerical solutions with a θ coordinate
model which agree more closely with Long's solution for the case
$Nh/\bar{u}=0.74$. However, in those simulations the lateral boundaries were
located at ±72a (where a is the mountain half-width) to remove all
lateral boundary influences; by contrast, the boundaries in the current simulations are located much closer to the mountain at ±9a.

These sensitivity tests suggest that the accurate simulation of
vertically propagating, highly nonlinear mountain waves requires a
very thick wave absorbing layer. The sensitivity to changes in the
height of "shallow" absorbing layers does not seem to be peculiar to
this model; similar behaviors have been observed in the numerical
models of Klemp and Lilly (1978) and Clark and Peltier (1977) (Clark,
personal communication, 1983), although they use somewhat different
formulations for the wave absorber. In addition, when an entirely different numerical radiation boundary condition (Klemp and Durran, 1983)

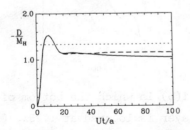

Fig.2 As in Fig.1 but in which the bottom of a $4.5\lambda_z$
thick wave-absorbing layer is located at $1.25\lambda_z$
(solid line) or $1.5\lambda_z$ (dashed line), and
$Nh/\bar{u} = 0.8$.

was used to replace the wave absorbing layer in this model, a similar sensitivity remained.

In most practical applications, it is simply not feasible to devote 80% of the numerical domain to the wave-absorbing layer. This difficulty can be avoided when simulating a real atmospheric flow by ensuring that the computational domain explicitly includes those regions which are primarily responsible for wave absorption or trapping. The effectiveness of this approach is illustrated in Fig. 3 which describes a situation identical to that in Fig. 1(b) except that the mean wind speed decreases linearly from 10 to 6 m s^{-1} between the heights of 3 and 5 km, so that the local inverse Froude number ($Nh/\bar{u}$) increases with height from 0.8 to 1.3. In this case, the wave breaks as it enters the region of decreasing wind speed, and its energy is largely dissipated by mixing so that the disturbance which reaches the top of the domain has relatively low amplitude and can be properly radiated by a "shallow" ($1.5\lambda_z$ thick) absorbing layer. Note, however, that the low level structure of the insensitive solution is at least as nonlinear as that in the sensitive case (compare the surface) wave drags in Figs. 1(b) and 3).

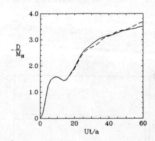

Fig.3 As in Fig. 1(b) in which the bottom of a $1.5\lambda_z$ wave absorbing layer is located at $1.25\lambda_z$ (solid line) or $1.5\lambda_z$ (dashed line), and a decrease in the mean wind speed produces a region of wave breaking in the computational domain.

Boundary condition sensitivities aside, it is also important to explicitly include the region in which the waves are absorbed because it can have an important impact on the wave structure. As an example,

966

note that the linear solutions to the cases shown in Figs. 1(b) and 3 should be identical in the layer between the ground and 3 km, since the mean Richardson number is large throughout the domain (Klemp and Lilly, 1975). However, as shown by the streamline fields in Fig.4, the nonlinear solution is strongly influenced by the changes in the upper-level wind speed. Thus, it seems important to include the region of stratospheric wave breakdown in most simulations of actual mountain waves. In the Rocky Mountain region of the United States, this will generally require that the modeling region extend to a height of 15 to 20 km.

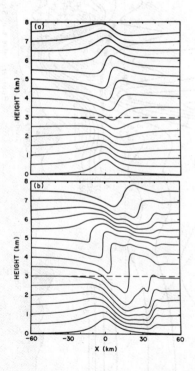

Fig.4 Streamlines for two flows, which have identical linear solutions between the heights of 0 and 3 km, obtained from numerical simulations at $ut/a = 40$. (a) $u = 10$ m s^{-1}, constant with height; (b) u decreases gradually from 10 m s^{-1} to 6 m s^{-1} above a height of 3 km.

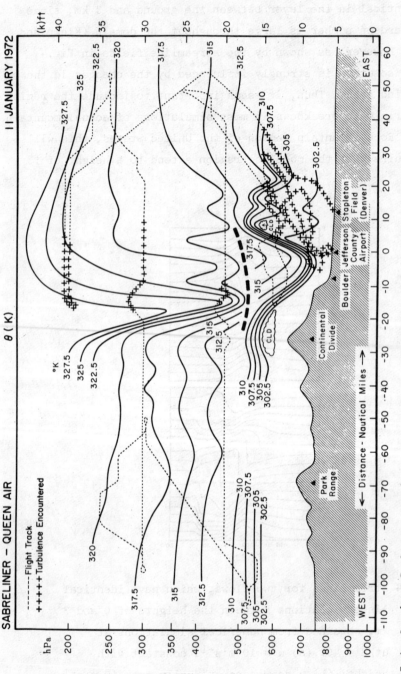

Fig. 5 Analysis of the potential temperature field (solid lines) from aircraft flight data and sondes taken on 11 January 1972. The dashed lines show aircraft track, with periods of significant turbulence shown by pluses. The heavy dashed line separates data taken by the Queen Air at lower levels before 2200 GMT from that taken by the Sabreliner in the middle and upper troposphere after 0000 GMT (12 January). The aircraft flight tracks were made along an approximate 130°–310° azimuth, but the distances shown are along the east–west projection of those tracks. Cloud locations and sizes are approximate. (From Lilly, 1978).

III. THE 11 JANUARY 1972 WINDSTORM IN BOULDER, COLORADO

On 11 January 1972, a severe downslope windstorm and mountain
wave event occurred in Boulder, Colorado. An unusually good documentation of the air flow on this day was provided by two NCAR aircraft
which took measurements in the wave. An analysis of the wave structure
revealed by their observations appears in Fig.5 which is reproduced
from Lilly and Zipser, 1972 and Lilly, 1978.

This case is simulated with a computational domain containing 128
points in the horizontal and 83 grid levels in the vertical; the
damping layer occupies the top 39 levels. The horizontal grid interval
are Δx = 1000 m and Δz = 341; the large and small time steps are 5 and
2.5 s. The mountain contour is specified by (3) with a = 10 km,
h = 2 km, and the upstream wind speed and temperature profiles are
specified by those observed on 11 January 1200 GMT in Grand Junction,
Colorado (see Fig.6).

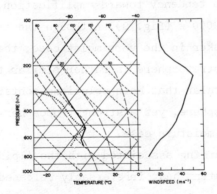

Fig.6 Upstream sounding used for the 11 January 1972
windstorm simulations. Moisture profile is from
the 11 January 1972 1200 GMT Grand Junction
sounding.

Three experiments were conducted. In the first, the air was
assumed to be completely dry. In the second, the moisture distribution

reported in the 11 January 1200 GMT Grand Junction sounding was used to initialize the humidity profile. In the third, a saturated layer containing 0.2 g kg^{-1} of liquid water between 700 and 500 hPa was added to the upstream flow.

The solutions obtained in the first two cases (completely dry and observed moisture) are quite similar. Because of their similarity, only the results from the observed moisture case are presented in Figs. 7(a) and 8(a), which show the streamlines and horizontal velocity fields. The cloud distribution is also shown in Fig.7(a). The model clouds are similar, though somewhat larger, than those actually observed during the windstorm and indicated schematically in Fig.5. The streamlines and horizontal velocity fields for the third case (low level cloud) are shown in Fig. 7(b) and 8(b). The time dependent behavior of the surface pressure drag is shown in Fig. 9 for all three cases.

As is evident, the additional moisture weakens the mountain wave considerably. The maximum downslope windspeed is reduced from almost 45 m s^{-1} to less than 25 m s^{-1} (Fig. 8); and the surface pressure drag, which shows no tendency toward amplification, is reduced by a factor of 6 at $t = 9000$ s (Fig. 9). Although the tropospheric wave response is much weaker in the very moist case, the wave amplitude above 12 km in the stratosphere is rather similar to that in the dry case. Note in particular that both solutions contain a breaking wave in the lower stratosphere, yet a strong tropospheric response is only produced in the low moisture case.

The lee side warming associated with the Alpine foehn is often attributed to the latent heat irreversably released in the low level air which ascends the windward slope moist adiabatically and then descends dry adiabatically. However, in this instance, the effect of irreversible heating on the lee side temperature is dominated by dynamical processes. Precipitation occurs only in the very moist case (from the cap cloud at a maximum rate of 0.2 cm h^{-1}), yet the lee side temperatures are several degrees lower than those obtained in the drie: nonprecipitating flow. The most important factor which influences the

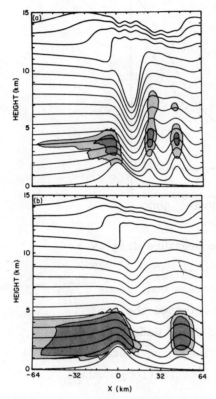

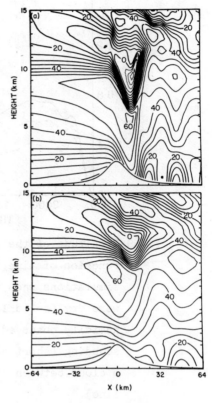

Fig.7 Streamlines obtained by
numerical simulation of the
11 January 1972 windstorm
in which the upstream
humidity (a) is determined
from the Grand Junction
sounding and (b) includes a
cloud layer with 0.2 g kg^{-1}
of liquid water between 700
and 500 hPa. Cloudy regions
are shaded: dark shading
represents liquid water concentrations exceeding 0.2 g
kg^{-1}.

Fig.8 As in Fig.7, except that
the fields plotted are the
horizontal winds speed
(m s^{-1}).

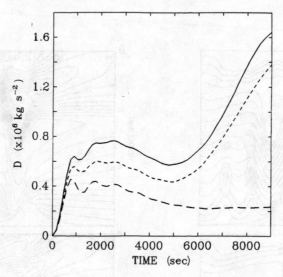

Fig.9 Magnitude of the surface pressure drag as a
function of time from the 11 January 1972
simulations when the upstream conditions
were dry (solid line), contained the
humidity observed in the Grand Junction
sounding (short dashed line), and contained
a cloud between 700 and 500 hPa (long dashed
line).

lee temperatures is the wave amplitude which is larger in the drier
case. A second simulation, in which rainwater was not allowed to form,
produced results very similar to those obtained with rain, suggesting
that (in this case) the irreversible heating associated with precipitation does not have a major impact on the wave structure. As noted by
Smith and Lin (1982) the influence of precipitation on the flow dynamics
can be much greater in tropical situations where the air holds more
water vapor.

Perhaps it is not suprising that the addition of moisture reduces
the wave amplitude in a situation when the dry atmosphere is already
favorable for the generation of strong mountain waves. However, the
results presented here, together with the previous studies of Barcilon
et al. (1979) and Durran (1981), suggest that the presence of low level
moisture does tend to decrease the wave response in a variety of situa-

972

tions. It may also be possible that the presence of moisture can amplify the wave response in cases which are otherwise unfavorable for the development of strong waves. According the linear theory of hydrostatic mountain waves presented by Klemp and Lilly (1975), such amplification may occur if the longer vertical wavelengths associated with saturated waves produce a phase shift across the troposphere which is closer to one-half vertical wavelength than the phase shift for the corresponding dry flow.

REFERENCES

Barcilon, A., J.C. Jusem and P.G. Drazin, 1979: On the two-dimensional hydrostatic flow of a stream of moist air over a mountain ridge. Geophys. Astrophys. Fluid Dyn., 13, 125-140.

Barcilon, A., J.C. Jusem and S. Blumsack, 1980: Pseudo-adiabatic flow over a two-dimensional ridge. Geophys. Astrophys. Fluid Dyn., 16, 19-33.

Clark, T.L., 1977: A small scale dynamic model using a terrain following coordinate transformation. J. Comput. Phys., 24, 186-215.

Durran, D.R., 1981: The effects of moisture on mountain lee waves. Ph.D. thesis, MIT, (NTIS PB 82156621).

Durran, D.R., and J.B. Klemp, 1982: The effects of moisture on trapped mountain lee waves. J. Atmos. Sci., 39, 2490-2506.

Durran, D.R., and J.B. Klemp, 1983: A compressible model for the simulation of moist mountain waves. Mon. Wea. Rev., 111, 2341-2361.

Fraser, A. B., R.C. Easter and P.V. Hobbs, 1973: A theoretical study of the flow of air and fallout of solid precipitation over mountainous terrain. Part I: Airflow model. J. Atmos. Sci., 30, 801-812.

Klemp, J.B. and D.K. Lilly, 1975: The dynamics of wave-induced downslope winds. J. Atmos. Sci., 32, 320-339.

Klemp, J.B. and D.K. Lilly, 1978: Numerical simulation of hydrostatic mountain waves. J. Atmos. Sci., 35, 78-106.

Klemp, J.B. and D.R. Durran, 1983: An upper boundary condition permitting internal gravity wave radiation in numerical mesoscale models. Mon. Wea. Rev., 111, 430-444.

Lilly, D.K., 1978: A severe downslope windstorm and aircraft turbulence event induced by a mountain wave. J. Atmos. Sci., 35, 59-77.

Lilly, and E.J. Zipser, 1972: The front range windstorm of 11 January 1972-A meteorological narrative. Weatherwise, 25, 56-63.

Lilly, and J.B. Klemp, 1979: The effects of terrain shape on nonlinear hydrostatic mountain waves. J. Fluid Mech., 95, 241-261.

Smith, R.B., and Y. Lin, 1982: The addition of heat to a stratified airstream with application to the dynamics of orographic rain. Quart. J. Roy. Meteor. Soc., 108, 353-378.

---

[1] A new radiation upper boundary condition (Klemp and Durran, 1983), in which the pressure at the upper boundary is determined from the Fourier transformed vertical velocity, has recently been incroporated in this model. It appears to significantly improve the model efficiency without sacrificing accuracy. However, since the model results we published earlier (Durran and Klemp, 1982) were obtained using an absorbing layer, and since we wish to discuss the model's sensitivity to the upper boundary using the simpler and more thoroughly understood absorbing layer formulation, the simulations described in this paper do not use this new radiation boundary condition.

A NUMERICAL MODEL OF SLOPE-WIND CIRCULATION REGIMES AND DISPERSION AND TRANSPORT OF POLLUTANTS IN A VALLEY

Wen Tang
University of Lowell

Li Peng
University Space Research Association

A numerical model is utilized to simulate 2-dimensional day and night slopewind circulation regime in a typical V-shaped valley under a specified prevailing wind velocity on the upper boundary. We assume the mountain-valley terrain (Fig.1) that the height of the mountain ridges and the width of the valley are 500m and 5700m, respectively. With the Boussinesq approximation and assumption of little or no change of all meteorological variables in the valley axis direction, the set of governing differential equations, which includes the vorticity equation, the equation of motion along the valley axis direction, and equation for the potential temperature, are obtained. The boundary

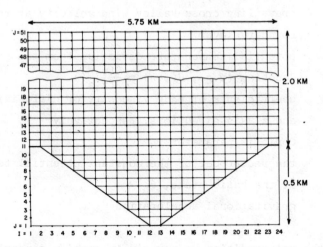

Fig.1. Mountain-valley profile and grids (M=22, N=50)

conditions are assumed as follows: At the ground surface we assume
that the stream function is zero. The no-slip condition is also assumed.
The lateral boundary conditions are assumed to be cyclic.

$$\frac{\partial \eta}{\partial t} = -J(\psi, \eta) - f\frac{\partial v}{\partial z} + \alpha_0 g \frac{\partial \theta'}{\partial x} + K_h \frac{\partial^2 \eta}{\partial x^2} + K \frac{\partial^2 \eta}{\partial z^2} \tag{1}$$

$$\frac{\partial v}{\partial t} = -J(\psi, v) + f\frac{\partial \psi}{\partial z} - fU_g + K_h \frac{\partial^2 v}{\partial x^2} + K \frac{\partial^2 v}{\partial z^2} \tag{2}$$

$$\frac{\partial \theta'}{\partial t} = -J(\psi, \theta') + K_h \frac{\partial^2 \theta'}{\partial x^2} + K \frac{\partial^2 \theta'}{\partial z^2} \tag{3}$$

where $\eta = \dfrac{\partial^2 \psi}{\partial x^2} + \dfrac{\partial^2 \psi}{\partial z^2}$

$$u = -\frac{\partial \psi}{\partial z}, \quad w = \frac{\partial \psi}{\partial x} \tag{4}$$

ψ stream function

η vorticity component in the negative y axis direction
parallel to the horizontal valley axis

u, v, w velocity components in cross-valley, valley axis and
vertical directions, respectively

U_g prevailing cross-valley wind velocity at the top of
boundary layer

θ' potential temperature deviation from the reference state

J Jacobian (e.g., $J(\psi, \eta) = \dfrac{\partial(\psi, \eta)}{\partial(x, z)}$)

K_h, K coefficients of eddy viscosity or diffusivity in the x
and z axis directions, respectively

f Coriolis parameter

α_0 the reciprocal of the constant potential temperature
at the basic state, θ_0

g gravitational acceleration

As to thermal condition we may simply specify a nearly constant
heat flux as usually assumed over homogeneous ground based on some
observations in a mountain-valley study.

$$\frac{\partial \theta}{\partial z} \text{ surface layer} = \begin{cases} -0.5^\circ \text{ K/50 m for day} \\ 1.0^\circ \text{ K/50 m for night} \end{cases}$$

The initial condition for potential temperature deviation is assumed to be linear function of height and the initial conditions for horizontal velocity components are roughly estimated based upon the Ekman boundary layer solution. The initial stream function field is determined from the initial conditions of the horizontal velocity field. The stream functions will satisfy all boundary conditions and vorticity equation.

$$\psi = \overline{\psi} = \text{constant}, \quad v = 0, \quad \theta = 6 \text{ K and}$$

$$\eta \cong \{-U_g - \frac{(\overline{\psi}-\psi_{i,N})}{\Delta z}\}/\Delta z,$$

$$w = \begin{cases} 0 & \text{for } z \geq D \\ u \dfrac{dz_g}{dx} \cos \dfrac{\gamma(z-z_g)}{2} & \text{for } z_g < z < D \end{cases}$$

As to numerical method used we use the central difference at interior grid and the one-side difference at upper and lower boundaries for the gradient and diffusion terms. Arakawa's scheme is modified and applied for the advection term. An area ratio factor is devised to allow the kinetic energy and vorticity to be conserved when we apply the method to an area which is not a rectangle. For the time difference, the "Simulated backward time difference scheme" is adopted in order to avoid high frequency oscillation from the solution. The Poisson equation of stream function is solved through the adjustment matrix.

$$J_{i,j}(\psi,a) = \frac{1}{4\Delta x \Delta z}[(\psi_{i+1,j} - \psi_{i,j+1})a_{i+1,j+1}$$
$$+ (\psi_{i-1,j}-\psi_{i,j-1})a_{i-1,j-1} - (\psi_{i-1,j}-\psi_{i,j+1})a_{i-1,j+1}$$
$$- (\psi_{i+1,j} - \psi_{i,j-1})a_{i+1,j-1}]$$

A constant K value is used since exact numerical value and distribution of K in any valley and its exact nature are not certain, and since variable forms of K over homogeneous horizontal surface

adapted for valley terrain have little real justification, as well as
for the sake of computational simplicity. Three cases of thermally
induced slope wind and the application to vehicular exhaust gas
dispersion in a valley highway are computed. The first case is a pure
thermally induced slope wind in midday without large scale prevailing
wind. The flow is evidently symmetrical with respect to the center of
the valley. The stream function field (Fig.2) exhibits a two cell

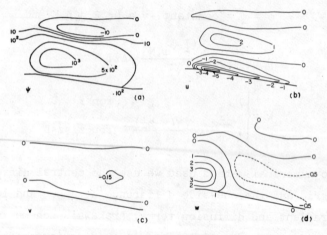

Fig.2. Computed circulation field for midday condition
without prevailing wind.
(a) Stream function, (b) u component, (c) v component,
(d) w component

vertical system. The cell just above the slope is the manifestation of
convection directly driven by the imput heat flux from slope surface.
The corresponding maximum upslope and downslope horizontal speeds are
5m s^{-1} and 2m s^{-1} respectively. The upslope wind covers the vertical
region from the sloping ground to about 500m above, in general agreement with linear theory. The center of maximum horizontal velocity
near the surface is situated near the top of the mountain ridge. The
velocity component parallel to the valley axis is very weak and anti-
symmetrical with respect to valley axis. The vertical velocity field

indicates upward motion above the ridge line, confined to a narrow horizontal region of 1 km except near slope, extending vertically 2 km level. The vertical velocity field agrees with cumulus cloud development observed on the top of a mountain range on a sunny day.

The second case is the case with a prevailing wind of 5m s^{-1} in the cross-valley direction in the daytime. A large separated cell, 2.5 km long and 700m thick, is formed above the lee slope with its center

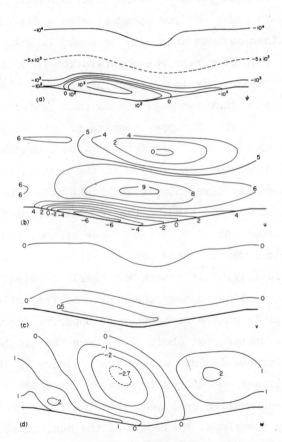

Fig.3. Computed circulation field for typical midday
condition with prevailing crossvalley wind
velocity of 5s^{-1}. (a) Stream function,
(b) u component, (c) v component, (d) w component

300m (Fig.3) above the middle of the surface. The reverse flow covers
most of the lee slope surface. The center of maximum speed of reverse
flow has a magnitude of approximately 7m s^{-1} at about 50m above the
middle of the lee slope. The reverse flow layer is 300m deep. Above
this layer the wind direction is essentially in the prevailing large-
scale wind direction. The center of maximum cross-valley horizontal
wind field, having a speed of 9m s^{-1}, is centered 900m directly above
the valley center. Next to the surface, except near the valley center
above the windward slope and lee slope, winds remain in an upslope
direction above most parts of both valley slopes. The velocity along
the valley axis has a maximum speed at 350m above the middle of the
lee slope with a maximum intensity of about 1m s^{-1}. The layer varies
with a depth from 300m near the mountain top to 900m above the center
of maximum v. Above this layer the wind reverses the flow direction.
Computations show that two separate maximum centers of upward velocity
about 2m s^{-1} are over the upper portion of both slopes and a maximum
velocity center of 2.7m s^{-1} is formed 1400m above the valley floor.

The third case of computation is the case with prevailing wind
in the cross-valley direction in the night time. The imposed condition
is a large scale 5m s^{-1} wind on the top with vertical potential
temperature ascendant of 0.02 $K \cdot m^{-1}$ near ground. A dramatic change
from the typical daytime case can be seen at once. The computed stream
function field (Fig.4) is smooth and nearly horizontal. The structure
becomes more complicated near the center of the valley, where one
elongated cell is on top of another of opposite circulation. The
longer one is centered at about 100m above the middle windward slope
and the smaller one is just within the vicinity of the center of the
valley. The longer one wedges in between the upper horizontal streamlines and lower smaller cell such that the flows over both slopes are
predominately downslope. The height of the maximum slopewind above the
ground is about 50m or less, comparable to results obtained independently by Prandtl, Defant, and Tang. The maximum speed is slightly
less than 1m s^{-1}. The wind fields in the vertical and valley-axis
direction are small because of stable stratification.

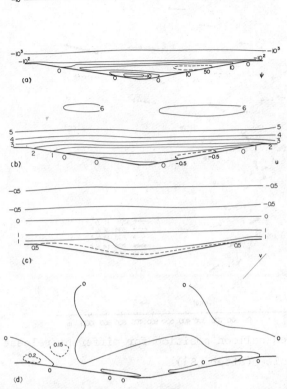

Fig.4. Computed circulation field for typical nighttime condition with prevailing crossvalley wind velocity of 5m s⁻¹.
(a) Stream function, (b) u component, (c) v component,
(d) w component

The computed results are in good qualitative agreement with many
interesting observations around the world especially in Innsbruck,
Austria, where many observations of mountain-valley circulations have
been made (e.g. Ficker, 1913; Jelinek, 1938; Kanitscheider, 1939;
Moll, 1935). These computed results are compared favorable with recent
observations made in the U.S. also. (e.g. Banta et al., 1981; Davidson,
1963; Tang, 1960, 1976, 1983) (see Figs.5, 6,7,8,9,10,11)

The numerical solution of the circulations are used to study the
dispersion and transport of vehiclar exhaust gas along a valley highway.
The highway is considered as a line source, situated on the lee slope

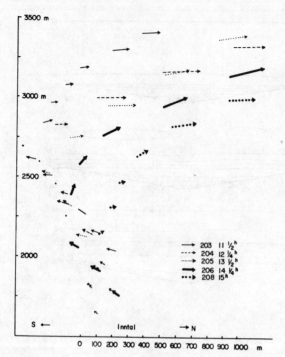

Fig.5. Observed balloon position for different release
in the lee (after [8])

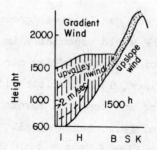

Fig.6. Illustration of the wind system observed in In-
nsbruck-Hafelekar area on 20 September 1935. I
Innsbruck, H Hungerburg, B Bodensteinalm, S
Seegrube, K Hafelekar (after [7])

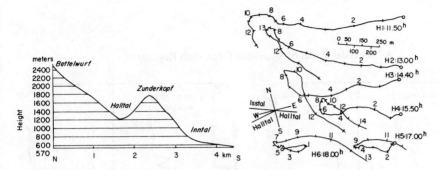

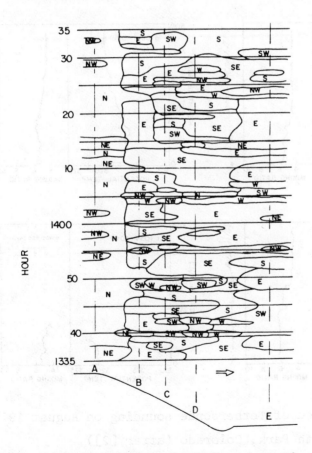

Fig.7. Left: Profile of the relief, right: Balloon trajectories Halltal, 1280 m, 26 July 1933. The numbers
marked along the side of each trajectory are the
elevations of the balloon in 100m (after [9])

Fig.8. Wind direction isopleth for given time interval
(after [15])

A Numerical Model of Slopewind Circulation Regimes

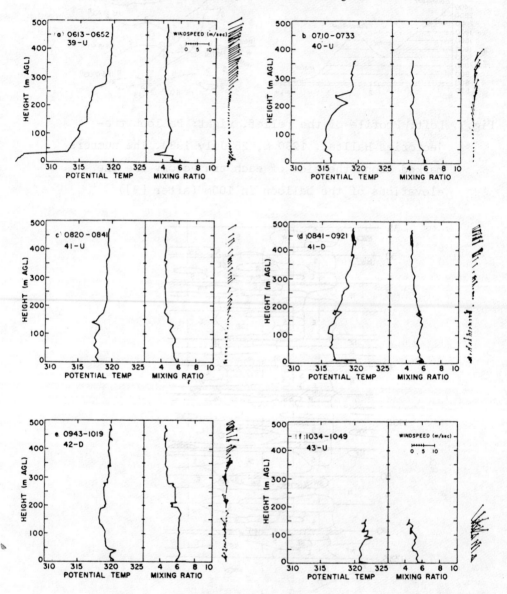

Fig.9. Sequence of Tethersonde sounding on August 1977 in South Park, Colorado (after [2])

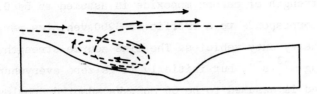

Fig.10. Observed balloon trajectory over the lee slope
(after [6])

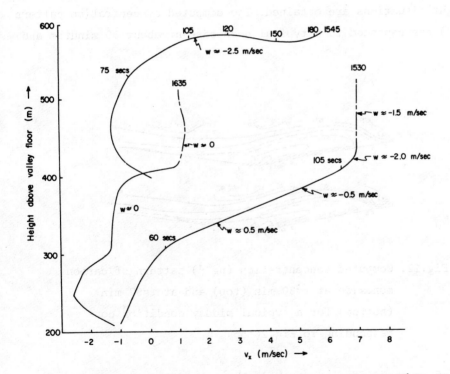

Fig.11. Three profiles of v_x component of velocity
observed from slope releases, 18 August,
1959 (after [4])

at 375m from the valley center and 50m above the valley floor. The source emission strength of carbon monoxide is assumed to be 0.019 $g \cdot s^{-1} \cdot m^{-1}$, which corresponds to a traffic of 6200 vehicles per hour, 5% of them being heavy duty vehicles. The line source strength per unit area is $3.06 \mu g \cdot m^{-3} \cdot s^{-1}$, but initially it is zero everywhere. The average route speed is assumed to be 50 mph. We adopted Crowley's fourth order conservative scheme and the splittling up method for the advection of pollutants.

Based on the forementioned conditions and the simple form of diffusion equation

$$C_1 + uC_x + wC_2 = S + (K_{p_1}C_x)_x + (K_{p_2}C_2)_2$$

and the boundary condition for C, $\frac{\partial C}{\partial n} = 0$

then the computed concentration of carbon monoxide for both typical day and night situations are obtained. The computed concentration pattern (Fig.12) corresponded to a typical noon case at about 30 minutes and

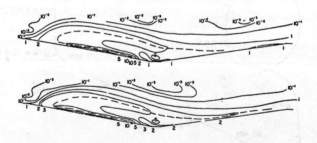

Fig.12. Computed concentration (μg^{-3}) pattern of carbon monoxide at t=30 min (top) and at t=57 min (bottom) for a typical midday condition corresponding to Fig.2(top)

1 hour after the diffusion started. The plume first follows the return flow of the separated cell and then bends back to follow the general flow direction above the valley. The difference between these two concentration patterns is rather insignificant. The centers of maximum

concentration are found at the place where the highway is located.
The rising portion of the plume is thin compared with the thickness of
the bent back portion. The concentration in general increases with
time even when the stability condition is favorable for dispersion,
although the rate of increase gradually levels off. The reason is that
the circulation pattern has a feed-back effect, so that part of the
pollutants re-enter into the source, producing a rather low dispersion
rate.

For the very stable nighttime case, the concentration is much
larger than in the noontime case. The magnitude is five times larger
at the source. Pollutants are generally restricted to the lower part
of the valley (Fig.13) because of low wind speed and stable thermal

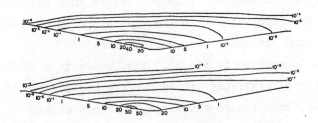

Fig.13 Computed concentration (μg^{-3}) patterns of
carbon monoxide at t=30 min (top) and at
t=53 min (bottom) for a typical night
condition corresponding to Fig.2 (bottom)

conditions. Thus, this computation indicates that the valley would be
a very unfavorable place for a highway if the assumed conditions in
the model are met. It appears that until the flow pattern is changed,
it will be difficult for the pollutant from the highway to be transported out of the valley.

In conclusion, the computation of the non-linear interaction
between the large scale flow and the thermally induced flow over
mountain-valley terrain has produced interesting results. For the
convection condition during the day, the return flow from the separated

987

cell in the lee of the ridge due to friction is enhanced by upslope
motion from differential heating, so that the wind becomes very strong
in the boundary layer. For the stable condition during the nighttime,
the friction-driven separated cell counteracts the drainage flow above
the lee slope leading to weakening of the separated cell.

On the windward slope, there is a little tendency to produce a
separated cell due just to the frictional effect, but the thermal effect
due to heating and cooling becomes a dominating factor. Over this
slope the flow situation is opposite to that over the lee slope. With
strong convective conditions during the day, no separated cell is
developed, whereas under strong stable conditions at night or in the
early morning, a weak elongated cell is developed. They show good
qualitative agreement with observations over many mountainous terrains,
and agree overall with the flow pattern obtained from analytical
solutions reported elsewhere. It is our belief that the results of the
present study have contributed to the understanding of the basic physics
governing the wind structure over valley terrain.

The model is applied to the pollution study in a valley. In the
typical day with a relatively strong surface heating the plume at first
inclines to lee slope toward the ridge, then curves back to the downwind direction with the prevailing flow; in the typical night with
stable conditions the dispersion showed no plume pattern but rather
diffused upward and horizontally.

REFERENCES

[1] Arakawa, A.: Computational Design for Long-Term Numerical
    Integration of the Equations of Fluid Motions: Two-Dimensional
    Flow. Part I. J. Comput. Phys. 1, 119-143 (1966).
[2] Banta, R., Cotton, W.R.: An Analysis of the structure of Local
    Wind Systems in a Broad Basin. J. Appl. Met. 20, 1255-1266 (1981).
[3] Crowley W.P.: Numerical Advection Experiments. Mon. Weath. Rev.
    96, 1-11 (1968).
[4] Davidson, b.: Some Turbulence and Wind Variability Observations

in the Lee of Mountain Ridges.J. Appl. Meteor. 2, 463-472 (1963).

[5] Defant. F.: Local Winds. Compendium of Meteorology. Boston: Amer. Met. Soc., pp. 1334, 1951.

[6] Ficker.H.V.: Wirbelbildung bei Ballonfahrten in Gebirge. Meteorol. Z., 48, 243-245 (1913).

[7] Jelinek, A., Riedel, A.: Uber die Schichtdicke der periodischen Lokalwinde im Inntal. Beitr. Phys. fr. Atm. 24 205-215 (1938).

[8] Kanitscheider, R.: Beitrage zur Mechanik des Fonns III. Beitr. Phys: fr. Atm. 25, 49-58 (1939).

[9] Moll, E.: Aerologische Untersuchung periodischer Gebirgswinde in V-formigen Alpentalern. Beitr. Phys. fr. Atm. 22, 177-199 (1935).

[10] Prandtl, L.: Fuhrer durch die Stromungslehre, 383 pp. Braunschweig: F. Vieweg & Sohn 1942.

[11] Tang, W.: The Diurnal Variation of Temperature and Wind Over Sloping Terrain. Ph.D. Thesis, Dept. of Meteorology and Oceanography, New York University 1960.

[12] Tang, W.: Theoretical Study of Cross-Valley Wind Circulation. Arch. Met. Geoph. Biokl., Ser. A 25, 1-18 (1976).

[13] Tang, W. Peng, L.: A Numerical Model of Slopewind Circulation Regimes in a V-Shaped Valley, Arch, Met. Geoph. Biokl., Ser. B, 361-380 (1983).

[14] Tang, W. Peng, L.: Dispersion and Concentration Pattern of Vehicular Exhaust Gases from a Valley Highway, Arch. Met. Geoph. Biokl., Ser. B, 33, 11-18 (1983).

[15] Yoshino, M.: The Structure of Surface Winds Crossing Over a Small Valley. J.Met. Soc.Japan, 35, 34-45 (1957).

TURBULENT EFFECTS IN LARGE SCALE FLOW OVER OROGRAPHY

Geoffrey K. Vallis and John O. Roads
Scripps Institution of Oceanography A-030
University of California

1. INTRODUCTION

The simplest models used to describe stationary eddies are the stationary linear models linearized around a zonal state (e.g., Saltzman, 1968). The primary justification for these models comes from the observation that the stationary eddies are very much weaker than the zonal flow-the eddy stationary energy is an order of magnitude smaller than the zonal stationary energy (Holopainen, 1970). Only minor modifications are to be expected if the stationary eddy components are allowed to nonlinearly interact (Ashe, 1980).

The effect of the transient eddies has usually been regarded as implicit through their effect on the zonal flow and intermediate to small scales. Indeed, many linear models do produce stationary anomalies in qualitative agreement with some observations (e.g., Alpert et al., 1983). However, some observational studies suggest that the transient eddies (which have more energy than the stationary eddies) have an important influence on the stationary eddies in linear models (e.g., Youngblut and Sasamori, 1980; Opsteegh and Vernekar, 1981). Other evidence lies in the instability properties associated with the stationary eddies. If these eddies are unstable then they are likely to give rise to perturbations that may draw energy from the stationary flow in an organized manner, eventually leading to a different time averaged field. A number of simple models of this interaction have been proposed (e.g., Frederiksen, 1978; Lin, 1980; Sasamori and

Youngblut, 1981).

The direct effect of the transients on the stationary eddies has received little attention, partly because it is very difficult, in the real atmosphere, to isolate their effects. They will affect the stationary flow by direct energy transfer and by stochastic perturbations. How important these processes are, which are neglected in linear models, and concomitantly, how realistic the results from such models are, are not well understood.

One step toward understanding such problems is to rigorously compare linear theory with the time averages in a nonlinear time dependent model (e.g., Roads, 1980; Phillips, 1982). The linear model can be identical with the full model in all respects other than the omission of nonlinearities. In particular it can use the same stationary forcings and finite differencing of the turbulent model. Further, the linear model can use the time averaged state of the turbulent model. By comparing the time averaged (stationary) response of the turbulent model with the linear model, linearized around the mean zonal flow of the full model, the effects of the transient eddies can be isolated.

In this paper we report the results of such a study with a turbulent (i.e., time dependent, highly nonlinear) quasi-geostrophic model on a beta plane with specified orography. The turbulent model (often denoted the "full" model below) is compared to the response in various abridged models-a linear model, a linear model with nonlinear stationary fluxes, a linear model with nonlinear stationary fluxes and transient thermodynamic or vorticity fluxes. Aside from the omission of one or more of the above processes, the models are identical. Energetics of the models are decomposed into stationary and transient components to examine the energy flow. Finally the stability of the solutions is examined. Here the aim is to see whether the variance of the full model is related to the unstable eigensolutions of the time averaged flow.

Overall, we are interested in how stationary orography is related to the time averaged fields in the turbulent model, in the hope of ultimately understanding the time averaged response through the use of simple models, such as that of White and Green (1982), in which turbulent processes are parameterized.

The basic model is described in Section 2. A description of the simulations and their energy budgets can be found in Section 3. Section 4 contains a comparison with linear theory. Section 5 is a description of the instability analysis. Section 6 summarizes and concludes.

2. MODEL

The two-level, beta-plane, quasi-geostrophic channel model comprises upper and lower level vorticity equations, a thermodynamic equation and boundary conditions. The relevant equations may be written, in standard notation, as

$$\frac{\partial}{\partial t}\nabla^2\psi_1 + J(\psi_1, \nabla^2\psi_1 + \beta_y) = f\frac{\omega}{\Delta p} - D_1 \tag{1}$$

$$\frac{\partial}{\partial t}\nabla^2\psi_3 + J(\psi_3, \nabla^2\psi_3 + \beta_y) = -f\frac{\omega}{\Delta p} + f\frac{\omega_3}{\Delta p} - D_3 \tag{2}$$

$$\lambda^2\frac{\partial r}{\partial t} + \lambda^2 J(\psi, r) = \lambda^2 F_r + \frac{f\omega}{2\Delta p} \tag{3}$$

The vertical pressure velocity is zero at the upper boundary and at the surface, ω, is given by

$$\frac{f\omega_3}{\Delta p} = -\frac{\rho_3 g f}{\Delta p}J(\psi_3, h') - K'f\frac{\rho_3}{\Delta p}\nabla^2\psi_3 = -J(\psi_3, h) - K\nabla^2\psi_3 \tag{4}$$

ψ_1 and ψ_3 are the upper and lower level streamfunctions. ψ and r are the barotropic and thermal stream-functions defined by $\psi = (\psi_1 + \psi_3)/2$, $\hat{r} = (\psi_1 - \psi_3)/2$. h' is the dimensional orographic height and F_r is the thermodynamic forcing. The flow is governed by these parameters as well as the value of β, the meridional derivative of the Coriolis parameter; the inverse deformation radius, λ, which is a measure of the static stability; the surface friction coefficient, K; and the dimensions of the periodic channel (zonal extent, L_x, 16,000 km and meridional extent, L_y, 8,000 km). D_1 and D_3 are high order diffusion operators designed to remove enstrophy and keep the energy spectra smooth at high wavenumbers. They have a negligible effect on low wavenumbers and are parameterized by $D_i = \nu\nabla^6\psi_i$. The numerical values of some of the parameters are

$\beta = 1.5\times10^{-11}s^{-1}m^{-1}$

$\Delta p = 400hPa$

$\lambda^2 = 3.16\times10^{-12}m^{-2}$ (corresponding to a nondimensional wavenumber 4.5)

$\nu = 1.4\times10^{16}m^4s^{-1}$

$K = (1/2.9)$ days^{-1}

By eliminating the vertical velocity the equations may be combined into equations representing the conservation of potential vorticity at the upper and lower levels.

$$\frac{\partial}{\partial t}q_1 + J(\psi_1,q_1) = -2\lambda^2 F_r - D_1 \tag{5}$$

$$\frac{\partial}{\partial t}q_3 + J(\psi_3,q_3) = +2\lambda^2 F_r - D_3 \tag{6}$$

where

$q_1 = \nabla^2\psi_1 + \lambda^2(\psi_3-\psi_1) + \beta y$

$q_3 = \nabla^2\psi_3 + \lambda^2(\psi_1-\psi_3) + \beta y + h$

and D_3 includes surface drag.

The vertical velocity is obtained from an omega equation. The channel is periodic in the x-direction and zero flow through the boundaries along the northern and southern edges. An appropriate spectral expansion of the stream-function is then:

$$\psi(x,y,t) = \sum_{l=1}^{N-1} b_{0l}(t)\cos ly + \sum_{l=1}^{N-1}\sum_{\substack{k=-(N-1)\\k\neq0}}^{k=N-1} b_{kl}(t)\sin ly\, e^{ikx} \tag{7}$$

Here x and y are nondimensional coordinates defined by $x = x^* \frac{2\pi}{L_x}$; $y = \frac{y^*\pi}{L_y}$; where* denotes a dimensional variable. Note that no eddy activity is allowed on the boundary and a temperature gradient can exist across the domain. Evaluation of the nonlinear Jacobian terms follows Vallis (1984) and Orszag (1971). Truncation occurs at N=16.

The form of the thermodynamic forcing is

$$F_r = (\frac{A_k}{2} + C_{r_k})$$

C gives a radiative damping, with a time scale of about 23 days (2×10^6 seconds). The zonal forcing component (A_{01}) is set to a value corresponding to 0.9 K day. In the absence of eddies this radiative forcing yields a symmetric radiative-equilibrium zonal wind of about 35 m/s at the upper level and an almost zero but negative lower level wind (Fig.1). The lower level wind would be exactly zero but for the high order diffusion operators (the enstrophy removers).

We note here that it is not our intention to perform a complete parameter study of the equations (5) and (6). We propose only a study of the effects of a localized range of mountains and a localized heat source. To this end experiments (denoted M1, M2) were performed with a strip of mountains through the whole meridional extent of the domain, and of 2000 km zonal extent. The total height in the strip is 2 km (M2) or 4 km (M1). The spectral amplitudes of orography are illustrated in Figure 4. Note the spectral expansions show a zero at wavenumber 8. The parameter range of the forcings is similar to that of Kalnay and Merkine (1982) and other idealized studies.

3. NUMERICAL EXPERIMENTS

The full (i.e., time dependent, nonlinear) model was integrated for 120 days (after a warm-up period of 30 days) with mountains (M1 and M2), and with no surface features (C1, the control). M1 and C1 were further integrated for 120 more days to give some measure of variance. The climatology of these simulations will first be described.

3.1. PHYSICAL SPACE RESPONSE

The time and zonally-averaged zonal wind for the various experiments is illustrated in figure 1. Note that the shear is reduced by about a factor of two from the equilibrium value, and the surface wind has the typical easterly, westerly, easterly variation. Note in particular that the main features of the zonal wind are not altered by the inclusion of orography.

The streamfunctions, minus the zonal mean, are displayed in Figure 2. The response shows a wave train propagating downstream of the mountains, with a stationary high north and west of the mountains, and a low directly east. The wave-train extends about halfway round the domain, and resembles, for example, that generated by the barotropic model of Kalnay and Merkine (1982). The flow pattern is shifted eastward somewhat at the lower level, and lowered in amplitude The difference map (M1-M2) shows some reduction in amplitude, and some slight phase shifts.

The nonlinearity inherent in the solutions is most noticeable by comparing the linear solution with the full solution. The linear solution uses the time averaged zonal wind from M1 and the same stationary topographic forcings. The amplitudes of the linear solution are about three times as large as the full solutions, and have larger responses near the critical latitudes in the heating cases. The comparison with linear theory is developed more fully in Section 4.

3.2. ENERGETICS

The model energetics are decomposed spectrally into kinetic and available potential energy for both stationary and transient components. The unforced, inviscid model conserves energy exactly. The equations for the kinetic energy budget are obtained by taking the time average of (1) and (2), multiplying by $[\psi_{1k}]$ and $[\psi_{3k}]$ and adding the resulting expressions. A square bracket denotes a time average, a prime a deviation therefrom and a subscript k the kth spectral mode. The potential energy budget is similarly obtained from (3).

The budget equations for the stationary flow are, for each wave-number k,

$$\frac{1}{2}k^2 \frac{d}{dt}([\psi_k]^2 + [r_k]^2) = T_k([K],[K]) + T_k([K],K')$$
$$+ C_k([P],[K]) + S_k([K]) \tag{8}$$

$$\lambda^2 \frac{d}{dt}[r_k]^2 = T_k([P],[P]) + T_k([P],P') - C_k([P],[K])$$
$$+ S_k([P]) \tag{9}$$

The left-hand sides of (4) and (5) are the total rates of change of the time averaged kinetic and potential energies. The terms on the right-hand side are

T([K],[K]) (transfer of stationary kinetic energy)
$$= [\psi_1] J([\psi_1],[\nabla^2 \psi_1]) + [\psi_3] J([\psi_3], \nabla^2[\psi_3] + h)$$

T([K],K') (transfer of stationary to transient kinetic energy)
$$= [\psi_1] [J(\psi', \nabla^2 \psi'_1)] + [\psi_3] [J(\psi'_3, \nabla^2 \psi'_3)]$$

S([K]) (frictional sink of kinetic energy)
$$+ [\psi_3] [D_3] + [\psi_1] [D_1]$$

C([P],[K]) (conversion of energy from potential to kinetic)
$$= -\frac{f}{2\Delta_p} [\omega] [r]$$

T([P],[P]) (transfer of stationary potential energy)
$$= -\lambda^2[r] J([\psi],[r])$$

T([P],P') (transfer of stationary to transient potential energy)
$$= -\lambda^2[r] [J(\psi',r')]$$

The transient energy budget is similar. The energy spectra are graphed in Figure 3 and the budgets are graphed by zonal wavenumber in Figures 4 and 5. (The energy in the stationary field in the control experiment is negligible).

The total energy budget and spectra (Figs. 3 and 4) depend only slightly on the inclusion of orography, and in all cases the slope and magnitudes are entirely reasonable. The total energy budgets for all cases are typical of quasi-geostrophic turbulence (Vallis, 1983; Boer and Shepherd, 1983): potential energy is transferred by baroclinic instability of the mean flow to the waves mainly between wavenumbers 3 and 7. This is balanced by a (local) conversion to kinetic energy where it is transferred upscale to be ultimately dissipated by friction. There are only quantitative differences between M1 and M2 so only those for M1 are shown. The energy in the zonal flow (both stationary and total) is very similar for all experiments M1, M2, and C1. Instability analysis shows its value to be about 50% higher than that required for baroclinic instability, implying that nonlinear equilibration mechanisms, rather than a reduction of the zonal APE, are responsible for equilibration.

The maintenance of the stationary energies for M1 is illustrated in Fig.5. Note the general direction of the energy cycle: Energy enters the (stationary) zonal flow through direct forcing creating zonal available potential energy. Most of it is transferred to transient potential energy, then to transient kinetic where it is dissipated. A Significant fraction is, however, transferred to stationary eddy potential energy. This is balanced by further transfer to transient potential energy and to stationary kinetic. There it is further converted to transient energy or dissipated. In M1 (and M2) note that the stationary energy budget is dominated by transfers at wavenumber 3. Over 70% of the conversion between zonal and eddy potential energy occurs here, suggesting that wavenumber 3 is linearly resonant with the zonal flow.

Note too that the stationary kinetic energy in M1 and M2 is not maintained by direct orographic forcing (which would be represented by a conversion from zonal kinetic to stationary eddy kinetic energy) but by transfer from potential energy (see also Yao, 1980). The balance is maintained partly by direct dissipation, and partly by the dissipative effects of the transient eddies. The upscale transfer of energy is accounted for, too, almost entirely in the transients even at low wavenumbers. The transfer of energy by the purely stationary flow shows no preferred direction.

4. STATIONARY MODELS

In this section we discuss the extent to which the stationary
fields of the full model are described by linear or stationary non-
linear models and to what extent the turbulent terms contribute.

4.1. TIME-AVERAGED EQUATIONS

The time mean asymmetric equations may be written

$$\frac{\partial}{\partial t}[\hat{q}] + G_w[\hat{\psi}] = [\hat{F}] \tag{10}$$

where at the upper level

$$[F] = \lambda^2 A - J([\hat{\psi}_1],[\hat{q}_1]) - [J(\psi_1',q_1')] \tag{11}$$

and at the lower level

$$[F] = -\lambda^2 A - J([\hat{\psi}_3],[\hat{q}_3]) - [J(\psi_3',q_3')] - J([\bar{\psi}_3],\hat{h}_3) \tag{12}$$

A bracket denotes a time average and a prime a deviation therefrom.
An overhead bar denotes a zonal average and a carat a deviation there-
from. An integration time of 4 months is sufficient to make $\frac{\partial}{\partial t}[\hat{q}]$
negligible. The terms on the RHS in (11) and (12) are the diabatic
heating, the fluxes due to the stationary eddies, the turbulent fluxes
due to the transient eddies, and the linear orographic forcing.
Including only the terms linear in $[\hat{\psi}]$ (i.e., radiative damping,
friction and interaction with the zonal flow which all are included as
$G_w[\hat{\psi}]$) and the non-homogeneous terms (i.e., those not involving ψ) in
[F] defines the linear model. Including the stationary nonlinear terms
in [F] using fields from the full model defines the stationary non-
linear model. The transient thermodynamic or vorticity fluxes may also
be incorporated. G_w is a matrix composed of the frictional terms and
terms derived by linearizing the asymmetric equations about the zonal
flow. For an inviscid linear model

$$G_\omega[\hat{\psi}] = J([\bar{\psi}], [\hat{q}]) + J([\hat{\psi}], [\bar{q}])$$

Setting up the matrix required use of interaction coefficients. The stationary linear solution is then given by $[\hat{\psi}] = G_\omega^{-1}[\hat{F}]$.

4.2. PHYSICAL SPACE COMPARISONS

The physical space amplitudes from the linear model are approximately three times too big (Fig.2), although some of the qualitative features of the nonlinear solutions are found. Nor is the linear model a good model of the difference (or anomaly) experiment for the differences M1-M2. That is, the increase of M1 from M2 and is much less than that given by linear theory.

Quantitative comparisons of the phases of the different fields are given by the correlation coefficient $\sigma = \int \psi_1 \psi_2 / (\int \psi_1^2 \int \psi_2^2)$ where the integrals are over the domain. For M1 the correlation between the full and linear models of about 50%. For the difference experiments, M1-M2, the correlations are similar. The point, though, is that as well as simply damping the response, transients are affecting the phase.

4.3 SPECTRAL AMPLITUDES AND ENERGETICS

The energy cycles for the linear models (not shown) are similar to the cycles of the stationary solutions of the full model in that the kinetic energy of the flow is maintained by conversion from potential energy, which in turn is maintained by transfer from the zonal flow. They differ in that the amplitudes are much higher and the orographic cases display a great sensitivity to the basic state. In the linear solution the balance in the stationary kinetic energy is maintained primarily by conversion to zonal kinetic energy, rather than to transients. Again direct orographic forcing is unimportant (i.e., terms directly involving the orography have a small impact on the energetics).

The kinetic energy spectrum for the models is shown in Figure 6. In each figure we plot the stationary kinetic energy achieved from an integration of the full model. Additionally we plot the energy from the linear model, this model plus stationary nonlinear forcings, and plus transient thermodynamic forcings or plus the transient vorticity forcings. That is, we include the nonlinear terms as non homogeneous terms on the right hand side of (10). The models are successive improvements to a basic linear model by the incorporation of additional forcings. If both transient vorticity and thermodynamic fluxes are included, we achieve again the results of the full model.

The abridged models have much higher energies in the stationary fields than the full model. Surprisingly, perhaps, the largest relative discrepancies occur at lower wavenumbers. The linear models are also extremely sensitive to the basic state-note the large differences in M1 in Fig.6 for two slightly different zonal winds from the two 120 day integrations. The very large amplitude response at wavenumber 3 is due to a simple resonance with the zonal wind. This also occurs in the full model, although its amplitude is greatly reduced by transfer of energy to the transient flow (Fig.5). In baroclinic flow, the stationary Rossby wave for the first zonal mode at the upper level occurs here near $k_x=3$. The large response at wavenumber 5 does not seem to be a simple resonance. However, for small surface easterlies resonance can occur near to $k_x^2 = \lambda^2$ (Egger, 1976). (In this model $\lambda \sim$ wavenumber 4.5).

The addition of stationary nonlinear terms, $J([\psi],[q])$, results in a negligible improvement over the linear model, consistent with the energetics which show very small transfers between stationary modes. Sensitivity to the basic state is displayed here, also. It is, therefore, the transient terms which are damping the linear response. The thermodynamic transients [the terms of the form $\lambda^2[J(\psi',r')]$] are dominant at lower wavenumbers, and the vorticity forcing is dominant at higher wavenumbers. For $k^2 >> \lambda^2$, inspection of (5) and (6) shows the layers to be effectively decoupled and the transient fluxes of temperature scale out of the problem; for $\lambda^2 >> k^2$ the temperature field is passively advected by the velocity fields and the fluxes of relative vorticity are small compared with the thermodynamic fluxes in the potential vorticity equation.

5. INSTABILITY PROPERTIES

Having shown that the transient fluxes reduce the amplitude of
the stationary flow, and greatly affect the phase, we examine in this
section the instability of the climatological flow to small perturbations. Our aim here is to see to what extent the flow variance may be
understood in terms of simple stability calculations.

5.1. EIGENVALUE EQUATIONS

The time dependent model may be written as

$$H\frac{\partial}{\partial t}\psi' + G\psi' = F' \tag{13}$$

In (13) G is the linear matrix operator derived by linearizing all
Jacobians about an equilibrium state-taken here to be the time averaged
non-zonal state from the nonlinear time dependent model. F' is composed
of the term $-J(\psi',q')'$. The linear stability properties of the above
equation is obtained by setting F'=0. If we assume $\psi'=\tilde{\psi}e^{\sigma t+i\omega t}$, then
we have the eigenvalues and eigenvectors of (13). Give the fundamental
frequencies and spatial structure of the time dependent system so long
as $J(\psi',q')'$ is negligible.

5.2. WAVE MEAN FLOW INSTABILITY

A simple problem is the stability of a basic state consisting of
the mean zonal flow, plus that due to the interaction of the zonal
flow with the asymmetric forcing. It is the simplest problem which
explicitly considers the effects of the asymmetric forcing. We
performed two sets of calculations. In the first we take the mean zonal
state from the full model, calculate the linear response to orography,
and then calculate the instability associated with this mean state.
The other instability calculation we perform uses the mean state given
by the full model integration for both the zonal flow and wave and
uses the corresponding spectral forcing.

The idealized wave mean flow interaction problem displays greater instability than the linear baroclinic problem (i.e., the problem of the instability of the purely zonal state), especially for low and high wavenumbers (Fig.7a). The phase speeds are also changed, generally being slower. We note here that linear resonance is evident, in both orography and heating cases (although more in the former). This is evidenced by the peak in stationary (linear) response just below wavenumber 3 which also is the slowest moving wave. However, this calculation itself is somewhat unrealistic, as can be seen by comparing it with the results from the true mean state (Fig.7b). Now the growth rates are very similar to those given by the purely zonal problems, suggesting baroclinic instability is the main contributor at smaller mean state amplitudes. However, the phase speeds are considerably reduced, especially at wavenumber 3.

The main point we wish to make here is that the idealized wave-mean flow problem is unrealistic as an indicator of the flow stability properties of the climatological flow because the linear amplitude of the waves is generally too high, and the system is too unstable. It does, however, suggest why in this model, and perhaps the atmosphere, linear behavior is not observed. That is, the linear state is likely to be highly unstable. Reduction of the orography and heating by 1/2 tends to also decrease the growth rates by 1/2 and therefore it is likely that linear theory is likely to work well only for topographic amplitudes an order of magnitude smaller than the ones we have chosen for this study. At high wavenumbers the topographic flow is stable. In spite of the flow being highly turbulent here, the stability is ensuring that linear theory performs fairly well.

5.3. THREE-DIMENSIONAL INSTABILITY

Finally we calculated the eigenvalues for the problem of the instability of the complete, time averaged, flow for experiment M1. The system was truncated at zonal wavenumber 8, for computational reasons, but all meridional modes were retained. Three unstable eigenvalues are present, two with large growth rates and fairly rapid oscillations, and a third, more slowly amplifying and oscillating mode.

The barotropic streamfunction eigenvectors for the orographic case, M1, are given in Figure 8. Plotted are the time averaged root mean square over one frequency cycle, ignoring the growth, and the eigenvector at t=0. The most unstable modes are similar to those given by the standard, linear, baroclinic instability calculations and the variance tends to have little zonal structure, indicating zonally propagating waves. The third mode has zonal structure associated with it. The minimum variance is over the mountain and the maximum is away from the mountain. Despite a nonzero phase speed, this mode acts like a standing mode in that its maximum amplitude stays more or less stationary.

Plotted in Figure 9 are the root mean squares (given by $\delta=(\frac{1}{T}\int\psi^2 dt)$) for the time integration of the full model. The low passed fields have been obtained by subjecting the time series of the stream function to a Hanning window which filters out periods shorter than five days (a ten day low pass filter showed little difference). Comparing the eigenfunctions (Fig.8) with the high and low passed variance fields of the numerical model (Fig.9), one can see a certain amount of resemblance. The fast oscillating modes tend to correspond to the high passed fields and the more slowly varying mode qualitatively corresponds to the low passed field. The agreement is only qualitative, though, which may be due to the choice of a time averaged basic state, which may not be relevant to the growth of instabilities.

6. SUMMARY AND CONCLUSIONS

This study has been concerned with the extent to which stationary features of flow over orography are the result of stationary linear dynamics and to what degree nonlinear dynamics, stationary and turbulent, contribute. We integrated a quasi-geostrophic model with idealized topographic forcing for a period of several months and compared the time averaged results with the results of linear theory. The addition of the stationary nonlinear, thermodynamic and transient vorticity fluxes successively brings the linear model to the full turbulent model.

The turbulent simulations display atmospheric-like features. The shape of the energy spectra, the direction and magnitude of the energy transfers and the physical space amplitudes are realistic. The mean zonal state of the model is little altered by the presence of orography. The variation in the zonal state with and without orography is no more different than the variation between long-term integrations started from different initial conditions.

The energetics of the total flow, transient plus stationary, is typical of geostrophic turbulence. Most of the upscale transfer of kinetic energy occurs in the transient flow. In all cases transfer of energy between stationary eddy modes has a negligible contribution. Transfer of energy from the stationary to the transient flow seems responsible for reducing the amplitude of the stationary flow, and in particular of the resonant structures which, nevertheless, can still be detected in the turbulent simulations. The energy cycles of the linear solutions are larger but in the same direction as the full solution, except that the main sink of stationary kinetic energy is in the zonal flow, rather than the transients.

The barotropic, time averaged stream function for M1 has a Rossby wave train propagating downstream of the mountain with a maximum wave-number response for $k_x=3$ corresponding to a simple resonant wavenumber. The linear responses were qualitatively similar except that the linear amplitudes were much too large, especially at the resonant wavenumbers, and somewhat out of phase.

The topographically forced flow was found highly unstable, although less so at high wavenumbers. The stability properties of the mean field give some indication of the flow variance. In particular, eigenfunctions are present which can be identified with the variance of the high-passed or low-passed time series. The presence of orography itself is non-negligible, and acts to stabilize the flow.

In summary, we conclude that linear theory will overestimate the response of flow over topographic features because it neglects the damping effects of the transients which arise because of the instability of the flow which would be produced in their absence. An important implication is that parameterization theories of the transient baroclinic eddies must incorporate the interaction of such eddies with the asymmetric stationary flow, in order that the stationary flow not be of unrealistically high amplitude even with the correct zonally-averaged flow.

ACKNOWLEDGMENTS

The order of authors is arbitrary. The research was supported by National Science Foundation Grant ATM82-10160 and by NASA Grant G-NASA-NAG5-236. G. Johnston text-edited the manuscript and F. Crowe and his group drafted the figures. Useful comments were received by R. C. J. Somerville.

BIBLIOGRAPHY

Alpert, J. C., M. A. Geller, and S. K. Avery, 1983: The response of
    stationary planetary waves to tropospheric forcing. J. Atmos.
    Sci., 40, 2467-2483.

Ashe, S., 1979: A nonlinear model for the time-averaged axially
    asymmetric flow induced by orography and diabatic heating J. Atmos.
    Sci., 36, 109-126.

Boer, G. J., and T. G. Shephered, 1983: Large scale two dimensional
    turbulence in the atmosphere. J. Atmos. Sci., 40, 164-184.

Charney, J. G., and J. G. Devore, 1979: Multiple flow equilibra in the
    atmosphere and flocking. J. Atmos. Sci., 36., 1205-1216.

Egger, J., 1976: The linear response of a hemispheric two-level
    primitive equation model to forcing by orography. Mon. Wea. Rev.,
    104., 351-364.

Frederiksen, J. S., 1978: Instability of planetary waves and zonal
    flows in two-layer models on a sphere. Quart. J. Roy. Meteor.
    Soc., 104, 841-872.

Holopainen, E.O., 1966: An observational study of the energy balance
    of the stationary disturbances in the atmosphere. Quart. J. Roy.
    met. Soc., 96, 626-644.

Lin, C. A., 1980: Eddy heat fluxes and stability of planetary waves.
    Part I and Part II. J. Atmos. Sci., 37, 2353-2380.

Opsteegh, J. D., and A. D. Vernekar, 1982: A simulation of the January
    standing wave pattern including the effects of transient eddies,
    J. Atmos. Sci., 39, 734-744.

Orszag, S., 1971: Numerical simulation of incompressible flow within
    simple boundaries. Stud. Appl. Math., L, 293-327.

Phillips, T. J., 1982: On the interaction of surface heating anomalies
    with zonally symmetric and asymmetric atmospheric flows. J. Atmos.
    Sci., 39., 1953-1971.

Roads, J. O., 1981: Linear and nonlinear aspects of snow albedo
    feedbacks in atmospheric models. J. Geophys. Res., 86, 7411-7424.

Saltzman, B., 1968: Surface boundary effects on the general circulation and manoclimate: A review of the theory of the quasi-stationary perturbations in the atmosphere. The causes of climatic change. Meteor. Monogr., 30, Boston, Amer. Meteor. Soc., 4-19.

Sasamori, T., and C. E. Youngblut, 1981: The nonlinear effects of transient and stationary eddies on the winter mean circulation. Part II: The stability of statationary waves. J. Atmos. Sci., 38, 87-96.

Vallis, G. K., 1983: On the predictability of quasi-geostrophic flow: the effects of beta and baroclinicity. J. Atmos. Sci., 40, 10-27.

Vallis, G. K., 1984: On the spectral integration of the quasi-geo-strophic equations for doublyperiodic and channel flow. To be submitted.

White, A. A., and J. S. A. Green, 1982: A nonlinear atmospheric long wave model incorporating parameterizations of transient baroclinic eddies. Quart. J. Roy. Met. Soc., 108, 55-85.

Yao, M. S., 1980: Maintenance of quasi-stationary waves in a two-level quasi-geostrophic spectral model with orography. J. Atmos. Sci., 37, 29-43.

Youngblut, C., and T. Sasamori, 1980: The nonlinear effects of transient and stationary eddies on the winter mean circulation, I. Diagnostic analysis. J. Atmos. Sci., 37, 1944-1957.

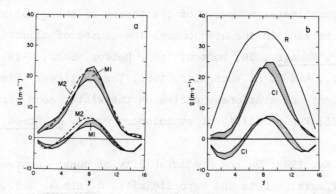

Figure 1   Time and zonally averaged zonal winds at the upper and lower
           levels for (a) M1 and M2; (b) C1. The shaded regions indicate
           two different 120-day integrations (M1a and M1b, and C1a
           and C2b), thereby giving some measure of the variability of
           the runs. In (b) the curve R is the radiative-equilibrium
           zonal wind.

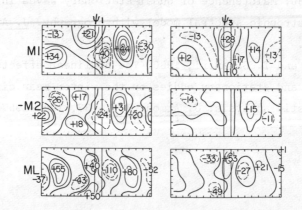

Figure 2   Upper and lower level stream functions, for the mountain
           cases M1 and M2. (a) is the time averaged results from the
           full model for M1 (b) displays the difference field M1-M2,
           and (c) shows the fields predicted by linear theory using
           the zonal wind of the full model for M1. Units are arbitrary
           with 10 units corresponding to a geopotential height of
           approximately 40 meters.

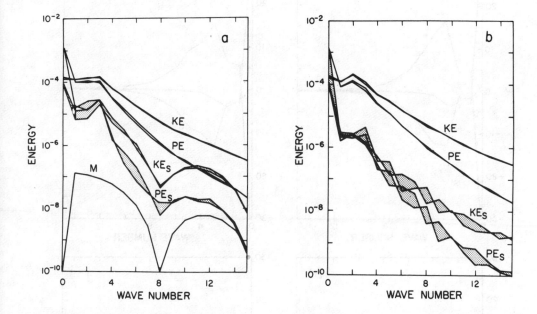

Figure 3   Kinetic (KE) and potential (PE) energy spectra for (a) M1;
            and (b) C1. Shaded regions denote integrations from different
            initial conditions. Curves without a subscript show the
            total, time-averaged, energies. A subscript indicates the
            energy is that in the stationary field. The curve M shows
            the amplitude of the stationary forcing-the mountains in
            (a). Units are arbitrary.

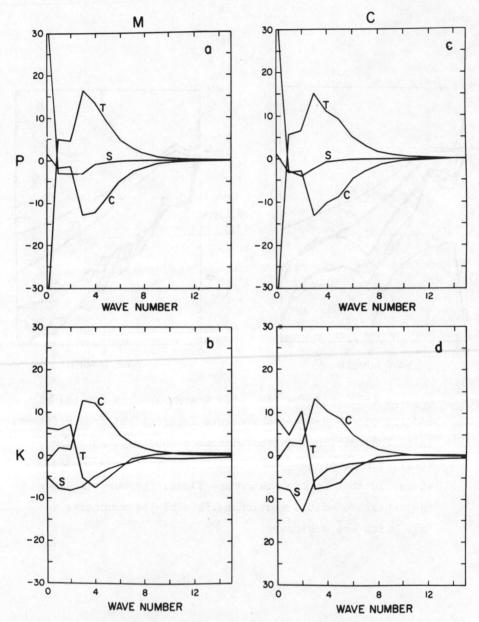

Figure 4 Total energy budgets for M1 (a and b), and C1 (c and d).
Units are arbitrary. The upper row is for potential energy,
the lower for kinetic. The labels denote: S, diabatic source
or frictional sink; C, conversion between potential and
kinetic; T, transfer.

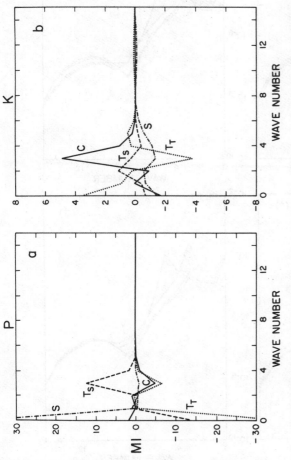

Figure 5 Energy budgets for the stationary field for M1 (a) shows the
budget of potential energy, (b) the kinetic energy budget.
The labels on the curves denote: S, diabatic source or
frictional sink; C, conversion between stationary potential
and stationary kinetic energy; T_3 contribution from/to other
stationary modes; T_T contribution to/from transient modes.
Note the different scales for the kinetic energy budget.
Units are same as Figure 4.

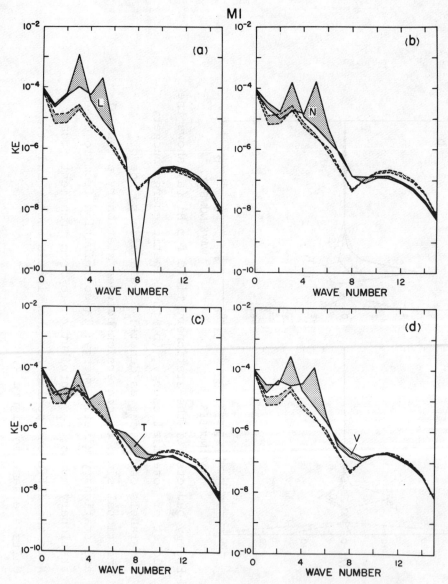

Figure 6 Comparison of stationary kinetic energy spectra for M1
between full model and various abridgements: (a) Spectra
from linear model; (b) stationary nonlinear model spectra;
(c) spectra from nonlinear model plus thermodynamic
transients, (d) spectra from nonlinear model plus vorticity
transients. The spectra from the full model is always shown
dashed. Units are arbitrary.

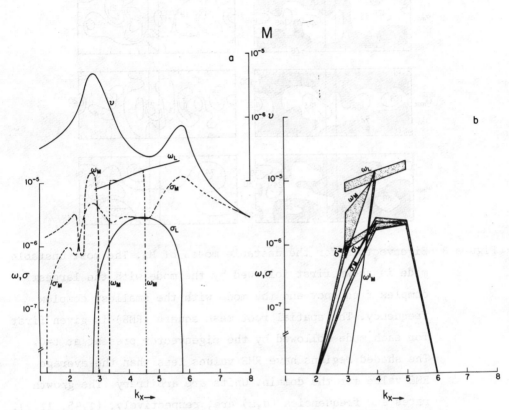

Figure 7   Instability for case M1 for the standard baroclinic in-
           stability problem and for the wave mean flow instability
           problem. σ denotes the growth rate and ω the complex frequency
           in s^{-1}. Subscript 1 denotes the standard baroclinic in-
           stability problem and subscript m the wave-mean flow problem.
           (a) Shows the response for the idealized problem in which
           the amplitude of the wave is taken from the linear model.
           The sum of the squares of the stationary linear amplitudes
           for the idealized problem is denoted V. (b) Shows the
           response when the amplitude of the wave is taken from the
           time averaged model. σ'_m and ω'_m refer here to the wave mean
           flow instability when the orographic feedback is neglected.

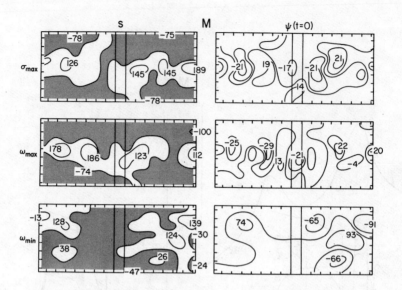

Figure 8   Eigenvectors for the unstable modes of M1. The most unstable
mode is shown first followed by the mode with the largest
complex frequency and the mode with the smallest complex
frequency. The spatial root mean square (RMS) is given first
for each mode followed by the eigenvector present at t=0.
The shaded regions have RMS values less than the average
RMS value for the domain. Units are arbitrary. The growth
rates and frequencies (σ,ω) are, respectively, (1.45, 11.2),
(1.35, 17.4), (0.43, 0.14) × 10^{-6} s^{-1}.

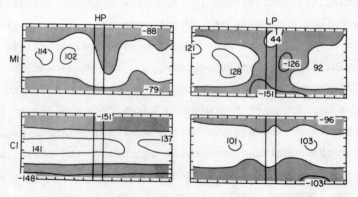

Figure 9   High and low passed RMS's from the full model for the
orographic case, M1 and the control run, C1. Units are
arbitrary.

A POSSIBLE EFFECT OF THE TIBETAN PLATEAU ON THE MID-LATITUDE WESTERLY SYSTEMS IN PRODUCING RAIN OVER THE BRAHMAPUTRA VALLEY IN WINTER

P.N. Sen

Meteorological Office, Pune-411, 005, India

I. INTRODUCTION

The main weather producing systems during the winter months over the northern India are the systems in the extra tropical westerlies. These systems generally originate to the west of India and move from west to east over the northwest India either as low pressure systems at the sea level or as cyclonic circulations or troughs in the upper air mainly in the lower troposphere. But their extension to the middle and upper troposphere are quite common. They are known as 'Western Disturbances' in India because they approach the country from the west. These systems after coming over to India get modified i.e. lose their frontal characteristics. In some rare cases the frontal characteristics could be detected. These systems give good amount of rain or snow over Jammu and Kashmir, Himachal Pradesh, Hills of West Uttar Pradesh; and rain over Rajasthan, Punjab, Haryana and Delhi and rest of Uttar Pradesh. Sometimes these systems affect the meteorological sub-divisions further south also. These systems can be detected over the Indian region on the weather charts till they move across the western Himalayas over the Tibetan Plateau when they can only be detected sometimes at the middle and upper tropospheric levels. Generally five to six such disturbances affect per month during the winter season. Sometimes a series of western disturbances affects the northern Indian region in succession like the cyclone families in the extratropics. It is quite often seen that after the passage of western disturbances across the western Himalayas over the Tibetan Plateau, Brahmaputra Valley gets good amount of rainfall. Even with disturbances which do not affect northwest India south of 30°N, Brahmaputra Valley gets **precipitation** a day or two after the western disturbance has caused weather over the western Himalayas. The connection between the rainfall over the Brahmaputra Valley and the movement of the western disturbances has been investigated in this paper.

II. DATA UTILIZED

The meteorological data for two months January and February for the period 1981 to 1984 collected from the Indian weather charts and satellite imageries from the Environmental Satellite Imagery (NOAA, Prepared by NESS) have been utilised for this study. To bringout the salient features only two cases have been presented in this paper, as it has been found the conditions are more or less similar for the entire period of study.

III. PHYSIOGRAPHY OF NORTHERN INDIA

The great Himalayas extend in a vast arc, convex to the south from 35°N, 74°E to about 30°N, 95°E. The height of the ranges are above 4 km and in many places above 6 km. To the south of the great Himalayas lie the lesser Himalayas; the Pirpanjal and the Siwalik ranges whose heights are between 2 and 4 km and less than 2 km respectively. To the north of the Himalayas lies the 'Tibetan Plateau'. 'The Tibetan Plateau has roughly the shape of an ellipsoid. The major axis is over 3000 km long and its minor axis over 1400 km' (Staff members, Academia Sinica, 1957). The average height of this Plateau is over 4 km. In the north-eastern part of India we have the Brahmaputra valley. This Valley has the Himalayan range and the Tibetan Plateau to the north, Garo-Khasi-Jaintia and Naga Hills to the south and the mountains of Yunnan to the east. The average height of these barriers are 4 km, 1 to 1.5 km and 3 km respectively. The river Brahmaputra flows along the

whole length of the plain enclosed by the hills. The Brahmaputra Valley
is 550 km long and 70 km wide. The orientation of the Valley as well
as the river Brahmaputra is more or less east to west. The Valley also
slopes down from east to west.

IV. MEAN FLOW PATTERN OVER INDIA DURING DECEMBER
AND JANUARY

The mean flow pattern over India at 900, 850, 700, 500, 300 and
200 hPa levels for the 10 day periods 1-10, 11-20, 21-31 December and
1-10, 11-20, 21-31 January have been presented in Figures 1-6.

If we look at the wind pattern over India in the lower levels the
winds are westerlies over the north Indian region excepting over the
Brahmaputra Valley where the winds are easterlies. The winds over the
Brahmaputra Valley are also easterlies at 0.3 km and 0.6 km levels
(Mean charts not shown). From the pressure distribution the westerlies
over the northern India are quite logical in the lower levels but the
easterlies over the Brahmaputra Valley in the lower levels are not in
general consistent with the pressure distribution. After the monsoon
recedes the sky over the Tibetan Plateau starts losing heat rapidly
and becomes a cold region. The Siberian high causes winter monsoon
over China. The Tibetan Plateau, the Himalayas to the north and the
mountains of Yunnan and Burma to the east prevent the easterlies from
reaching India (Byers, 1959). It follows that the prevailing wind
over the Brahmaputra Valley is the result of super-imposition of local
factor on the air flow resulting from the existing pressure distribution. Mukherjee and Ghosh (1965) explained the causes of these local
winds. It is well known that there is a downslope or katabatic wind on
a diurnal scale resulting from the cooling of hills or mountains
during night. The easterlies over the Brahmaputra Valley are katabatic
winds on a seasonal scale as no other local wind system is conceivable
from the geography of the place. The Himalayan range and the Tibetan
Plateau having the highest orography cools down excessively during
winter and katabatic wind starts flowing from them. As the Valley

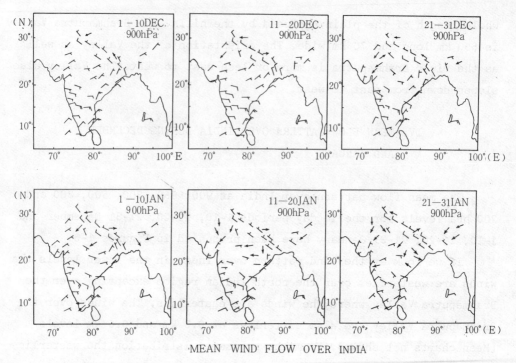

MEAN WIND FLOW OVER INDIA

Fig.1

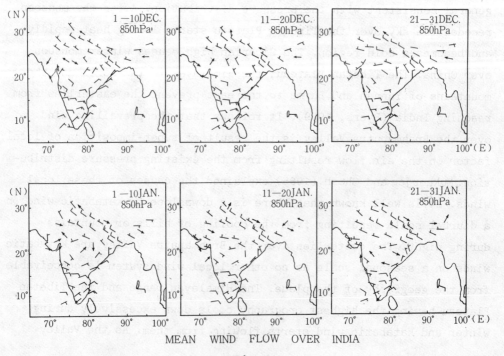

MEAN WIND FLOW OVER INDIA

Fig.2

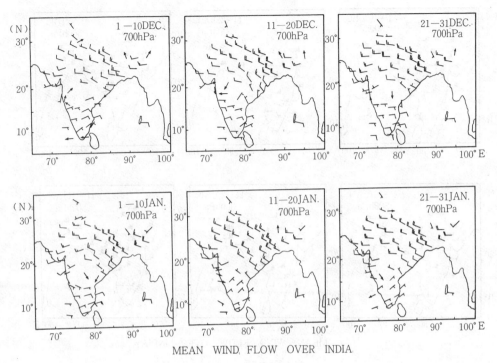

MEAN WIND FLOW OVER INDIA

Fig.3

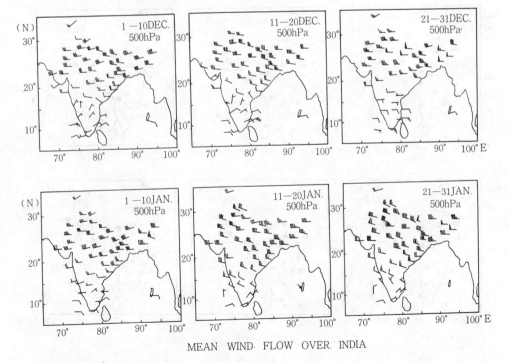

MEAN WIND FLOW OVER INDIA

Fig.4

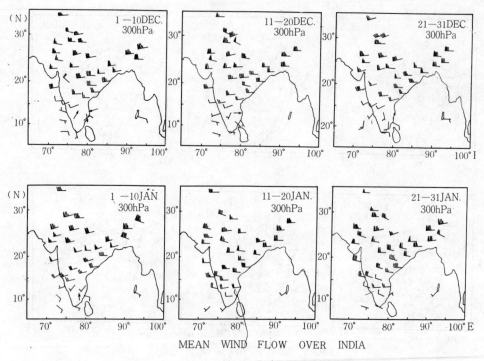

MEAN WIND FLOW OVER INDIA

Fig.5

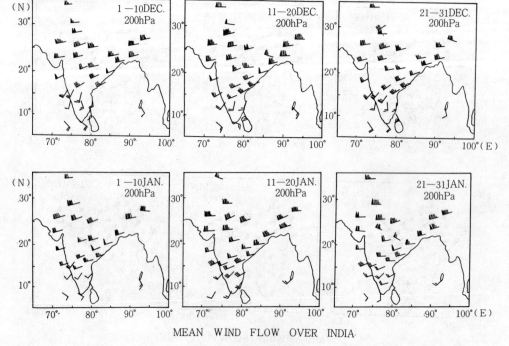

MEAN WIND FLOW OVER INDIA

Fig.6

slopes down from east to west, the winds should be easterlies. These
easterlies should be stronger at the lower levels and gradually weaken
as we go up. This is what has been observed in Figures 1 to 6. Easterlies should have blown over the Valley upto the mean height of the
Tibetan Plateau. In absence of Tibetan Plateau and the Himalayas
ranges the northwesterlies would have blown over the Valley as the
pressure distribution suggests. Thus the westerlies, will try to exert
themselves against the local wind system and that is why the westerlies
appear at 700 hPa level as the katabatic wind is expected to be weak
at higher levels. As in all mountains wind systems there should be a
counter current in this also. The katabatic winds are easterlies and
the counter current blowing at the southern edge of the Tibetan Plateau
should be westerlies and should blow upto the mean height of the
Tibetan Plateau. The normal westerlies over the Valley are likely to
get laterally mixed with this westerly counter current, whereas the
westerlies below the height of the Tibetan Plateau are opposed by the
katabatic winds. The counter currents accentuate the westerlies in the
mid-tropospheric levels over the vicinity of the Brahmaputra Valley.
The counter currents in a mountain wind system are expected to be
shallow and the effect of accentuation should vanish after a certain
height. That feature has been revealed in Figures 1 to 6.

It has been noticed that the commencement of rainfall over the
Brahmaputra Valley during the passage of the western disturbances
across the western Himalayas over the Tibetan Plateau is either
preceeded by or coincides with the change over of the easterlies to
westerlies over the Brahmaputra Valley particularly over Gauhati,
which is located at 26° 11'N and 91° 45' E on the southern bank of the
river Brahmaputra, in the lower tropospheric level upto 0.9 km i.e.
about 900 hPa level. This was noticed by the author when he was
engaged in forecasting work in the Brahmaputra Valley. The commencement
of the westerly over Gauhati upto 900 hPa level takes place even when
the western disturbance is over the Tibetan Plateau at a much northerly
latitude. This interesting phenomenon has encouraged the author to
investigate further into this aspect.

V. PRESENTATION OF RESULTS

The period chosen for this study is 1981 to 1984 and the months December and January. The number of western disturbances affecting northwest India during this period is shwon in the following Table.

Table: Number of Western Disturtances moved over N.W. India.

	1981-82		1982-83		1983-84	
	Dec.	Jan.	Dec.	Jan.	Dec.	Jan.
Number of disturbances	7(2)	7(3)	7(2)	7(2)	6(2)	6(4)

The number of occasions when the Brahmaputra Valley received rainfall during the passage of western disturbances across the western Himalayas during the period has been shown in brackets.

In all the cases when the rainfall occurred over the Brahmaputra Valley after the passage of western disturbances, westerly wind appeared over Gauhati at 0.9 km level. In no case rainfall over the Brahmaputra Valley was accompanied by easterlies alone. Whenever rain-fall was accompanied by easterlies some disturbances to the south of the Brahmaputra Valley moved northward over the Valley and the rainfall was not the effect of the westerly system. For the illustration of the above observations two cases have been presented here.

1. Western Disturbance of 17-19 December 1981

A western disturbance moved over Jammu and Kashmir on 17 December as an upper air system in the lower troposphere with a trough aloft. It moved away eastward across the western Himalayas by 12 GMT of 18 December. The associated upper air trough at 5.8 km a.s.l. also moved eastward and lay along 95°E at 00 GMT of 19 December. At 3.1 km a.s.l. a cyclonic circulation with its center at 27°N, 102°E was seen. The wind at 0.9 km level over Gauhati was westerly on 17 December because of the passage of an earlier western disturbance. It changed over to easterly on the same day by 06 GMT as revealed in the vertical time section (Figure 10) and persisted till 12 GMT of 18

December. It again changed over to westerly at the same level by 18 GMT of 18 December and continued to be so on 19 December. The movement of this system has been shown in Figures 7 to 9. This western disturbance did not cause any precipitation over Jammu and Kashmir and adjoining areas. But the Brahmaputra Valley received rainfall on 19 December. Thus the rainfall over the Brahmaputra Valley was preceeded by westerly over Gauhati at 0.9 km a.s.l. level. The vertical time section of Gorakhpur (Figure 11) which is located to the west of Gauhati more or less to in the same latitude belt (26° 45'N, 83° 22'E) did not show any significant change in the upper winds. The visible cloud pictures of the system have been shown in Figures 12 to 15.

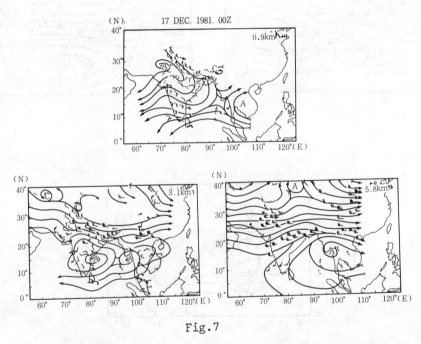

Fig.7

2. Western Disturbance of 28-31 December 1982

A western disturbance which could be traced from over Iran and Afghanistan lay on 28 December 1982 over Jammu and Kashmir and adjoining Punjab as an upper air system. It moved eastward initially slowly and then rapidly across the western Himalayas. The sequence of movement of the system has been shown in Figures 16 to 19. The vertical time section of Gauhati (Figure 20) indicates easterly wind at 0.9 km a.s.l. till 00 GMT of 29 December. Thereafter it changed over to

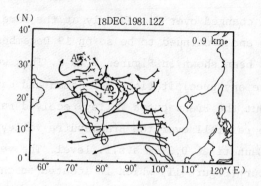

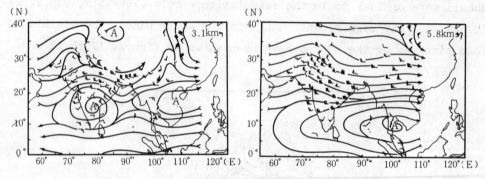

Fig.8

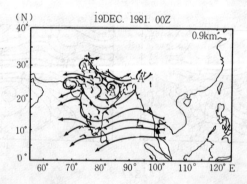

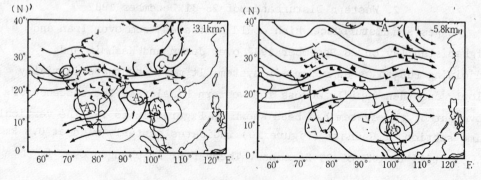

Fig.9

VERTICAL TIME SECTION

GAUHATI

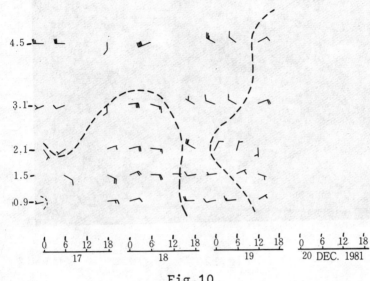

5.8
4.5
3.1
2.1
1.5
0.9
0 6 12 18 0 6 12 18 0 6 12 18 0 6 12 18
17
18
19
20 DEC. 1981

Fig.10

VERTICAL TIME SECTION

GORAKHPUR

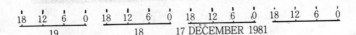

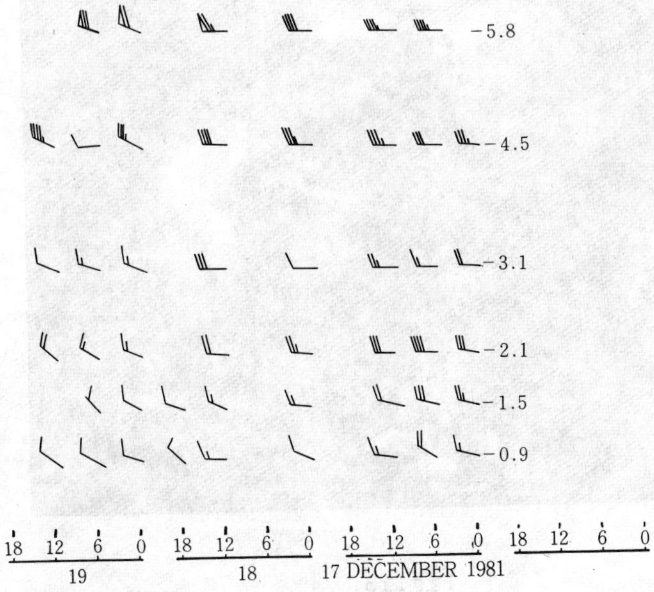

5.8
4.5
3.1
2.1
1.5
0.9
18 12 6 0 18 12 6 0 18 12 6 0 18 12 6 0
19
18
17 DECEMBER 1981

Fig.11

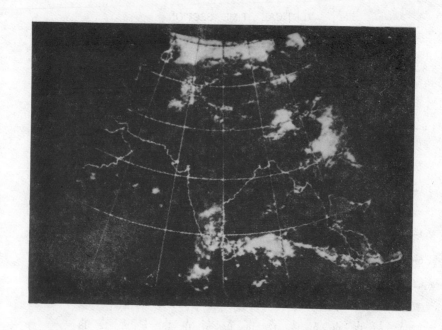

Fig.12

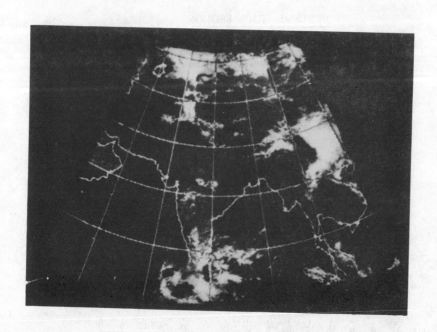

Fig.13

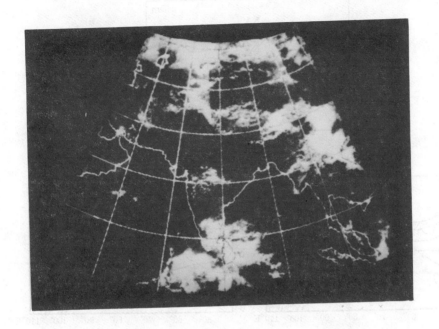

Fig.14

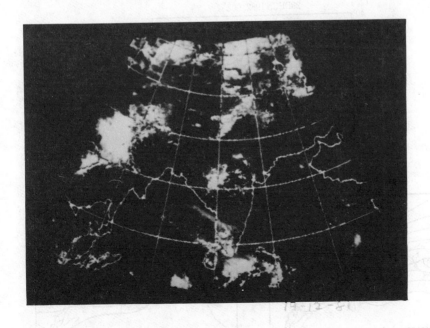

Fig.15

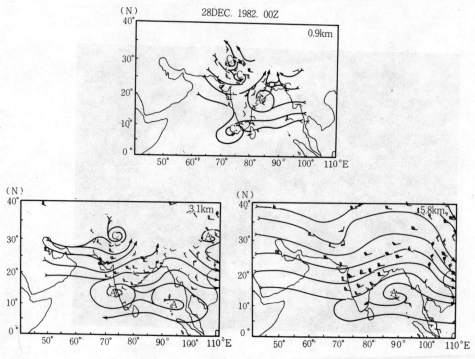

Fig.16

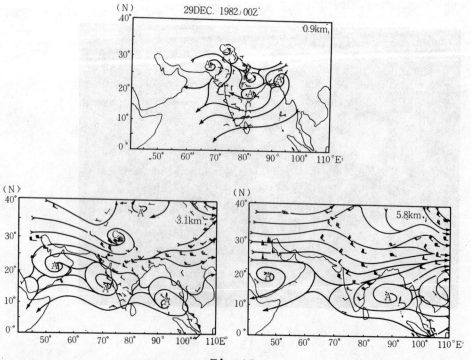

Fig.17

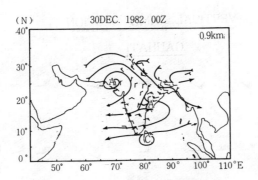

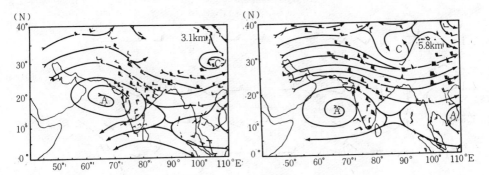

Fig.18

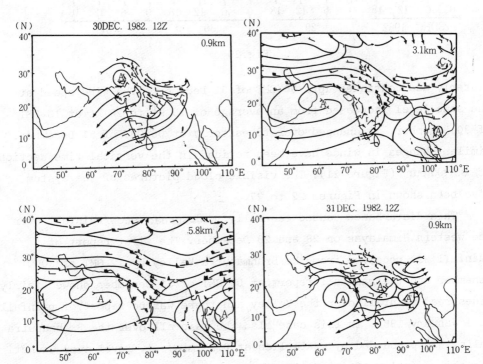

Fig.19

VERTICAL TIME SECTION

GAUHATI

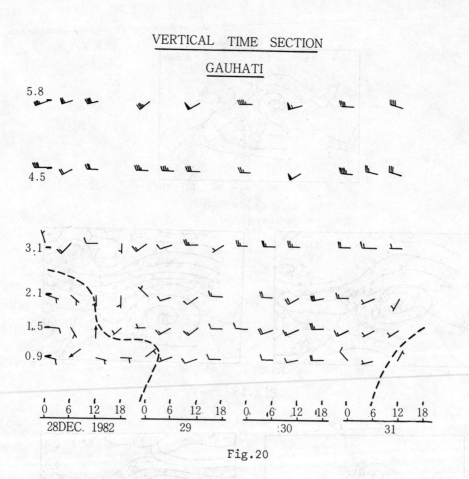

Fig.20

westerly and continued upto 06 GMT of 31 December. The wind speed at
the same level increased from an average of 10 kt to 20 kt at 18 GMT
of 30 December. It changed over to easterly at 06 GMT of 31 December.
Similar changes in winds have been noticed in the vertical time section
of Gorakhpur (Figure 21). The visible cloud pictures of the system
have been shown in Figures 22 to 27.

This disturbance caused generally widespread snow/rainfall over
the western Himalayas on 28 and 29 December. The first report of
rainfall was received from the Brahmaputra Valley by 12 GMT of 30
December and the rainfall report at 03 GMT of 31 December showed fairly
widespread rainfall over the Valley and there was no report of rainfall
on 1 January 1983. In this case also the rainfall over the Brahmaputra
Valley was preceeded by westerlies at 0.9 km a.s.l. level over Gauhati
and the cessation of rainfall took place as soon as the westerly

GORAKHPUR

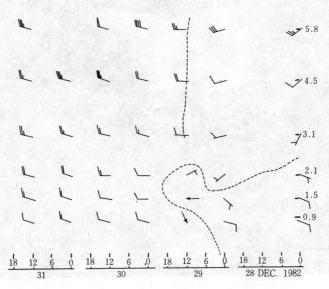

Fig.21

changed over to easterly.

These two cases clearly show that the movement of the disturbances over the Tibetan Plateau region has been much faster compared
to that over northwest India. The orography i.e. the Tibetan massif
might have exerted some influence causing this faster movement over the
area. The two cases discussed above also bring out the fact that the
onset of rainfall over the Brahmaputra Valley is definitely associated
with the appearance of westerlies at Gauhati in the lower levels.
With the onset of westerlies in the lower level with the local moisture
available from the river Brahmaputra, the orography might have
caused lifting leading to condensation and precipitation over the
Valley. Another interesting feature is that the cessation of rainfall
took place as soon as westerlies changed over to easterlies in the
lower levels. It can also be noticed that the westerlies at 500 hPa
level over the Brahmaputra Valley have strengthened with the appearance
of the western disturbance over the Tibetan Plateau.

Let us now try to find out the reasons for the occurrence of
westerlies over Gauhati in the lower levels. If the hypothesis of

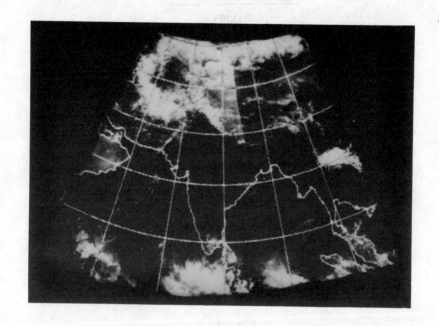

Fig.22

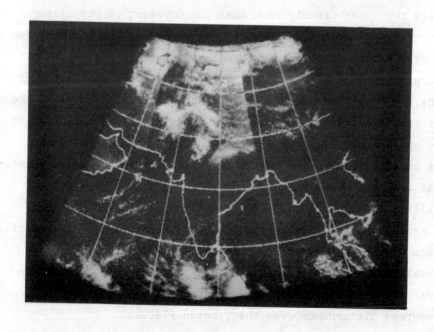

Fig.23

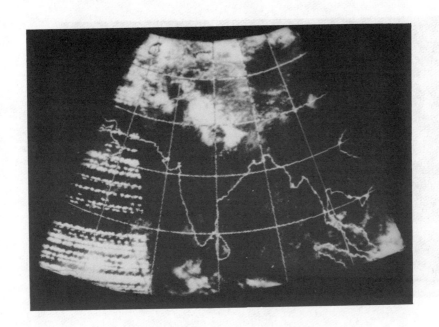

Fig.24

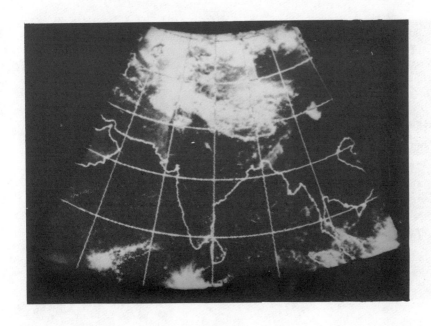

Fig.25

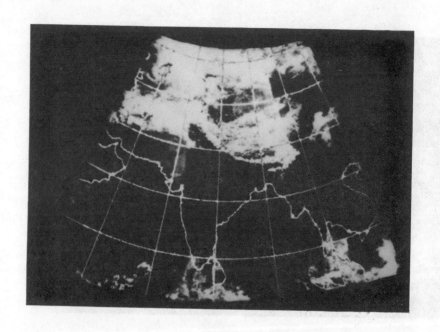

Fig.26

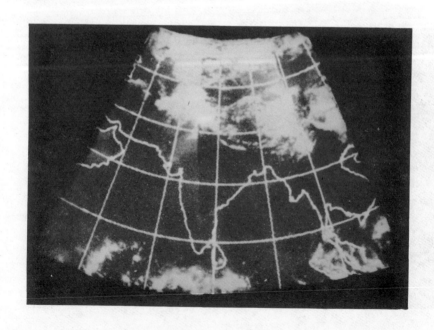

Fig.27

Mukherjee and Ghosh is accepted then the local easterlies caused due to katabatic wind must have been inhibited by some mechanism.

Due to the movement of the western disturbance over the Tibetan Plateau there has been transport of westerly momentum to the south of it which is confirmed by the strengthening of westerlies over the Brahmaputra Valley at the 5.8 km level. This transport of westerly momentum at the midtropospheric level may be responsible for the perturbation in the lower level flow over the Brahmaputra Valley and the rainfall over the region. It may, therefore; be argued that orography has some effect on the induction of westerlies over Gauhati. The actual transfer mechanism and the advection of vorticity to the south of the Tibetan Plateau region could probably be explained with extensive upper air data from the Tibetan Plateau region.

As the distrubances moved across the western Himalayas from over northwest India to the Tibetan Plateau the cloud mass also moved over the area (as revealed by the satellite picture). The satellite cloud pictures suggest that the cloudiness has increased over the Tibetan Plateau region. Convective clouds can also be seen over the Tibetan Plateau region. The convective heating over the plateau might have caused the increase in temperature in the middle and upper troposphere over the southern part of the Tibetan Plateau causing thereby the weakening of the seasonal katabatic wind over the Brahmaputra Valley. The heating might have been enchanced due to the warm air advection from the warm sector of the front. This can only be confirmed with extensive Radio sonde data over the Tibetan Plateau region.

VI. CONCLUDING REMARKS

It can be seen from the foregoing sections that the Tibetan Plateau plays an important role for the weather phenomena over the Brahmaputra Valley during winter. The plateau must be exerting some influence in causing the western disturbances to move at a much faster rate over the plateau region than over northwest India. The precipitation over the Brahmaputra Valley also seems to be due to the influence exertedon the Tibetan Plateau. Further study in this direction with extensive data over the Plateau is called for.

ACKNOWLEDGEMENT

The author is grateful to Dr. A.K. Mukherjee and Dr. A.A. Ramas-astry for helpful discussions and comments. He is also indebted to Mr. N.C. Biswas for the unhesitating help he rendered during the course of investigations.

REFERENCES

[1] Byers, H.R., General Meteorology, 285, McGraw Hill Book Company, New York, 1959.

[2] Mukherjee, A.K. and S.K. Ghosh, Orographical influence on the air flow over Brahmaputra Valley-Indian, J.Met and Geophys., 16, 429, 1965.

[3] Staff Members, Academia Sinica, On the general Circulation over Asia-Tellus, 9, 432, 1957.